PETROLEUM GEOLOGY OF IRELAND

Also published by Dunedin Academic Press

The Geology of Ireland Second Edition (2009)
Edited by Charles H. Holland and Ian S. Sanders
9781903765715 (HB)
9781903765722 (PB)

Glacial Geology and Geomorphology:
The Landscapes of Ireland (2008)
Marshall McCabe
9781903765876

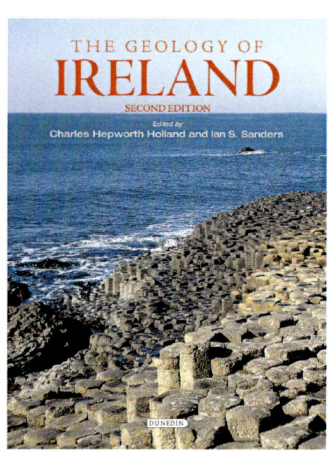

PETROLEUM GEOLOGY OF IRELAND

David Naylor and Patrick M. Shannon

DUNEDIN

Published by
Dunedin Academic Press Ltd
Hudson House
8 Albany Street
Edinburgh EH1 3QB
Scotland

ISBN: 978-1-906716-13-4

British Library Cataloguing in Publication Data
A catalogue record for this book is available from the British Library

Typeset by Makar Publishing Production, Edinburgh
Printed and bound in Poland, produced by Hussar Books

Contents

Acknowledgements

We are grateful to the many persons, institutions and commercial companies who have helped us in different ways during the preparation of this book. Clearly, a work of this type is made possible only through the labours of the many academic, government and oil industry scientists who have studied and published on the Irish region over the past few decades. However, we wish to make clear that the views and opinions expressed herein are those of the authors, and are not necessarily shared by authors or institutions whose data are employed or quoted in the book.

We acknowledge permissions to publish some of the material granted by the Petroleum Affairs Division, DCENR (Department of Communications, Energy and Natural Resources) and by the Director of the Geological Survey of Northern Ireland. Peter Croker and Michael Hanrahan have been very helpful in providing access to the Petroleum Affairs Division data, while Koen Verbruggen of the Geological Survey of Ireland and Charise McKeon of the Marine Institute kindly provided access to data and permission to use material from the Irish National Seabed Survey. Colleagues in the oil industry have been supportive throughout, and we express our gratitude in particular to Lansdowne Oil and Gas Plc, Providence Resources Plc, and TOG Ltd for permission to publish proprietary seismic data.

Of the numerous academic colleagues who have willingly provided material, help and advice we are particularly grateful to Professor Geoff Clayton, Professor George Sevastopulo and Dr John Graham of the Department of Geology, Trinity College Dublin. We are greatly in debt to John Kennedy, UCD School of Geological Sciences at University College Dublin who laboured long and hard in preparing text figures for this volume. Elaine Cullen, Department of Geology, Trinity College Dublin, also gave invaluable help with some of the figures. Also, several diagrams in the book are a reprise of figures prepared by Elaine for an earlier publication (*The Geology of Ireland*, also Dunedin Academic Press).

The references used in the preparation of each chapter are cited, but we recognise that many important papers are not referenced. To the authors of these papers, who have contributed important details to the story, we extend our apologies. We also express our gratitude to the many oil industry geologists who have worked unsung on the background data over the years.

Finally, our deepest thanks and gratitude go to Verney Naylor and Adrienne Shannon for their support and understanding during the preparation of this book. They did not complain when the dining room tables were covered in books, diagrams maps – probably because they were too busy walking the dogs, cutting the lawn and all the other chores that we should have been doing. Thank you both very much.

Preface

Fifty years have now elapsed since the start of petroleum exploration in Ireland. In that time there have been periodic phases of enthusiastic exploration, both onshore and offshore, interspersed with periods of inactivity and reflection. While the results of the exploration have been modest if measured solely in terms of discoveries of oil and gas, they have yielded an enormous amount of new knowledge regarding the extent and evolution of the sedimentary basins and their underlying basement rocks. A vast amount of new geological and geophysical information has been gathered and analysed, and has led to new models for basin formation and the development of new petroleum plays. In addition, petroleum exploration in the offshore has led to a better understanding of the recent and modern history of deep-water basins and to a realisation of the interplay between the geosphere, hydrosphere and atmosphere. It is interesting to compare the current understanding of the basins, their internal sedimentary architecture and controls, with the knowledge and ideas that were current at various times in the past 50 years.

The large volume of new information, acquired and published in the past couple of decades, has been the major stimulus for the writing of this book. In it we hope to convey the present understanding of the petroleum geology and prospectivity of the onshore and especially the offshore parts of Ireland, and to set these in the context of our understanding of basin development, and broader geological development of the region. We hope that the reader will gain an insight into how increased new geological and geophysical data have influenced basin development and exploration models through time.

The first two chapters set the geological framework for the basin development by outlining the pre-Permian history in a regional, plate tectonic framework, and then by focusing upon the more local development. The nature and history of the basement rocks that underlie the sedimentary basins played a major role in controlling the zones of rifting that facilitated the later formation of basins and petroliferous structures. In particular,

the Caledonian, Variscan and Alpine orogenies, together with the crustal thinning and ultimate rupture and the onset of seafloor spreading in the North Atlantic, shaped basin development.

The third chapter describes the history of exploration. This shows how exploration developed, guided initially by results from onshore drilling, then moved into the shallow offshore areas south and east of Ireland and later into the deeper and more remote regions of the North Atlantic where the risks but potential rewards are likely to be larger. Exploration in these frontier areas has not only provided information on the nature and composition of the basins but has also helped shape geological thinking on the development of passive margins that will be of value in exploring in other passive margin regions of the world. Exploration is always encouraged by commercial discoveries, and unfortunately the successes to date in the Irish region have been, at best, very modest. However, the offshore region is vast – 90% of Ireland's acreage lies offshore – and these deep-water areas are very lightly explored and still hold significant exploration potential. They need a combination of innovative exploration ideas and good luck. While success encourages exploration, no discoveries will come without exploration, and without the acquisition of seismic data and the drilling of wells. These, in turn, are facilitated by a robust legal framework and by internationally competitive exploration and development terms. Chapter 3 charts the development of the exploration licensing framework of Ireland and also outlines the current position regarding the delimitation of the offshore region under the United Nations Convention on the Law of the Sea (UNCLOS III).

Exploration commenced initially in the onshore sedimentary basins, and the geological setting of these basins is described in Chapter 4. The information from the onshore geology guided the initial understanding of the offshore geological framework. The onshore region provided much information on the Palaeozoic and older strata of the region, and this is provided in the fourth chapter. As such the treatment of this chapter is somewhat different from the others. It addresses the structure, stratigraphy and petroleum habitat of the onshore area in a single, early chapter rather than waiting until the later chapters of the book, following description of the regional stratigraphic succession, to describe the basin development and petroleum prospectivity, as is done for the offshore basins. This approach is aided by the fact that onshore Ireland is comprised almost entirely of Palaeozoic and older rocks.

Extension of our knowledge of the basement framework into the offshore regions has been facilitated by a scattering of deeper exploration wells that have penetrated the pre-Permian section (Chapter 5). The following four chapters (6–9) describe the detailed stratigraphy of the Permo-Triassic, Jurassic, Cretaceous and Cenozoic successions respectively, with reference to the key data and wells from which the information is gleaned. Exploration wells, by their very nature, are drilled in anomalous settings – typically on anticlinal structural highs, on the crests of fault footwalls or close to stratigraphic pinchouts on basin floors or slopes – and consequently alone do not give a representative picture of the geology of a region. Therefore the well data are supplemented, where possible, by geophysical information such as that provided by seismic profiles, and where available by seabed and shallow stratigraphic boreholes, in order to facilitate regional correlation and extrapolation.

The next three chapters (10–12) describe each of the three major groups of sedimentary basins in the offshore region. The first of these encompasses the Celtic Sea basins, south of Ireland, where offshore exploration first began and which to date are host to the majority of discoveries, some mature and almost completely depleted, others small but with hopes of production. The second group of basins contains the large, typically deep-water basins of the North Atlantic, which represent challenging and frontier exploration, and which has met with some success in locating both gas and oil accumulations. These basins are considered to be the only ones with the potential for giant discoveries in both structural and especially stratigraphic plays. The third group consists of a number of relatively small, and typically Permo-Triassic, basins that lie onshore in Ulster and in the shallow waters of the North Channel and the Irish Sea. There have been no petroleum discoveries of great significance to date in these basins, but they do provide an important insight both into the Mesozoic evolution of the region and into the challenges in trying to understand the petroleum habitat of Ireland.

The final chapter (perhaps fittingly Chapter 13), looks to the future of petroleum exploration in Ireland. It speculates upon the remaining potential in each of the main groups of basins, highlighting some possible plays, prospects and associated risks. Ireland's position at the western margin of the energy pipeline makes it particularly vulnerable to any fluctuation in supply. This could be brought about in the short term by political issues, and in the medium term by resource depletion in producing countries. This underlying uncertainty emphasises the importance for Ireland of discovering indigenous supplies in the coming years. This chapter also examines briefly the possible potential in unconventional petroleum resources such as shale gas, coalbed methane and methane hydrates. Finally, it reviews the resources that the onshore and offshore geology may hold in terms of reservoirs for sequestered carbon dioxide. Carbon capture and storage/sequestration (CCS) is currently seen to hold significant potential, albeit at probably considerable cost, to help in reversing increasing emissions, and may serve as an important transitional technology to allow for the continued use of fossil fuels (oil, gas and coal) until cleaner and more sustainable forms of energy can be provided in sufficient quantities and with acceptable reliability into the future.

We have written the book in a manner than ensures that each chapter can stand alone, without the reader having to refer excessively to other chapters. Each chapter is therefore accompanied by its own set of references. This approach means that there are some elements of overlap and duplication, particularly with regard to the stratigraphic descriptions and to parts of the exploration history. However, we have tried to keep this to a minimum and have only included such elements where necessary for the cohesion and completeness of the various chapters. It is hoped that the book will serve as a repository of information on the geology of Ireland, both onshore and offshore, and especially on the petroleum geology of the region. In documenting both the incremental accumulation of information and also the historical context for the development of the petroleum industry, it is hoped that the book will provide a sense of the dynamic nature of both geological exploration in general and petroleum exploration in particular. We hope it will serve as a marker for the past 50 years of petroleum exploration, and will provide some context and pointers towards greater geological understanding and exploration in the coming decades.

Chapter 1

Structural framework

Introduction

Ireland lies towards the western edge of the European continent (Figure 1.1), a region where the crust has been affected by multiple episodes of tectonism over the course of geological time. These events, and their resultant structures, played a major role in influencing the location, orientation and history of the set of Permo-Triassic to Cenozoic sedimentary basins that lie in the offshore region surrounding Ireland. This relationship between basement geological structure and the later basin evolution is complex. This is due, at least in part, to the fact that relatively little is yet known about the precise location and the nature of the boundaries of the various basement blocks and structural terranes in the region. However, it is clear that the younger sedimentary basins in the offshore region owe their existence and their evolution to the complex, largely tensional, tectonic regimes that operated through latest Palaeozoic and Mesozoic times. The products of sedimentation in the rift and consequent thermal subsidence post-rift basins were periodically modified, especially in Cenozoic times, by a series of compressional and inversion events. The driving forces for these have been linked to a combination of many factors, the main ones being (a) Alpine orogenic pulses, (b) far field effects of North Atlantic seafloor spreading rates and (c) plate readjustments and mantle thermal structures linked with crustal breakup. The complex nature of this interaction, and the role played by older structural units and lineaments, is well illustrated in the Irish offshore region.

In this chapter the nature and development of the regional basement structure is explored. The chapter summarises the early crustal and basement evolution, up to Late Palaeozoic time. In particular, it documents the Caledonian and Variscan structural framework that played a key role in controlling the location and early history of the Mesozoic and Cenozoic sedimentary basins. It also reviews the location, orientation and formation of the major lineaments in the region, as well as providing a background to the evolution of ideas concerning the way in which crustal structure developed during Mesozoic extension.

Basement in the region comprises a pastiche of rocks of different ages, compositions, histories and rheologies. A simplified version of the broad basement terrane structure is shown in Figure 1.2. The present basement pattern and structure reflects successive phases of crustal extension, intrusion, suturing and accretion, that resulted in a very complex basement geology. In particular, over Phanerozoic time the region was affected by several orogenic episodes, notably the Caledonian (Grampian/Acadian) orogenic cycle of Early Ordovician to Early Devonian age, and the Variscan orogenic cycle with the peak of deformation taking place from latest Carboniferous to early Permian times. These, together with earlier Precambrian deformation episodes, led to the initiation and later reactivation of structural weaknesses and crustal fabrics, and resulted in a complex set of deep-seated crustal lineaments. In turn, these had a major impact on the location and the tectono-stratigraphic development of the overlying sedimentary basins.

Precambrian Framework

The Precambrian is generally taken to cover the vast period before the start of the Cambrian at approximately 542 Ma (Gradstein *et al.*, 2004). It is subdivided into the Proterozoic (542–2500 Ma) and the older Archaean (2500–4567 Ma) history. This part of geological history is relatively poorly constrained in Ireland, and therefore this vast period of time is dealt with very briefly. It is included not just for completeness but because it laid the original foundations for the later structures and events that were to be of profound importance in shaping the formation of the younger sedimentary basins. Precambrian rocks make up approximately 10% of the surface area of Ireland and are described in detail in Daly (2009). These are predominantly seen along the west and northwest

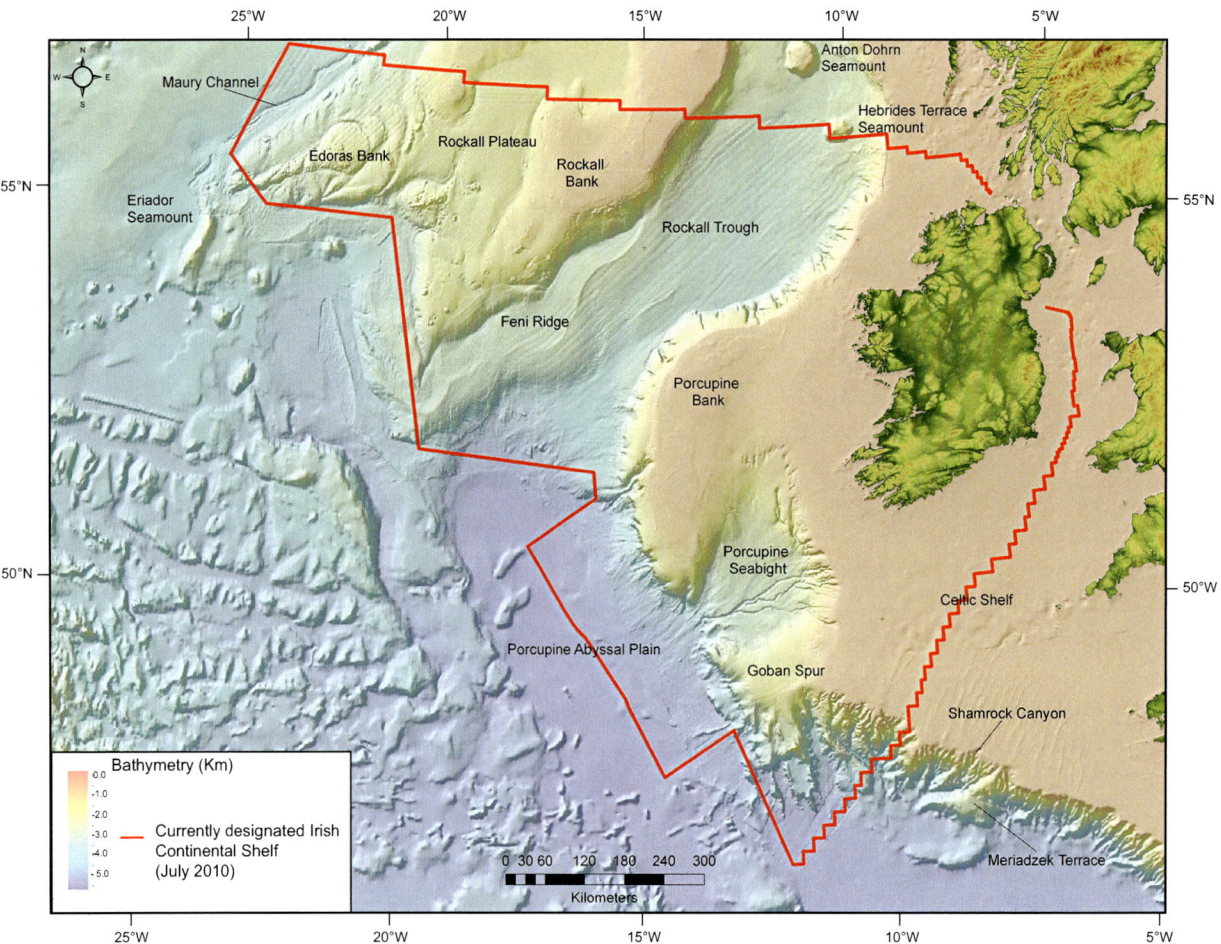

Figure 1.1. Regional swath bathymetry map showing the location of Ireland and its offshore region in the context of the North Atlantic Ocean. Basement blocks and Cenozoic igneous seamounts typically define the shallow waters in the Atlantic region, while the bathymetric troughs overlie major sedimentary basins. (Image from the Irish National Seabed Survey (INSS) courtesy of the Geological Survey of Ireland and the Marine Institute).

Atlantic fringes of Ireland, and also in the extreme south-east of the country. The oldest known Precambrian rocks are the Palaeoproterozoic syenitic orthogneisses on Inishtrahull off Malin Head in County Donegal, which are dated at 1779 ± 3 Ma (Daly *et al.*, 1991). No Archaean rocks (i.e. older than 2500 Ma) are known from Ireland. However, Lu–Hf isotopic data from zircons in granitic rocks from the Tyrone Central Inlier suggest the presence of Archaean rocks at depths (Flowerdew *et al.*, 2008). It is therefore likely that Archaean lithosphere may be present at depth beneath NW Ireland.

In a more regional context, the nearest Archaean rocks to Ireland, dating back to 2900 Ma, are the Lewisian Gneisses that characterise the Outer Hebrides (e.g. Glennie, 1990). Archaean crustal rocks are also suggested to occur west of the Hebrides, beneath the

northern part of the Rockall Trough, in an area lying north of the Anton Dohrn Transfer Fault Zone (Figure 1.2) within the Laurentian continental terrane.

Much of the rest of the Precambrian basement rocks seen in the classical exposures of the Scottish Highlands comprise the Moine and Dalradian supergroups and inform our understanding of regional Precambrian development. They collectively span a period of approximately 600 Ma within Proterozoic time. The Moine Supergroup sediments of Scotland were deposited in the time period 1200–1000 Ma. These Mesoproterozoic rocks are generally preserved between the Moine Thrust and the Great Glen Fault and were subjected to several phases of deformation and metamorphism. East of the Great Glen Fault the Moines are subdivided by some workers into the older Central Highlands and the younger Grampian

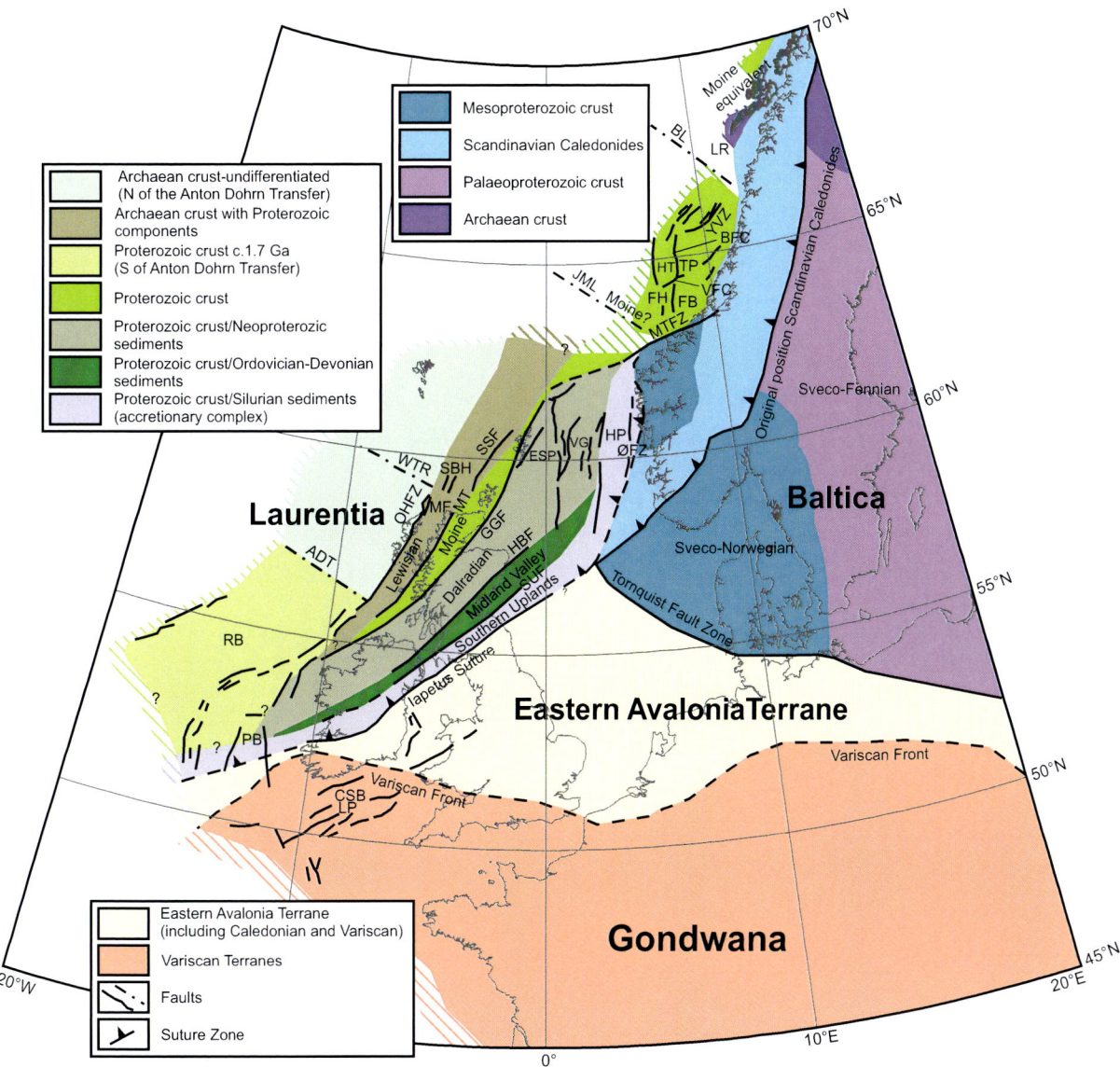

Figure 1.2. Simplified structural and terrane map of the NW European tectonic margin (from Štolfova and Shannon, 2009). BL, Bivrost Lineament; CSB, Celtic Sea Basins; ESP, East Shetland Platform; FB, Froan Basin; FH, Frøya High; GGF, Great Glen Fault; HBF, Highland Boundary Fault; JML, Jan Mayen Lineament; HP, Horda Platform; HT, Halten Terrace; LP, Labadie Bank-Pembrokeshire Ridge; LR, Lofoten Ridge; MF, Minch Fault; MTFZ, Møre-Trøndelag Fault Zone; MT, Moine Thrust; ØFZ, Oygarden Fault Zone; OHFZ, Outer Hebrides Fault Zone; PB, Porcupine Basin; RB, Rockall Basin; SBH, Solan Bank High; SSF, Shetland Spine Fault; SUF, Southern Upland Fault; TP, Trøndelag Platform; VFC, Vingleia Fault Complex; VG, Viking Graben; WTR, Wyville-Thompson Ridge; YVZ, Ylvingen Fault Zone. Dashed areas represent uncertain terrane occurrence or boundary. Based on compilation of previously published work (e.g. Roberts *et al.,* 1999; Coward, 1990, 1995; Plant *et al.,* 2003; Doré *et al.,* 1997; Park *et al.,* 2002).

divisions (Barr *et al.,* 1988). These are separated by a combination of an unconformity and a tectonic slide. However, in this region it has been suggested that the only distinction between them is the degree of deformation and metamorphism. The Grampian Group is now considered by many geologists to be the basal division of the Dalradian, separated unconformably from

the overlying Appin Group. The younger Dalradian Supergroup rocks are largely Neoproterozoic in age, although it is argued by various workers, summarised by Glennie (1990), that Dalradian deposition probably began about the same time as the Moines of the Northern Highlands (<1200 Ma) and continued until the Late Precambrian Grampian Orogeny (~600 Ma). The Moine

Supergroup rocks can generally be considered as being confined to the areas northwest of the Great Glen Fault, while the Dalradian Supergroup succession lies entirely in the area to the southeast. The deformed Dalradian rocks are intruded by the Late Precambrian Ben Vuirich Granite of 590 Ma age (Rogers *et al.*, 1989). Their latest period of metamorphism occurred in the period between the Early Silurian emplacement of the Moine Thrust and the intrusion of Caledonian granites in Early Devonian time. Glennie (1990) considers the Dalradian rocks to be the products of depositional environments that were replicated in a number of separate locations along the southeastern margin of Laurentia.

As mentioned above, the oldest dated rocks in Ireland are the Palaeoproterozoic orthogneisses of the Rhinns Complex in Donegal. Younger Palaeoproterozoic rocks are found in the Annagh Gneiss Complex of NW Mayo, as are some Mesoproterozoic (younger than 1.6 Ga) orthogneisses (Daly, 2009). This rock suite in NW Ireland shows the effects of multiple phases of deformation, including the Grenvillian (approximately 1 Ga), marking the onset of the Neoproterozoic. However, most of the Precambrian rocks exposed in Ireland are supracrustal rocks (i.e. rocks with sedimentary and volcanic protoliths) belonging to the Dalradian Supergroup. This thick succession of Neoproterozoic (younger than 1 Ga) rocks is exposed in a number of inliers in western and northwestern Ireland. Dalradian deposition probably continued into the Middle Cambrian in Ireland, but because it is predominantly Precambrian in age it is treated as part of the Precambrian history. In addition to the Precambrian rocks preserved along the Atlantic coastline of western and northwestern Ireland, a small inlier of Precambrian rocks is also preserved in the southeast of Ireland. In contrast to the Laurentian terrane rocks of the Atlantic margin, the Rosslare Complex contains Neoproterozoic rocks that formed on the edge of the Avalonian continental margin. These rocks preserve evidence of a different history to those of the Laurentian margin rocks and contain the products of three orogenic events (Tietzsch-Tyler, 1996). The oldest of these is the Avalonian Orogeny, dated as approximately 620 Ma. This was followed by the Monian Orogeny (485 Ma) in early Ordovician time, while the youngest was the Caledonian Orogeny at approximately 430 Ma, in Silurian time.

By Late Precambrian time a wide region of oceanic crust, the Iapetus Ocean, had opened. This separated the large Laurentian continent from the small Avalonian microcontinent. The northwest of Ireland and Scotland, including the Grampian Highlands, lay on the eastern edge of Laurentia, while southeast Ireland and the southeast of the UK lay on the trailing northwestern edge of Avalonia. Scandinavia lay on the Baltica continent to the north of Avalonia and to the east of Laurentia. The Iapetus Ocean began to close during the Cambrian, heralding the start of the Caledonian orogenic cycle.

Caledonian Framework

On a regional scale the broadly linear Arctic–North Atlantic Caledonides bifurcate at their southern end into the Scottish–Irish–Appalachian and the North German–Polish Caledonides (Ziegler, 1982). Together with the Mid-European Caledonides and the Ligerian–Moldanubian Cordillera, these fold belts enclose a number of microcontinents (Figure 1.3). The Ligerian Cordillera possibly grade westwards into the northern Appalachian zone of incomplete Late Caledonian consolidation. The northern sector (Arctic–North Atlantic) Caledonides are relatively well understood, with orogenic fronts marking the collision between the Laurentia and

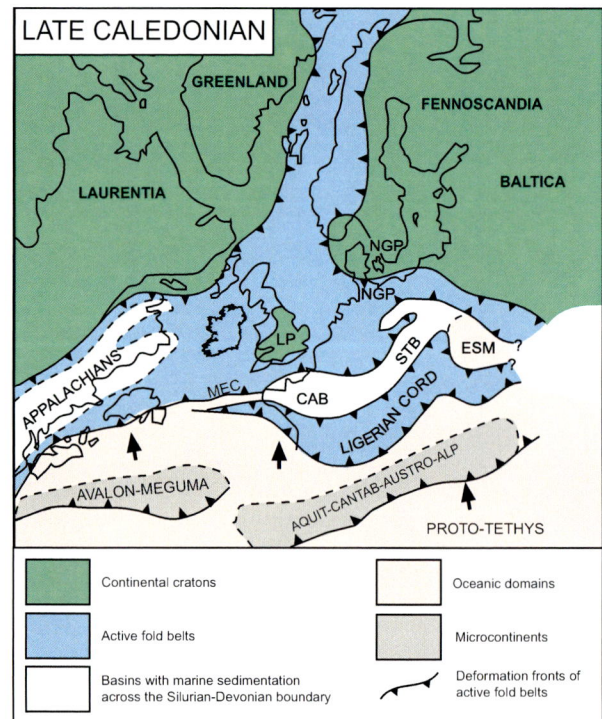

Figure 1.3. Regional Late Caledonian tectonic framework of the North Atlantic region (modified largely from Ziegler, 1982). NGP: North German-Polish Caledonides. LP: London Platform. MEC: Mid-European Caledonides. CAB: Central Armorican Basin. STB: Saxothuringian Basin. ESM: East Silesian Massif.

Baltica continents, but the overall geometry and extent of the southern part of the Caledonides belt is complex and poorly constrained. However, faunal provinciality and palaeomagnetic data suggest that during Cambrian and Early Ordovician times a number of microcontinental fragments spawned off the northern margins of Gondwana, drifted northwards and were successively accreted, during Late Ordovician to Late Silurian times, against the southern margin of the northern continent (Perroud and Bonhommer, 1981). However, the dimensions of the oceans that were subducted during the northward migration of these continents, and the location of the suture zones between the cratonic fragments, are less clearly understood. Nevertheless, the overall evidence is clear that during the Caledonian orogenic cycle, the Laurentia and Baltica continents collided in an east–west direction, and that the oceanic Proto-Tethys plate (Rheic Ocean) converged with the Laurentia and Baltica continents in a broadly north–south direction, thereby giving rise to a three-armed orogenic system (Ziegler, 1982). This indicates that the early Caledonian development involved the convergence of four plates. These were the large Laurentia and Baltica continental plates and the Iapetus and Proto-Tethys (with its micro-continents) oceanic plates. A gross simplification of the Caledonian development model is indicated in Figure 1.4.

From mid-Cambrian time onwards the eastern margin of the Laurentian continental plate carrying parts of Scotland and northwest Ireland lay in a near-equatorial setting and became the site of carbonate platform deposition. The presence of serpentinites, gabbros and basic volcanic rocks in the Highland Border Complex is suggestive of the presence of oceanic crust in the region. By Early Ordovician time the western part of Ireland

and Scotland lay at a latitude of approximately 10° south of the equator, while SE Ireland and England were at a latitude of some 60° S, separated by the wide Iapetus Ocean floored by oceanic crust. By Early Silurian time the Iapetus Ocean had narrowed considerably as a result of subduction, and the margins of the ocean containing the fragments of Ireland and the UK were approximately 500 km apart and straddled 20° S (Cocks and Fortey, 1982). The Tornquist Sea had undergone subduction and had closed by the end of the Ordovician. Although it did not give rise to a major fold belt due to oblique convergence, it resulted in the formation of a major suture, the Tornquist Line (also known as the Tornquist–Teisseyre Line), which was periodically reactivated through time as a major NW–SE deep-seated structural feature in the Danish and Dutch onshore and offshore regions. Its location corresponds to that of the North German–Polish Caledonides (Ziegler, 1982). Other fractures associated with this closure later resulted in the formation of the Ringkøbing-Fyn High, a major structural feature in the evolution of the North Sea.

The Caledonian orogenic cycle, previously typically referred to as the Caledonian Orogeny, encompasses the set of tectonic events that occurred during the closing of the Iapetus Ocean between the continental masses of Laurentia, Avalonia and Baltica (Figure 1.5). In Ireland the cycle comprises two main phases, each associated with a collisional event. The earlier of the two is known as the Grampian Orogeny (Chew, 2009) and took place in Early to Middle Ordovician times (460–475 Ma). Subduction led to the collision of an intra-oceanic island arc system that formed by subduction of oceanic crust within the Iapetus Ocean, with the northern (Laurentian) continental margin. Rocks affected by the

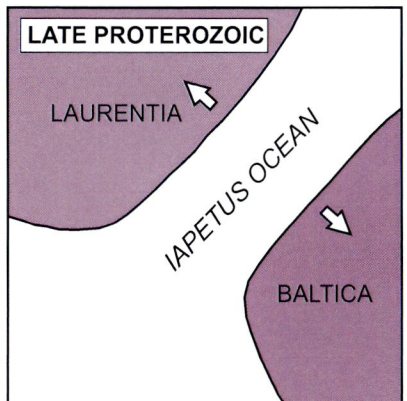
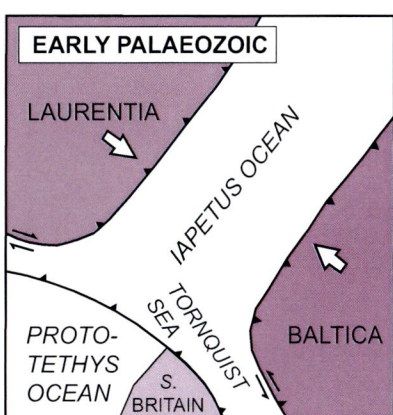
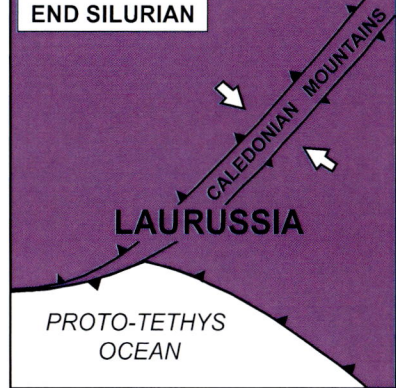

Figure 1.4. Cartoon figures (not to scale) showing the major relative plate movements that occurred from latest Proterozoic through to the end of Early Palaeozoic time (from Glennie, 1990 and Ziegler, 1982).

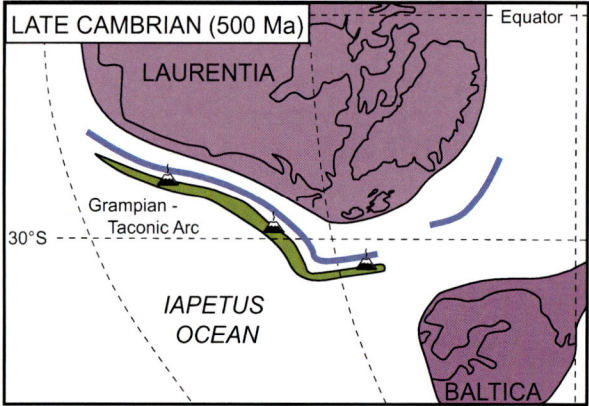

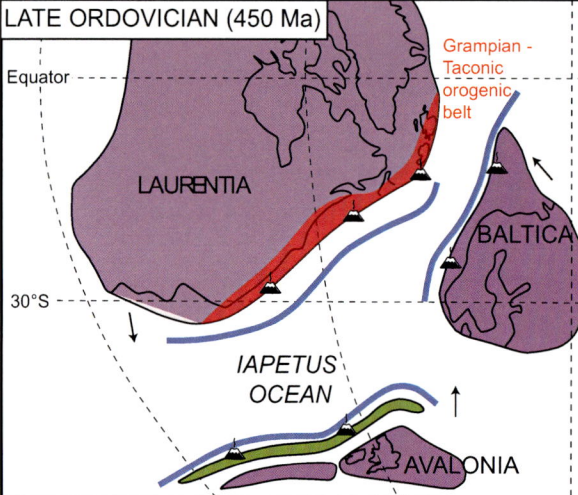

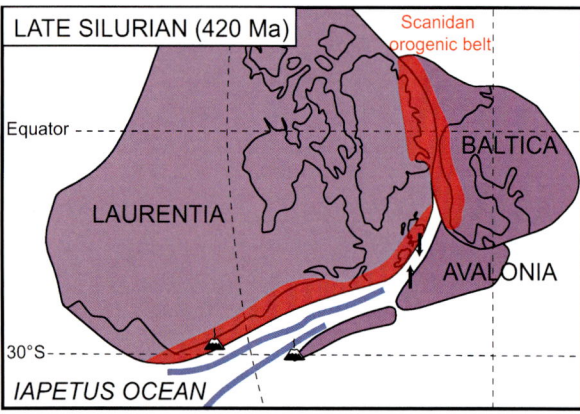

Figure 1.5. Schematic regional evolution of the Caledonian orogenic cycle that documents the closure of the Iapetus Ocean and the multi-continent collision and suturing of Laurentia, Baltica and Avalonia (from Chew, 2009).

(Chew, 2009), itself the continuation of the Highland Boundary Fault of Scotland into Ireland.

Following this collision, northward-directed subduction of oceanic crust beneath the Laurentian continental margin continued through to Late Silurian time. At the same time, southward oceanic subduction occurred beneath Avalonia. By latest Silurian time the oceanic crust between the continents had been subducted and a complex multi-continent collision occurred. The collision between Laurentia and Baltica caused the Scandian Orogeny in the Greenland/Scandinavian region in latest Silurian times. Major N–S striking thrust structures resulted, including westward thrusting of the Moine over the Lewisian basement. The collision also involved the obduction of the Unst ophiolites onto the Dalradian rocks of Scotland (Dewey and Shackleton, 1984). Further south, the more oblique collision between Laurentia and Avalonia, seen in Ireland and Scotland, occurred slightly later, in Early Devonian time (405 Ma), and gave rise to the Acadian Orogeny which produced a more NE–SW general strike of tectonic structures than in the northern Iapetus collision zone. This was the final closure of the Iapetus Ocean and resulted in the formation of the Laurussia megacontinent. The boundary between Laurentia and Avalonia is termed the Iapetus Suture, a broad band that can be mapped trending in a NE–SW direction across Ireland from north of Dublin to south of the estuary of the River Shannon. Overall, most of the major strike-slip faults that bound the various Caledonian terranes of the region formed zones of weakness that were periodically reactivated during the later history of the region.

On a regional scale the Caledonian orogenic cycle resulted in the assembly of a series of major structural massifs in northwest Europe. The major ones are the Laurentian, Baltica and Welsh–Brabant massifs (Coward, 1995). The Laurentian craton is cross-cut by the Anton Dohrn Transfer Fault Zone, a major terrane boundary between Archaean gneisses in the north, and Proterozoic crust in the south (Dickin, 1992) (Figure 1.2). Northwest of the Hebrides, mid-crustal structures are horizontal or gently dipping to the southeast and were generated by NW-verging thrusts followed by NW–SE extension in the Middle Proterozoic (Coward, 1995). Reactivation of these structures occurred during Late Palaeozoic–Mesozoic times (Stoker *et al.*, 1993; Hitchen *et al.*, 1995; Doré *et al.*, 1997). Laurentian crystalline basement is commonly overlain by Torridonian, Permo-Triassic and Jurassic strata (e.g. Hitchen *et al.*, 1995). Magmatic arc

Grampian Orogeny are restricted to the area northwest of the Iapetus Suture. The boundary between the rocks on the Laurentian margin and those of the colliding arc terrane is marked by the Fair Head–Clew Bay Line accretion in the Middle Proterozoic resulted in the formation of Baltica (Coward, 1995). Roberts *et al.* (1999) suggested that major terrane boundaries in Scandinavia include the Møre–Trøndelag Fault Zone, the Caledonian Front and the Tornquist Line, which separates the Baltic and Eastern Avalonia terranes (Figure 1.2). Basement structures within the Baltica basement rotate from NE–SW in the south to a more N–S orientation in the north. A well-established major fault boundary separates the Caledonian fold belt from its foreland in Greenland and North America. Within the Archaean and Proterozoic foreland of Greenland and eastern Canada, older terrane boundaries (e.g. Grenville and Ketilidian fronts) trend at a high angle to the frontal thrust zone of the Caledonides. The Outer Isles Thrust, west of Scotland, has been traditionally correlated with the Caledonian Front, but Roberts *et al.* (1999) suggested, on the basis of deep seismic data, that the Caledonian Front, linking into major terrane boundaries on Newfoundland and North America, would be more appropriately placed at the eastern margin of the Rockall Trough, and controlled the location of the later Mesozoic and Cenozoic basin.

On a more local scale, Late Caledonian (Acadian) deformation affected four principal pre-Devonian crustal blocks in Ireland. These are referred to (Chew and Stillman, 2009) as the Grampian, Midland Valley, Longford-Down and Leinster terranes (Figure 1.6). These were of different ages, lithologies and rheologies, and their tectonic response to orogeny was different. The Grampian terrane comprises largely Dalradian metasediments, and underwent multiphase ductile deformation prior to granite intrusion and further deformation. It was cut by a series of major crustal lineaments, the most pronounced of which are the Leannan Fault (a major splay of the Great Glen Fault of Scotland), and the Fair Head–Clew Bay Line which forms the southeastern limit to the Dalradian outcrop in Ireland. The Midland Valley terrane consists of Ordovician rocks, typically deep-water clastic sediments and volcanics. The rocks in this terrane underwent several phases of deformation, dominated by sinistral transpression. Both the Grampian and Midland Valley terranes show a pronounced strike swing from NE–SW to E–W towards the Atlantic. The Longford-Down terrane lies to the southeast of the Grampian and Midland Valley terranes and consists of deep marine Lower Palaeozoic clastics deposited on the northwest margin of the Iapetus Ocean. It continues into Scotland as the Southern Uplands terrane, and represents an accretionary prism succession formed above a NW-directed subduction of the Iapetus oceanic slab beneath the Laurentian continent. The Iapetus Suture forms the southern margin of the Longford-Down terrane. To the south lies the Leinster terrane. Regional deformation in this area is largely thought to be Acadian in age. A series of deformation phases occurred before and after major granite intrusion.

Following the latest Caledonian orogenic phase, the tectonic setting of the general region underwent a major change. The compressional setting of the North German–Polish and the Mid-European Caledonides terminated during Early Devonian time and gave way to a tensional regime (Figure 1.7). However, subduction of the Proto-Tethys along the southern margin of the Ligerian–Moldanubian Cordillera persisted and culminated in the Middle Devonian Ligerian diastrophism (Ziegler, 1982). In addition, contemporaneous tectonic events are also present in the Innuitian fold belt of the Canadian Arctic Islands and northern Greenland (Trettin, 1973; Trettin and Balkwill, 1979).

Variscan Framework

To the south of Ireland, the Proto-Tethys oceanic plate subducted northwards beneath the new Late Caledonian magacontinent. This subduction commenced during the Caledonian orogenic cycle and by Late Devonian time the Proto-Tethys Ocean was narrowing. Gondwana approached the Laurussia megacontinent, which was now undergoing internal sinistral oblique movement focused on the general vicinity of the Caledonian suture zone. By latest Carboniferous to earliest Permian times, continued subduction had resulted in the collision of Gondwana against Laurussia, with the complete subduction of the oceanic crust of the Proto-Tethys, the formation of a generally east–west orogenic Variscan orogenic fold/thrust belt and the formation of the Pangaean Supercontinent. This now provided the structural template for the location of the younger sedimentary basins upon a complex basement of annealed terranes.

The Variscan (also commonly referred to regionally as Hercynian or Armorican) orogenic belt can be traced across western Europe to the Atlantic and further west to the eastern seaboard of the United States (*see* Figure 2.1 in the next chapter). South of the main orogenic front, extending from Belgium to the west of Ireland, the overall pattern of the fold belt and accreted terranes is highly arcuate. It changes trend from WNW–ESE to ENE–WSW and to broadly north–south between Newfoundland and Iberia (Roberts *et al.*, 1999).

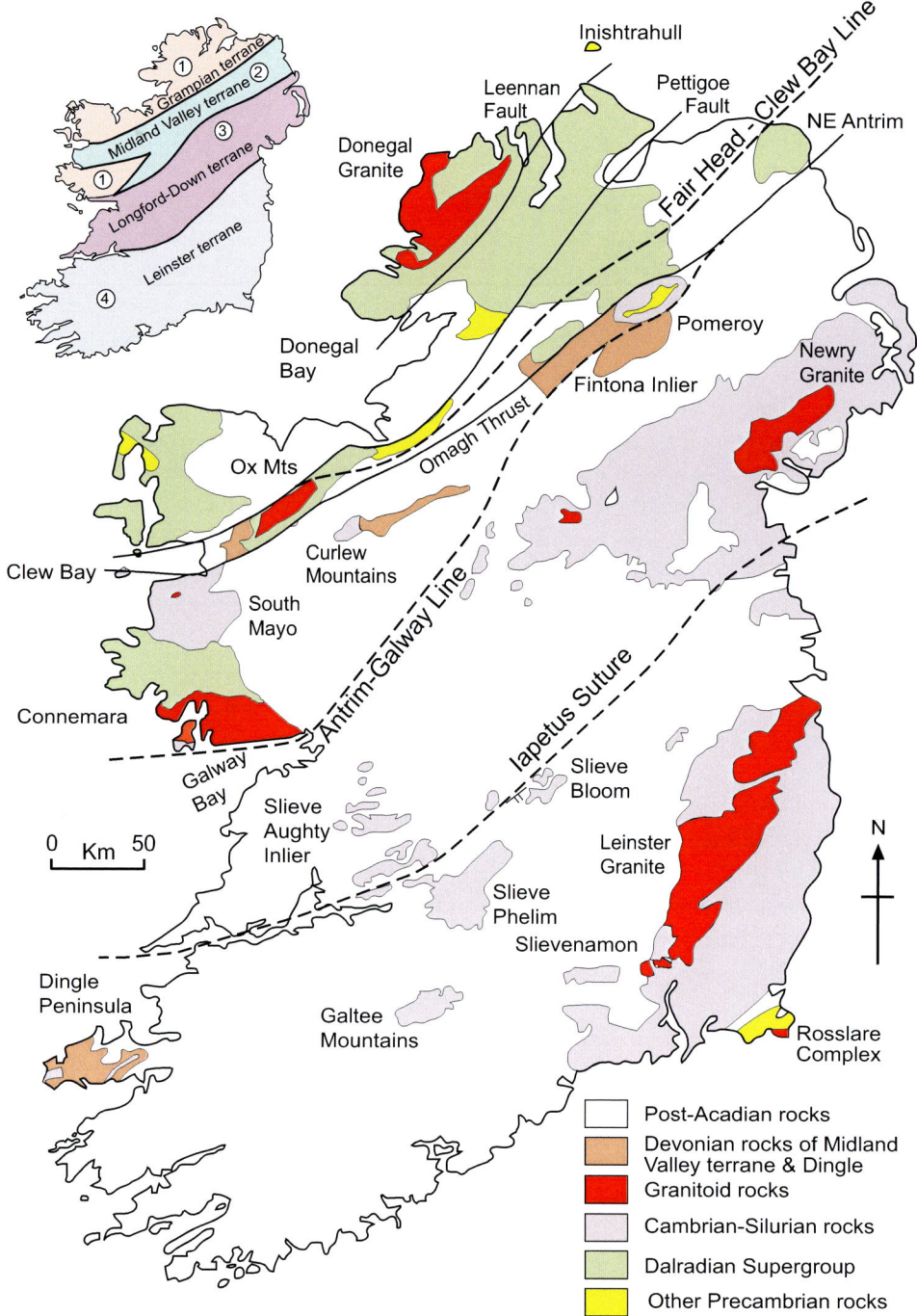

Figure 1.6. Map of Ireland showing the location of major Caledonian structures and terranes (from Chew and Stillman, 2009).

Variscan structures are excellently exposed in the south of Ireland and also locally further north in the country. Major changes in the style of Variscan deformation occur from south to north (Graham, 2009). Tight folds and thrusts in the south of Ireland give way to broad gentle folds, with some fault reactivation further north. Gill (1962) and Cooper *et al.* (1986) used the nature and

style of the deformation to define various Variscan zones. However, the boundaries between these zones are often neither clear nor sharp. A major feature, often referred in the literature as the 'Dingle–Dungarvan Line' has been labelled by many authors as the Variscan Front, the northern boundary between the Rheno-Hercynian zone of major thrusting and folding and the foreland

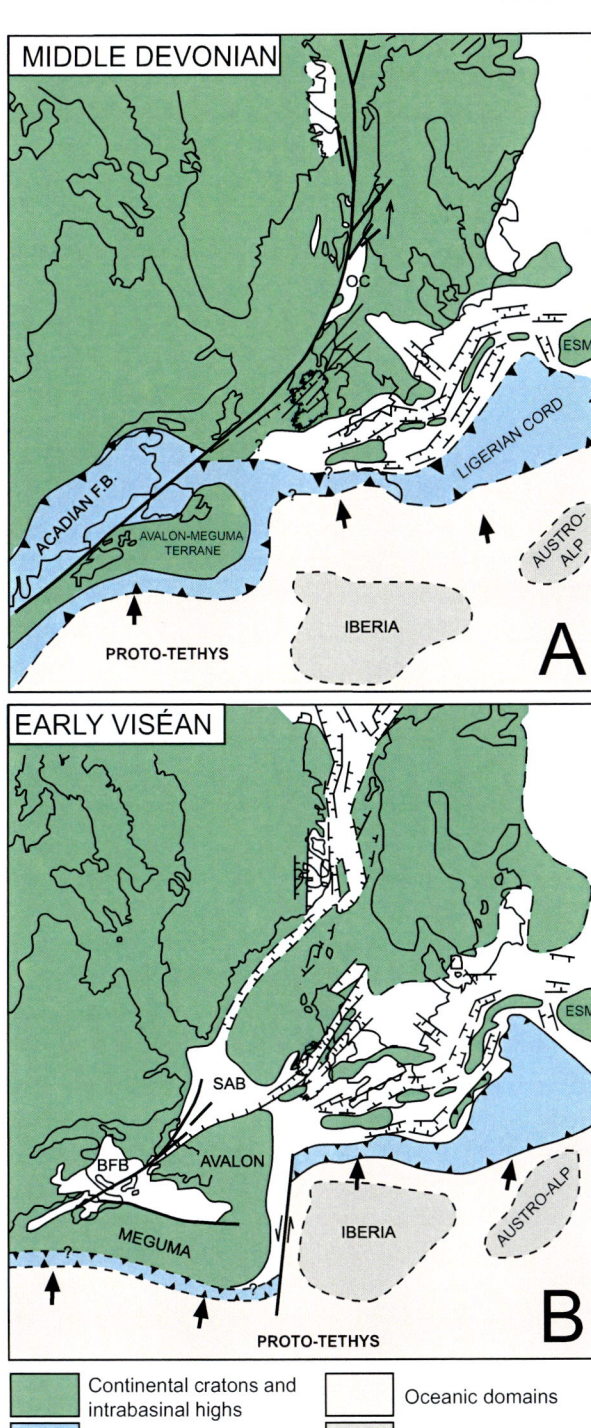

Figure 1.7. (A) Regional Middle Devonian tectonic framework and (B) Early Carboniferous tectonic framework of the North Atlantic region (modified from Ziegler, 1982). OC: Orcadian Basin. ESM: East Silesian Massif. SAB: Saint Anthony Basin. BFB: Bay of Fundy Basin.

zone where deformation is not as severe. In the southwest of the country this is marked by a major fault, the Killarney–Mallow Fault, but the boundary is less obvious further east where there does not appear to be a major change in either original depositional facies or in the strain gradient (Naylor and Sevastopulo, 1979; Bresser, 2000). Geophysical evidence, summarised in Graham (2009), indicates that Variscan deformation onshore in Ireland is likely to be thin-skinned in nature, and does not involve the entire crust. Detachments within the crust, combined with the likely presence of subsurface Caledonian granites, are likely to have played a role in the formation of the overall geometry of the Variscan structural configuration. The Variscan orientation in the south of Ireland was likely influenced by inherited Caledonian structures and features. In the south of Ireland the Caledonian structures swing from NE–SW to a more east–west trend as they approach the main thrust front. However, it is difficult to trace the line of the latter into the offshore, and it is possible that either the two major orogenic trends merge or that the Variscan Front dies out towards the west into the offshore. Further discussion of Variscan structure in southernmost Ireland is presented in Chapter 4 where the onshore succession and basins are described, while the details of the influence of Caledonian/Variscan structure on the development of Mesozoic basins in the Celtic Sea region can be found in Chapter 10.

The Devonian to early Carboniferous evolution of the Variscan basinal system of west-central and Northwest Europe reflects an interplay between tensional and compressional tectonics. From the late Early Carboniferous onwards, tensional effects were minimal and compressional structures became the dominant feature of the regional development. By Moscovian (late Westphalian) time formation of the Variscan fold belt was essentially complete. Thereafter it provided the structural template for the formation of the younger post-orogenic basins and ultimately for the extensional basins that developed as the Pangaean supercontinent disintegrated and crustal extension and ultimate crustal rupture occurred.

The Variscan orogenic development was the response to the progressive northward subduction of the Proto-Tethys oceanic crust beneath the Laurussia megacontinent (Ziegler, 1982; Glennie, 1990). Accretion of the Avalon–Meguma and the Austro-Alpine microcontinents against the southern margin of Laurussia megacontinent during the Middle Devonian was followed by collision of the Aquitaine–Iberian microcontinent

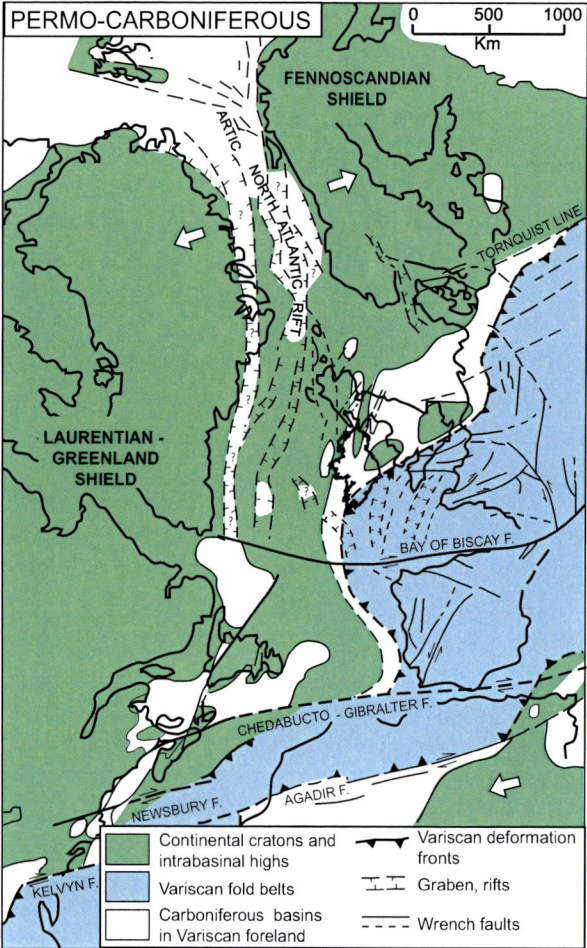

Figure 1.8. Latest Variscan to early Permian tectonic framework of the Atlantic-Arctic region showing major reactivated lineaments. Arrows indicate the major plate movement (from Ziegler, 1982).

during the late Early Carboniferous (Figure 1.7). By Late Viséan time the western parts of the Proto-Tethys Ocean had been subducted and continent–continent collision between Laurussia and Gondwana commenced, resulting in major deformation and orogeny. North-directed thrusting overrode the back-arc basin system, causing downwarping of the northern foreland, a process that was accentuated by loading with the erosional products of the Caledonian Mountains in the north and the rising Variscan Mountains to the south. This gave rise to the deposition of thick accumulations of Serpukhvian and Pennsylvanian (late Namurian–Westphalian) Coal Measures. The Asturian compressive phase, of Moscovian (Late Westphalian) to Early Stephanian times, represents the last orogenic pulse that affected parts of the Variscan fold belt. In the late stages of the Variscan Orogeny the rigid basement margins of the

Laurentian and Baltic components of Laurussia formed a re-entrant into which the orogen was squeezed from the south. This resulted in NE–SW right-lateral wrench faults. The arcuate shape of the Variscan fold belt resulted from draping of mobile fold belts around the more rigid accreting microcratons, as well as being controlled by the location and shapes of the earlier Devonian and Early Carboniferous rifts. Following the consolidation of the Variscan fold belt in latest Carboniferous to earliest Permian times, the north–south convergence between Gondwana and Laurussia changed to a more east–west convergence, resulting in the development of a right-lateral transform system. This is represented by a set of major wrench fault systems. These include the major Kelvyn–Agadir, Chedabusco–Gibralter and Bay of Biscay fault systems, together with the slightly oblique Tornquist–Teisseyre lineament, reactivated from an inherited Caledonian structure (Figure 1.8). These were the sites of major wrench, compressional and inversion structures, and were locally accompanied by volcanism in early Permian time.

Crustal Structure

While the current crustal structure of the region is reasonably well documented from deep seismic and other geophysical studies, the pre-Permian crustal make-up of the northwest European region is still poorly understood. The thickness of the crust probably varied regionally due to the existence of the Variscan orogen and remnants of the Caledonian orogen. The crustal thickness of the basement massifs, which were little affected by later lithospheric extension during Mesozoic basin formation, may serve as a proxy for estimating probably pre-Permian crustal thickness. The crustal thickness beneath the Irish massif ranges from 29 km to 33 km (Landes *et al.*, 2003; 2005), the Cornubian Massif from 30 km to 32 km (Tesauro *et al.*, 2008), and the present thickness of the crust beneath the East Shetland Platform is approximately 34 km (Chadwick and Pharaoh, 1998). The thickness of the Baltica crust in onshore areas, in the mid-Norway region, varies from 27 km to 37 km (Faleide *et al.*, 2008). Comparison with the nearest younger orogenic system (e.g. Pyrenees), where the crustal thickness is well constrained (Díaz and Gallart, 2009), suggests that the Variscan crust was thinned to 28–35 km from an original thickness of 40–60 km during the Early Permian in Northwest Europe. A somewhat thicker crustal root, a residual of the Caledonian orogenic accretion, could also be anticipated at the time in the western UK and part of the Norwegian offshore. In the East Shetland Basin region the initial Permian crustal thickness is suggested to have varied from approximately 35 km in the platform areas to less than 30 km beneath the basin (Kinck et al.,1991; Odinsen et al., 2000). This is slightly thicker than normal crust and is thought to reflect residual thickening following Caledonian continental suturing.

The thickness of the crust beneath Europe at the present time is quite variable (Tesauro et al., 2008). It is thickest in the central European and Scandinavian shield region (40–50 km). It is also thick in elongate belts in the Alpine and Pyrenees regions, reflecting crustal thickening beneath the young mountains. In general, the crust is approximately 30 km thick beneath much of the Irish, UK, French, Dutch and Iberian regions but thins significantly beneath the offshore basins, especially those of the Atlantic margins.

Seismic refraction studies (Lowe and Jacob, 1989) indicate the presence of rocks with velocities greater than 6 km s^{-1} at a relatively shallow depth beneath the Palaeozoic cover in central Ireland. These have generally been assumed to represent crystalline Precambrian basement. However, some constraints on crustal structure and composition beneath Ireland can be obtained from wide-angle shear wave seismic data and from lower crustal xenoliths (Hauser et al., 2008). The results indicate that the Irish crust in the eastern Avalon terrane within the Iapetus Suture Zone is unusually felsic in bulk composition. Variations with Poisson's ratio within the upper 5 km of crust are greatest and can be correlated with siliclastic and carbonate sediments in Devonian and Carboniferous sedimentary basins that formed during the early part of the Variscan orogenic cycle. The more uniform variation in the mid crust down to 15 km depth is consistent with a granite/granodiorite or metagreywacke composition of greenschist to amphibolite facies mineralogy basement. U–Pb zircon dating indicates that the xenoliths comprise sedimentary, volcanic and granitic protoliths of Palaeozoic age (Daly and Van den Berg, 2004) and thus are more likely to represent high metamorphic grade Caledonian rocks, rather than Precambrian basement. The lower crust was therefore probably largely derived by the accretion of sedimentary materials derived from oceanic, island arc, and continental margin sources during Caledonian collision.

Within the past couple of decades a significant amount of petroleum industry and academic information, mostly in the form of deep and wide-angle seismic data, supplemented by a large volume of gravity, magnetic, seismic reflection and some petroleum exploration boreholes, has provided a general picture of the crustal thickness and broad structure of the Irish offshore. However, the location, orientation and influence of the inherited basement fabrics still remain uncertain and loosely constrained. The crust beneath onshore Ireland is approximately 30 km thick, with little evidence of stretching or thinning. Beneath the Irish and Celtic Sea regions it is approximately 25 km thick, with evidence of crustal decoupling to accommodate the slight thinning beneath the basins (O'Reilly et al., 1991). The nature and the thickness of the crust beneath the large outboard Atlantic basins such as the Porcupine and Rockall basins has been the subject of considerable debate during the years (see Smythe, 1989; Shannon et al., 1999). However, robust geophysical evidence from the RAPIDS (Rockall And Porcupine Irish Deep Seismic) profiles now points to the presence of severely attenuated continental crust beneath the largest of the basins (Figure 1.9), intruded in places by mantle serpentinites or volcanics (Makris et al., 1988, 1991; Hauser et al., 1995; O'Reilly et al., 1996, 2006; Reston et al., 2001). Beneath the centres of the Rockall and Porcupine basins the continental crust is 2–5 km thick and has been modelled to indicate differential stretching, with greater upper and middle crustal extension facilitated by rheologically-controlled detachments at the top of the lower crust. The uppermost mantle is serpentinised where the stretching is most severe, and this may have provided a degree of added strength to the lithosphere to prevent rupture.

While the thickness of the crust on the continental shelf has been well constrained by deep-seismic profiles (Morewood et al., 2005; O'Reilly et al., 2006), little is known about the composition or detailed nature of the crust or its age. Naylor and Shannon (2005) identified several distinct basement (pre-Permian) provinces in the Porcupine–Rockall region, based on seismic character and on the structure of the overlying basin sediments. The European Atlantic margin was interpreted by Doré et al. (1999) as an oblique re-opening of the Caledonian suture and fold system, and thus the present western Rockall margin may follow an earlier basement lineament. Recent data from the Hatton region (Hitchen, 2004) demonstrate that the area has a thick cover of Cenozoic to possibly Palaeozoic sedimentary rocks. Rockall Bank is thought to belong to the Islay (Rockall Bank basement) terrane, with the boundary against Lewisian rocks lying to the north, in UK waters (Hitchen et al., 1997). Similar rocks may underlie the

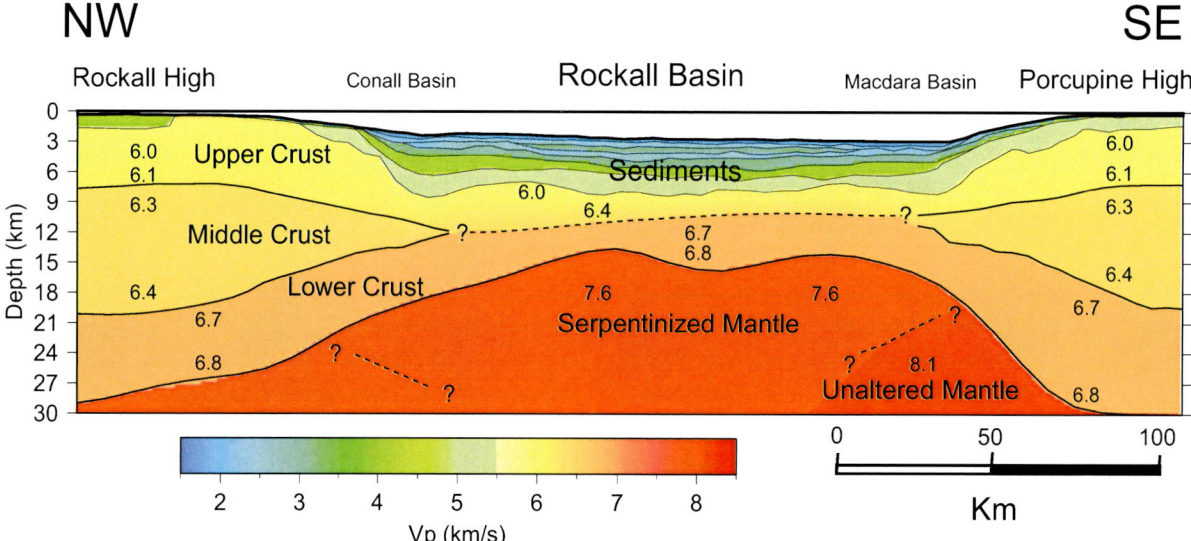

Figure 1.9. Crustal velocity structure across the Rockall Basin, based on wide-angle seismic reflection studies (after Morewood *et al.,* 2005). The location of the figure is shown on Figure 11.2.

younger Hatton Bank succession. The Rockall High comprises a distinct province of Precambrian rocks, with no evidence of Palaeozoic cover. This suggests that the extension that formed Rockall Basin took place along an earlier basement structure that had either controlled Palaeozoic deposition, or at the very least controlled the pattern of later denudation. Tyrrell *et al.* (2007) speculated on the nature of basement terranes and terrane boundaries on the Atlantic margin based upon the geochemistry of the few basement samples taken in the offshore region, integrated with the onshore basement of Ireland, the UK, Greenland and Canada. Their work on Pb isotopic domains suggested the existence of five major domains (an Archaean, two Proterozoic, an Avalonian and a Variscan domain), extending through the offshore region. While some of these are suggested to extend along strike from onshore Ireland and the UK into the Porcupine and Rockall regions, the paucity of *in situ* offshore basement samples places a high degree of uncertainty on their precise boundaries, and indeed, on whether other unsampled basement terranes may exist in the region.

Major Lineaments and Structures

What is clear from the available geological and geophysical data is that the continental shelf of NW Europe, and especially offshore Ireland, bears the imprints and structures resulting from Variscan, Caledonian and older orogenic events. Reactivation of some of these basement structures clearly influenced the location, orientation and large-scale structures of the sedimentary basins. On a regional scale, closure of the Iapetus Ocean in late Silurian to earliest Devonian times led to the docking and suturing of distinct basement terranes. This imposed a general north–south orientation on pre-Mesozoic basement in Norway, changing southwards to NE–SW through Scotland and Ireland. The Mesozoic basins and the continent–ocean margin in these regions typically parallel the Caledonian structures. These are predominantly NE–SW to N–S in the Norwegian and Scottish regions overlying Caledonian and older basement, whereas further south, overlying Variscan basement, structures commonly have an E–W strike. The Møre–Trøndelag Fault Zone, the Great Glen Fault and the southeastern margin of the Rockall Basin are NE–SW orientated, commonly regarded as being of Late Caledonian age (Rumph *et al.,* 1993; Coward 1995; Doré *et al.,* 1999), that affected post-Carboniferous basins. Some structures within the Møre–Trøndelag Fault Zone may also have a Proterozoic origin and such structures were reactivated periodically during the Palaeozoic (Grønlie and Roberts, 1989). The N–S trend seen west of the Trøndelag Platform, in the Viking Graben and in the Porcupine Basin (Figure 1.2) is also regarded as being an inherited orientation. Færseth *et al.* (1995) suggested a Proterozoic origin of the N–S structural orientation in the Norwegian offshore.

A NW–SE structural fabric, which offsets Mesozoic–Cenozoic basins, is observed along the entire Northwest European Atlantic margin. Examples include the Bivrost and the Jan Mayen lineaments and the Anton Dohrn Transfer Fault Zone (Figure 1.2). This trend influenced basin geometry west of Shetlands and in the Faroes (Rumph *et al.*, 1993). It probably represents Caledonian and older structural fabrics, prominent in the basement on both sides of the Caledonian orogenic front (Doré *et al.*, 1999; Brekke, 2000). Structures within the Baltica basement swing from NE–SW in the south to a more N–S orientation in the north. Basement structures within the Variscan region are broadly E–W to ENE–WSW.

Focusing on the slightly smaller scale, the control exerted by basement structure can be seen clearly in the orientation of many of the Irish basins and in many of the basin boundaries. The early Mesozoic basins in the North Channel area follow the NE–SW structural fabric of the Southern Uplands, while the Central Irish Sea Basin has a general NE–SW orientation, parallel to the Caledonian lineaments in the basement rocks of both SE Ireland and Wales/Anglesey (*see* Chapter 12). Onshore in Ireland the NE–SW Caledonian fabric in the north of the country changes to an ENE–WSW alignment towards the south coast, so that the Caledonian and Variscan fabrics are sub-parallel. The North Celtic Sea Basin has a more ENE–WSW orientation (*see* Figure 2.15 in the next chapter), reflecting either a strike swing in the Caledonian fabric, or a Variscan structure (or possibly a combination, with a Caledonian structure that was reactivated in Variscan times). Within the Celtic Sea region, deep seismic profiles suggest that the steep northern margin of the basin is controlled in places by extensional reactivation of a southward-dipping Variscan structure that detaches in the lower crust (Shannon, 1991, and *see* Chapter 10). The ENE–WSW fabric is seen again in the eastern part of the Goban Spur province and in the orientation of the Porcupine Fault at its northern margin.

A number of different interpretations have been proposed for the extension of major faults and lineaments westwards from Ireland across the Atlantic Continental Platform, based in the main on gravity and magnetic data (Young and Bailey, 1973; Riddihough, 1975; Max *et al.*, 1982; Masson *et al.*, 1985). Nevertheless there is general agreement that the Great Glen and Fairhead–Clew Bay fault systems trend southwestwards towards the north of the Porcupine Basin. Reactivated splays from these systems may have played a role in the segmentation of the Slyne–Erris and North Porcupine basins into discrete sub-basins.

Klemperer (1989) and Klemperer *et al.* (1991) considered that the Great Glen and Clew Bay faults, and possibly the Variscan Front, can be identified on deep seismic profiles. They suggested that north-dipping reflectors west of the Shannon estuary on the BIRPS WIRE 1 profile are structures related to the Iapetus Suture. Any westward extension of the Variscan Front or the Iapetus Suture could be anticipated to cross the Porcupine Basin, but these have not been identified and there are no clear effects of any such deep structures on the orientation or development of the overlying Mesozoic–Cenozoic basin.

There is a major swing in strike of the eastern margin of Rockall Basin at approximately 54°N. To the north, the margin parallels the NE–SW caledonoid strike of the narrow Slyne–Erris basin system. At 54°N the strike changes almost to E–W. The strike change lies due east of a similar strike swing in the Caledonian of NW Ireland documented by Hutton and Alsop (1996), which they believe to have a deep pre-Caledonian (>600 Ma) structural origin. A possible extension of the important Donegal Lineament of Hutton and Alsop (1996) could have influenced the trend of the Rockall margin at this latitude. The NE–SW faults of the set of small basins in the footwall of the main Rockall Basin (the Colm, Fursa and Macdara basins of Naylor *et al.*, 1999) along the 54°N sector are oblique to, and intersect, the margin in much the same way as the Leannan and related Caledonian faults are shown as intersecting the older lineaments onshore. Bends in the Rockall Trough in the UK sector were also thought by Musgrove and Mitchener (1996) to be controlled by the geometry of the Caledonian foreland thrust belt. South of 54°N the eastern Rockall margin is aligned N–S parallel to the axis of the Porcupine Basin. In the Goban Spur province the NW–SE trend is parallel to the main margin faults and to the orientation of the continent–ocean boundary.

References

Barr, D., Strachan, R.A., Holdsworth, R.E,. and Roberts, A.M. (1988) Summary of the Geology. *In*: Allison, I., May, F., and Strachan, R.A. (eds), *An Excursion Guide to the Moine Geology of the Scottish Highlands.* Scottish Academic Press, Edinburgh, 11–38.

Brekke, H. (2000) The tectonic evolution of the Norwegian Sea Continental Margin with emphasis on the Vøring and Møre Basins. *In*: Nøttvedt, A. (ed.), *Dynamics of the Norwegian Margin*. Geological Society, London, Special Publications, **167**, 327–378.

Bresser, G. (2000) *An integrated structural analysis of the SW Irish Variscides.* Aachener Geowissenschaftliche Beiträge, 190pp.

Chadwick, R.A. and Pharaoh, T.C. (1998) The seismic reflection Moho beneath the United Kingdom and adjacent areas. *Tectonophysics*, **299**, 255–279.

Chew, D.M. (2009) Grampian Orogeny. In: Holland, C.H. and Sanders, I.S. (eds), *The Geology of Ireland.* Dunedin Academic Press, Edinburgh, 69–102.

Chew, D.M. and Stillman, C.J. (2009) Late Caledonian orogeny and magmatism. In: Holland, C.H. and Sanders, I.S. (eds), *The Geology of Ireland.* Dunedin Academic Press, Edinburgh, 143–173.

Cocks, L.R.M. and Fortey, R.A. (1982) Faunal evidence for oceanic separations in the Palaeozoic of Britain. *Journal of the Geological Society of London*, **139**, 465–478.

Cooper, M.A., Collins, D., Ford, M., Murphy, F.X., Trayner, P.M., and O'Sullivan, M. (1986) Structural evolution of the Irish Variscides. *Journal of the Geological Society of London*, **143**, 53–61.

Coward, M.P. (1990) The Precambrian, Caledonian and Variscan framework to NW Europe. In: Hardman, R.F.P. and Brooks, J. (eds), *Tectonic Events Responsible for Britain's Oil and Gas Reserves.* Geological Society, London, Special Publications, **55**, 1–34.

Coward, M.P. (1995) Structural and tectonic setting of the Permo-Triassic basins of northwest Europe. In: Boldy, S.A.R. (ed.), *Permian and Triassic rifting in Northwest Europe,* Geological Society, London, Special Publications, **91**, 7–39.

Daly, J.S. (2009) Precambrian. In: Holland, C.H. and Sanders, I.S. (eds), *The Geology of Ireland.* Dunedin Academic Press, Edinburgh, 7–42.

Daly, J.S. and Van den Berg, R. (2004) Ion microprobe U–Pb zircon evidence from lower crustal xenoliths for Palaeozoic crust formation and crustal development within the Caledonian suture zone in central Ireland. *Geological Society of America Abstracts with Programs*, **36**, No. 5, 48.

Daly, J.S., Muir, R.J., and Cliff, R.A. (1991) A precise U–Pb zircon age for the Inishtrahull syenitic gneiss, Co. Donegal, Ireland. *Journal of the Geological Society of London*, **148**, 639–642.

Dewey, J.F. and Shackleton, R.M. (1984) A model for the evolution of the Grampian tract in the early Caledonides and Appalachians. *Nature*, **312**, 115–121.

Díaz, J. and Gallart, J. (2009) Crustal structure beneath the Iberian Peninsula and surrounding waters: a new compilation of deep seismic sounding records. *Physics of the Earth and Planetary Interiors*, **173**, 181–190.

Dickin, A.P. (1992) Evidence for an Early Proterozoic crustal province in the North Atlantic region. *Journal of the Geological Society of London*, **149**, 483–486.

Doré, A.G., Lundin, E.R., Fichler, C., and Olesen, O. (1997) Patterns of basement structure and reactivation along the NE Atlantic margin. *Journal of the Geological Society of London*, **154**, 85–92.

Doré, A.G., Lundin, E.R., Jensen, L.N., Birkeland, Ø, Eliassen, P.E., and Fichler, C. (1999) Principal tectonic events in the evolution of the northwest European Atlantic margin. In: Fleet, A.G. and Boldy, S.A.R (eds), *Petroleum Geology of Northwest Europe: Proceedings of the 5th Conference.* Geological Society, London, 41–61.

Færseth, R.B., Gabrielsen, R.H., and Hurich, C.A. (1995) Influence of basement in structuring of the North Sea Basin, offshore southwest Norway. *Norsk Geologisk Tidsskrift*, **75**, 105–119.

Faleide, J.I., Tsikalas, F., Breivik, A.J., Mjelde, R., Ritzmann, O., Engen, Ø., Wilson, J., and Eldholm, O. (2008) Structure and evolution of the continental margin off Norway and the Barents Sea. *Episodes*, **31**, 82–91.

Flowerdew, M.J., Daly, J.S., Chew, D.M., Millar, I.L., and Horstwood, M.S.A. (2008) In situ Hf isotopic measurements of complex zircons from Irish granitoids reveal hidden Palaeoproterozoic and Archaean sources at depth. *Abstracts Volume, 2008 Highland Workshop 24th, 25th April*, 24.

Gill, W.D. (1962) The Variscan Fold Belt in Ireland. In: Coe, K. (ed.), *Some Aspects of the Variscan Fold Belt.* Manchester University Press, 44–64.

Glennie, K.W. (1990) Outline of North Sea History and Structural Framework. In: Glennie, K.W. (ed.), *Introduction to the Petroleum Geology of the North Sea.* Blackwell Scientific Publications, London, 34–77.

Gradstein, F.M., Ogg, J.C., and Smith, A.G. (2004) *A geological timescale 2004.* Cambridge University Press, 589pp.

Graham, J.R. (2009) Variscan deformation and metamorphism. In: Holland, C.H. and Sanders, I.S. (eds), *The Geology of Ireland.* Dunedin Academic Press, Edinburgh, 295–310.

Grønlie, A. and Roberts, D. (1989) Resurgent strike-slip duplex development along the Hitra–Snåsa and Verran Faults, Møre–Trøndelag Fault Zone, Central Norway. *Journal of Structural Geology*, **11**, 295–305.

Hauser, F., O'Reilly, B.M., Jacob, A.W.B., Shannon, P.M., Makris, J., and Vogt, U. (1995) The crustal structure of the Rockall Trough: differential stretching without underplating. *Journal of Geophysical Research*, **100**, 4097–4116.

Hauser, F., O'Reilly, B.M., Readman, P.W., Daly, J.S., and Van den Berg, R. (2008) Constraints on crustal structure and composition within a continental suture zone in the Irish Caledonides from shear wave wide-angle reflection data and lower crustal xenoliths. *Geophysical Journal International*, **175**, 1254–1272.

Hitchen, K. (2004) The geology of the UK Hatton–Rockall margin. *Marine and Petroleum Geology*, **21**, 993–1012.

Hitchen, K., Morton, A.C., Mearns, E.W., Whitehouse, M., and Stoker, M.S. (1997) Geochemical implications from geochemical and isotopic studies of Upper Cretaceous and Lower Tertiary igneous rocks around the northern Rockall Trough. *Journal of the Geological Society London*, **154**, 517–521.

Hitchen, K., Stoker, M.S., Evans, D., and Beddoe-Stephens, B. (1995) Permo-Triassic sedimentary and volcanic rocks in basins to the north and west of Scotland. In: Boldy, S.A.R. (ed.), *Permian and Triassic rifting in Northwest Europe.* Geological Society, London, Special Publications, **91**, 87–102.

Hutton, D.H.W. and Alsop, G.I. (1996) The Caledonian strike-swing and associated lineaments in NW Ireland and adjacent areas: sedimentation, deformation and igneous intrusion patterns. *Journal of the Geological Society, London*, **153**, 345–360.

Kinck, J.J., Husebye, E.S., and Lund, C.E. (1991) The south Scandinavian crust: Structural complexities from seismic reflection and refraction profiling. *Tectonophysics*, **189**, 117–133.

Klemperer, S.L. (1989) Seismic reflection evidence for the location of the Iapetus suture west of Ireland. *Journal of the Geological Society, London*, **146**, 409–412.

Klemperer, S.L., Ryan, P.D., and Snyder, D.B. (1991) A deep seismic reflection transect across the Irish Caledonides. *Journal of the Geological Society, London*, **148**, 149–164.

Landes, M., O'Reilly, B.M., Readman, P.R., Shannon, P.M. and Prodehl, C. (2003) VARNET-96: three-dimensional upper crustal velocity structure of SW Ireland. *Geophysical Journal International*, **153**, 424–442.

Landes, M., Ritter, J.R.R., Readman, P.W., and O'Reilly, B.M. (2005) A review of the Irish crustal structure and signatures from the Caledonian and Variscan Orogenies. *Terra Nova*, **17**, 111–120.

Lowe, C. and Jacob, A.W.B. (1989) A north–south seismic profile across the Caledonian suture zone in Ireland. *Tectonophysics*, **168**(4), 297–318.

Makris, J., Egloff, R., Jacob, A.W.B., Mohr, P., Murphy, T., and Ryan, P. (1988) Continental crust under the southern Porcupine Seabight west of Ireland. *Earth and Planetary Science Letters*, **89**, 387–397.

Makris, J., Ginzburg, A., Shannon, P.M., Jacob, A.W.B., Bean, C.J., and Vogt, U. (1991) A new look at the Rockall region. *Marine and Petroleum Geology*, **8**, 410–416.

Masson, D.G., Miles, P.R., Max, M.D., Scrutton, R.A., and Inamdar, D.D. (1985) *A free-air gravity anomaly map of the Irish Continental Margin and a new gravity model across the southern Porcupine Seabight*. Geological Survey of Ireland, Report Series **RS 85/4**, 8pp.

Max, M.D., Inamdar, D.D., and MacIntyre, T. (1982) *Compilation magnetic map: the Irish Continental Shelf and adjacent areas*. Geological Survey of Ireland, Report Series **RS 82/2**, 7pp.

Morewood, N.C., Mackenzie, G.D., Shannon, P.M., O'Reilly, B.M., Readman, P.W., and Makris, J. (2005) The crustal structure and regional development of the Irish Atlantic margin region. *In*: Doré, A.G. and Vining, B. (eds), *Petroleum Geology: North-West Europe and Global Perspectives – Proceedings of the 6th Petroleum Geology Conference*. Geological Society, London, 1023–1034.

Musgrove, F.W. and Mitchener, B. (1996) Analysis of the pre-Tertiary rifting history of the Rockall Trough. *Petroleum Geoscience*, **2**, 353–360.

Naylor, D. and Sevastopulo, G.D. (1979) The Hercynian 'Front' in Ireland. *Krystalinikum*, **14**, 77–90.

Naylor, D. and Shannon, P.M. (2005) The structural framework of the Irish Atlantic Margin. *In*: Doré, A.G. and Vining, B.A. (eds), *Petroleum Geology: North-West Europe and Global Perspectives – Proceedings of the 6th Petroleum Geology Conference*, Geological Society, London, 1009–1021.

Naylor, D., Shannon, P.M., and Murphy, N. (1999) *Irish Rockall Basin region – a standard structural nomenclature system.*

Petroleum Affairs Division, Dublin, Special Publication **1/99**, 42pp.

O'Reilly, B.M., Shannon, P.M., and Vogt, U. (1991) Seismic studies in the North Celtic Sea Basin: implications for basin development. *Journal of the Geological Society, London*, **148**, 191–195.

O'Reilly, B.M., Hauser, F., Jacob, A.W.B., and Shannon, P.M. (1996) The lithosphere below the Rockall Trough: wide-angle seismic evidence for extensive serpentinisation. *Tectonophysics*, **255**, 1–23.

O'Reilly, B.M., Hauser, F., Ravaut, C., Shannon, P.M., and Readman, P.W. (2006) Crustal thinning, mantle exhumation and serpentinisation in the Porcupine Basin, offshore Ireland: Evidence from wide-angle seismics. *Journal of the Geological Society of London*, **163**, 775–787.

Odinsen, T., Reemst, P., van der Beek, P., Faleide, J.I., and Gabrielsen, R.H. (2000) Permo-Triassic and Jurassic extension in the northern North Sea: results from tectonostratigraphic forward modelling. *In*: Nøttvedt A (ed.), *Dynamics of the Norwegian Margin*, Geological Society, London, Special Publications, **167**, 83–103.

Park, R.G., Stewart, A.D., and Wright, D.T. (2002) The Hebridean terrane. *In*: Trewin, N.G. (ed.), *The Geology of Scotland*. Geological Society, London, 45–80.

Perroud, H. and Bonhommer, N. (1981) Palaeomagnetism of the Ibero-Armorican arc and the Hercynian orogeny in Western Europe. *Nature*, **292**, 445–448.

Plant, J.A., Reeder, S., Salminen, R., Smith, D.B., Tarvainen, T., De Vivo, B., and Petterson, M.G. (2003) The distribution of uranium over Europe: geological and environmental significance. *Applied Earth Science (Transactions of the Institution of Mining and Metallurgy B)* **112**, 221–238.

Reston, T.J., Pennell, J., Stubenrauch, A., Walker, I., and Perez-Gussinye, M. (2001) Detachment faulting, mantle serpentinization, and serpentinite-mud volcanism beneath the Porcupine Basin, southwest of Ireland. *Geology*, **29**, 587–590.

Riddihough, R.P. (1975) *A magnetic map of the continental margin of west of Ireland involving part of the Rockall Trough and the Faeroe Plateau*. Dublin Institute for Advanced Studies, Geophysical Bulletin **33**, 1p. and Enclosure.

Roberts, D.G., Thompson, M., Mitchener, B., Hossack, J., Carmichael, S., and Bjørnseth, H-M. (1999) Palaeozoic to Tertiary rift and basin dynamics: mid-Norway to the Bay of Biscay – a new context for hydrocarbon prospectivity in the deep water frontier. *In*: Fleet, A.J. and Boldy, S.A.R. (eds), *Petroleum Geology of Northwest Europe: Proceedings of the 5th Conference*, Geological Society, London, 7–40.

Rogers, G., Dempster, T.J., Bluck, B.J., and Tanner, P.W.G. (1989) A high precision U–Pb age for the Ben Vuirich granite: implications for the evolution of the Scottish Dalradian Supergroup. *Journal of the Geological Society, London*, **146**, 789–798.

Rumph, B., Reaves, C.M., Orange, V.G., and Robinson, D.L. (1993) Structuring and transfer zones in the Faeroe Basin in a regional tectonic context, *In*: Parker, J.R. (ed.), *Petroleum Geology of Northwest Europe: Proceedings of the 4th Conference*, Geological Society, London, 999–1009.

Shannon, P.M. (1991) The development of Irish offshore

sedimentary basins. *Journal of the Geological Society, London,* **148**, 181–189.

Shannon, P.M., Jacob, A.W.B., Makris, J, O'Reilly, B., Hauser, F., Readman, P.W., and Makris, J. (1999) Structural setting, geological development and basin modelling in the Rockall Trough. *In*: Fleet, A.G. and Boldy, S.A.R (eds), *Petroleum Geology of Northwest Europe: Proceedings of the 5th Conference.* Geological Society, London, 421–431.

Smythe, D.K. (1989) Rockall Trough – Cretaceous or Late Palaeozoic? *Scottish Journal of Geology,* **25**, 5–43.

Stoker, M.S., Hitchen, K., and Graham, C.C. (1993) *The Geology of the Hebrides and West Shetland shelves, and adjacent deepwater areas.* United Kingdom Offshore Regional Report, British Geological Survey, 149pp.

Štolfová, K and Shannon, P.M. (2009) Permo-Triassic development from Ireland to Norway: basin architecture and regional controls. *Geological Journal,* **44**, 652–676.

Tesauro, M., Kaban, M.K., and Cloetingh, S.A.P.L. (2008) EuCRUST-07: a new reference model for the European crust. *Geophysical Research Letters,* **35**, L05313.

Tietzsch-Tyler, D. (1996) Precambrian and early Caledonian orogeny in south-east Ireland. *Irish Journal of Earth Sciences,* **15**, 19–39.

Trettin, H.P. (1973) Early Palaeozoic evolution of the northern parts of the Canadian Archipelago. *In*: Pitcher, M.G. (ed.), *Arctic Geology.* American Association of Petroleum Geologists, Memoir, **19**, 57–75.

Trettin, H.P. and Balkwill, H.R. (1979) Contributions to the tectonic history of the Innuitian Province, Arctic Canada. *Canadian Journal of Earth Sciences,* **16**, 748–769.

Tyrrell, S., Haughton, P.D.W., and Daly, J.S. (2007) Drainage reorganization during breakup of Pangea revealed by in-situ isotopic analysis of detrital K-feldspar. *Geology,* **35**, 971–974.

Young, D.G.G. and Bailey, R.J. (1973) A reconnaissance magnetic map of the continental margin west of Ireland. *Communication, Dublin Institute for Advanced Studies, Series D. Geophysical Bulletin* **29**.

Ziegler, P.A. (1982) *Geological Atlas of Western and Central Europe.* Shell Internationale Petroleum Maatschappij, B.V. 130pp.

Chapter 2

Regional development

Introduction

The basement geology, described in Chapter 1, provided the initial structural template for the development of the various offshore sedimentary basins that surround Ireland. This chapter builds upon that with a description of the post-Variscan regional development of the North Atlantic. The regional evolution was shaped to a large extent by the opening of the North Atlantic Ocean. This occurred in a complex manner, with pulsed northward rupture of the crust and the onset of seafloor spreading coeval with southward progression of rupture from the Arctic domain. The two sets of opposing crustal ruptures overlapped to the west of the Norwegian mainland in early Cenozoic time, resulting in the final separation of Europe from Greenland and the North American continent and the development of the passive European margin. Various extensional and compressional events preceded and succeeded the crustal rupture to provide a complex, structurally linked basin system. The major Mesozoic and Cenozoic tectonic events, important in the development and modification of traps and of reservoir fairways, are discussed in a regional context. The overall regional geological framework of the Irish offshore is outlined, with a focus on the major structural features that influenced the location, orientation and evolution of the major basin systems.

Development of the North Atlantic
Variscan Collapse

The end of the Carboniferous saw the grouping of most of the world's continents to form the Pangaean supercontinent. This followed the closure of the proto-Tethys ocean and the development of the Alleghanian–Variscan fold belt running in a general east–west direction across France, the south of England and the south of Ireland. Further east the crustal suturing of West Siberia and Laurussia along the Urals fold belt took place in Late Permian to Early Triassic times (Scotese, 1987) further annealing the supercontinent. However, supercontinents

formed by orogenic suturing are inherently unstable and begin to break up shortly after formation. This facilitates early sedimentation and the onset of crustal rifting and basin formation. Supercontinents typically undergo uplift, collapse and breakup as a variety of factors interplay. Heat buildup beneath the low conductivity lithosphere of the supercontinent may result in uplift and tensional stresses due to mantle convection cells (Nance *et al.*, 1988). These tensional stresses are an integral part of the Wilson cycle (Wilson, 1966), whereby continents commonly split to form new oceans along the lines of inherited old orogens. There is thus a repetitious opening and closing of oceans, with the formation and disintegration of continents and the formation and deformation of sedimentary basins. The formation and collapse of orogenic belts, such as the Caledonian and Variscan fold belts, is central to such tectonic cycles. Ryan and Dewey (1997) argued that the presence of eclogite-facies roots of partially collapsed orogens are important to the cycle, and that such roots weaken the orogenic lithosphere and provide preferred locations for later rifting and basin development. The evolution of the collapse and breakup of the Pangaean supercontinent is punctuated by a series of major and semi-regional unconformities and uplift events, which may be accounted for by phases of slab drop-off as the subduction zone material breaks and sinks into the mantle.

The Variscan fold belt of western Europe comprises a patchwork of annealed terranes that influenced the location and configuration of the later Permian to Cenozoic basins (Figure 2.1). The collapse of the Variscan orogenic belt was shown by Praeg (2004) to have expanded northwards over time, and to have occurred in three main stages. The earliest, in the late Viséan to Moscovian (previously mid-Westphalian; 335–310 Ma) interval, took place in a relatively narrow (<500 km) zone due to the collapse of the central internides, facilitated by NW–SE extension, and resulted in the passive infill of residual

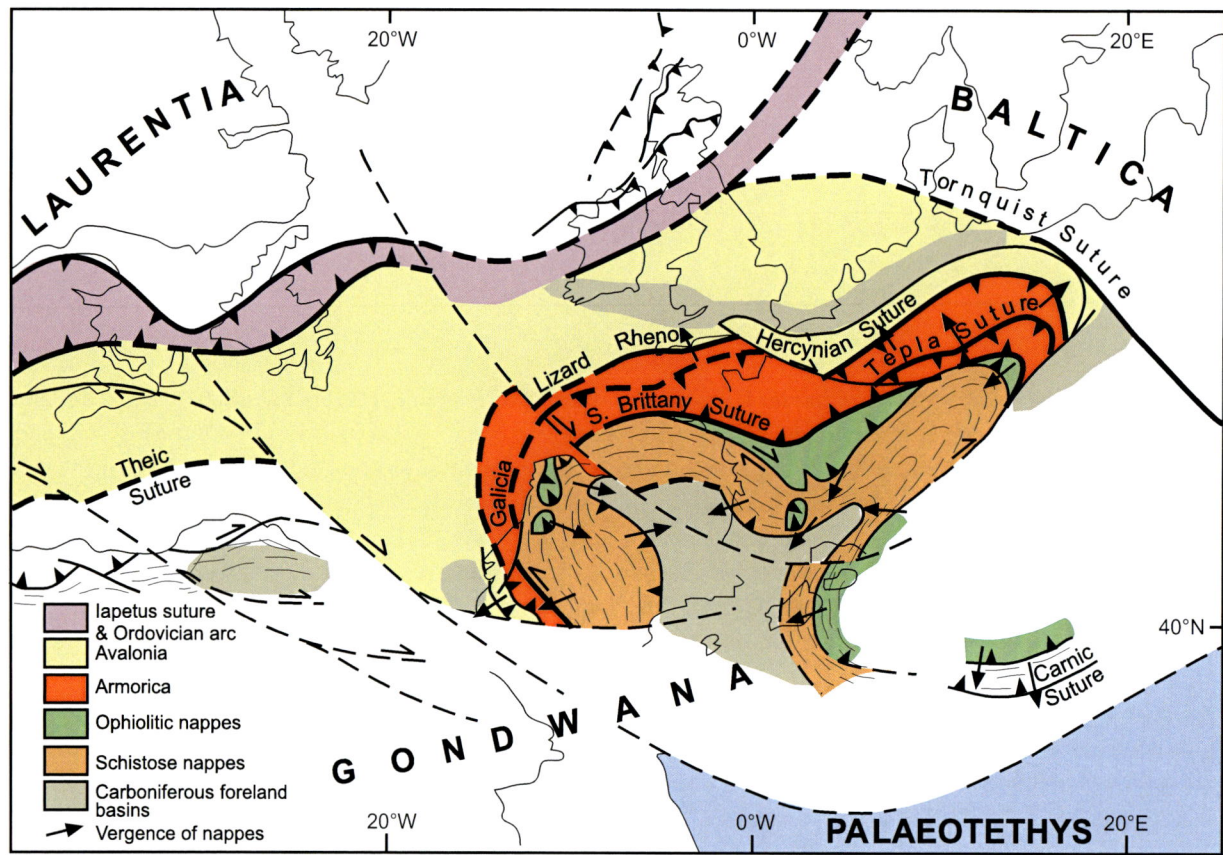

Figure 2.1. The Variscan fold belt of western Europe, in early Permian time, at the onset of post-Pangaean basin development (from Graham, 2009 and Matte, 2001). The Gondwana, Baltica and Laurentia basement terranes were annealed through a complex system of collisional suture zones.

shallow basins inherited from extension during the closure of the Rheic Ocean. The second stage was the re-orientation and expansion of the collapse, in Moscovian (mid-Westphalian) to Autunian times (310–300 Ma). An orthogonal rotation of the extension direction in the central internides was accompanied by episodic basin formation in the northern internides, thrust propagation in the externides and the foreland, while basins began to form in the central internides and expanded to the externides, coeval with final nappe emplacement across the orogenic front. The final stage was the collapse of the foreland in the late Stephanian to Early Permian (300–290 Ma), when kilometre-scale uplift and erosion of the foreland took place prior to widespread Early Permian basin formation. Praeg (2004) related the three stages to successive detachments of negatively buoyant litho-spheric material; a collisionally-thickened orogenic root and two Rheic subducted oceanic slabs that lay beneath the orogen (southward subducted) and the foreland (northward subducted). This evolutionary development is shown in Figure 2.2.

Permo-Triassic extension and incipient rifting

Basin development and thick sediment accumula-tion became more regionally extensive on the collaps-ing Pangaean supercontinent in Early Permian time. Although dating and correlation of the successions are difficult due to the largely continental red-bed nature of the strata, and later structuring has resulted in partial removal of many of the depositional margins of the basins, the overall pattern of early post-orogenic basin formation suggests a variety of local depocentres in a set of disconnected basins. A wide variation of exten-sion vectors is recorded from studies (Coward, 1995). Generally, the Permo-Triassic basins were wider than the later Jurassic rift basins. They reflect a complex develop-ment of orogenic collapse basins, wide-rift basins and narrower rift basins controlled by reactivated crustal structures (Štolfová and Shannon, 2009). In broad terms, several swaths of Permo-Triassic basins developed in the North Atlantic region (Doré *et al.*, 1999), shown in Figure 2.3. Through much of the future North Atlantic region these followed the trend of the Caledonian fold

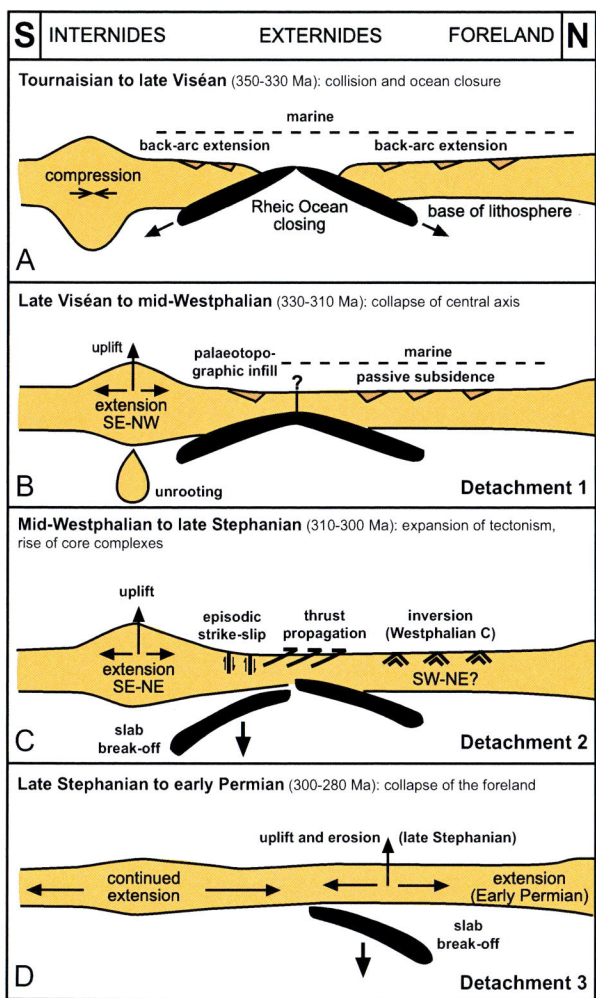

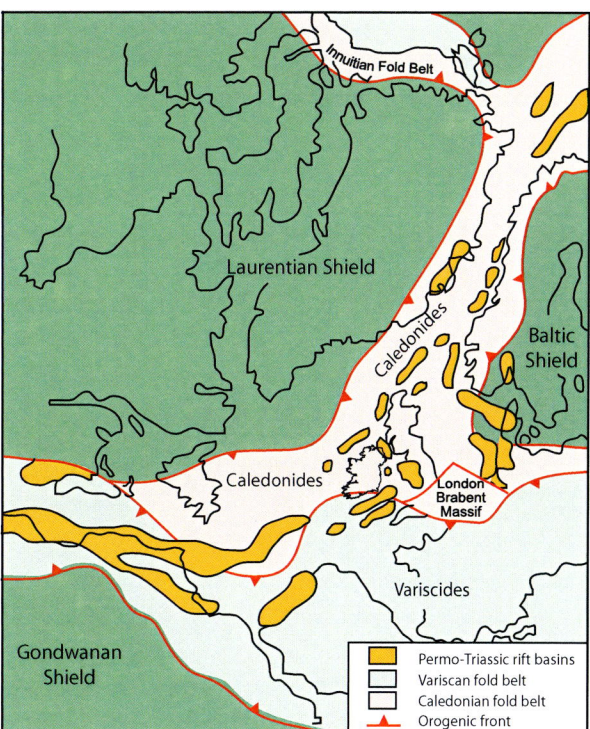

Figure 2.3. Location of main Permo-Triassic remnant basins in the context of the Pangaean orogenic fronts (after Doré *et al.*, 1999). Note the close relationship between the main structural fabrics (Caledonian and Variscan) and the basin size and orientation. Smaller basins follow the Caledonian NE-SW fabric while the larger basins are broadly parallel to the Variscan (and Innuitian) E-W structural fabrics.

Figure 2.2. A model of Variscan late-orogenic collapse in response to multiple slab detachments (from Praeg, 2004). (A) Early Carboniferous culmination of collision and ocean closure; (B) Late Viséan to mid-Westphalian (Moscovian) loss of thickened orogen root resulting in uplift, with the weight of the subducted Rheic oceanic slabs maintaining subsidence and marine conditions across the externides and foreland; (C) Mid-Westphalian (Moscovian) to late Stephanian rise of core complexes, expansion of tectonism through faulting, thrust propagation and inversion caused by break-off of the Rheic slab beneath the internides; (D) Late Stephanian to Early Permian collapse of the foreland and onset of extension following break-off of the Rheic slab beneath the foreland.

belt, while to the north they paralleled the Innuitian fold belt of Arctic Canada (Doré *et al.*, 1999) and the Uralides in the eastern Barents Sea. To the south of Ireland, Permo-Triassic extension in the Celtic Sea region was influenced by the Variscan structural fabric (Shannon, 1991), while along strike to the west, on the eastern seaboard of America and northwest Africa the post-orogenic

rift basins followed the lines of the Variscan–Alleghanian Orogen (Doré *et al.*, 1999).

Jurassic rifting and extension

The onset of the Jurassic saw a major change in the nature, organisation and geometries of basins in Northwest Europe. These changes were largely driven by changing plate tectonic stresses as incipient seafloor spreading took place in the Tethys to the southeast of the North Atlantic and in the proto-Central Atlantic to the southwest (Doré *et al.*, 1999). Rifting breached Pangaea, resulting in marine flooding of the typical red-bed sequences of the Triassic. Rifting then became more focused and resulted in the development of elongate half-graben depocentres such as the Slyne, Erris and the Sea of Hebrides basins in latest Early Jurassic times. In early Middle Jurassic time seafloor spreading began in the Central Atlantic but appears to have been restricted to the region south of the Azores–Gibraltar Fracture Zone. Rifting occurred to the north of this deep-seated major structure, with extensional stresses largely east–west directed. This produced a series of generally north–south oriented basins such as the Porcupine Basin, Viking Graben, East Greenland Rift and Halten Terrace. Local volcanism is also recorded, possibly in response to the development of mantle plume hot spots rather than crustal thinning decompression melting. The igneous activity and pronounced uplift in Middle Jurassic time in the North Sea is well documented (e.g. Underhill and Partington, 1993), while Shannon (1995) suggested that the presence of Middle Jurassic igneous rocks and the general lack of late Middle and Upper Jurassic strata in the Fastnet Basin at the southwest end of the Celtic Sea may also reflect the presence of a broadly coeval mantle plume. Extensional crustal stresses were predominantly east–west. Rifting became more pronounced in the region in Late Jurassic time, with the development of a series of disconnected, broadly elongate north–south basins developed perpendicular to the predominant extensional direction (Figure 2.4). Older structures (e.g. east–west Variscan structures) still played a pronounced role in the continued development of basins in the Celtic Sea region, with these basins oblique to the general north–south trend of the major basin systems in the Porcupine, North Sea and mid-Norway regions to the north. By Late Jurassic times, early rift onset warp basin development had given way to fault-controlled rift basins containing major facies differences ranging from fluvial and lacustrine sands and muds to basin floor sandy turbidites (Sinclair *et al.*, 1994). The rift basins became larger, more elongate and more focused than in Middle Jurassic times. Some of the older and smaller Early Jurassic rift basins, such as the Slyne and West Shetland basins, became less active as extension was preferentially focused in the larger basins. In these smaller basins, while Upper Jurassic strata were deposited, there is little evidence of the major synsedimentary growth faulting that typified the major rift phase of basin development.

Cretaceous extension and seafloor spreading

By Early Cretaceous time the Tethyan seafloor spreading that had influenced Late Jurassic extensional stresses had ceased and was replaced by northward subduction on the northern margin of the ocean (Ziegler, 1988). Further west, seafloor spreading and oceanic crust had propagated into the Bay of Biscay and the Grand Banks region by Aptian times. However, no spreading had progressed north of the Charlie-Gibbs Fracture Zone into the Rockall region or west of Greenland. As

Figure 2.4. Late Jurassic plate reconstruction showing the location and orientation of rift basins (modified from Doré *et al.,* 1999). The main plate motion is east-west, resulting in a series of generally north-south basins.

observed by Lundin (2008) geodynamic changes at the Jurassic–Cretaceous boundary resulted in a fundamental but often overlooked event. Although Jurassic and Cretaceous rift events were not separated by a significant amount of time, their distribution and orientation differed markedly. As a consequence of the cessation of seafloor spreading in the Tethys, the north–south Jurassic rift system was overprinted and sometimes 'beheaded' by a set of large NE–SW oriented Early Cretaceous rifts that accumulated very thick, generally marine, strata. This is seen between the Porcupine and Rockall basins, and further north between the northern North Sea and the Møre Basin, and between the Haltenbanken and the Vøring Basin. Doré *et al.* (1999) also described east–west extension in Late Jurassic time changing to NW–SE extension in Early to Mid-Cretaceous time, resulting in a broad zone of crustal extension and subsidence extending northwards from the southern end of the Rockall Basin through to the Barents Sea. A second, slightly younger, rift arm extended northwards through the Labrador Sea between Labrador and Greenland (Figure 2.5).

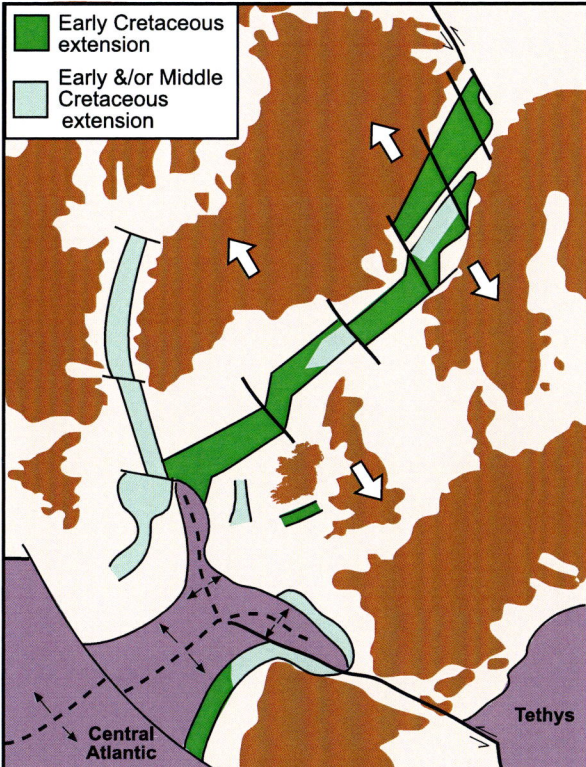

Figure 2.5. Mid-Cretaceous plate reconstruction showing the location and orientation of basins (from Doré *et al.,* 1999). The main plate motion is NW-SE, resulting in a series of NE-SW east-west large basins that transect the smaller and older Jurassic rift basins.

Oceanic crustal propagation into the North Atlantic is marked by the M0 anomaly (mid-Aptian) seen off Iberia and in the Grand Banks (Figure 2.6). Spreading had reached Galicia Bank by the late Aptian and shortly afterwards a triple junction had formed in the Bay of Biscay, where spreading continued until Campanian (Late Cretaceous) time. This was followed, until the Eocene, by the northwest rotation of Iberia with respect to Eurasia, producing subduction of the Bay of Biscay oceanic crust beneath Iberia, and in turn resulting in the formation of the Pyrenees (Lundin, 2002). Seafloor spreading propagated northwards from the Bay of Biscay triple junction, reaching the Goban Spur by Middle to Late Albian times (De Garciansky and Poag, 1985). The spreading axis had reached the Charlie-Gibbs Fracture Zone by the Santonian (Chron 34). Although it is not thought likely that seafloor spreading propagated into either the mouth of the Porcupine Basin or the southern mouth of the Rockall Basin, Early Cretaceous igneous activity is suggested for both regions. The Porcupine Median Volcanic Ridge and the Barra Volcanic Ridge systems are thought

to be broadly coeval with spreading in the Goban Spur region and of approximately Chron 34 age.

The onset of seafloor spreading in the Labrador Sea is interpreted by Roest and Srivastava (1989) and Srivastava and Roest (1999) as being of Early Campanian (Chron 33) age (Figure 2.7). However, Chalmers and Laursen (1995) argued that crust of this age is transitional crust and that true oceanic crust was not formed until Paleocene (Chron 27) time. Nevertheless, severe attenuation leading to seafloor spreading along this axis represented a change in direction in the northward propagation of the Atlantic, which would ultimately lead to the separation of Greenland from North America.

Seafloor spreading is generally thought to have taken place in the Arctic Canada Basin during Mid- to Late Cretaceous (M0-A34) time (Weber and Sweeney, 1990), although the age of the crust in the region is poorly constrained. This matches the general age of magmatism on Franz Josef Land, Svalbard and in the Sverdrup Basin. Lundin (2008) suggested that the opening of the Canada Basin was influenced by a plume track in the region. Opening of the Canada Basin occurred in response to counterclockwise rotation of Eastern Siberia and the North Slope of Alaska away from the Canadian Arctic Islands (Grantz *et al.*, 1990). Seafloor spreading also appears to have propagated as far north as Baffin Bay by latest Cretaceous to earliest Cenozoic times (Figure 2.6).

Cenozoic rifting and seafloor spreading

The early Cenozoic rifting culminated in the final phase of Pangaean supercontinent breakup. It is characterised by major volcanism, resulting in the development of a classic volcanic passive margin, in contrast to the Mesozoic non-volcanic nature of the margin to the south. Early Cenozoic (Paleocene) rifting was associated with the formation of the North Atlantic Igneous Province (NAIP), one of the world's largest igneous provinces (Coffin and Eldholm, 1992). There was widespread igneous activity across a broad region, with a diameter of more than 2000 km. This included lavas, including seaward-dipping reflectors along the continental–oceanic margin, possible magmatic underplating at the base of the crust, and sills and dykes within the crust and sedimentary succession. The NAIP is thought to reflect the development of the Iceland plume, with major pulsed volcanism, most pronounced during Paleocene time but beginning in the Late Cretaceous and continuing in a pulsed manner throughout the Palaeogene (O'Connor *et al.*, 2000). Plume development is interpreted to have

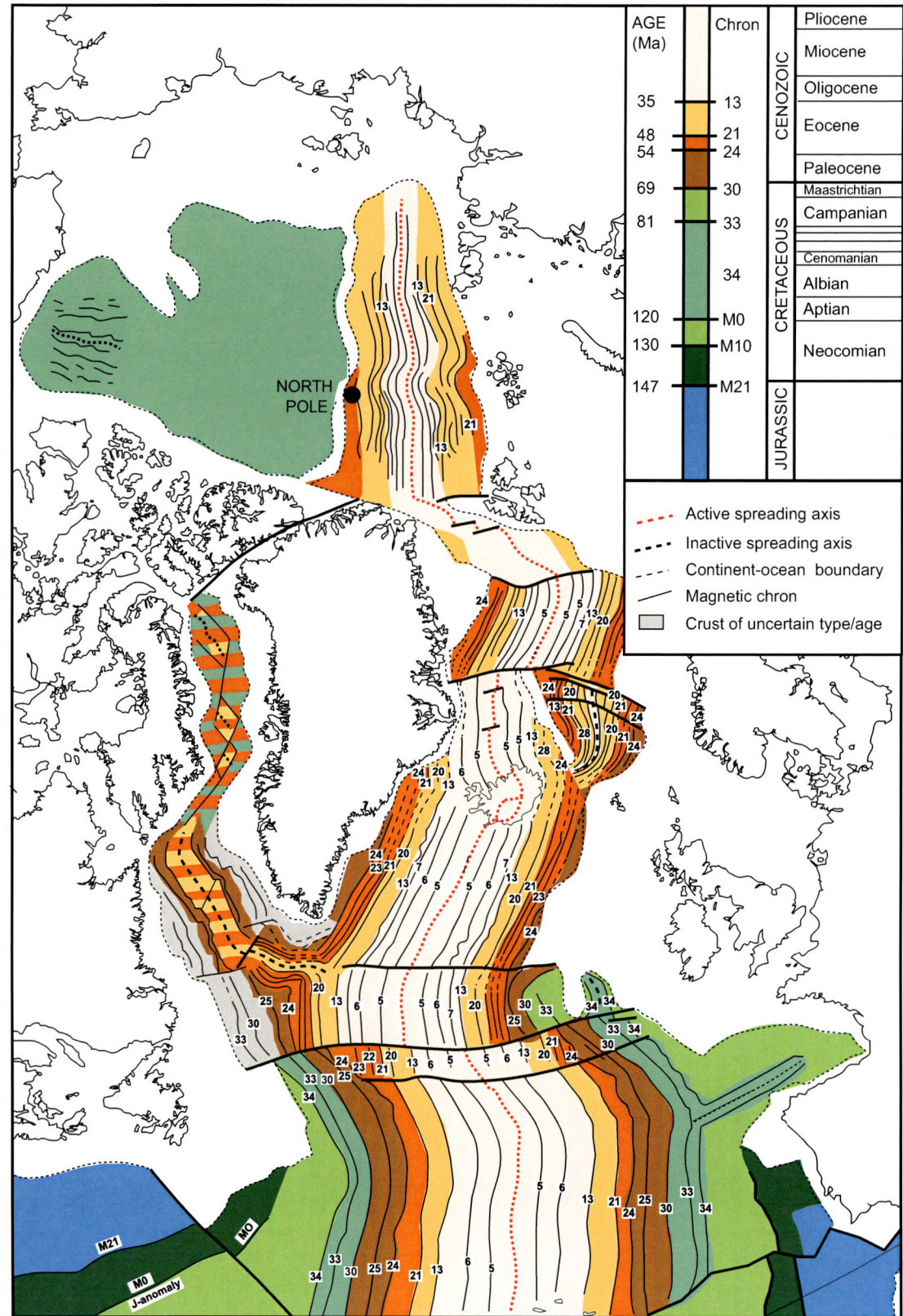

Figure 2.6. Simplified map of the seafloor in the North Atlantic to Arctic region (from Lundin, 2002). The ages and magnetic anomalies are indicated on the inset.

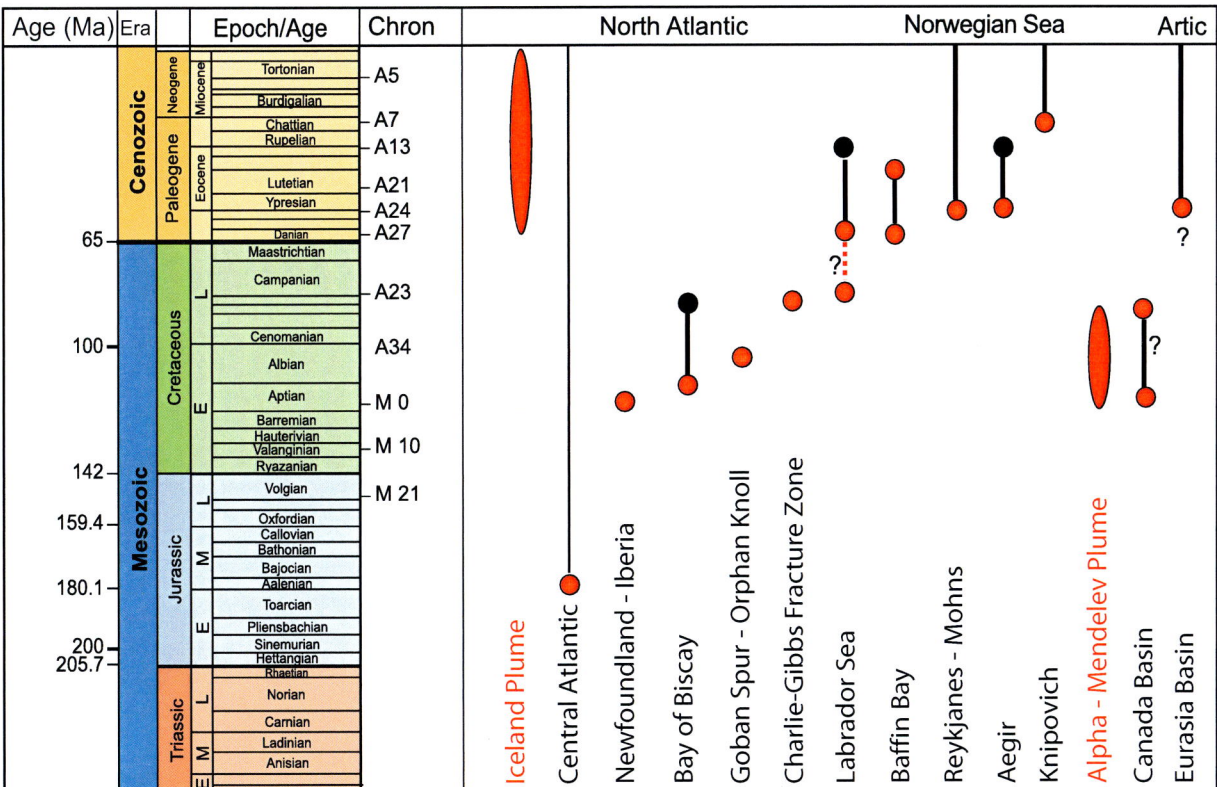

Figure 2.7. Chronological chart indicating the ages and durations of the main seafloor spreading events in the North Atlantic and Arctic region (from Lundin, 2002).

led to uplift and episodic influx of clastics from continental margins into the adjacent sedimentary basins such as the North Sea, Porcupine Basin and the Norwegian offshore basins (White and Lovell, 1997). The arrival of the plume in the North Atlantic region was probably a decisive factor in causing continental breakup. Previous episodic rifting since Permo-Triassic times had not resulted in plate separation in the North Atlantic. However, the arrival of the plume resulted in elevated asthenospheric temperatures combined with high rates of crustal extension. This increased melt generation (Brown and White, 1995), leading to acceleration of crustal rupture.

Cenozoic rift style is noticeably different to that of the Mesozoic. In the latter, fault-control and synsedimentary faulting is a characteristic feature, while in the Cenozoic extensional structures are poorly expressed along the entire length of the North Atlantic margin. While this may be partly the result of masking of such features beneath sills, lava flows and seaward-dipping reflectors, there does appear to be a major discrepancy between the amount of extension calculated from seismically-observed faults and the amount of extension required to explain the thermal subsidence (e.g.

Roberts *et al.*, 1997; Walker *et al.*, 1997). This suggests that the effect is real rather than an artefact created by poor data quality beneath the Cenozoic igneous cover. It appears likely that the lack of faulting is a response to depth-dependent or differential stretching (e.g. Davis and Kusznir, 2004). In a number of basins (e.g. Rockall and Porcupine) the amount of stretching appears to be greater in the upper/mid crust than in the lower crust/mantle lithosphere (Hauser *et al.*, 1995; O'Reilly *et al.*, 1996), with the differential crustal stretching accommodated by a major crustal detachment. However, closer to the continental margin, where rupture was successful (e.g. Hatton Continental Margin, Norwegian Atlantic margin basins), whole crustal thinning exceeded that of the upper and middle crust. This may be explained by preferential removal of the lower crust and mantle, probably by flow into the oceanic domain. However, the processes responsible for the physical removal of the lower crustal and mantle material are still poorly understood. The overall effect of such differential stretching is to produce large-scale regional subsidence, generally devoid of the synsedimentary faulting that characterises cold crustal 'normal' rifting.

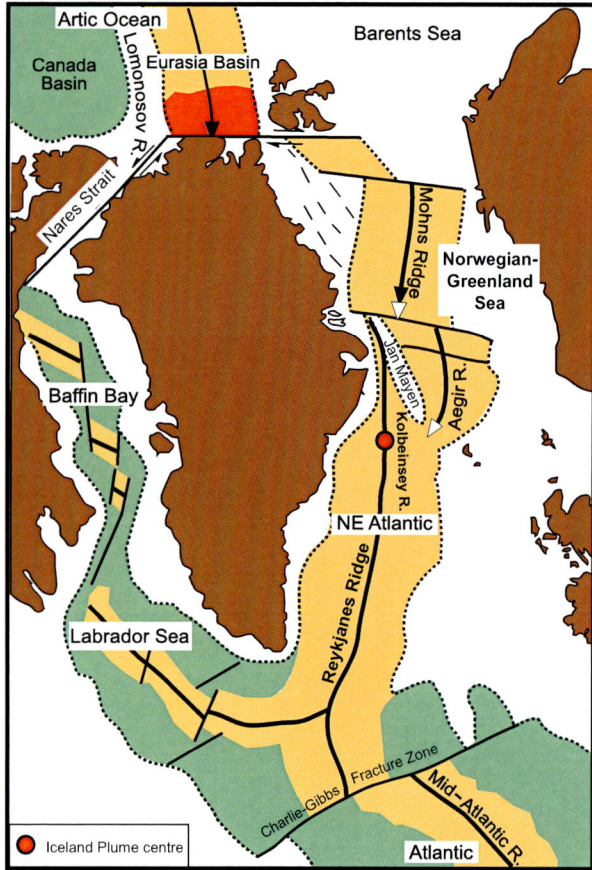

Figure 2.8. Simplified Late Palaeogene (Chron 13) plate reconstruction (from Lundin, 2002). Northward crustal propagation in the NE Atlantic overlaps with southward crustal propagation in the Norwegian-Greenland Sea, while a complex propagation arm has spread through the Labrador Sea to Baffin Bay.

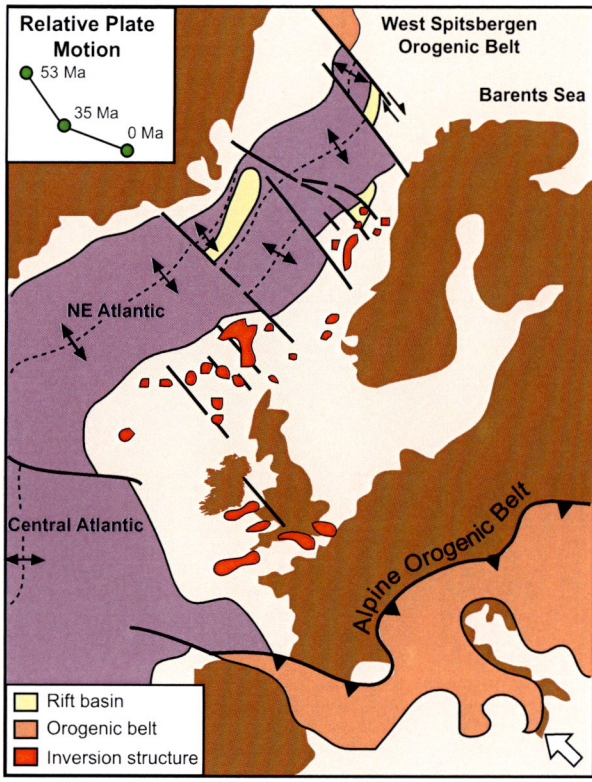

Figure 2.9. Mid-Cenozoic (Oligocene-Miocene) plate reconstruction showing the propagation of seafloor spreading in the NE Atlantic west of Ireland to Norway (from Doré *et al.,* 1999). A series of clustered inversion structures, of generally Miocene age, occur along the eastern continental European margin.

Early Eocene (Chron 24) marked the onset of seafloor spreading in the Northeast Atlantic and also in the Eurasian Basin in the Arctic (Figure 2.7). Spreading still continued in the Central Atlantic and Labrador Sea arms, resulting in the development of a short-lived triple junction. The north–south arm was the major active segment. However, this was a complex system, with the interplay of the northward propagating systems along the Reykjanes Ridge with the southward propagating Mohns Ridge (Figure 2.8). Lundin (2008) suggested that the Aegir Ridge represents the southern tip of a southward propagating Arctic rift system, while the Kolbeinsey Ridge is the northern tip of a northward propagating Atlantic rift system (Figure 2.8). These opposing systems overlapped as they attempted to link with the major systems of the Reykjanes and Mohns ridges respectively. This link made further seafloor spreading along the Aegir Ridge redundant. The timing of this overlap also coincided

with the cessation of spreading in the Labrador–Baffin Bay segment in earliest Oligocene times (Chron 13), and is likely to have been the cause of this cessation, as linkage between the Arctic and Atlantic systems had been achieved. Chron 13 also marked a major change in regional plate motions (Figure 2.9). The Eurasian plate started to move northeastwards with respect to the North American plate. Seafloor spreading was initiated along the Knipovich Ridge from Chron 7 (Late Oligocene). This led to the opening of the Arctic seaway and completed the North Atlantic oceanic system. No further changes took place in the spreading system thereafter.

Cenozoic Tectonism

The onset of seafloor spreading and the successful plate separation of the European and North American plates would have been expected to result in tectonic quiescence along the North Atlantic passive margins. In fact, these regions experienced post-rift and post-drift periodic vertical movements of up to kilometre scale. Praeg *et*

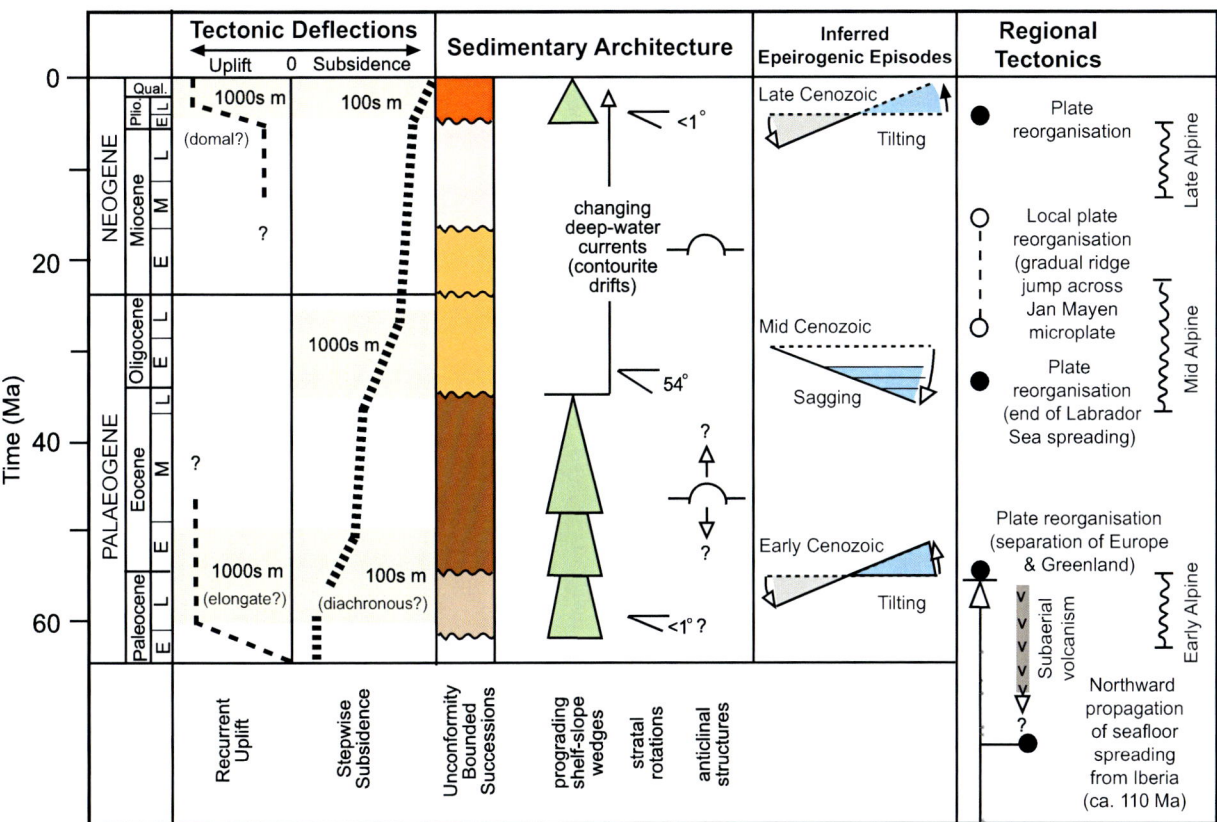

Figure 2.10. Summary of uplift/subsidence patterns, sedimentary architecture, epeirogenic episodes and regional tectonics in the NE Atlantic in Cenozoic times (from Praeg *et al.*, 2005). Early-, mid- and late Cenozoic km-scale epeirogenic movements of different form occurred and had a profound influence on the Cenozoic evolution of the NE Atlantic region.

al. (2005) identified three periods of epeirogenic movement, in the early, mid- and late Cenozoic respectively, together with at least one major phase of compressive tectonism. Two distinct types of epeirogenic movement were recognised: tilting (coeval uplift and subsidence with rotations of up to 1° over distances of hundreds of kilometres), and sagging (differential subsidence with rotations of up to 4° over distances of up to 100 km). Each episode involved rapid (<10 Ma) kilometre-scale movements that resulted in characteristic sedimentation patterns. Paleocene–Eocene tilting resulted in basinward progradation of shelf-slope wedges from elongate uplifts along the inner continental margin and the offshore structural highs. Mid-Cenozoic sagging of late Eocene to early Oligocene times terminated the wedge progradation and initiated the onset of contourite deposition in the deep-water basins. Late Cenozoic (early Pliocene to present) tilting caused further basinward progradation of shelf-slope sediment wedges from uplifted regions along the inner margin and from offshore structural highs (Figure 2.10).

The basins of the NW European margin exhibit an excess of post-rift subsidence, exceeding that predicted from models of lithospheric stretching and cooling. This excess subsidence has been related to a variety of Cenozoic syn-rift processes, either in a mantle plume context (e.g. Clift and Turner, 1998) or as a consequence of depth-dependent lithospheric stretching (Davis and Kusznir, 2004). However, this Cenozoic subsidence along the Atlantic Ocean margin was accompanied by coeval periodic uplift and erosion in the onshore and shelf regions (Japsen, 1997; Japsen and Chalmers, 2000). These have been constrained by a range of thermal history methods based on apatite fission track and vitrinite reflectance (Green *et al.*, 1999, 2002) as well as from studies of compaction and velocity information (e.g. Japsen, 1997). These indicate kilometre-scale uplift in early Palaeogene (late Paleocene) and in late Neogene (early Pliocene) times (Figure 2.11). Both events correspond to periods of sudden stepwise subsidence and were accompanied by clastic sedimentation in the offshore basins. Additionally, a period of sudden stepwise

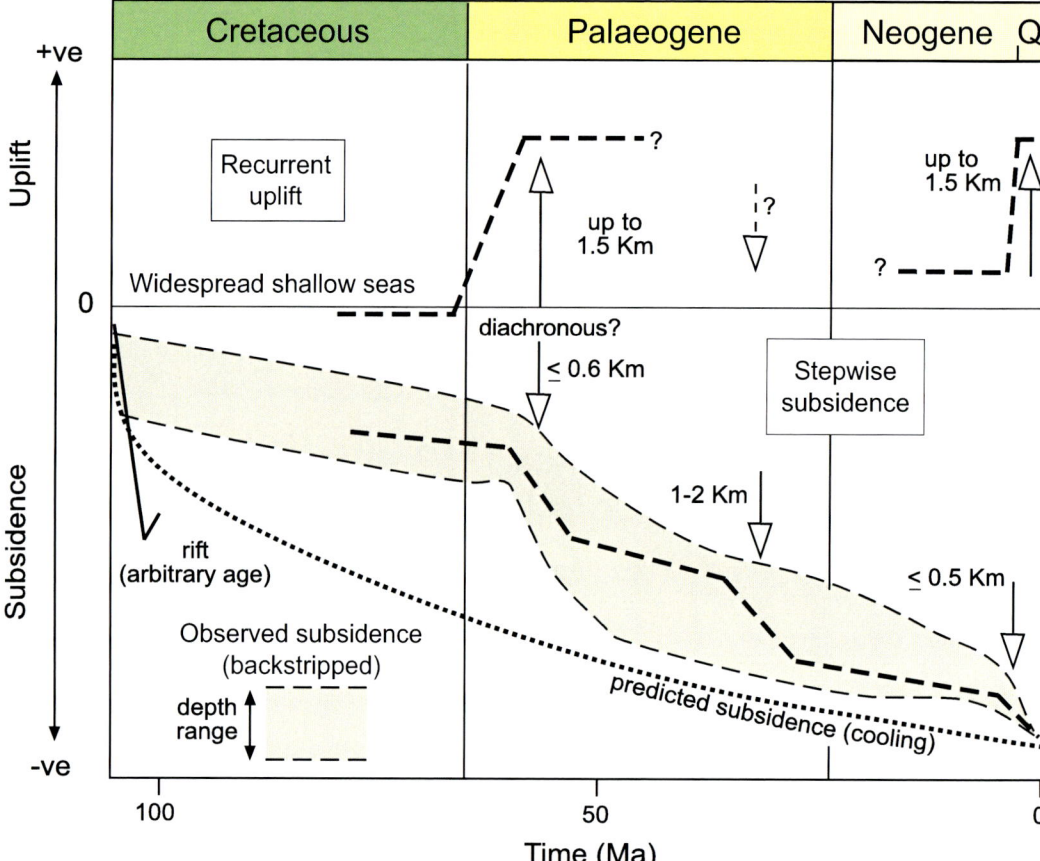

Figure 2.11. Kilometre-scale Late Cretaceous to Neogene uplift and subsidence patterns of the NE Atlantic region (from Praeg *et al.,* 2005). Based on various data from onshore and offshore, they show a pattern of recurrent uplift and stepwise subsidence that differs from the predicted post-rift subsidence (cooling) patterns.

sagging and subsidence occurred in earliest Oligocene time, resulting in rapid deepening and the onset of contourite sedimentation in the large basins (Rockall and Porcupine).

The three epeirogenic episodes coincide with Atlantic plate reorganisations (Figure 2.10). The latest Paleocene tilting is coeval with the separation of Greenland from Europe and follows upon the early Alpine movements. The mid-Cenozoic sagging is coincident with plate re-organisation, the end of Labrador Sea spreading and the onset of the main Alpine orogenic phase. The earliest Pliocene tilting coincides with a change in the Pacific plate motion (Pollitz, 1986), with the formation of the Isthmus of Panama by final closure of the seaway between the Pacific and Atlantic oceans (Lear *et al.*, 2001), a phase of increased spreading rates (Géli, 1993) and with the termination of the last phase of late Alpine orogeny. However, Praeg *et al.* (2005) argued for a causal mechanism in mantle–lithosphere interactions rather than a direct response to either plate reorganisation or to more localised intra-plate stresses. They argued that the kilo-metre-scale movements are too large to be accounted for as flexural deflections due to intra-plate stress variations and explained them as dynamic topographic responses to changing small-scale convective flow in the upper mantle. The tilting is explained as upwelling and down-welling above an edge-driven convection cell, while the sagging results from the loss of dynamic support above a former upwelling.

Another characteristic feature of the North Atlantic margin is the presence of Cenozoic domal structures, interpreted as compressional/inversion features. These have been the subject of significant discussion in the literature, both in terms of their origin and their effects on Cenozoic development and on the formation and reorganisation of trapping structures (Boldreel and Andersen 1993, 1998; Doré and Lundin, 1996; Doré *et al.*, 2008; Stoker *et al.*, 2005). They occur, often as groups of elongate structures, along the Atlantic margin from the Barents Sea to the west of Ireland, although their

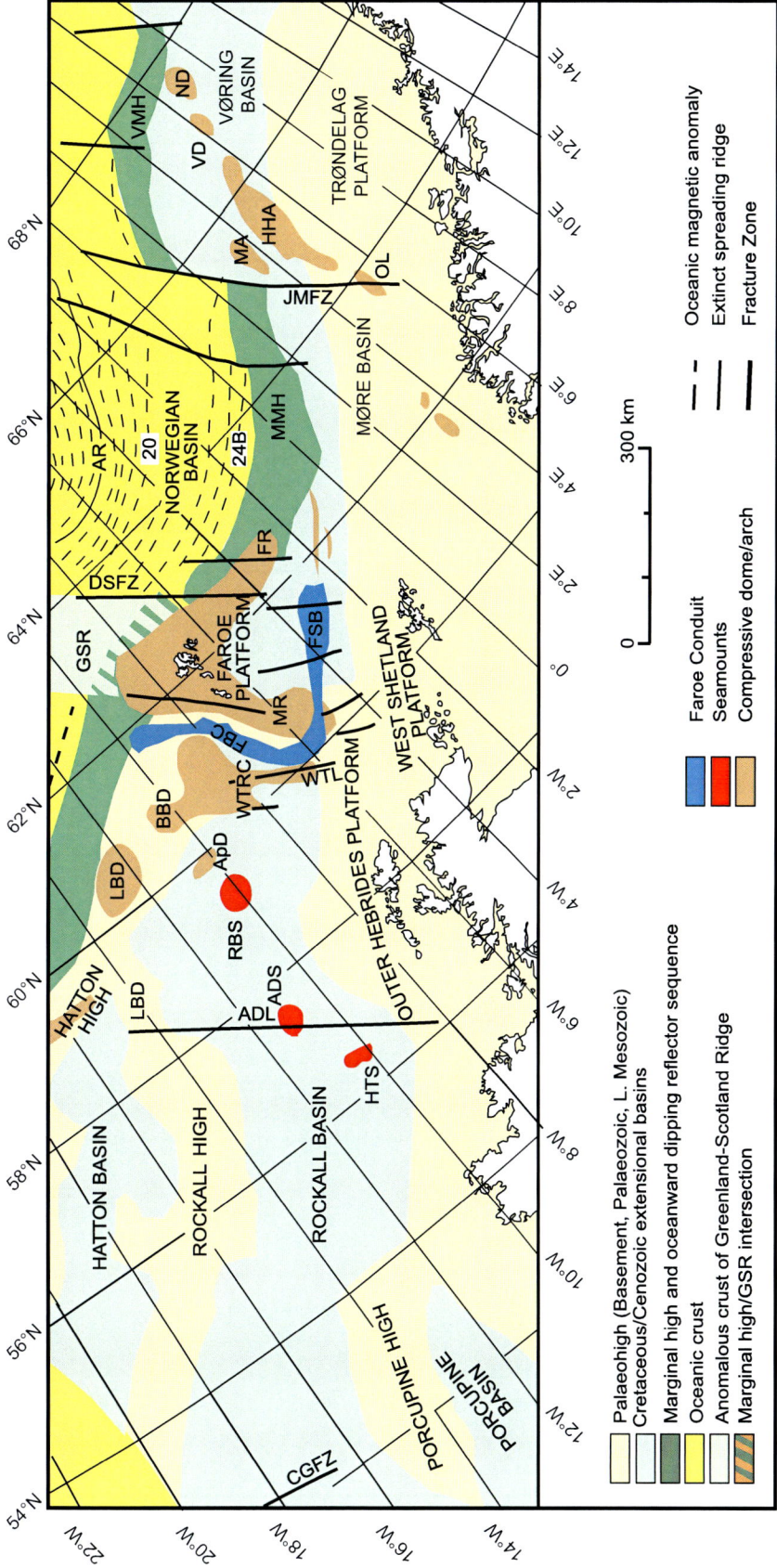

Figure 2.12. Mid-Miocene structural elements of the NW European Atlantic margin (modified after Doré *et al.,* 1999) showing the major groups of compressional domes and arches (from Stoker *et al.,* 2005). Abbreviations: ADL, Anton Dohrn Lineament; ADS, Anton Dohrn Seamount; ApD, Alpin Dome; AR, Aegir Ridge; BBD, Bill Bailey's Dome; BFZ, Bivrost Fracture Zone; CGFZ, Charlie-Gibbs Fracture Zone; DSFZ, Denmark Strait Fracture Zone; FBC, Faroe Bank Channel; FBD, Faroe Bank Dome; FR, Fugloy Ridge; FSB, Faroe-Shetland Basin; GBD, George Bligh Dome; GSR, Greenland-Scotland Ridge; HHA, Helland-Hansen Arch; HTS, Hebrides Terrace Seamount; JMFZ, Jan Mayen Fracture Zone; LBD, Lousy Bank Dome; MA, Modgunn Arch; MMH, Møre Marginal High; MR, Munkagrunnur Ridge; ND, Naglfar Dome; OL, Ormen Lange Dome; RBS, Rosemary Bank Seamount; VD, Vema Dome; VMH, Vøring Marginal High; WTL, Wyville-Thomson Lineament; WTRC, Wyville-Thomson Ridge Complex.

distribution is non-uniform. While relatively uncommon in the Irish offshore region, they are spectacularly developed in clusters in the Wyville-Thomson Lineament/Faroe Platform region and in the Møre and Vøring basins (Figure 2.12). The features often occur as elongate anticlinal structures of kilometre scale, with axes of structures sometimes traced for hundreds of kilometres

(Boldreel and Andersen, 1998). The major phase of inversion appears to have taken place in Early Miocene times, although the features frequently show an extensive multiphase history with significant inversion also occurring in latest Eocene time (Figure 2.13). Numerous potential driving mechanisms have been suggested for their formation. The contractional phases are variously

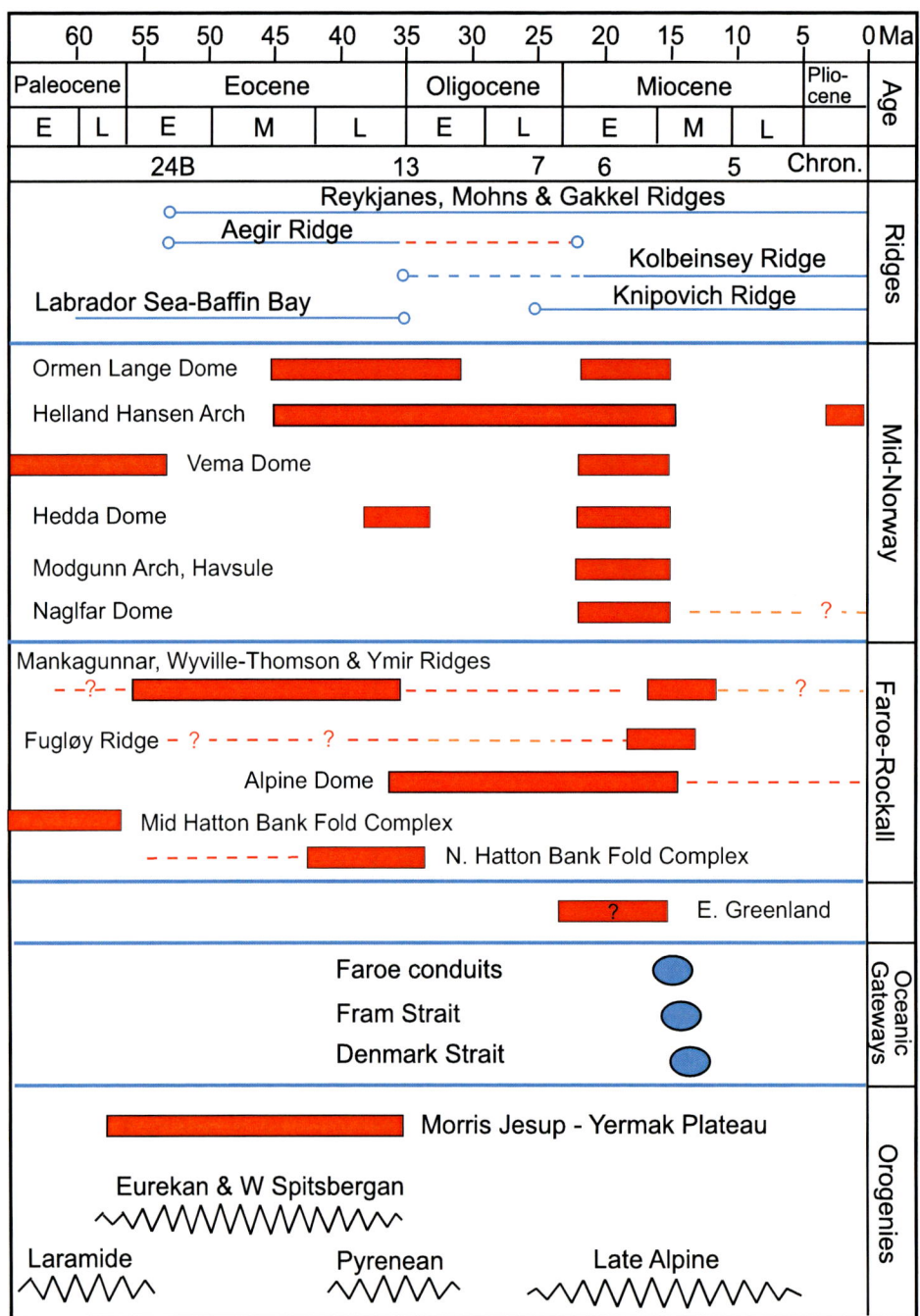

Figure 2.13. Chronological diagram showing the approximate timing of inversion structures along the NE European Atlantic margin and their relationship to seafloor spreading and orogenic episodes (from Doré *et al.*, 2008).

interpreted to reflect changes in compressive stresses due to Alpine orogeny, changes in seafloor spreading geometries and rates, ridge-push and mantle drag forces, variations in basement and crustal strength, and even differential loading by sediment wedges. Stoker *et al.* (2005) suggested that the early Neogene compression, responsible for the structures, created the Faroe Conduit, thereby establishing the first true deep-water connection between the Atlantic Ocean and the Nordic Seas. This enabled the accelerated exchange of deep waters across the Greenland–Scotland Ridge and the rapid expansion of contourite drift formation in the region.

Irish Offshore Geological Framework

In this section the physiography of the extensive continental shelf is outlined and geophysical evidence from the continental shelf and the continental–oceanic boundary is discussed. Finally, the distribution and broad outlines of the offshore sedimentary basins in the Irish offshore region are briefly outlined. This framework provides the background for the stratigraphic development that follows in the succeeding chapters.

The Continental Shelf

The waters east and south of Ireland are typically shallow, ranging from approximately 30 m to the east and northeast of Ireland to little more than 100 m south of Ireland. There is no bathymetric expression in these areas of the presence of the thick sedimentary basins that lie beneath the seabed. These include the North Channel, Kish Bank and Central Irish Sea basins to the northeast and east of Ireland, and the elongate Celtic Sea basins to the south of Ireland. This lack of noticeable bathymetric expression is indicative of the sediment-filled nature of the basins and/ or of the current isostatic balance between the crust and sediment column in these basins. The lack of bathymetric expression is also thought to be due in large part to Late Mesozoic and Cenozoic inversion and subsequent erosion of the basins.

The continental shelf west of Ireland is much more variable in its nature and physiography. It extends out more than 300 km, and considerably more than this in the region of the Rockall Bank (Figure 2.14). In the southwest, the Goban Spur, a remote plateau area on the continental margin south of the Porcupine Seabight, is located approximately 250 km southwest of Ireland. Bathymetrically it comprises a smooth platform sloping gently westwards away from the Celtic Shelf to depths of 2000 m. South of the Goban Spur the continental edge

curves eastwards, and the continental slope is cut by a series of deep canyons. These are of probable Cenozoic age, and were modified and further incised during Quaternary time. North of the Goban Spur the foot of the continental slope swings westwards from Ireland around the Rockall Bank. The Porcupine Seabight and Rockall Trough are deep-water embayments within the shelf that separate a number of higher plateau areas. The Porcupine Seabight is a large north–south trending deep-water area, overlying the thick Late Palaeozoic to Cenozoic successions of the Porcupine Basin. It opens southwestwards onto the Porcupine Abyssal Plain (Figure 2.14). Water depths increase from about 350 m in the north of the Seabight to more than 4 km in the south. The Porcupine Seabight is bounded on three sides by bathymetrically shallow platforms. The Irish Mainland Shelf lies to the east, the Slyne Ridge to the north, whilst the western boundary is formed by the Porcupine Ridge, extending southwards from the Porcupine Bank. The existence of Mesozoic and older rocks on the Irish Atlantic margin was first indicated by dredge samples reported by Cole and Crook (1910). The Rockall Trough, directly overlying the sediments infilling the Rockall Basin, is bounded to the west by the Rockall Bank, an elongate shallow bank marked at its shallowest point by Rockall islet. Further west lie the Hatton Trough, Hatton Bank and Continental Margin, west of which the slope deepens rapidly into the Iceland oceanic basin where water depths are of the order of 4 km. Again, the bathymetry reflects the underlying geology, with a major sedimentary depocentre (Hatton Basin) underlying the bathymetric Hatton Trough, while the bathymetrically shallow region (Hatton Bank) mirrors the underlying, structurally pronounced Hatton High. The narrow, elongate Slyne, Erris and Donegal sedimentary basins inboard of the eastern margin of the Rockall Basin and to the northeast of the Porcupine Basin, do not have a pronounced bathymetric expression and lie in a region of westward increasing water depth. As with the basins to the northeast, east and south of Ireland, this is largely a reflection of the major uplift and erosion that they underwent in late Mesozoic times, resulting in the removal of a substantial thickness of strata.

Geophysical evidence (gravity, magnetic and seismic data) shows that the continent–ocean boundary (COB), formed during the latest Cretaceous in the region south of the Charlie-Gibbs Fracture Zone, and during Paleocene time in the region west of the Hatton Continental Margin, coincides approximately with the 4 km isobath.

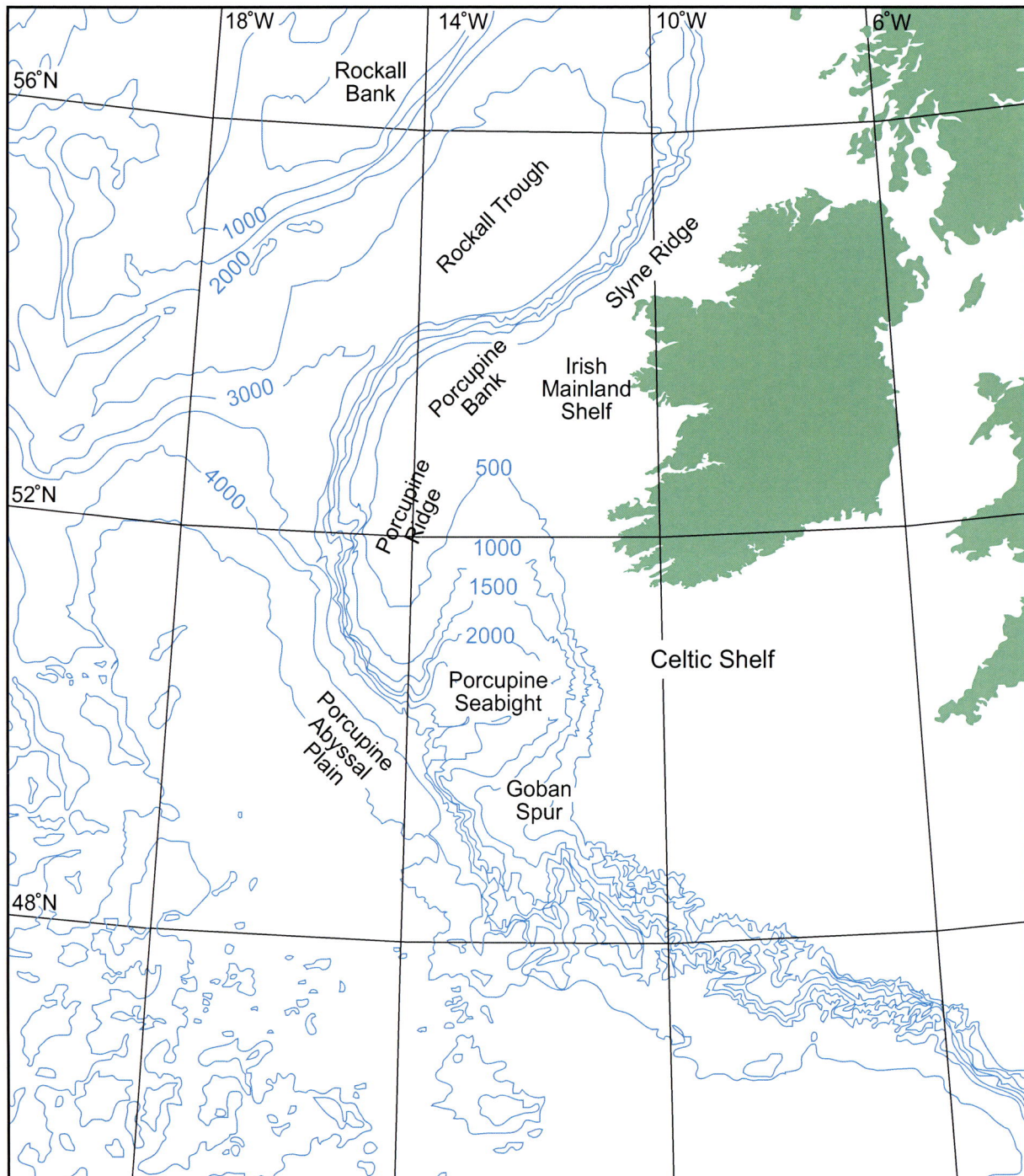

Figure 2.14. Bathymetric map showing the major physiographic features of the Atlantic shelf and margin of Ireland (from Naylor and Shannon, 2009 and Naylor *et al.*, 1999). Water depth is in metres.

The COB lies west of the Goban Spur and the mouth of the Porcupine Seabight, swings east–west, parallel to the Charlie-Gibbs Fracture Zone, immediately south of the Rockall Trough and the Rockall Bank, and then turns into a NE–SW orientation to the west of the Hatton Continental Margin. The COB is clearly seen on magnetic data (e.g. Srivastava and Verhoef, 1992) along much of its length, although it is more diffuse and less certain in the southwest of the Hatton region, where the boundary is complicated by the presence of igneous intrusions, seaward-dipping reflections from lava flows and areas of thin crust. The boundary is seen on deep seismic profiles at the mouth of the Porcupine Seabight (Makris *et al.*, 1988), the mouth of the Rockall Trough (Hauser *et al.*, 1995) and to the west of the Hatton Continental Margin (Vogt *et al.*, 1998). The southern and western continental–oceanic boundaries are different, with the southern boundary, parallel to the Charlie-Gibbs Fracture Zone, reflecting a tectonic (probably transpressional) aspect with a thickened ridge, while the western boundary is marked by a large high-density Lower Crustal Body. This is a regionally extensive, complex feature that is seen along the contiguous boundary west of the UK and Norway (e.g. White, 1992). It has been traditionally interpreted as an underplated body that was added to the base of the crust during early Cenozoic opening of the North Atlantic. However, its uneven thickness, complex nature and detailed morphology suggest that this interpretation may not be valid for the entire length of the feature. Some parts of it may be linked to serpentinisation of exhumed mantle adjacent to the COB. Other parts may represent fragments of a Caledonian orogenic root and consist of high-grade metamorphic rocks (Ebbing *et al.,* 2006).

Distribution and Development of Offshore Sedimentary Basins

A set of Permo-Triassic to Cenozoic sedimentary basins of different sizes, shapes and ages surround Ireland, extending onshore only in the northeast of the country (Figure 2.15). These are part of a regional system of basins extending along strike for several thousand kilometres from North Africa and Iberia to the Barents Sea, and covering a width of several hundred kilometres from the continental shelf west of Ireland, the UK and Norway, through to the North Sea in the east. Similar basins on the eastern margin of Canada (especially offshore Newfoundland and Nova Scotia) form part of the same basin system that shared a linked history until Late Cretaceous time (Sinclair *et al.*, 1994) when crustal separation occurred. The basins' development history reflects the response to the instability and breakup of the Pangaean Supercontinent and to the later development of the North Atlantic Ocean, as described in the earlier sections of this chapter.

In general, the Irish offshore basins can be grouped into three broad categories, each with distinct size, shape and age profiles. A set of small basins lies between Ireland and the UK in the Irish Sea region. The Rathlin, Lough Neagh, Larne and North Channel basin complex extends between Northern Ireland and Scotland. These basins are generally elongate parallel to the Caledonian structural fabric of the region. Further south in the Irish Sea, the Kish Bank, Central Irish Sea and St George's Channel basins have variable orientations. While the Kish Bank Basin is largely circular to slightly elongate, the basins further south generally follow the Caledonian orientation of the basement rocks in the adjacent Irish and Welsh mainlands. This first group of basins includes the smallest and generally the least elongate of the offshore basins and they reflect development during the initial tectonic instability and incipient breakup of Pangaea. They are the oldest of the basins, containing a predominantly Permo-Triassic basin fill, and typically have a thin to absent Cretaceous and Cenozoic succession as a result of early intra-Mesozoic or early Cenozoic inversion, uplift and erosion.

A second group of basins lies to the south of Ireland (Figure 2.15). These basins are elongate, and are larger and typically have a younger basin-fill than the Irish Sea basins. They occur in the Celtic Sea and run parallel to the south coast of Ireland, and are slightly oblique to the major onshore Variscan structures of Ireland. The Fastnet and North Celtic Sea basins lie to the north of an intermittent basement structure (the Pembrokeshire–Labadie basement high), while the Cockburn and South Celtic Sea basins lie to the south of the ridge. At their eastern end they merge, with a major strike swing, into the Irish Sea basins. These basins may have begun their development as a series of small, isolated basins, coeval with the Permian basins of the Irish Sea. However, they contain only rare pockets of Permian sediments, but have extensive Triassic strata. The main phase of basin development was in the Jurassic to Early Cretaceous, leading to thick deposits of generally clastic sediments. The basins typically underwent a number of periods of regional inversion in the early Cenozoic, so that a post-Mesozoic succession is typically only patchily developed.

The third group of basins contains the largest and generally the youngest basins in the Irish region (Figure 2.15). The basins lie exclusively in the Atlantic region and are the furthest west of the three basin groups. Most of the basins follow a general NNE–SSW trend and form part of a chain of structurally-linked basins that extends from mid Norway to west of Iberia (Doré *et al.*, 1997a, b). The notable exceptions to the NNE–SSW basin orientation of this third group of basins are the Porcupine Basin that trends north–south, and the Goban Spur Basin, lying south of the Porcupine Basin and west of

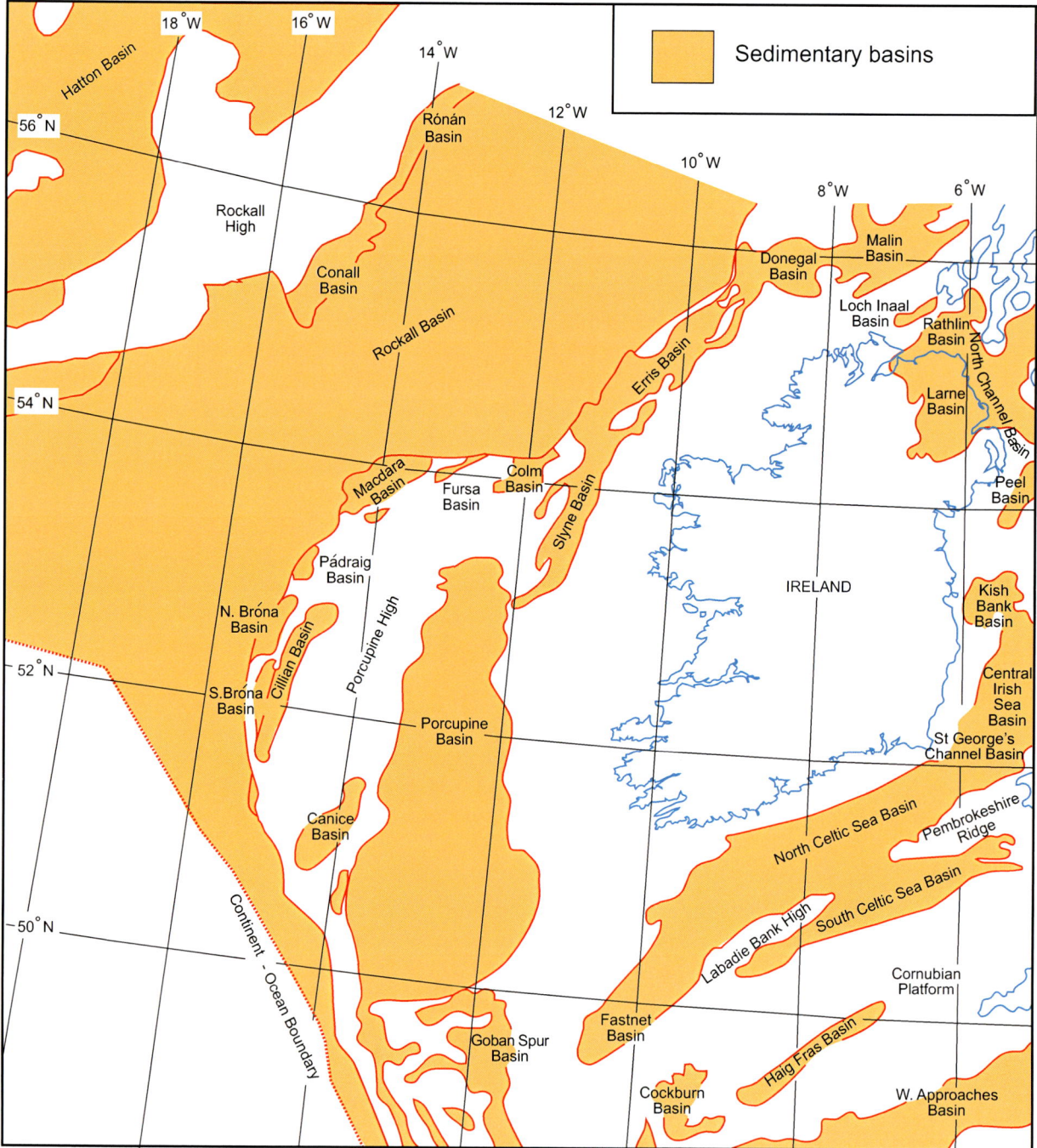

Figure 2.15. Location of sedimentary basins around Ireland (from Naylor and Shannon, 2009 and Naylor *et al.*, 2002).

the Fastnet–Celtic Sea basins system. Within the Irish Atlantic offshore region a band of narrow (inboard) basins includes the Slyne, Erris and Donegal basins. These lie landward of a set of larger (outboard) basins that includes the Porcupine, Rockall and Hatton, basins. A number of small, elongate, probably early Mesozoic basins are located in the footwalls of the main Cenozoic

Rockall Basin (Naylor *et al.*, 1999), often referred to as 'perched' basins. The main phase of sedimentation in the large basins was in Cretaceous and especially in Cenozoic times. Unlike the two other basin types, the largest of the basins underlie significant bathymetric depressions, reflecting rapid subsidence and under-sedimentation from mid-Cenozoic time, following the regional sagging event described by Praeg *et al.* (2005) and outlined earlier in the chapter.

The overall progression in size, shape and age of the offshore basins reflects a westward progression of extensional strain towards the area of continental breakup and seafloor spreading. Initial sedimentary basin development took place in a series of localised Permian basins in to the foothills of the Variscan mountains, and sometimes in depocentres controlled by the gravitational collapse of Variscan orogenic structures. Early to mid-Triassic depocentres were either partly infilled topographic lows or rift induced depressions, with the basin geometries representing the transition from late Variscan collapse basins, through to early wide-rift systems that developed over warm and thick crust (Štolfová and Shannon, 2009). Through Late Triassic and into Early Jurassic times, localised fault-controlled depocentres developed, but deposition was largely in broad and shallow thermal subsidence depocentres. Major classical narrow-rift development occurred during Middle Jurassic to Early Cretaceous times. Rifting during the Middle Jurassic appears to have followed caledonoid-oriented crustal fabrics and structures (e.g. the Celtic Sea and the Slyne–Erris basins). The onset of the north–south elongate rifting in the Porcupine Basin was probably somewhat later (Late Jurassic). A major Base Cretaceous unconformity, or grouped set of unconformities, is seen in most of the basins and corresponds to a period of plate reorganisation. Residual rifting continued in places (e.g. the Porcupine Basin), with localised inversion in others (e.g. the Fastnet Basin). A local phase of rifting occurred during the Aptian–Albian and thereafter regional thermal subsidence, accompanied by a eustatic sea level rise, occurred in all the basins, giving rise to major Upper Cretaceous chalk deposition with very little terrigenous input. The Cenozoic marked a major change with a combination of basin inversion, return to clastic deposition, and marked igneous activity especially in the Atlantic margin basins. Various Cenozoic plate reorganisations led to localised major sediment wedges, inversion events and differential basin restructuring. The internal structure of the basins was complicated by igneous activity associated with crustal breakup, and by localised inversion due to plate readjustments.

References

Boldreel, L.O. and Andersen, M.S. (1993) Late Paleocene to Miocene compression in the Faroe–Rockall area. *In*: Parker, J.R. (ed.), *Petroleum Geology of NW Europe: Proceedings of the 4th Conference.* The Geological Society, London, 1025–1034.

Boldreel, L.O. and Andersen, M.S. (1998). Tertiary compressional structures on the Faroe–Rockall Plateau in relation to northeast Atlantic ridge-push and Alpine foreland stresses. *Tectonophysics*, **300**, 13–28.

Brown, J.W. and White, R.S. (1995) Effects of finite extension rate on melt generation at rifted continental margins. *Journal of Geophysical Research*, **100**, B9, 18011–18029.

Chalmers, J.A. and Laursen, K.H. (1995) Labrador Sea: the extent of continental and oceanic crust and the timing of the onset of seafloor spreading. *Marine and Petroleum Geology*, **12**, 205–217.

Clift, P.D. and Turner, J. (1998) Paleogene igneous underplating and subsidence anomalies in the Rockall–Faeroe–Shetland area. *Marine and Petroleum Geology*, **15**, 223–243.

Coffin, M.E. and Eldholm, O. (1992) Volcanism and continental break-up: a global compilation of large igneous provinces. *In*: Storye, B.C., Alabaster, T., and Pankhurst, R.J. (eds), *Magmatism and the Causes of Continental Break-up.* Geological Society, London, Special Publications, **68**, 17–30.

Cole, G.A.J. and Crook, T. (1910) On rock specimens dredged from the floor of the Atlantic off the coast of Ireland and their bearing on submarine geology. *Memoir Geological Survey of Ireland*, 84pp.

Coward, M.P. (1995) Structural and tectonic setting of the Permo-Triassic basins of northwest Europe. *In*: Boldy, S.A.R. (ed.), Permian and Triassic rifting in Northwest Europe. Geological Society, London, Special Publications, **91**, 7–39.

Davis, M. and Kusznir N.J. (2004) Depth-dependent lithospheric stretching at rifted continental margins. *In*: Karner, G.D. (ed.), *Proceedings of NSF Rifted Margins Theoretical Institute.* Columbia University Press, 92–136.

De Garciansky, P.C. and Poag, C.W. (1985) Geological history of the Goban Spur, northwest Europe continental margin. *In*: De Garciansky, P.C. *et al.* (eds), *Initial Reports of the Deep Sea Drilling Project*, **80**, U.S. Printing Office, Washington, D.C., 1187–1212.

Doré, A.G. and Lundin, E.R. (1996) Cenozoic compressional structures on the NE Atlantic margin: nature, origin and potential significance for hydrocarbon exploration. *Petroleum Geoscience*, **2**, 299–311.

Doré, A.G., Lundin, E.R., Fichler, C., and Olesen, O. (1997a) Patterns of basement structure and reactivation along NE Atlantic margin. *Journal of the Geological Society, London*, **154**, 85–92.

Doré, A.G., Lundin, E.R., Jensen, L.N., Birkeland, Ø, Eliassen, P.E., and Jensen, L.N. (1997b) The NE Atlantic Margin: implications of late Mesozoic and Cenozoic events for hydrocarbon prospectivity. *Petroleum Geoscience*, **3**, 117–131.

Doré, A.G., Lundin, E.R., Jensen, L.N., Birkeland, Ø, Eliassen, P.E. and Fichler, C. (1999) Principal tectonic events in the evolution of the northwest European Atlantic margin. *In*: Fleet, A.G. and Boldy, S.A.R (eds), *Petroleum Geology of Northwest Europe: Proceedings of the 5th Conference.* Geological Society, London, 41–61.

Doré, A.G., Lundin, E.R., Kusznir, N.J., and Pascal, C. (2008) Potential mechanisms for the genesis of Cenozoic domal structures on the NE Atlantic margin: pros, cons and some new ideas. *In*: Johnson, H., Doré, A.G., Gatliff, W.R., Holdsworth, R., Lundin, E., and Ritchie, I.D. (eds), *The Nature and Origin of Compression in Passive Margins.* Geological Society, London, Special Publications, **306**, 1–26.

Ebbing, J., Lundin, E., Olesen, O., and Hansen, E.K. (2006) The mid-Norwegian margin: a discussion of crustal lineaments, mantle intrusions, and remnants of the Caledonian root by 3D density modeling and structural interpretation. *Journal of the Geological Society, London*, **163**, 47–59.

Géli, L. (1993) Volcano-tectonic events and sedimentation since Late Miocene times at the Mohnes Ridge, near 72N, in the Norwegian-Greenland Sea. *Tectonophysics*, **222**, 417–444.

Graham, J.R. (2009) Variscan deformation and metamorphism. *In*: Holland, C.H. and Sanders, I.S. (eds), *The Geology of Ireland.* Dunedin Academic Press, Edinburgh, 295–310.

Grantz, A., May, S.D., Taylor, P.T., and Lawver, L.A. (1990) Canada Basin. *In*: Grantz, A., Johnson, L., and Sweeney, J.F. (eds), *The Arctic Ocean region: The Geology of North America. Vol. 1.* Geological Society of America, Boulder, Colorado, 379–402.

Green, P.F., Duddy, I.R, Hegarty, K.A., and Bray, R.J. (1999) Early Tertiary heat flow along the UK Atlantic margin and adjacent areas. *In*: Fleet, A.J. and Boldy, S.A.R. (eds), *Petroleum Geology of Northwest Europe: Proceedings of the 5th Conference.* Geological Society, London, 348–357.

Green, P.F., Duddy, I.R. and Hegarty, K.A. (2002) Quantifying exhumation from apatite fission-track analysis and vitrinite reflectance data: precision, accuracy and latest result from the Atlantic margin of NW Europe. *In*: Doré, A.G., Cartwright, J.A., Stoker, M.S., Turner, J.P. and White, N. (eds), *Exhumation of the North Atlantic Margin: Timing, Mechanisms and Implications for Petroleum Exploration.* Geological Society, London, Special Publications, **196**, 331–354.

Hauser, F., O'Reilly, B.M., Jacob, A.W.B., Shannon, P.M., Makris, J., and Vogt, U. (1995) The crustal structure of the Rockall Trough: differential stretching without underplating. *Journal of Geophysical Research*, **100**, 4097–4116.

Japsen, P. (1997) Regional Neogene exhumation of Britain and the western North Sea. *Journal of the Geological Society, London*, **154**, 239–247.

Japsen, P. and Chalmers, J.A. (2000) Neogene uplift and tectonics around the North Atlantic: overview. *Global and Planetary Change*, **24**, 165–173.

Lear, C.H., Rosenthal, Y., and Wright, J.D. (2001) The closing of a seaway – ocean water masses and global climate change. *Earth and Planetary Sciences*, **210**, 425–436.

Lundin, E. (2002) Atlantic–Arctic seafloor spreading history. *In*: Eide, E.A. (coord.) *BATLAS – Mid Norway plate reconstructions atlas with global and Atlantic perspectives.* Geological Survey of Norway, 40–47.

Lundin, E.R. (2008) *Late Mesozoic and Cenozoic tectonic evolution of the Mid-Norwegian margin: a NE Atlantic perspective.*

Published dissertation, Faculty of Mathematics and Natural Sciences, University of Olso, No. **801**.

Makris, J., Egloff, R., Jacob, A.W.B., Mohr, P., Murphy, T. and Ryan, P. (1988) Continental crust under the southern Porcupine Seabight west of Ireland. *Earth and Planetary Science Letters*, **89**, 387–397.

Matte, P. (2001) The Variscan collage and orogeny (480–290 Ma) and the tectonic definition of the Armorica microplate: a review. *Terra Nova*, **13**, 122–128.

Nance, R.D., Worsely, T.R., and Moody, J.B. (1988) The supercontinent cycle. *Scientific American*, July, 44–51.

Naylor, D. and Shannon, P.M. (2009) Geology of offshore Ireland. *In*: Holland, C.H. and Sanders, I.S. (eds), *The Geology of Ireland.* Dunedin Academic Press, Edinburgh, 405–460.

Naylor, D., Shannon, P.M., and Murphy, N. (1999) *Irish Rockall Basin region – a standard structural nomenclature system.* Petroleum Affairs Division, Dublin, Special Publication **1/99**, 42pp.

Naylor, D., Shannon, P.M., and Murphy, N. (2002) *Porcupine-Goban region – a standard structural nomenclature system.* Petroleum Affairs Division, Dublin, Special Publication **1/02**, 65pp.

O'Connor, J.M., Stoffers, P., Wijbrans, J.R., Shannon, P.M., and Morrissey, T. (2000) Evidence from episodic seamount volcanism for pulsing of the Iceland plume in the past 70 Myr. *Nature*, **408**, 954–958.

O'Reilly, B.M., Hauser, F., Jacob, A.W.B., and Shannon, P.M. (1996) The lithosphere below the Rockall Trough: wide-angle seismic evidence for extensive serpentinisation. *Tectonophysics*, **255**, 1–23.

Pollitz, F.F. (1986) Pliocene change in Pacific plate motion. *Nature*, **320**, 738–741.

Praeg, D. (2004) Diachronous Variscan late-orogenic collapse as a response to multiple detachments: a view from the internides in France to the foreland in the Irish Sea. *In*: Wilson, M., Neumann, E.-R., Timmerman, M.J., Heeremans, M. and Larsen, B.T. (eds), *Permo-Carboniferous Magmatism and Rifting in Europe.* Geological Society, London, Special Publications, **223**, 89–138.

Praeg, D., Stoker, M.S., Shannon, P.M., Ceramicola, S., Hjelstuen, B.O., and Mathiesen, A. (2005) Episodic Cenozoic tectonism and the development of the NW European 'passive' continental margin. *Marine and Petroleum Geology*, **22**, 1007–1030.

Roberts, A.M., Lundin, E.R., and Kusznir, N.J. (1997) Subsidence of the Vøring Basin and the influence of the Atlantic continental margin. *Journal of the Geological Society*, **154**, 551–557.

Roest, W.R. and Srivastava, S.P. (1989) Seafloor spreading in the Labrador Sea: a new reconstruction. *Geology*, **17**, 1000–1003.

Ryan, P.D. and Dewey, J.F. (1997) Continental eclogites and the Wilson Cycle. *Journal of the Geological Society, London*, **154**, 327–442.

Scotese, C.R. (Co-ordinator) (1987) *Phanerozoic plate tectonic reconstructions. Palaeoceanographic Mapping Project.* Institute of Geophysics. University of Texas Technical Report **90**.

Shannon, P.M. (1991) The development of the Irish offshore sedimentary basins. *Journal of the Geological Society, London*,

148, 181–189.

Shannon, P.M. (1995) Permo-Triassic development of the Celtic Sea region, offshore Ireland. *In*: Boldy, S.A.R. (ed.), *Permian and Triassic Rifting in Northwest Europe*. Geological Society, London, Special Publications, **91**, 215–237.

Sinclair, I.K., Shannon, P.M., Williams, B.P.J., Harker, S.D., and Moore, J.G. (1994) Tectonic control on sedimentary evolution of three North Atlantic borderland Mesozoic basins. *Basin Research*, **6**, 193–218.

Srivastava, S.P. and Roest, W.R. (1999) Extent of oceanic crust in the Labrador Sea. *Marine and Petroleum Geology*, **16**, 65–84.

Srivastava, S.P. and Verhoef, J. (1992) Evolution of Mesozoic sedimentary basins around the North Central Atlantic: a preliminary plate kinematic solution. *In*: Parnell, J. (ed.), *Basins on the Atlantic Seaboard: Petroleum Geology, Sedimentology and Basin Evolution*. Geological Society, London, Special Publications, **62**, 397–420.

Stoker, M.S., Hoult, R.J., Nielsen, T., Hjelstuen, B.O., Laberg, J.S., Shannon, P.M., Praeg, D., Mathiesen, A., van Weering, T.C.E., and McDonnell, A. (2005) Sedimentary and oceanographic responses to early Neogene compression on the NW European margin. *Marine and Petroleum Geology*, **22**, 1031–1044.

Štolfová, K. and Shannon, P.M. (2009) Permo-Triassic development from Ireland to Norway: basin architecture and regional controls. *Geological Journal*, **44**, 652–676.

Underhill, J.R. and Partington, M.A. (1993) Jurassic thermal doming and deflation in the North Sea: implications of the sequence stratigraphic evidence. *In*: Parker, J.R. (ed.), *Petroleum Geology of Northwest Europe: Proceedings of the 4th Conference*. Geological Society, London, 337–346.

Vogt, U., Makris, J., O'Reilly, B.M., Hauser, F., Readman, P.W., Jacob, A.W.B., and Shannon, P.M. (1998) The Hatton Basin and continental margin: Crustal structure from wide-angle seismic and gravity data. *Journal of Geophysical Research*, **103**, 12545–12566.

Walker, I.M., Berry, K.A., Bruce, J.R., Bystol, L., and Snow, J.H. (1997) Structural modeling of regional depth profiles in the Vøring Basin; implications for the structural and stratigraphic development of the Norwegian passive margin. *Journal of the Geological Society, London*, **154**, 537–544.

Weber, J.R. and Sweeney, J.F. (1990) Ridges and basins in the Central Atlantic Ocean. *In*: Grantz, A., Johnson, L., and Sweeney, J.F. (eds), *The Arctic Ocean region; The Geology of North America. Vol. 1*. Geological Society of America, Boulder, Colorado, 305–336.

White, N. and Lovell, B. (1997) Measuring the pulse of a plume with the sedimentary record. *Nature*, **387**, 888–891.

White, R.S. (1992) Crustal structure and magmatism of North Atlantic continental margins. *Journal of the Geological Society, London*, **149**, 841–854.

Wilson, J.T. (1966) Did the Atlantic close and then re-open? *Nature*, **211**, 676–681.

Ziegler, P.A. (1988) Evolution of the Arctic–North Atlantic and the western Tethys. *AAPG Memoir*, **43**, 198pp.

Chapter 3

History of oil and gas exploration

Onshore exploration

Before considering the history of exploration for oil and gas in Ireland it is instructive to summarise the early British experience. Onshore oil exploration has been carried out in parts of mainland Europe since the nineteenth century and had resulted in the discovery of many onshore fields. In Britain, natural seepages of oil have been known for centuries in different places and in rocks of different ages. Collieries in the north of England frequently encountered small flows of oil, some of which were locally exploited. The demand for petroleum in the First World War (1914–18) prompted the search for indigenous oil supplies. This resulted in the drilling of a number of wells in the period preceding the Second World War (1939–45) and in the immediate post-War period, with modest success. Many of these small fields (Lees and Taitt, 1945; Falcon and Kent, 1960) occur in the Carboniferous, particularly in Pennsylvanian coal measures. However, interesting oilfields were also found in Jurassic rocks in Dorset. The discovery of the large Wytch Farm field near Swanage in 1973 led to a wave of further interest in onshore exploration in Britain. Wytch Farm is the largest onshore oilfield in Western Europe, and the sixth largest oilfield in the UK (onshore and offshore), with recoverable reserves estimated at 480 million barrels in Jurassic and especially Triassic reservoirs.

Petroleum exploration in Ireland has a quite different history. There was no equivalent of the exploration carried out onshore Britain during the period up to 1950. While the major political events in Ireland during the first part of the nineteenth century played some part in this, geological factors were also significant deterrents. No oil or gas seeps, or even significant staining, were known in Ireland. Pennsylvanian coal measures had been largely eroded from the Irish mainland, and mines in the remaining small coalfields did not encounter oil seeps. The fact that the coals of the Leinster coalfields are of anthracitic rank was a further negative indicator. More significantly, from the point of view of systematic exploration by the oil industry, thick Mesozoic sequences are found only in the northeast portion of the island. For these various reasons Ireland remained completely unexplored for hydrocarbons throughout the first half of the nineteenth century.

Onshore exploration finally began through an Irish-American initiative in 1959. A lively and entertaining account of the early years of Irish oil exploration is found in Collins (1976). The agreement in 1959 between the Minister for Industry and Commerce and Ambassador Irish Oil Ltd gave the latter an Exploration Licence for 20 years from 29th March, 1960. The area licensed to the company was the whole onshore area of the Republic of Ireland and any seas under Irish jurisdiction, subject to the surrender of 25% of the original area every five years. This licence was issued under the Petroleum and Other Minerals Development Act of 1960. The apparent generosity of the licence terms has to be judged against the lack of previous exploration, and against a general climate of low expectation in the period before any North Sea development had taken place (the first commercial offshore gas discovery in the southern North Sea was not made until 1965).

Conoco and Marathon Petroleum Ireland Ltd farmed into the Ambassador licence in 1961. The three-company consortium then drilled six exploration wells in the main onshore Carboniferous basins during 1962 and 1963. The first reflection seismic data within the Republic had also been acquired across the prospective areas, with S.S.L. as contractor, using vibroseis for the first time outside the United States. Two wells were drilled in 1962 – at Trim, County Meath and Doonbeg, County Clare (Figure 3.1). A third well, drilled at Dowra, Co. Cavan in the Northwest Carboniferous Basin during 1963, flowed gas at a rate of about 31,000 standard cubic feet per day (scfd) from Lower Carboniferous sandstones (Sheridan, 1977). Two further wells, drilled at Meelin, County Kerry and Ballyragget, County Kilkenny, were dry. However, a sixth, Macnean-1, drilled about 12 km

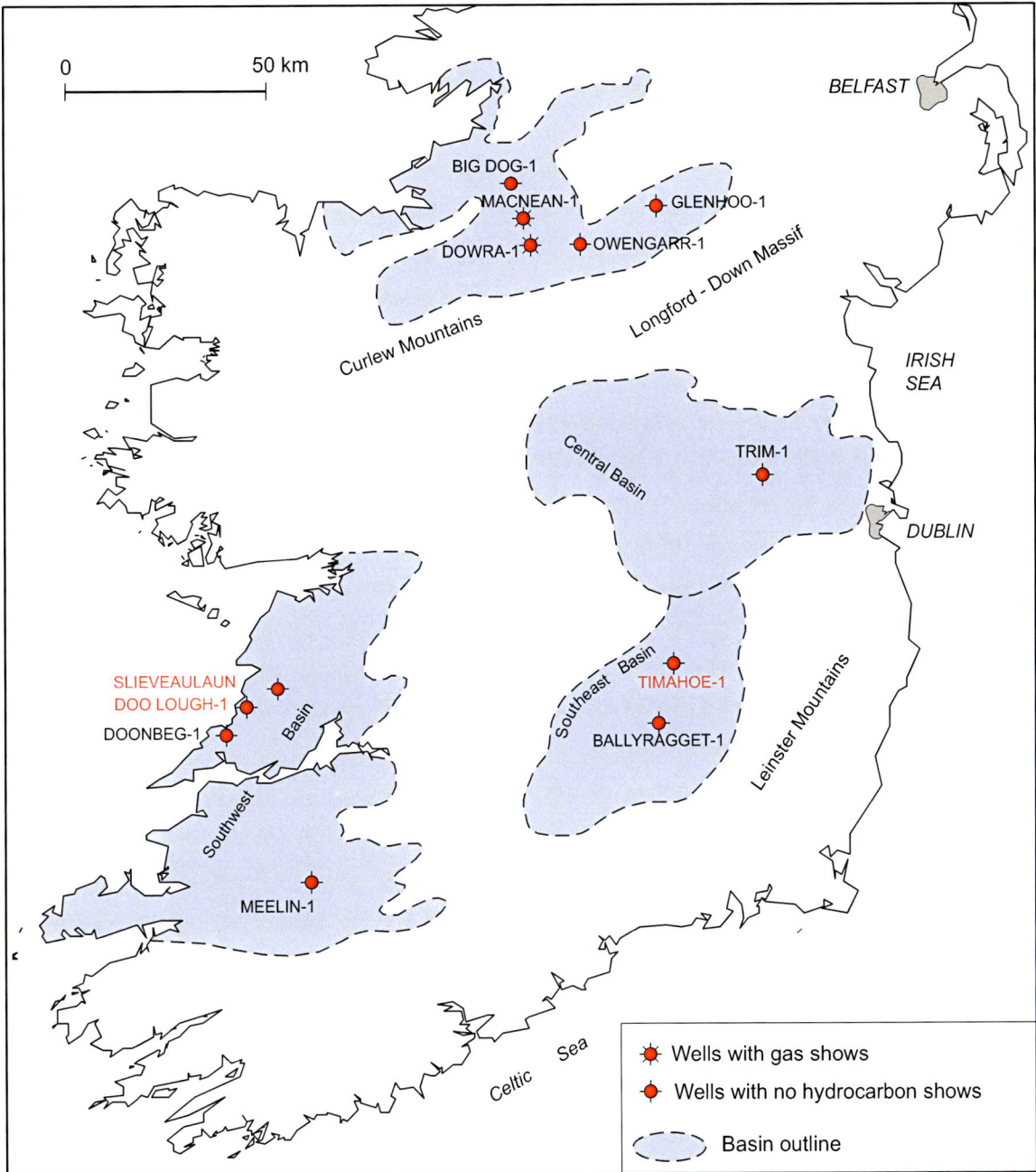

Figure 3.1. Onshore wells drilled in Ireland by Ambassador Irish Oil Company and partners, 1962-1966. The outlines of the Carboniferous basins identified and named by the consortium are also shown (simplified after Sheridan, 1977). Later wells, Slieveaulaun and Doo Lough (Irish Petroleum Prospecting Ltd) and Timahoe-1 (Tullow Oil) are also shown.

from Dowra-1 and close to the border with Northern Ireland, also tested gas at 15,000 scfd (Figure 3.1). The outline results of the exploration programme were given by Sheridan (1972, 1977), and details of individual wells were released in other publications. This exploration programme produced the first regional overview of the Irish Carboniferous in terms of basin development, regional isopachytes and reservoir potential.

The shows of gas in the Northwest Carboniferous Basin encouraged the consortium to apply for an Exploration Licence over a contiguous area in Northern Ireland. A licence covering 3500 km² was issued to Marathon Petroleum (UK) Ltd in May 1965 under the Petroleum (Production) Act (Northern Ireland), 1964 (Griffith, 1983). Three wells (Figure 3.1) were drilled for Carboniferous targets in the border area of Northern

Ireland (Big Dog-1, Owengarr-1 and Glenoo-1), the first two with gas shows, whilst Glenoo-1 was dry (Reay, 2004).

In the Republic, after withdrawal of two partners (Conoco in 1964 and Ambassador in 1966), and a first relinquishment of area (50% in 1965), Marathon was left holding the licence and the whole of the remaining concession. By this time Marathon was negotiating with the licensing authorities with regard to the offshore areas and took no further part in onshore exploration, which remained at reduced levels during the 1970s. Drilling in this period was restricted to two exploratory coreholes by Irish Petroleum Prospecting Ltd in 1980 to test the Carboniferous in County Clare – Slieveaulaun to 1760 ft (536.4 m), and Doo Lough to 3320 ft (1011.9 m), both without recorded shows (Figure 3.1).

After this period of low onshore activity, licences were taken up in 1980 over the Northwest Carboniferous Basin and adjoining areas by a consortium comprising Aran Energy, Marinex Petroleum and North West Oil & Gas. This group, operated by Marinex, re-entered and attempted to stimulate the earlier Dowra-1 well through hydrofracturing (with a resultant flow rate of 289,000 scfd from the Dowra Sandstone), and carried out a regional seismic programme. Santa Fe joined the group and, with Aran Energy now as operator, four wells were drilled during 1984 and 1985. One well was drilled in County Leitrim (Drumkeeran South-1), one in County Cavan (Macnean-2), and two in the immediate border area of County Fermanagh, Northern Ireland (Kilkoo Cross-1 and Slisgarrow-1). All were plugged and abandoned after encountering only minor gas shows. More recently (1996) licences have been held over this area by Evergreen Resources and some re-appraisal undertaken. During 2001 the company drilled Dowra-2 – a twin to the Dowra-1 well drilled by Ambassador in 1963 – and a further well, Thur Mountain-1. The same year the group drilled four further wells in County Fermanagh, Northern Ireland. The reservoir intervals in the wells were hydraulically fractured and subjected to extended well tests during 2002, but were plugged and abandoned in 2003. However, only low gas flows were achieved (<100,000 scfd) and the exploration programme failed to prove commercial quantities of gas in the area (Reay, 2004). During the past number of years acreage has been held under onshore Petroleum Prospecting Licences over a large portion of the Northwest Carboniferous Basin, most recently by Finavera Ltd until November 2008, but no further wells have been drilled. An account of the geological results of drilling in the Northwest Carboniferous Basin is given in Chapter 4 (*see* Figures 4.7 and 4.8).

A licence was also held by Tullow Oil in the 1980s in east central Ireland over the crop of the Leinster Coalfield. A slimhole exploration test, drilled in December 1987 at the northern end of the coalfield (Timahoe-1), was terminated at less than 1000 ft (305 m) due to mechanical difficulties.

Onshore oil exploration in the Republic of Ireland has therefore met with little success to date. An analysis of the reasons for this is presented later in the book. In consequence of the perceived low potential for oil and gas exploration, there has been no significant exploration activity in the Republic in recent years. However, in March 2010 the Irish licensing authorities announced the first onshore licensing round, closing for bids in June 2010. This covers large tracts of the Northwest Carboniferous Basin in counties Cavan, Donegal, Leitrim, Mayo, Monaghan, Roscommon and Sligo, as well as acreage in the Clare Basin covering the Carboniferous succession in counties Clare, Cork, Limerick and Kerry. A Licensing Option of up to 24 months' duration will be awarded to successful bidders, which can be converted to a Petroleum Exploration Licence with drilling commitments thereafter.

A quite separate onshore exploration province exists in the thick Permo-Triassic sequences of the Ulster Basin – a composite name encompassing a number of contiguous basins in the northeast of the island (Figure 2.15). The main play here is for hydrocarbon traps in porous Permo-Triassic sandstone reservoirs, sourced from underlying Carboniferous strata. The geology and prospectivity of the Ulster Basin are discussed in Chapter 12. The earliest exploration in this region was driven not by the search for oil and gas but instead was in an attempt to identify concealed coalfields beneath the Mesozoic cover (Griffith, 1983). Coal was formerly mined in the Tyrone Coalfield, comprising small outliers of Coal Measures west of Lough Neagh. Tests, by shaft and drill, were made prior to the Second World War west of the Lough, near the basin margin, in an unsuccessful attempt to locate more coal reserves beneath thin Permo-Triassic cover. The Department of Commerce for Northern Ireland then sponsored further deep boreholes (Table 3.1) in the sub-basins of the Ulster Basin between 1956 and 1980 in search of Carboniferous coal and Permo-Triassic halite and evaporites. Also, a deep geothermal test – Larne No 2 – was drilled at the east coast (Penn, 1981). These deep

Table 3.1. Wells and boreholes drilled in the Larne and Lough Neagh basins, east of Lough Neagh.

Name	Grid ref.	Date	Depth (m)	TD formation	Reference
Langford Lodge	J091748	1956	1524.6	Lower Palaeozoic	Manning et al., 1970
Ballyalton	J464728	1957	552.4	Lower Palaeozoic	Bazley, 1975
Belfast Harbour	J3786-7840	1978	523.0m	Dinantian	Smith, 1986
Larne-1	D4010-0239	1962	1283.5	Sherwood Sst. Group	Manning and Wilson, 1975
Newmill-1	J34603950	1971	1981.1	Permian breccia	Marathon/Shell well completion report
Ballymacilroy	J05749761	1977	2272.0	Tertiary intrusive	Thompson, 1979
Larne-2	D40700226	1981	2877.0	Permian 'brockram'	Penn, 1981
Ballytober-1	~ D33-05	1990	1282.0	?Lower Permian	Shelton 1997, Ruffell & Shelton, 1999
Annaghmore-1	H9966-8786	1993	1554.4	?Lower Permian	Naylor et al., 2003
Ballynamullan-1	H9990-8682	1994	1371.6	?Lower Permian	Naylor et al., 2003

tests, although not drilled in search of hydrocarbons, did confirm the nature of the thick Permo-Triassic succession, and opened up the possibilities of the presence of hidden oil and gas deposits.

None of the coreholes or wells drilled in the Ulster Basin prior to 1980 had been located using seismic surveys. Under an initiative from the Department of Commerce in 1981, C.G.G. shot two vibroseis lines across the coastal section of the Larne Basin, and a further line extending from Tertiary lavas onto Lower Carboniferous rocks to the west of Lough Neagh. These lines were made available to the oil industry (Illing, 1982). They showed in particular the presence of promising faulted structures in the Permo-Triassic. They also demonstrated that acquisition techniques had advanced sufficiently for reasonable seismic results to be obtained through the basalt that covers much of the basin. This latter point had been a source of technical concern. Interestingly, the first acquisition of reflection seismic in Ireland had been by the Geological Survey of Northern Ireland in 1959, using an S.S.L. dynamite crew to attempt, unsuccessfully, to image sub-basalt structure beneath part of the Rathlin Trough.

Four oil exploration wells have been drilled to date to test the Permo-Triassic sequence of the Larne and Lough Neagh basins. In 1968 Marathon obtained an exploration licence for an onshore area northeast of Belfast, and also covering the adjacent offshore. Some offshore seismic lines were shot, but there was no onshore seismic programme. An exploration well (Newmill-1, at

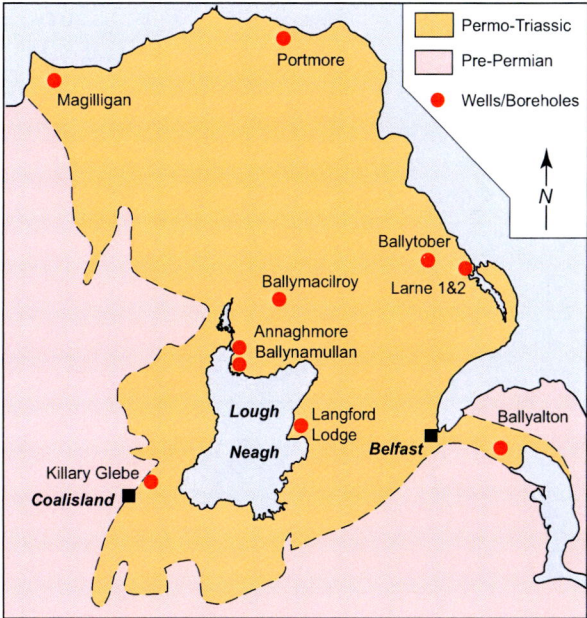

Figure 3.2. Location map of wells and boreholes in the onshore Ulster Basin mentioned in the text.

the head of Larne Lough) was drilled in 1971 by Shell under a farm-in arrangement with Marathon, but was dry and abandoned. A long interval then elapsed before three further petroleum exploration wells were drilled during the 1990s, each located using vibroseis surveys. The first, Ballytober-1, was drilled northwest of Larne by a consortium with Kirkland as operator on a seismically-defined structure, but only outline accounts have been published. The other two wells (Annaghmore-1 and

Ballynamullan-1) were drilled by Nuevo Energy (as operator for a group including Lough Neagh Exploration and Wiser) on the north shore of Lough Neagh (Figure 3.2). A discussion of the results is available in Naylor *et al.* (2003).

Offshore exploration

General summaries of the geology and exploration history of the Irish offshore basins can be found in Naylor and Shannon (1982), Naylor (1984, 1996), and Shannon (1993). Historical data are now also available on the Irish Petroleum Affairs Division website (www. pad.ie). Following the discovery of several large gas fields in Permian aeolian to shallow-marine sandstones in the southern part of the North Sea during the 1960s, the offshore exploration effort in that region gradually moved into the northern North Sea. This eventually led to the discovery in the early 1970s of a series of large oilfields in British and Norwegian waters in marine rocks of Mesozoic and Cenozoic age. This, in turn, directed

significant oil industry attention to the virgin areas west of Britain, and to Irish waters. However prior to this, exploration of the Irish offshore regions was already under way.

According to the original 1960 licence agreement Marathon Oil Company, as the remaining member of the original group of companies, was entitled to the entire continental shelf when this was designated by the government. After negotiations with the Irish Government, Marathon was granted three offshore tracts (Robinson and Riddihough, 1975; Naylor and Mounteney, 1975). In 1968 the first Irish offshore designation claimed an area slightly larger than that granted to Marathon (Figure 3.3). The following year a revised agreement was negotiated under which Marathon relinquished any exclusive claim to the Continental Shelf, except for the areas it already held. The company retained claims in the three areas already granted for the second term of the original licence (1965–1970). These were reduced by relinquishment for the third term (1970–1975), and again in March 1975 for the final term (1975–80).

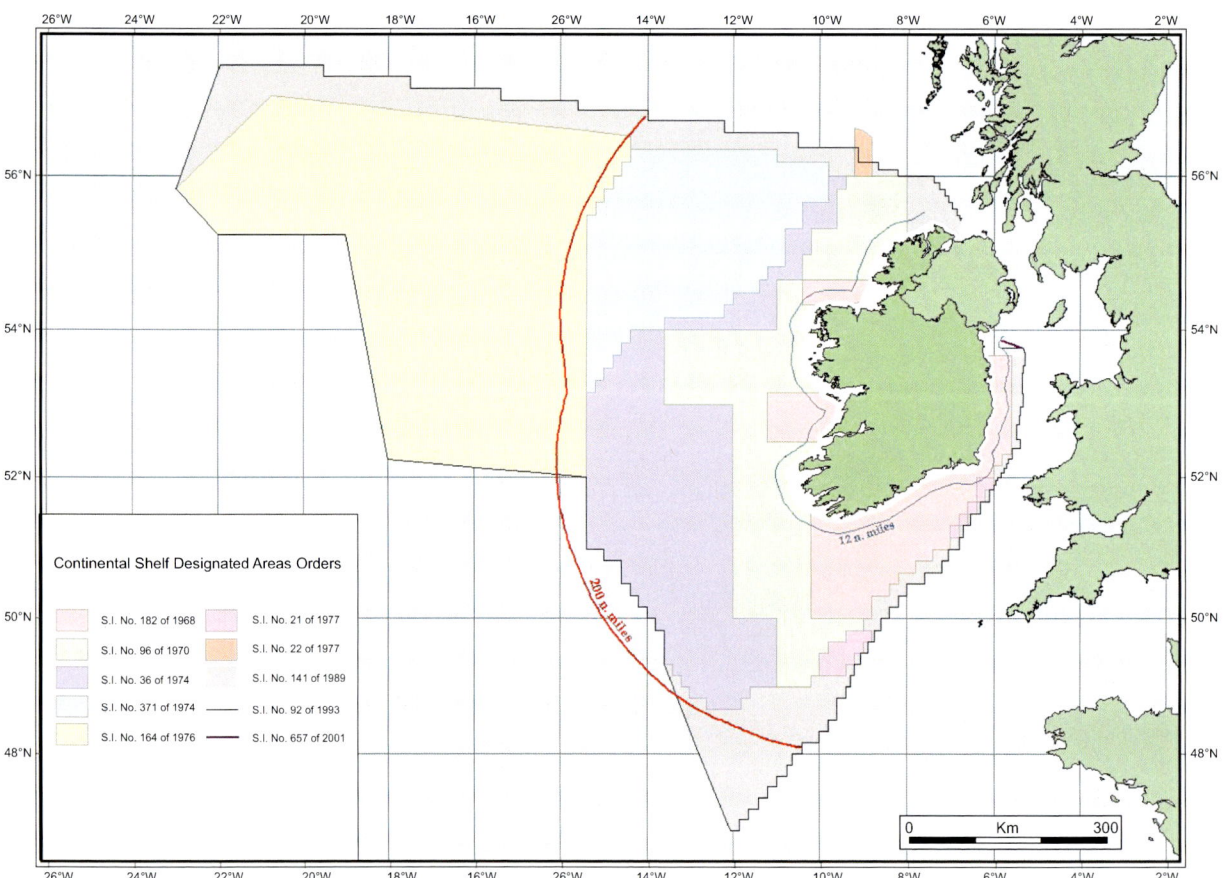

Figure 3.3. Successive stages in the designation of the Irish Continental Shelf (source: Petroleum Affairs Division, DCENR: website 2010).

As in United Kingdom waters, each 1° Latitude × 1° Longitude sector of the Irish offshore is numbered and divided into thirty licence blocks (this system is sometimes referred to as the Williams Grid). Each block measures 10' in latitude and 12' in longitude, and has an area of approximately 250 sq. km. Individual blocks are referred to by the sector number followed by the block number within the sector, thus – 50/1, 50/2 etc. Wells are numbered in the sequence of drilling within the block – 49/9-1, 49/9-2 etc.

The first seismic surveys were shot in the Celtic Sea in 1969. The discoveries in the southern North Sea had by this time imparted a considerable impetus to Irish exploration, as had the growing understanding of the extent of the Mesozoic sedimentary system around Britain and Ireland (Figure 2.15).

A series of further Irish offshore designations took place in 1970, 1974, 1976 and 1977, extending the area westwards over the Porcupine and south Rockall region (Figure 3.3). There was no agreed boundary between Ireland and the United Kingdom at this time, and it was not until 1990 that agreement was finally negotiated between the two countries. The problems and legalities relating to the definition of national offshore areas, and of the Irish offshore area in particular, are discussed later in this chapter.

In May 1970 the 48/25-1 well, with Marathon as operator, became the first petroleum exploration well to be drilled in the Irish offshore. The well, located in the North Celtic Sea Basin (Figure 3.4), targeted a shallow Cretaceous prospect and encountered gas shows before being plugged and abandoned. Marathon then drilled another unsuccessful well (50/11-1, with hydrocarbon shows), before returning to drill a second well in Block 48/25. This well, completed in late 1971, was the discovery well of the Kinsale Head Gasfield. The field, which produces from Lower Cretaceous sandstones, contained recoverable reserves of 1.5 trillion cubic feet (Tcf) of gas (Winn, 1994). It came on stream in 1978.

In 1971, Esso agreed a farm-in with Marathon by which Esso would earn a 50% interest in the western half of Marathon's Celtic Sea concession, excluding the 48/25 Kinsale Head Gasfield, in return for a work obligation involving the drilling of a number of exploration wells and further seismic surveys. At this time, no other concessions had been granted outside the Marathon areas, despite growing interest in the Irish offshore. By 1974, sixty two companies and groups had been granted non-exclusive Petroleum Prospecting Licences (Table 3.2), which allowed the companies to carry out exploration surveys on open acreage on the Irish Continental Shelf. A list of the companies is given in Robinson and Riddihough (1975) and reveals the substantial oil industry interest in the Irish offshore sector at that time.

Exploratory surveys and drilling were carried out by Marathon and Esso-Marathon on the North Celtic Sea Basin acreage throughout the 1970s (Elf later joined the farm-in). The two companies drilled a number of wells, most of which targeted shallow Kinsale Head look-alike structures. However, until relatively recently the seismic data quality in the basin has been poor, largely due to Cenozoic basin inversion resulting in the exposure of compacted Upper Cretaceous Chalk at the seabed. In consequence of the poor imaging of deeper targets, the early drilling results were generally disappointing and no further commercial discoveries were made. Despite this lack of success, the companies were granted a total of 20 petroleum leases – 17 to Marathon, 2 to Marathon-Esso, and one to Marathon-Esso-Elf), covering 37 blocks.

The oil crisis of 1973 led to a dramatic increase in oil prices, and spurred European and American countries into encouraging exploration for indigenous oil and gas. In 1975 the Irish Department of Industry and Commerce published a set of Terms and Conditions under which exclusive Exploration Licences (Table 3.2) would be awarded. The new terms, modelled on the Norwegian terms of the time, reflected a belief that large fields awaited discovery in the Irish offshore and the aspiration that Ireland should benefit by joint ownership of the fields (State Participation), as well as by the receipt of taxes and royalties. Applications were then invited for a First Round of Licensing, covering all blocks within Irish designated territory, with exception of those held by Marathon and Esso/Marathon, and 24 other blocks that at the time were being negotiated directly with a number of joint venture groups. The first Exploration Licences were granted in 1976 to 11 different groups, for a total of 43 blocks spread across the Fastnet, Porcupine, Slyne, Erris, Donegal and Kish Bank basins. These gave the holders exclusive rights to carry out exploration, including drilling, on the specified blocks. Licences were issued for an initial six year period where water depths were less than 600 ft (183 m), or nine years in deeper water. A 50% relinquishment was required after four or six years, respectively. Current licence types or authorisations (2009) are listed in Table 3.2.

Following on the First Round licence awards, a phase of drilling commenced in 1976. The first drilling took

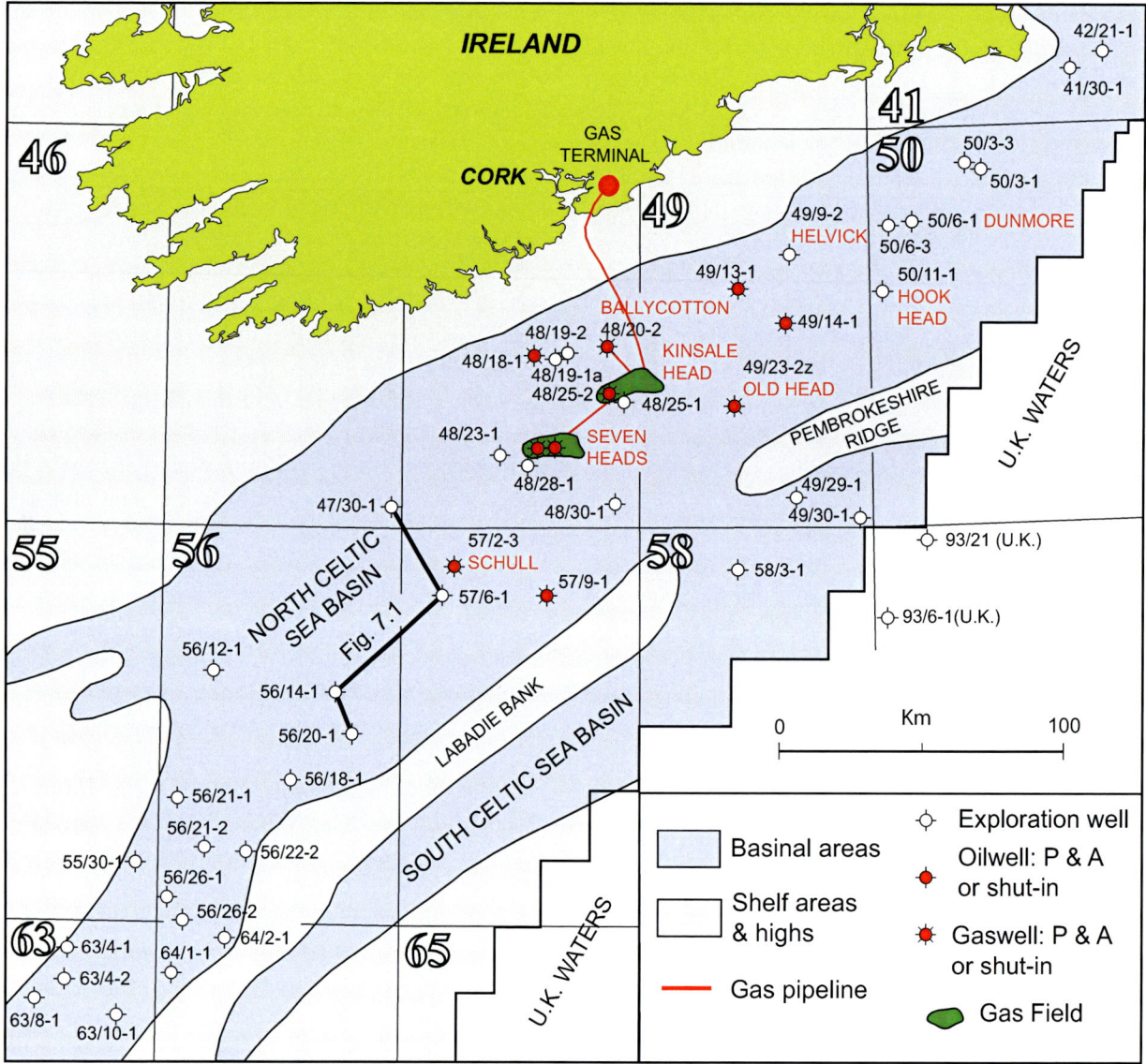

Figure 3.4. Fields, discoveries and wells in the Celtic Sea Basins, as mentioned in the text.

place in the Fastnet Basin, a western extension of the Celtic Sea basins (Figure 2.15), in about 130 m of water. The following year saw further drilling in the Fastnet Basin, and the first well in the Kish Bank Basin, east of Dublin. Drilling also commenced in 1977 in the deeper waters of the Porcupine Basin, off the west coast, when Shell drilled 35/13-1 in 482 m of water. Drilling reached a peak of activity in 1978 when 14 wells were drilled (Figure 3.5). One of these (Phillips 35/8-1) flowed 730 barrels of oil per day (bopd) on test (730 bopd) for the first time in Irish waters from Lower Cretaceous marine sandstones. This undeveloped discovery is now known as the Burren Prospect (Figure 3.6). Interest in the basin west of Ireland was fuelled by the major discoveries being

made at that time in the UK waters of the northern North Sea. The BP 26/28-1 well in 366 m of water in the northern part of the Porcupine Basin during 1979 discovered the Connemara oil accumulation (undeveloped), with proven reserves in place of 120 million barrels of 32–38° API oil (Macdonald *et al.*, 1987). The BP 26/28-1 discovery well flowed 5589 bopd from Middle and Upper Jurassic fluvial and shallow-marine sandstones. A follow-up well (BP 26/28-2), drilled in 1980, flowed 1500 bopd from a similar sequence. However, three further appraisal wells were dry, indicating a relatively small and complex structure (Earls, 1995). In 1981 Phillips Petroleum drilled the 35/8-2 well in 422 m of water and this flowed 4.85 MMcfd gas and 925 barrels per day of condensate

Table 3.2. Petroleum Authorisations for Exploration and Production Activities.

Various authorisations are issued by the Minister for Communications, Energy and Natural Resources under the Petroleum and Other Minerals Developmment Act, 1960.

1. Petroleum Prospecting Licence (issued under Section 9 (1) of the 1960 Act). This is a non exclusive licence giving the holder the right to search for petroleum in any part of the Irish Offshore which is not subject of a Petroleum Exploration Licence, Reserved Area Licence or Petroleum Lease granted to another party.

2. Licensing Option (issued under Section 7 (1) of the 1960 Act). This is a non exclusive licence giving the holder the first right, exercisable at any time during the period of the Option, to an Exploration Licence over all or part of the area covered by the Option.

3. Exploration Licence (issued under Section 8 (1) of the 1960 Act). There are three categories of Exploration Licence:-
  - Standard Exploration Licence for water depths up to 200m
  - Deepwater Exploration Licence for water depths exceeding 200m
  - Frontier Exploration Licence for areas so specified by the Minister.
For Standard and Deepwater Explorations Licences the holder is obliged to carry out a work programme that must include the drilling of a least one exploration well in the first phase. For a Frontier Exploration Licence the holder must commit to at least one exploration well in order to proceed to the second phase. The area of an Exploration Licence shall be expressed in terms of blocks and/or part blocks of the Williams Grid.

4. Lease undertaking (issued under Section 10 (1) of the 1960 Act). When a discovery is made in a licensed area and the licensee is not in a position to declare the discovery commercial during the period of the licence but expects to be able to do so in the foreseeable future, the licensee may apply for a Lease Undertaking. This is an undertaking by the Minister, subject to certain conditions, to grant a Petroleum Lease at a stated future date. The holder of a Lease Undertaking is required to hold a Petroleum Prospecting Licence which will govern activities under the Lease Undertaking.

5. Petroleum Lease (issued under Section 13 (1) of the 1960 Act). When a commercial discovery has been established it will be the duty of the authorisation holder to notify the Minister and apply for a Petroleum Lease with a view to its development.

6. Reserved Area Licence (issued under Section 19 (1) of the 1960 Act). A Petroleum Lease holder may apply for a reserved area licence in respect of an area adjacent to or surrounding the leased area and which is not subject of an authorisation other than a Petroleum Prospecting Licence.

Notes
Terms and conditions, including environmental provisions, are attached to the authorisations. These licensing terms are set out in the Department's *Licensing Terms For Offshore Oil And Gas Exploration, Development & Production 2007* which provide the operational framework for oil and gas exploration and production. They are the terms on which the Minister is prepared to issue the various authorisations.

For authorisations awarded prior to 1 January 2007 or Leases/Lease Undertakings awarded as a result of a discovery made under a licence awarded before 1 January 2007, the terms and conditions are set out in the Department's Licensing Terms for Offshore Oil and Gas Exploration 1992.

In addition Rules and Procedures for Offshore Petroleum Exploration Operations apply to all petroleum exploration and development operations in the internal waters of the State, the territorial waters or in the designated areas of the continental shelf under Irish jurisdiction.

from Jurassic sandstones – this prospect became known as 'Spanish Point'. The company's reserve estimate for Spanish Point was 1.1 Tcf gas and 112 MMbo, but these figures failed to pass the company's risk and economic threshold at that time and the acreage was relinquished as the oil price collapsed during the 1980s. It is now held under licence by Providence Resources (*see* Chapter 11 for further discussion).

Since the First Licensing Round the Irish Department of Energy (now the Department of Communications, Energy and Natural Resources) has, in contrast to the United Kingdom, operated an 'open door' policy with respect to licensing. Companies are free to enter into discussions with the Department regarding the work programme that would be required to obtain a licence over a block or blocks in open acreage. This policy has resulted in the granting of a small flow of licences over the succeeding years. Licensing Option agreements were first signed by exploration groups in 1978. Under the terms of an Option (Table 3.2), the company is required to carry out a work programme, normally a seismic survey, within a specified period. The company then has to decide whether any of the blocks held under the option are to be retained as an Exploration Licence with an agreed work programme.

The second oil crisis of 1979 stimulated a renewal of exploration interest by oil companies in the Irish offshore. The Second Licensing Round took place in 1981 following the required relinquishment of a large amount of acreage in the Celtic Sea region by Marathon, at a

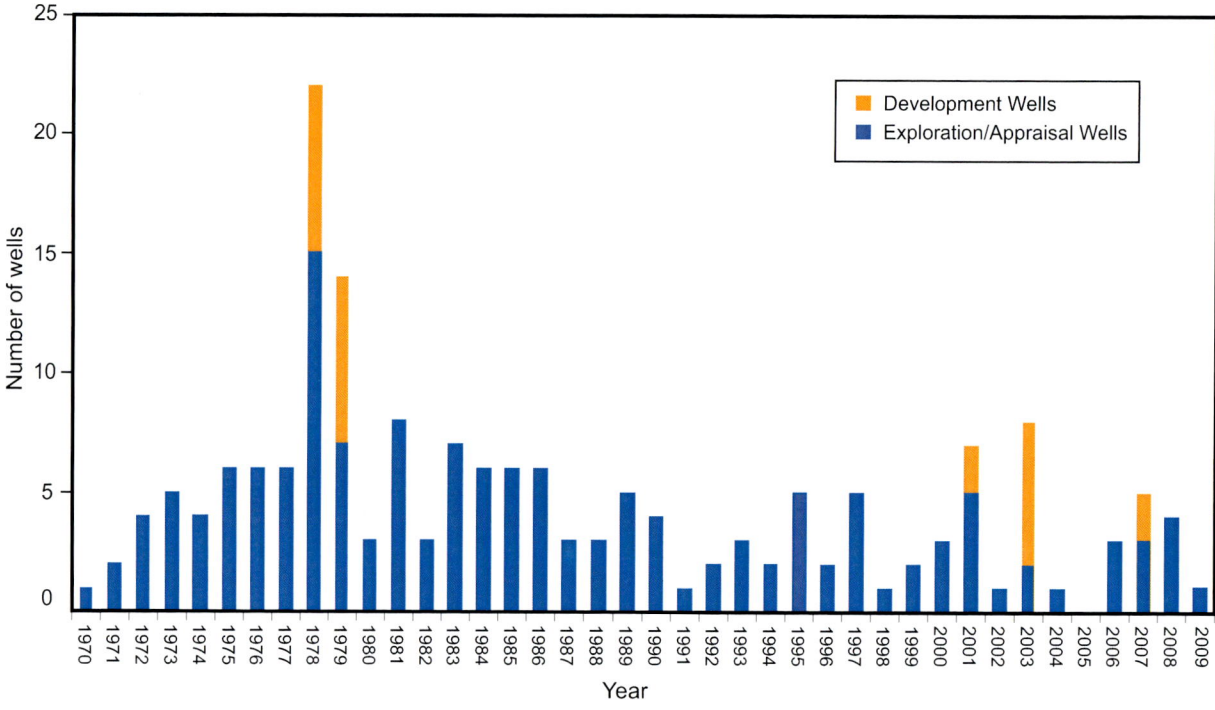

Figure 3.5. Histogram of the offshore wells drilled each year since 1971 (source: Petroleum Affairs Division, DCENR: website 2010).

time when seismic acquisition techniques were improving to allow better definition of deep, unexplored structures within the North Celtic Sea and Porcupine basins. A large number of companies new to the Irish offshore entered the Round and provided a boost to the falling number of wells being drilled, especially in the North Celtic Sea Basin. One hundred and eight blocks were on offer (45 in the North Celtic Basin, 46 in the Porcupine Basin, 7 in the Slyne Basin, 6 in the Donegal Basin and 4 in the Kish Bank Basin). Thirty-seven companies (21 new to offshore Ireland) were involved in applications. In 1982 a total of 24 blocks were awarded to 10 consortia (18 in the Celtic Sea basins, 4 in the Porcupine Basin, and 2 in the Kish Bank Basin). In a significant separate development, Esso took a 30-block seismic option over a portion of the Goban Spur on the Atlantic margin. After a phase of seismic acquisition the option was converted in 1981 to a 14-block exploration licence and in 1982 the company drilled the 62/7-1 well in 1075 m of water (Figure 3.6). The well reached a depth of 4661 m (15.293 ft; Cook, 1987) and remains the only well drilled in the Goban frontier province to date.

The Second Licensing Round led to an intensive phase of seismic acquisition, followed in 1983 by the drilling of further wells in the North Celtic Sea Basin. Improved

seismic techniques now allowed better imaging of the deeper plays, particularly in the Jurassic, which was judged to have considerable potential. The discovery of hydrocarbons in 1983 by a Gulf–Union–Atlantic Resources consortium within the Middle–Upper Jurassic section in Block 49/9 at the eastern end of the North Celtic Sea Basin (Figure 3.4), gave rise to considerable industry interest. The well tested at rates up to 6467 bopd of 44°API oil (aggregating 9911 bopd oil and 2.1 MMcfd gas), and demonstrated the potential for light oil accumulations in the Celtic Sea area. The Department of Energy suspended the 'open door' policy for licence acquisition in the discovery area, and in 1984 announced a Third Round of Licensing, encompassing most of the open acreage (76 blocks) in the Irish sector of the Celtic and Irish Seas. Twenty two companies were involved in the Round and in October 1985, the Minister announced the award of fifteen blocks to nine consortia. Thirteen of the blocks were in the Celtic Sea and two in the Irish Sea. Unfortunately, follow-up drilling of the 49/9 discovery showed that it was of small extent. This discovery – the Helvick Oilfield (Caston, 1995), now operated by Providence Resources plc., remains undeveloped. In 1985 a Gulf-operated group drilled a further discovery in the same region, when well 50/6-1 flowed

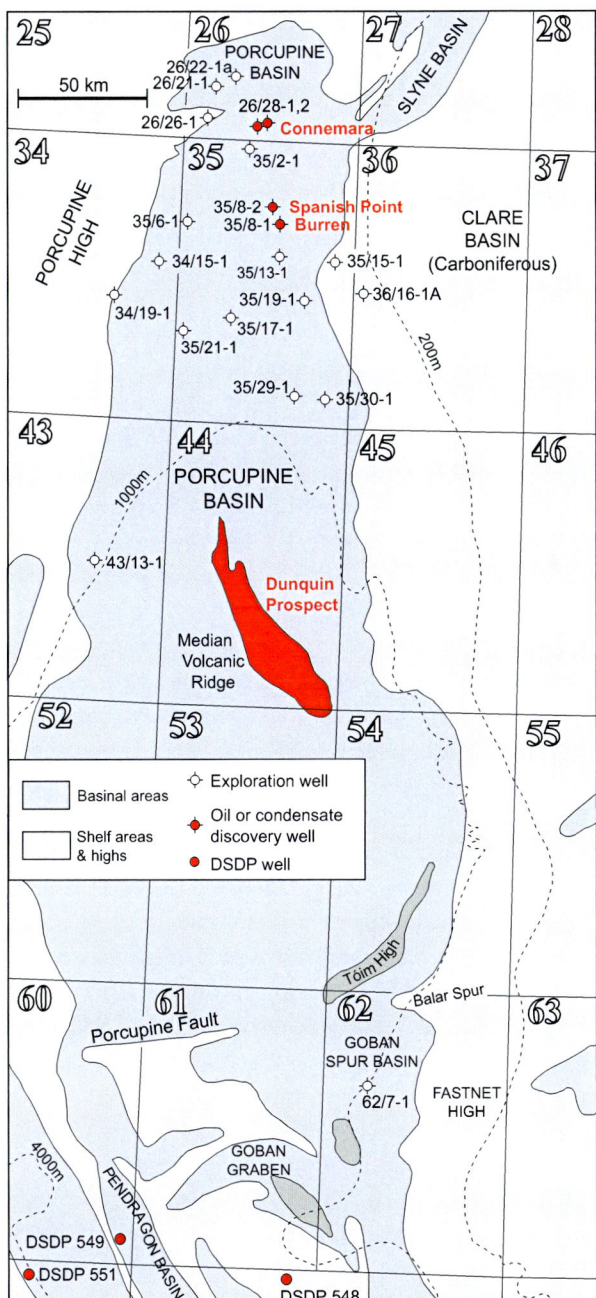

Figure 3.6. Fields, discoveries, wells and DSDP boreholes in the Goban Spur – Porcupine Basin region, as mentioned in the text.

A number of hydrocarbon accumulations, judged to be non-commercial at the time, were encountered in drilling within the Celtic Sea during the early phases of exploration. Some of these have undergone re-appraisal in recent years with the spur of increased product prices and improved technology (*see* later in the chapter). One was the Esso–Marathon Seven Heads prospect, on which two wells flowed hydrocarbons on test. Esso–Marathon 48/28-1 produced at a rate of 1550 bopd from thin Lower Cretaceous sandstones. A further well, 48/23-1, had oil and gas shows, but failed to flow on test. The final well on the prospect, 48/24-2, drilled in 1978, encountered both gas and oil zones. Reserve estimates by the operator at the time were about 100 bcf gas, and 1–2 million barrels of oil. A later well, Bula 48/19-2, drilled in 1992, west of the Ballycotton Gasfield, encountered a large accumulation of heavy biodegraded oil, similar to the Seven Heads structure, in Greensand and Wealden sandstones, sourced from the Liassic (Howell and Griffiths, 1995) or possibly from the Portlandian–Purbeckian. Further east, gas flow rates of 119 mcfd were recorded in the Marathon well 49/13-1, drilled on the 'Ardmore' prospect in 1974. An offset well, 49/14-1, flowed gas on test at a rate of 8.81 MMcfd, but at the time (1974–75) the accumulation was not regarded as commercial. Marathon also encountered hydrocarbons when drilling a large anticlinal structure in Block 50/11, known as the 'Hook Head' structure. The Marathon 50/11-1 discovery well in 1971 encountered oil and gas shows in five sandstones in the Lower Cretaceous. The well was not tested due to operational problems. A later well, Marathon 50/11-2, drilled in 1975, was a delineation well at the down-dip edge of the structure, following which the accumulation was deemed non-commercial.

By the mid-1980s the effects of the oil price rises of the late 1970s had begun to take a toll on the industrial countries and recession had begun to bite. The demand for oil and gas dropped and the oil industry began to look away from the high risk, unproven regions. Drilling results from the Irish offshore had been disappointing – some small, non-commercial discoveries but no indications of major or giant fields. In comparison to other countries, the Irish licensing terms had begun to look unattractive, especially for the types of small to marginal fields that were regarded as the most likely outcome in the North Celtic Sea and Irish Sea basins. Over the course of the next few years the Irish financial terms were modified several times in order to bring them into line with those of neighbouring competitors. Eventually the state

2074 bopd and 1.3 MMcfd gas probably from the pre-Cretaceous section (the 'Dunmore' prospect). In March 1989 Marathon drilled the discovery well (48/20-2) for the Ballycotton Gasfield (Figure 3.4), a small field with an estimated 89.9 bcf of gas in place (Winn, 1994). Using subsea completion technology, this small satellite field was linked into the Kinsale Head production facility in 1991 (Murray, 1995).

participation and royalty components were abolished, and the terms became significantly more attractive by international standards. The underlying exploration policy reflected the pragmatic view that oil and gas is only found by drilling wells, and this should be facilitated and encouraged by every means possible. By the early 1990s the pace of exploration had slowed in the Irish basins. Most of the obvious structures in the Celtic Sea and Irish Sea had been drilled and the results had been generally disappointing.

Oil industry attention has gradually turned to the Irish Atlantic margin basins. The results of exploration in the UK and Norwegian parts of the North Atlantic margin had proved encouraging. New seismic data in the Slyne, Erris and Rockall basins added to the encouragement and indicated the presence of interesting structures and potential plays. Exploration interest and activity in the deeper water under-explored basins of the Irish Atlantic margin has tended to fluctuate according to interaction between conflicting factors. The basins are recognised as having the potential for the large discoveries required to fund exploration and development in this environment. Also, successes offshore West Africa and South America have highlighted the potential of deep-water areas. On the other hand, the price of oil was low in the 1990s and has fluctuated in the general rise to a high in 2008, before subsiding to significantly lower levels in 2009. This makes planning for exploration and development in high-cost environments difficult for the oil industry. The longer period of the Frontier Exploration Licence (Table 3.2) was intended to assist in long term planning.

Although no further standard Licensing Rounds have been held, the Irish licensing authorities have sought since 1993 to promote exploration of the Atlantic Margin by holding a succession of *Frontier Licensing Rounds*. These have been:

i)   *Erris–Slyne*: closing date 15th December 1993. 128 blocks on offer.
ii) *Porcupine Basin*: closing date 15th December 1994. 172 blocks on offer.
iii) *Rockall Basin*: closing date 26th March 1997. 615 blocks and 35 part-blocks on offer.
iv) *South Porcupine Basin*: closing date 15th December 1998, 156 blocks on offer.
v)   *Porcupine Basin*: in three tranches, with closing dates 15th March 2003, 15th October 2003 and 15th March 2004.
vi) *Rockall Basin*: closing date 15th March 2005. 65 blocks and 12 part blocks on offer.

vii) *Slyne–Erris–Donegal*: closing date 15th March 2006. 73 blocks and 31 part blocks on offer.
viii) *Porcupine Basin*: closing date 18th December 2007. 229 full blocks and 3 part blocks on offer.
ix) *Rockall Basin*: closing date 22nd April 2009. 477 full blocks and 44 part blocks on offer (later extended in area).
x)   *Atlantic Margin*: closing date 31st May 2011. 996 full blocks and 58 part blocks on offer.

These Rounds have been generally successful in attracting major companies to the region and resulted in the issuing of a scatter of licence blocks in each basin. Because of the large size of the basins and the very early stage in the exploration process, it is not surprising that relatively few blocks have been licensed, compared with the total number on offer. The terms offered were seen by the oil industry as being realistic and pragmatic and afforded a relatively long lead-time to companies before a decision is required to drill or drop acreage. The First Frontier Licensing Round offered blocks covering the Slyne and Erris basins, and ten companies, grouped in five consortia, applied for 28 of the blocks. Five licences were awarded to the consortia led by Enterprise (2 licences), Kerr-McGee Oil, Statoil (UK) Ltd. and Texaco (Petroleum Affairs Division 1994).

The Second Frontier Licensing Round in 1994 (awards in 1995) offered blocks in the northern part of the Porcupine Basin and was again heavily subscribed. The Third Frontier Licensing Round of 1997 offered deep-water blocks in the eastern Rockall and Erris basins and was also successful in terms of the numbers of licence applications. A total of 58 blocks or part blocks were awarded to 11 consortia comprising 16 companies. A novel aspect of this round was the instigation of the Petroleum Infrastructure Programme (PIP), whereby the licensees contributed to a fund which was used to finance research on the Rockall region, largely in Irish academic and service institutions. Although the number of wells drilled under the Frontier Rounds has been relatively small, such drilling led to the discovery of the Corrib Gasfield by a group operated by Enterprise Oil (now Shell) in the north Slyne Basin during 1996 (Figure 3.7). Two wells were drilled in the Porcupine Basin during 1997 on blocks awarded in the 1995 Round: Total 35/17-1 (4305 m); and Marathon 35/30-1 well (5215 m) that was reported to have encountered multiple hydro-carbon shows.

The Fourth Frontier Licensing Round, in the south Porcupine Basin, took place in 1998. At the time of the

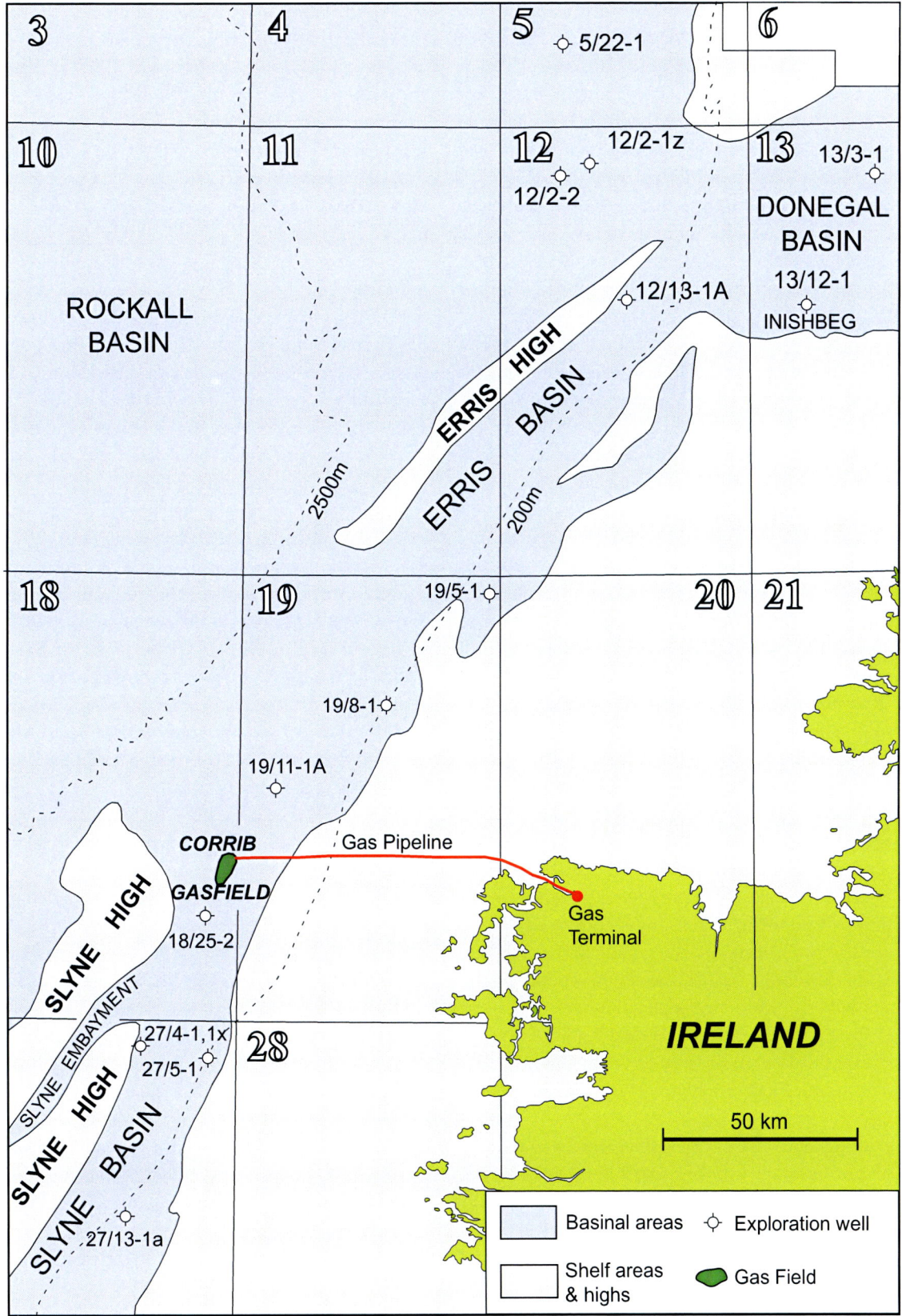

Figure 3.7. Fields, discoveries and wells in the basins of the northern Atlantic margin, as mentioned in the text.

announcement of the Round an optimistic air still prevailed in the international exploration industry. However, by the time the Round closed for bidding, major changes in the nature and confidence of the industry had taken place. The results of the appraisal of the Connemara oil accumulation in the Porcupine Basin by Statoil (who had acquired the licence from BP) had been disappointing. Statoil had drilled two extended reach appraisal wells during 1997, but these had only served to demonstrate the complexity of the geology and had failed to establish the commerciality of the field. The oil price had dropped significantly and was showing no sign of a base level. Company mergers were taking place, especially among the major oil companies, with resultant redundancies and cutbacks in exploration. Mergers were seen, in part, as a cost-effective way to acquire reserves without carrying out expensive exploration. Fewer companies meant a decrease in competition and a consequent decrease in the critical mass of exploration ideas that could be focused upon a particular area. Many companies began to feel over-exposed to the high risk, deep-water North Atlantic margin region. In consequence, only two exploration groups (Agip and Elf) were awarded acreage (11 blocks) in the region as a result of the Fourth Frontier Round. The results of the 2005 Rockall Licensing Round were also disappointing, with the award of licences to only two exploration groups. This reflected the erosion of confidence in the prospectivity of this, still largely undrilled, region. At the same time companies continued to take licence options on the Atlantic and Celtic Sea acreage. Appraisal drilling by Enterprise also continued on the Corrib Gasfield during 1999 and 2001. The 2006 Slyne/Erris/Donegal Licensing Round fared well, with five groups submitting applications and four licences awarded. In 2008, licences were granted to three groups under the Seventh Frontier Licencing Round in the Porcupine Basin, while two applications were received under the Eighth Frontier Licensing Round covering blocks in the Rockall Basin in 2009. Serica Energy (UK) Ltd were awarded a Frontier Exploration Licence in July, 2009.

Resulting in part from the Frontier Licensing rounds, and also from other licences, small numbers of wells continued to be drilled over the past decade, both in the Celtic Sea and the Atlantic margin basins. In 2000 Providence Resources drilled the 49/9-6z appraisal well which was flow tested at rates of *c.* 5200 bopd on the Helvick discovery (Block 49/9) in the North Celtic Sea Basin. In 2001 the first exploration well was drilled by Enterprise in the Irish sector of the Rockall Basin on the 'Errigal' prospect, targeting Lower Tertiary reservoirs on a structural high in Block 5/22. Unfortunately the well, drilled in 1623 m of water and at the time the deepest water well drilled on the NW European shelf, was plugged and abandoned (P&A) as a dry hole. Two further exploration wells drilled in 2001 were also P&A; Statoil 35/21-1 in the Porcupine Basin and EDC 63/4-2 in the Fastnet Basin.

In 2002 encouraging results were announced from the Enterprise 12/2-1 deep-water exploration well on the Dooish prospect on the eastern flank of the Rockall Basin, confirming the presence of a working petroleum system (Figure 3.7). The well was drilled in 1476 m of water and was located 125 km northwest of the Donegal coast. The discovery well was re-entered and deepened in 2003 and confirmed the existence of a substantial (214 m: *Offshore* magazine July 2009) gas condensate column. There were also encouraging results from renewed exploration and appraisal interest in the Celtic Sea. Marathon's Southwest Kinsale Head gas accumulation, confirmed with two deviated wells in 2001 and one in 2003, was brought on stream in July 2003, via a tie-back to the Kinsale Head Bravo platform. Ramco returned to Marathon's then sub-commercial Seven Heads gas field and drilled one appraisal well in 2001 and five in 2003. The field was declared commercial in 2002 and came on stream in late 2003, using a subsea tie back to the Kinsale Head Alpha platform.

Only one well (in the North Celtic Sea Basin) was drilled in 2004 and, for the first year since Irish offshore drilling began in 1970, no wells were drilled in 2005. Gas flows from Seven Heads Gasfield did not match expectations and Ramco sold their share, and the operatorship of the field, to Marathon in 2006. However, smaller independent oil companies were showing renewed interest in prospects that had tested hydrocarbons during earlier phases of exploration, but had been judged sub-economic at the time. Licences were awarded to Island Oil and Gas over a 622 km^2 area centred on the Connemara oil accumulation on Block 26/28, and to Providence Resources Limited over a 500 km^2 area around the Spanish Point condensate accumulation on Block 35/8 in the Porcupine Basin. 3D seismic data were subsequently acquired over the prospect.

Drilling activity resumed in 2006, with three wells drilled in the Irish offshore. Island Oil and Gas drilled a successful appraisal well at the southwestern end of the Seven Heads Gasfield, and also drilled a successful exploration well (30 m gas column) on the Old Head prospect (49/23-1), near the Kinsale Head Gasfield (reported by the *Oil & Gas Journal* of Nov. 26th, 2007 as containing median-case gross gas reserves of 28.9 bcfg). This was the first successful exploration well to be drilled since the discovery of the Corrib Gasfield in 1996. Island Oil and Gas and partners had returned to the Donegal Basin in the Frontier Licensing Round and identified a shallow gas-prone Triassic target – the Inishbeg prospect – and related structures. The prospect (in Block 13/12) was drilled in 2006, 75 km northwest of the Donegal coast, with Lundin Petroleum as operator, but the well was plugged and abandoned as a dry hole at a depth of 1533 m.

Drilling activity again rose in 2007, during which five wells were drilled and all encountered oil or gas. Two were appraisal/development wells drilled by Shell on the Corrib Gasfield in the Slyne Basin. The other three were drilled in the North Celtic Sea Basin and marked the continuing appraisal of earlier prospects or sub-commercial discoveries (Figure 3.4). Island Oil and Gas drilled a successful exploration/appraisal well on the Schull gas field (57/2-3) and a second appraisal/development well on the Old Head of Kinsale gas field (well 49/23-2). Tie-backs from the Old Head of Kinsale accumulation are planned to the Kinsale 'A' production platform. The Schull gas accumulation is reported by the *Oil & Gas Journal* (11 November, 2007) as containing median-case gross reserves of 27.7 bcf gas, with the possibility of development by tie-back to the Kinsale Head platform. The Old Head of Kinsale appraisal, tested through restricted choke, had an estimated absolute open flow potential of 47 MMcfd from Lower Cretaceous sandstone reservoir intervals. Providence Resources and partners drilled an appraisal well on the Hook Head structure (Block 50/11), discovered by Marathon in 1971, and encountered good quality oil. A further appraisal well, 50/11-4, was drilled in 2008 and is reported to have yielded disappointing results. Also in 2008 Providence Resources drilled 50/6-4 on the Dunmore prospect to test a Jurassic sandstone play. While the sandstones were thinner than anticipated, and were water-wet, the well is reported to have encountered a new hydrocarbon-bearing Jurassic carbonate reservoir succession and was suspended. In 2009 Serica announced that well 27/4-1 (on the Bandon structure) had encountered oil, but that further technical assessment of volumes was needed. The well, in the Slyne Basin, drilled through the Jurassic and Sherwood Sandstone Group intervals to a TD of 6233 ft (1899.7 m),

and is significant in confirming the presence of a working oil system in the basin to complement the gas system of the Corrib Gasfield.

There has been slow progress on the Corrib gas field development since discovery of the field in 1996. Shell E&P Ireland and partners Statoil Exploration (Ireland) Limited and Marathon International Petroleum Hibernia Limited drilled a number of successful appraisal wells. The gas will be produced via five subsea wells to be tied back to a central subsea manifold to connect to the main offshore pipeline, and could supply 75% of Ireland's peak needs for up to a decade. The pipeline will carry gas from the manifold to a landfall at Broadhaven, County Mayo, and then on to the onshore terminal. At the terminal the gas will be conditioned and exported to the BGE national grid. The connection to the grid will be via a pipeline supported by the project partners and operated by BGE. The tie-in to the national grid is planned for Craughwell near Galway. The development has been delayed due to environment and safety concerns and especially to disputes regarding the route of the onshore pipeline to the onshore processing terminal. The terminal received planning permission in April 2004 and the decision was upheld by An Bord Pleanala in October 2004. In November 2007 the project received a boost when the Environmental Protection Agency confirmed its earlier decision to grant a licence to Shell E&P Ireland Limited to develop the gas refinery and combustion facilities at its Bellanaboy Bridge site. Despite proposed changes to the route and modifications to the pipeline specifications, the routing of the onshore pipeline still remains to be finalised (2010).

Exploration Summary

From the brief historical review above it will be seen that offshore exploration in Irish waters has differed markedly from that of the North Sea. In the North Sea progressively expanding exploration and production followed as discoveries were made in different geological habitats. Viewed in detail, of course, or by region, this development did not follow a smooth unbroken curve. As a province, however, the North Sea has followed the classic development to a mature production province, now in the first stages of decline. Ireland, on the other hand, was distant from the northern North Sea oil play, and overspill of exploration activity into the Irish Atlantic Margin basins was delayed. The Porcupine Basin was shown to contain attractive Jurassic horst block structures comparable to the Northern North Sea, but initial drilling unfortunately failed to locate Jurassic sandstone reservoirs to match those of the North Sea province. Initial successes in both the Celtic Sea and Porcupine basins were not sustained. Furthermore, the more promising remaining prospects are in deeper water areas of the Atlantic margin that are costly to explore. For this variety of reasons oil and gas exploration in the Irish offshore region has followed an erratic path and has, as yet, met with only limited success.

More than 80 wells have been drilled in the Celtic Sea region over the past 20 years (see below), but the Kinsale Head, Ballycotton and Seven Heads gas fields remain the only producing fields. Contributing reasons for this lack of success have been the relatively poor seismic data quality, and the tectonic complexity of the area. A range of untried plays remain at pre-Cretaceous levels in the Celtic Sea basins (see Chapter 10). However, due to the structural complexity, individual prospects are likely to be modest in size. Improvements in seismic data quality has contributed to something of an exploration renaissance in the North Celtic Sea Basin in the past few years, especially by smaller exploration companies concentrating on the appraisal of earlier 'sub-commercial' discoveries, and recent drilling and appraisal results of the past couple of years have been encouraging.

Perhaps the cleanest and thickest reservoirs in the Irish offshore have been encountered in the wells drilled in the Irish Sea region (see Chapter 12). However, there has been very little encouragement by way of significant oil or gas shows, and no hydrocarbon flows have been recorded on test. The major impediment in this region appears to be the structural complexity, and especially the multiple inversion and structuring episodes in Cretaceous and Cenozoic times that appear to have breached most of the obvious large structures.

Most of the wells to date in the Atlantic margin basins have been drilled on the shallower flanks of the Porcupine Basin, with the Slyne Basin being the next most drilled basin in the region. In the latter basin, exploration has been stimulated to some extent during the past decade by the discovery of the Corrib Gasfield. However, the presence of shallow igneous flows, sills and dykes have made detailed imaging of the structures quite challenging. Although most of the wells in the Porcupine Basin encountered traces of hydrocarbons, the lack of thick Jurassic reservoirs, in particular, gave rise to disappointing and discouraging results, compounded by the significant water depths and harsh drilling conditions of the region. However, the basin is large and under-explored

and a number of prospects are currently being re-evaluated (see Chapter 11). Improved deep-water equipment and technology, together with the existence of proven petroleum systems operating along the Irish Atlantic margin, will lead to further exploration activity in the region.

In summary, despite offshore exploration since 1969, most of the prospective offshore sedimentary basins remain lightly explored, particularly in the deeper water areas off the west coast. At the time of writing (2010) a total of 156 exploration or appraisal wells have been drilled (Figures 3.4 and 3.5) in Irish waters, distributed as follows:

Celtic Sea basins	84
Fastnet Basin	12
Porcupine Basin	30
Rockall Basin	4
Goban Spur	1
Slyne Basin	10
Erris Basin	4
Donegal Basin	2
Kish Bank Basin	4
Central Irish Sea	5

Fiscal Terms

The main current legislation with respect to oil and gas exploration and production in Ireland is as follows:
◆ Petroleum and Other Minerals Development Act, 1960;
◆ Continental Shelf Act, 1968;
◆ Finance Act & Licensing Terms, 1992;
◆ Licensing Terms for Offshore Oil and Gas Exploration, Development and Production, 2007.

During the 1980s there were a number of revisions to the 1975 fiscal terms for exploration and development, including the announcement in 1987 of the abolition of royalties on production, and of the provision for State participation. The continuing low levels of exploration prompted a further radical review, which produced new petroleum taxation measures in the Finance Act, 1992. This provided a special low rate of 25% Corporation Tax – the only State 'take' from oil and gas production – on production under Petroleum Leases granted before certain future dates (2003–2013), dependent on the length of Petroleum Lease tenure, and a normal 40% Corporation Tax beyond those dates. There was provision for 100% allowances for exploration, development and operating expenses.

These fiscal terms were accompanied (November 1992) by revised Licensing Terms from the Minister for Transport, Energy and Communications (Petroleum Affairs Division 1994). These terms apply to all new authorisations and replace the amended 1975 Terms. Existing authorisations (i.e. pre-1992) continue to be subject to the 1975 Terms, but the operating companies could, in appropriate circumstances, opt for the new Terms. The full range of authorisations (*see* Table 3.2) were covered under the 1992 Terms and the rental fees restructured.

These terms remained in place until modified in 2007, when a new profit resource rent tax was introduced, whereby the total taxation level would increase to 40% in the case of the most profitable fields. The period for Deepwater Licences, previously 12 years, was reduced to 9 years, and the term of a Frontier Licence to 12 years. At the end of the first phase of all exploration licences, there will now be an automatic surrender of 50% of the acreage, with a further 50% surrender at the end of the second phase in the case of Deepwater and Frontier licences, regardless of drilling commitments.

The Law of the Sea
Historical
Maritime law and international maritime agreements apply to a wide range of subjects, mostly beyond the scope of this book. These are as widely varied as fishing rights, safe passage for vessels, seabed mining and many others. We are concerned here to summarise those aspects that allow, and also limit, the rights of a country to extend its territory from the coastline out across the adjoining shelf. We also examine the factors that influence the position of offshore boundaries between neighbouring states, particularly when these are in dispute. Here we concentrate mainly on the Irish situation as it impinges on hydrocarbon exploration – a full review of the range of issues relating to Ireland and the Law of the Sea can be found in Symmons (1993).

The legal position regarding the marine areas of the world during the early twentieth century was simple, but largely undefined. National offshore boundaries were limited to a narrow zone extending from the coastline, usually three nautical miles (based on the seventeenth century concept of a cannon shot range). Beyond the national limits the seas were considered to be international waters. Other than for maritime trade and freedom of innocent passage, the main economic interest was fishing, often taking place under historically-defined patterns.

Pressure on fish stocks was not yet a severe problem and offshore oil exploration had not yet begun. However, the situation became more confused as the century progressed. After the Second World War the USA, using the principle of national control over natural resources, extended that control to the whole of its continental shelf. Some nations followed this example, whilst others extended their territorial seas to 12 nautical miles. However, along the west coast of South America, Argentina, Chile, Peru and Ecuador extended their maritime limits to 200 nautical miles to protect fish stocks in their Humboldt Current fishing grounds.

The United Nations held the first Conference on the Law of the Sea (UNCLOS I) in Geneva 1956–58. This resulted in four separate treaties, including one dealing with the Continental Shelf. The 1958 Geneva Convention contained some provisions for the determination of national offshore areas. In May 1964 the U.K became the required twenty-second country to sign the Convention and thus bring it into force. Later in the same year the positions of the median lines between the countries around the North Sea were agreed, but not the boundaries along the south and west of the UK. This left the boundaries between France, the UK and Ireland unresolved. Under the provisions of the Geneva Convention a nation could exercise sovereignty over:

◆ the territorial area of the seabed and submarine zone adjacent to the coast; also
◆ outside the boundary of the territorial area up to a depth of 200 metres (657 ft);
◆ beyond this limit, up to a point at which the depth of the superjacent waters acts as a technical limit to the exploitation of the natural resources of the said areas;
◆ and the seabed and subsoil of submarine areas adjacent to the coasts or islands.

Although partially successful, UNCLOS I left open the width of the territorial waters and failed to place a defined limit on the extent of the continental shelf. Clearly, technical developments would inevitably extend the exploitable area. The continued development of improved offshore technology, offshore oil and gas discoveries and the increased demand of petroleum combined to pressure governments in the next decade to progressively designate extensions to their offshore areas. UNCLOS II was a brief six-week session in Geneva in 1960. Despite support of a Canadian proposal to fix the territorial limit at 12 nautical miles, the conference failed to produce any new agreements. One of the problems that had developed was block voting related to the Cold

War opposition between the USA and the Soviet Union and their various allies and dependants. The position in the 1960s regarding many maritime issues was therefore confused. Although UNCLOS I and II had made some progress, they had left many issues unresolved – notably width of territorial sea, fishing limits and definition and extent of the continental shelf. As a result, countries had adopted different limits – 3 to 12 nautical miles for the territorial zone and 12 to 200 nautical miles for an exclusive fishing limit (later called EEZ: Exclusive Economic Zone).

In an attempt to comprehensively resolve the growing number of problems and concerns regarding the extension of national sovereignty, protection of fish stocks, ownership of the deeper marine areas and many other topics, the third United Nations Convention on the Law of the Sea (UNCLOS III) was convened. This took place, with regular sessions in New York and Geneva, between 1973 and 1982. UNCLOS III was to examine all elements of the Law of the Sea – right of passage, fishing limits, continental shelf limits, deep seabed ownership. To prevent block voting, progress was to be achieved by gradual consensus rather than majority vote. As a result the countries (more than 160 participated) split into negotiating groups of countries with similar agendas (the Margineers or wide margin states (a group for which Ireland acted as Chair), the land-locked and geographically disadvantaged states (LL & GDS), the Arab group, Russia and allies, and others).

However, even whilst UNCLOS III was in session countries continued to expand their offshore areas by designation. As we have seen, Ireland made a series of offshore designations in 1968, 1970, 1974, 1976 and 1977, extending out westwards over the Porcupine and south Rockall region (Figure 3.3). At the time of the first Irish designation in 1968, most offshore exploration was limited to the shallow shelf up to 200 m. At the time of writing (2010) it is possible to drill and complete a well in almost 3000 m of water. It was this technological pressure that made it important that some solutions regarding the limits of national sovereignty emerge from the UNCLOS III deliberations. However, in the case of Ireland, other factors were involved in the westwards expansion into the Atlantic domain. In 1955 a landing party from HMS *Vidal* had taken possession the islet of Rockall for Britain. The British Admiralty subsequently placed a navigational beacon on top of the rock. By passing the Rockall Act 1972, the British Parliament incorporated Rockall into the County of Inverness (Scotland) and by doing so acquired the right to designate the areas around the rock. This took place at a time when there was no agreed offshore boundary between the two countries, and produced an angry response from the Irish press and politicians.

If any single feature symbolises the hostile operating environment west of Britain and Ireland, it is the precipitous shipping hazard of Rockall rising starkly from the stormy waters of the north Atlantic. The rock is 301.4 km (162.7 nmi) west of the (Scottish: now uninhabited) St Kilda archipelago, and 367 km (198 nmi) from the inhabited island North Uist. A number of ships have been lost on the granitic prominence, which is only 23 m in height and 27 m in diameter. It is very difficult to land on Rockall from the sea because of the steep cliffs and large swell. The first recorded landing was a small party from the frigate HMS *Endymion* in July 1810. However, it is almost certain that fishing boats from St Kilda and the Faroes had landed men there in the past. Even earlier, Saint Brendan the Navigator reputedly set sail from County Kerry around AD 520–530 on a seven year voyage to find the Blessed Isle. He probably visited the Faroes and Iceland en route to the New World. However, the description of his journey (*Navigatio Sancti Brendani*), written by an Irish monk more than three centuries later, is full of mythology and it is difficult to separate out the facts. Nevertheless, a few days out on the journey the account gives the description of a rock and the difficulties of landing thereon that reads remarkably like Rockall, and this gave the Irish a feeling that they had an earlier claim on the rocky prominence.

A number of arbitration cases in disputes between countries (notably UK and Norway, USA and Canada) had set precedents regarding issues such as the drawing of baselines, status of fringing islands, geometric construction of offshore boundaries, et cetera. The results of these cases could not be ignored and were incorporated into the UNCLOS III framework as it developed. For the Irish it was important that uninhabitable rocks such as Rockall should not generate their own shelf and economic zone. This was achieved when the final wording of the UNCLOS III text included the provision that 'Rocks which cannot sustain human habitation or economic life of their own shall have not exclusive economic zone or continental shelf'. UNCLOS III ended in 1982 and the convention came into force in 1994, one year after the required sixtieth country ratified the treaty. Ireland ratified the 1982 United Nations Convention on the Law of the Sea on 21st June 1996, and it entered into force with respect to Ireland on 21st July 1996. At the time of writing (2010) there are 22 countries that have signed but not yet ratified the treaty (most notably the United States), and some countries have not yet signed. The main points in the Convention relating to continental shelf definitions can be summarised as follows (Figure 3.8):

Internal waters: This zone includes all water and waterways on the landward side of the baseline. Normally, a sea baseline follows the low-water line, but when the coastline is deeply indented, has fringing islands/sandbanks or the coastline shifts, straight baselines may be used. The coastal state is free to set laws, regulate use, and develop any resource within the zone, and foreign vessels have no right of passage.

Territorial waters: Out to 12 nautical miles from the baseline, the coastal state is free to set laws, regulate use, and use any resource, with vessels given the right of innocent passage.

Archipelagic waters: When a string of islands extends parallel to the coast, the baseline is drawn between the outermost points of the outermost islands. All waters inside this baseline are designated Archipelagic Waters. The state has full sovereignty over this zone, but foreign vessels have right of innocent passage.

Contiguous Zone: Beyond the 12 nautical mile limit

there is a further zone 12 nautical miles in width, the contiguous zone, in which a state could continue to enforce laws regarding illegal activities.

Exclusive Economic Zone (EEZ): These zones were a new concept developed at UNCLOS III and extend 200 nautical miles from the baselines. Within the zone, the coastal nation has sole exploitation rights over all natural resources, including fishing rights and the exploitation of mineral or oil resources. Foreign nations have the freedom of navigation and overflight, subject to the regulation of the coastal states.

Continental Shelf: The continental shelf is defined as the natural prolongation of the land territory of the coastal state to the outer edge of the continental margin, or 200 nautical miles from the baseline, whichever is greater. However, although the continental shelf may extend beyond 200 nautical miles, the coastal state's claim may never exceed either: 350 nautical miles from the baseline; or it may never exceed 100 nautical miles beyond the 2500 metre isobath. Coastal states have the right to exploit minerals, including hydrocarbons, and non-living material in the subsoil of its continental shelf, and living resources 'attached' to the continental shelf, but this does not include exclusive fishing rights beyond the limit of the EEZ.

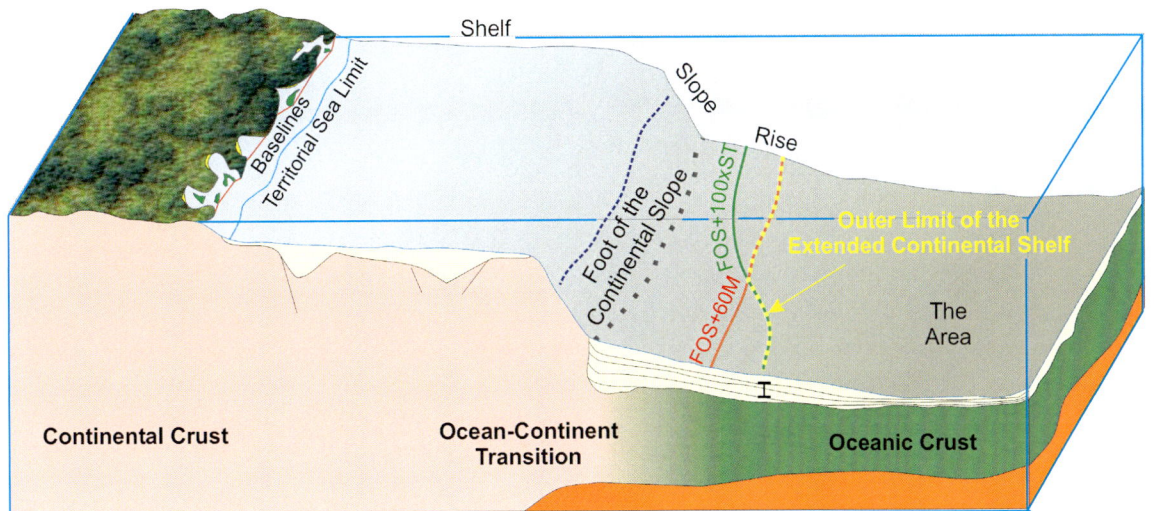

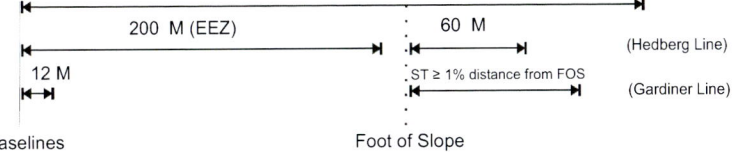

Figure 3.8. Continental Shelf delineation under Article 76 of the United Nations Convention on the Law of the Sea (source: Petroleum Affairs Division, DCENR: website 2010).

Within the maximum limit defined above, the coastal state can establish the outer limits of the continental shelf beyond 200 nautical miles by (Article 76, United Nations Convention on the Law of the Sea), either (Figure 3.8):

◆ a line (comprising straight line segments not longer than 60 nautical miles) linking the outermost fixed points at which the thickness of sedimentary rocks is at least 1 per cent of the shortest distance to the foot of the continental slope (Irish or Gardiner line); or

◆ a line (comprising straight line segments not longer than 60 nautical miles) linking fixed points not exceeding 60 nautical miles from the foot of the continental slope (Hedberg line).

In practice, one or other of these formulae will be furthest removed from the baseline, and the coastal state will use a mixture of the two to provide the most favourable result.

UNCLOS III established an International Seabed Authority (ISA) to authorise the exploitation of seabed resources, notably minerals, in the deeper water areas beyond the limits of national jurisdiction claimed by coastal nations, and to distribute the royalties arising from such activity. The Convention does not, however, provide authority to the United Nations to arbitrate in disputes between nations regarding the positioning of offshore boundaries between their territories. The seaward extent of national jurisdiction is determined by submission of suitable technical data to the Commission on the Limits of the Continental Shelf established by the Convention for this purpose. This is of great importance to Ireland, as one of the wide margin states, but can only be achieved where there is agreement on a mutual boundary between adjoining states. In 1988 Ireland and the United Kingdom agreed their mutual boundaries on the Continental Shelf in the Irish Sea–Celtic Sea and on the Atlantic margin. In both areas only a narrow stepped corridor existed between the previously designated areas of the two countries. However Ireland, in its 1977 designation in the Atlantic, with the issue of Rockall still unresolved, had counter-designated a small area designated by the UK to avoid *de facto* recognition of the UK designations limits. This was subsequently modified by the 1989 designation (Statutory Instrument No. 141 of 1989) that gave effect to the Agreement of 7th November 1988 between Ireland and the UK on delimitation of areas of the continental shelf between the two countries, and corrected the earlier designation limits. There were probably two factors that led to Ireland and the United Kingdom agreeing their offshore boundaries. The provision in the UNCLOS III text regarding offshore islands and rocks that cannot sustain habitation had eased Irish fears that Rockall could generate an independent EEZ, and the United Kingdom was moving towards an acceptance of this. There was also a realisation on the Irish side that the treatment of baselines in international disputes, such as that between the UK and Norway, combined with the UNCLOS III text, made it difficult for Ireland to make a serious case for the construction of a boundary in the Atlantic that trended north of Rockall. The second factor was that both countries recognised that their mutual interest was in strengthening the case for the natural extension of their territories across Rockall Trough out to Hatton Bank. Iceland and Denmark (on behalf of the Faroese) had given notice of their intent to make counter claims to this western region (Figure 3.9).

The southwestern extension of the Celtic Sea boundary agreed by Ireland and the United Kingdom enters into an area of contention with both France and Spain. Ireland therefore entered into quadrilateral negotiations, in the northwest with UK, Denmark and Iceland, and in the southeast with UK, France and Spain. At the same time data were being gathered towards a submission to the Commission on the Limits of the Continental Shelf. Because of the disputed areas, the Irish submission was divided into three parts (Figure 3.9) – *see* the Petroleum Affairs Division website (www.pad.ie) for details. A central area on the Porcupine Abyssal Plain (B on Figure 3.9) was judged not to prejudice the outcome of discussions to north and south, and to concern only Ireland (submitted 2005). In the event, the discussions between Ireland, UK, France and Spain ended in an agreement to collaborate in a joint submission to the Commission (19th May, 2006 – the first joint submission to the Commission). The unresolved boundaries between the countries will be determined after the decision of the Commission, on the basis of applicable principles of International Law. In both B and C submissions, a combination of the Hedberg and the Irish formulae was used within each zone to define different portions of the outer limit boundary submitted to the Commission, the approximate position of which is shown on Figure 3.9. The submissions have relied on seismic profiles across the margin acquired by the Petroleum Affairs Division for this purpose in 1995, and a bathymetric survey of the margin in 1996, allied with data from a complete bathymetric survey of the Irish shelf (Irish National Seabed Survey, which began in 1999). Quadrilateral negotiations have been ongoing since 2001 to resolve the problem of

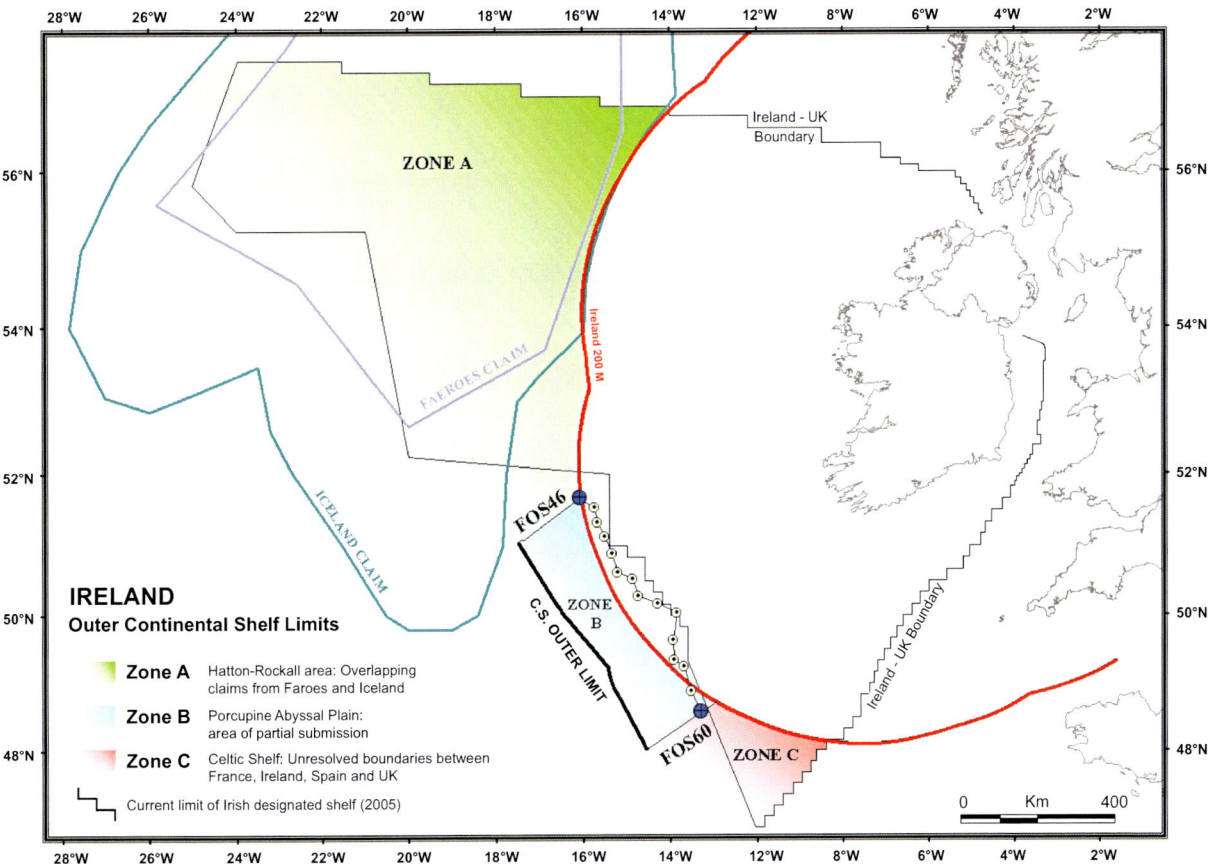

Figure 3.9. Zonation of the Irish Atlantic margin for the purpose of submission to the Commission on the Limits of the Continental Shelf. Also shown are the conflicting national claims in the region (source: Petroleum Affairs Division, DCENR: website 2010).

overlapping counterclaims in the northwest (Zone A on Figure 3.9) and no submission to the Commission on the Irish boundary in this sector has yet been made (2010). The outcome of these submissions and negotiations will eventually determine the area of Shelf under Irish jurisdiction (and for petroleum exploration). For the purpose of this book, in the Atlantic sector we have included discussion of the geology of the area south of the agreed boundary between the United Kingdom and Ireland and out to the western margin of Hatton Bank.

References

Bazley, R.A.B. (1975) The Tertiary igneous and Permo-Triassic rocks of the Ballyalton Borehole, County Down. *Bulletin of the Geological Survey of Great Britain*, **50**, 71–101.

Caston, V.N.D. (1995) The Helvick oil accumulation, Block 49/9, North Celtic Sea Basin. *In*: Croker, P.F. and Shannon, P.M. (eds), *The Petroleum Geology of Ireland's Offshore Basins*. Geological Society, London, Special Publications, **93**, 209–225.

Collins, C.B. (1976) *Wildcats and Shamrocks*. Mennonite Press Inc., Kansas. 100pp.

Cook, D.R. (1987) The Goban Spur: exploration in a deep-water frontier basin. *In*: Brooks, J. and Glennie, K.W. (eds), *Petroleum Geology of North-West Europe*. Graham and Trotman Ltd, London, 623–632.

Earls, T.C. (1995) Potential for development of the Connemara Field – Block 26/28. *In*: Croker, P.F. and Shannon, P.M. (eds), *The Petroleum Geology of Ireland's Offshore Basins*. Geological Society, London, Special Publications, **93**, 343.

Falcon, N.L. and Kent, P.E. (1960) Geological results of petroleum exploration in Britain 1945–1957. Geological Society of London Memoir, **2**, 56pp.

Griffith, A.E. (1983) The search for petroleum in Northern Ireland. *In*: Brooks, J. (ed.), *Petroleum geochemistry and exploration of Europe*. Blackwell Scientific Publications (for the Geological Society of London), 213–222.

Howell, T.J. and Griffiths, P. (1995) A study of the hydrocarbon distribution and Lower Cretaceous Greensand prospectivity in Blocks 48/15, 48/17, 48/18 and 48/19, North Celtic Sea Basin. *In*: Croker, P.F. and Shannon, P.M. (eds), *The Petroleum Geology of Ireland's Offshore Basins*. Geological Society, London, Special Publications, **93**, 261–275.

Illing, L.V. (1982) *An assessment of petroleum prospects near Larne, Northern Ireland.* The Department of Economic Development for Northern Ireland. 3pp.+ enclosure.

Lees, G.M. and and Taitt, A.H. (1945) The geological results of the search for oilfields in Great Britain. *Quarterly Journal of the Geological Society, London*, **101**, 255–317.

MacDonald, H., Allan, P.M. and Lovell, J.P.B. (1987) Geology of an oil accumulation in block 26/28, Porcupine Basin, offshore Ireland. *In*: Brooks, J.& Glennie, K.W. (eds), *Petroleum geology of Northwest Europe: Proceedings of the 3rd Conference.* Graham & Trotman, London, 643–651.

Manning, P.I. and Wilson, H.E. (1975) The stratigraphy of the Larne Borehole, County Antrim. *Bulletin of the Geological Survey of Great Britain*, **50**, 1–27.

Manning, P.I., Robbie, J.A. and Wilson, H.E. (1970) *Geology of Belfast and the Lagan Valley.* Memoirs of the Geological Survey. HMSO, 242pp.

Marathon Petroleum U.K. Ltd./Shell U.K. Ltd. *Newmill No.1, County Antrim, Northern Ireland. Final Well Report*, 15pp.

Murray, M.V. (1995) Development of small gas fields in the Kinsale Head area. *In*: Croker, P.F. and Shannon, P.M. (eds), *The Petroleum Geology of Ireland's Offshore Basins.* Geological Society, London, Special Publications, **93**, 259–260.

Naylor, D. (1984) Petroleum exploration in the Republic of Ireland: a review. *Energy Exploration and Exploitation*, **3**, 5–26.

Naylor, D. (1996) History of oil and gas exploration in Ireland. *In*: Glennie, K. and Hurst, A. (eds), *AD 1995 – NW Europe's Hydrocarbon Industry.* Geological Society, London, 43–52.

Naylor, D. and Mounteney, S.N. (1975) *Geology of the North-West European Continental Shelf. Vol. 1.* Graham and Trotman, London, 162pp.

Naylor, D. and Shannon, P.M. (1982) *The Geology of Offshore Ireland and West Britain.* Graham and Trotman, London, 161pp.

Naylor, D., Philcox, M.E., and Clayton, G. (2003) Annaghmore-1 and Ballynamullan-1 wells, Larne–Lough Neagh Basin, Northern Ireland. *Irish Journal of Earth Sciences*, **21**, 47–69.

Penn, I. E. (1981) Larne No.2 Geological Well Completion Report. Deep Geology Unit, *Institute of Geological Sciences Report*, **81/6**, 59pp.

Reay, D.M. (2004). Oil and Gas. *In*: Mitchell, W.I. (ed.), *The Geology of Northern Ireland: Our Natural Foundation.* Geological Survey of Northern Ireland, Belfast, 273–290.

Robinson, K.W. and Riddihough, R.P. (1975) Ireland – Oil and Gas Exploration. *Geological Survey of Ireland Information Circular*, **8**, 11pp.

Ruffell, A. and Shelton, R. (1999) The control of sedimentary facies by climate during phases of crustal extension from the Triassic of onshore and offshore England and Northern Ireland. *Journal of the Geological Society, London*, **156**, 779–789.

Shannon, P.M. (1993) Oil and gas in Ireland – exploration, production and research. *First Break*, **11**, 429–433.

Shelton, R. (1997) Tectonic evolution of the Larne Basin. *In*: Meadows, N.S., Trueblood, S.P., Hardman, M., and Cowan, G. (eds), *Petroleum Geology of the Irish Sea and Adjacent Areas.* Geological Society, London, Special Publications, **124**, 113–133.

Sheridan, D.J.R. (1972) Upper Old Red Sandstone and Lower Carboniferous of the Slieve Beagh Syncline and its setting in the Northwest Carboniferous Basin, Ireland. *Geological Survey of Ireland Special Paper*, **2**, 129pp.

Sheridan, D.J.R. (1977) The hydrocarbons and mineralization proved in the Carboniferous strata of deep boreholes in Ireland. *In*: Garrard, P. (ed.), *Proceedings of Forum on Oil and Ore in Sediments.* Imperial College London 1975, 113–145.

Smith, R.A. (1986) *Permo-Triassic and Dinantian rocks of the Belfast Harbour Borehole.* British Geological Survey Report, **18(6)**, 13pp.

Symmons, C.R. (1993) *Ireland and the Law of the Sea.* The Round Hall Press, Dublin, 255pp.

Thompson, S.J. (1979) Preliminary Report on the Ballymacilroy No.1 Borehole, Aloghill, Co. Antrim. *Geological Survey of Northern Ireland Open File Report*, **63**, 15pp.

Winn, R.D. (1994) Shelf sheet-sand reservoir of the Lower Cretaceous Greensand, North Celtic Sea Basin, offshore Ireland. *AAPG Bulletin*, **78**, 1775–1789.

Chapter 4

Onshore basins

Introduction

A comprehensive account of the onshore geology of Ireland (Figure 4.1) can be found in Holland and Sanders (2009). In this chapter we give a brief account of the development of the onshore sedimentary basins, but are concerned in the main with their hydrocarbon potential. Chapter 1 of the book reviewed the structural framework of Ireland and its continental shelf. As outlined in Chapters 1 and 2, and as we shall see throughout the book, the structural grain of the Carboniferous and older rocks has imposed considerable control on the development of the overlying younger basins, both in the onshore and especially in the offshore region.

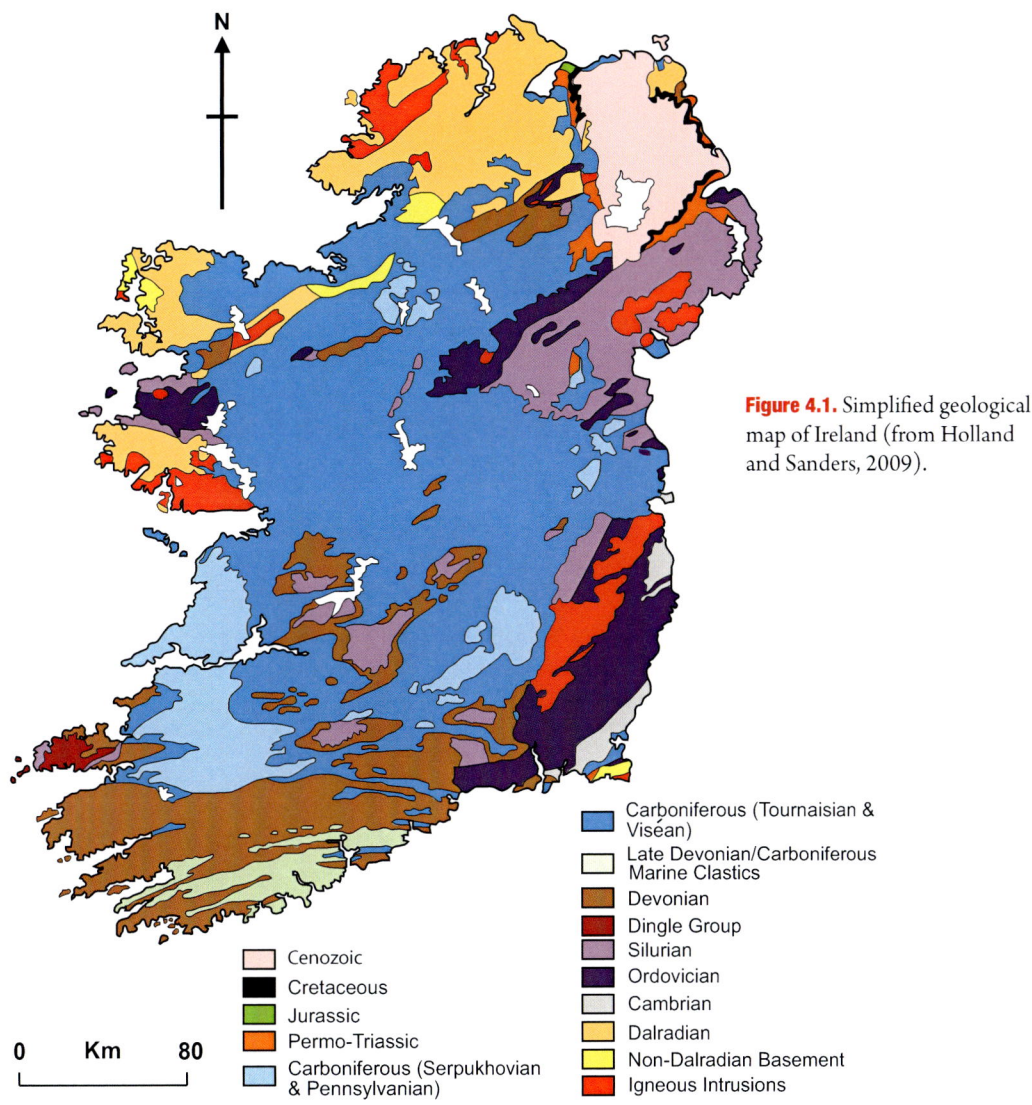

Figure 4.1. Simplified geological map of Ireland (from Holland and Sanders, 2009).

Legend:
- Cenozoic
- Cretaceous
- Jurassic
- Permo-Triassic
- Carboniferous (Serpukhovian & Pennsylvanian)
- Carboniferous (Tournaisian & Viséan)
- Late Devonian/Carboniferous Marine Clastics
- Devonian
- Dingle Group
- Silurian
- Ordovician
- Cambrian
- Dalradian
- Non-Dalradian Basement
- Igneous Intrusions

N

0 Km 80

The onshore geology of Ireland reflects its position at the western extremities of the Caledonian Orogen of northwest Europe and the Variscan Orogen of central and west Europe. Within the Caledonian Orogen, the Iapetus Suture Zone crosses the central part of the country (Figures 1.2 and 1.6) and marks the closure between the Northwestern (American) Plate and the Southeastern (European) Plate. Two zones can be discerned within the Northwestern Plate. An early Caledonian orogenic cycle formed the Orthotectonic Caledonides (c.700–460 Ma) to the northeast, while a younger cycle (c.580–400 Ma) formed the Paratectonic Caledonides to the southeast. South of the Iapetus suture the Caledonian rocks occur in the southeast of the country, generally strike northeast–southwest, and are cored by the early Devonian Leinster Granite. They also appear mainly in the cores of major Variscan folds of the south Central Midlands. These strata comprise Cambrian to Silurian marine sandstones, mudstones and volcanic sequences.

The Caledonian orogeny was followed by a period of uplift and erosion. While the sandstones and thick black mudstones would have originally held significant reservoir and source potential respectively, structural deformation and metamorphism has destroyed any such potential in the Lower Palaeozoic sequence. Devonian strata rest, generally with marked unconformity, on the eroded surface of the Caledonides.

Devonian and Carboniferous Basins

Devonian rocks are widespread in Ireland and are represented in the main by rocks of non-marine facies. As noted above, pre-Devonian rocks occupy large areas in the north and east of the country and also occur as inliers within the Upper Palaeozoic rocks of the Central Midlands. Further south, however, the Devonian and Carboniferous sequences are extremely thick and older rocks are not exposed.

Devonian

Three regions of Devonian rocks can be distinguished (see Clayton et al. (1980), Graham (2009)) and are shown on Figure 4.2:

◆ Thick red-bed clastic sequences in the Munster Basin in the southernmost part of the country.
◆ Old Red Sandstone of the Central Midlands that crops out as a rim around the Lower Palaeozoic inliers.
◆ Small, isolated Old Red Sandstone basins in the north of the country.

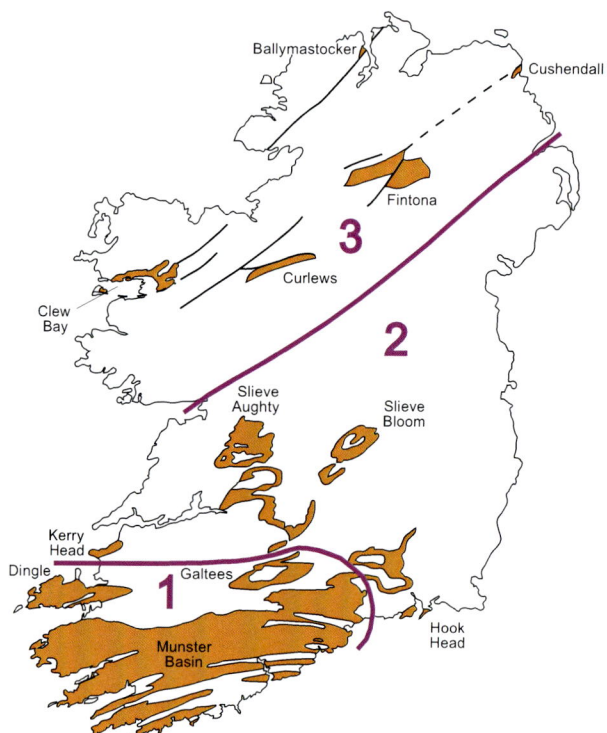

Figure 4.2. Distribution of Old Red Sandstone rocks in Ireland (modified after Graham, 2009). The demarked zones 1–3 are discussed in the text.

Munster Basin

The Munster Basin (Capewell, 1957) is an extensional half-grabenal basin that contains up to 6 km of strata, mainly Late Devonian in age (Figure 4.3). The basin has a general asymmetric profile, as demonstrated by gravity modelling and by outcrop geology (Naylor and Jones, 1967; Graham 1983, 2009). Sandstones and mudrocks predominate, with conglomerates locally important near the basin margins. Derivation directions, where measured, are all from the north, and there is a gross fining-upwards in the sequence (Graham et al., 1992; Figure 4.3). Biostratigraphical dating of the sequence is sparse, and based mainly on palynology and fish fragments. The available evidence (see Graham 2009 for summary) suggests that the lowest exposed beds are of lower Frasnian or possibly Givetian in age. Within the southern region there is a gradual and conformable upward passage into a thick shallow-marine sequence during the Fammenian (PL miospore zone, Clayton et al., 1986a). Evidence in Cork Harbour (Sleeman et al., 1978) shows that the first marine influence becomes progressively younger northwards (LN Subzone).

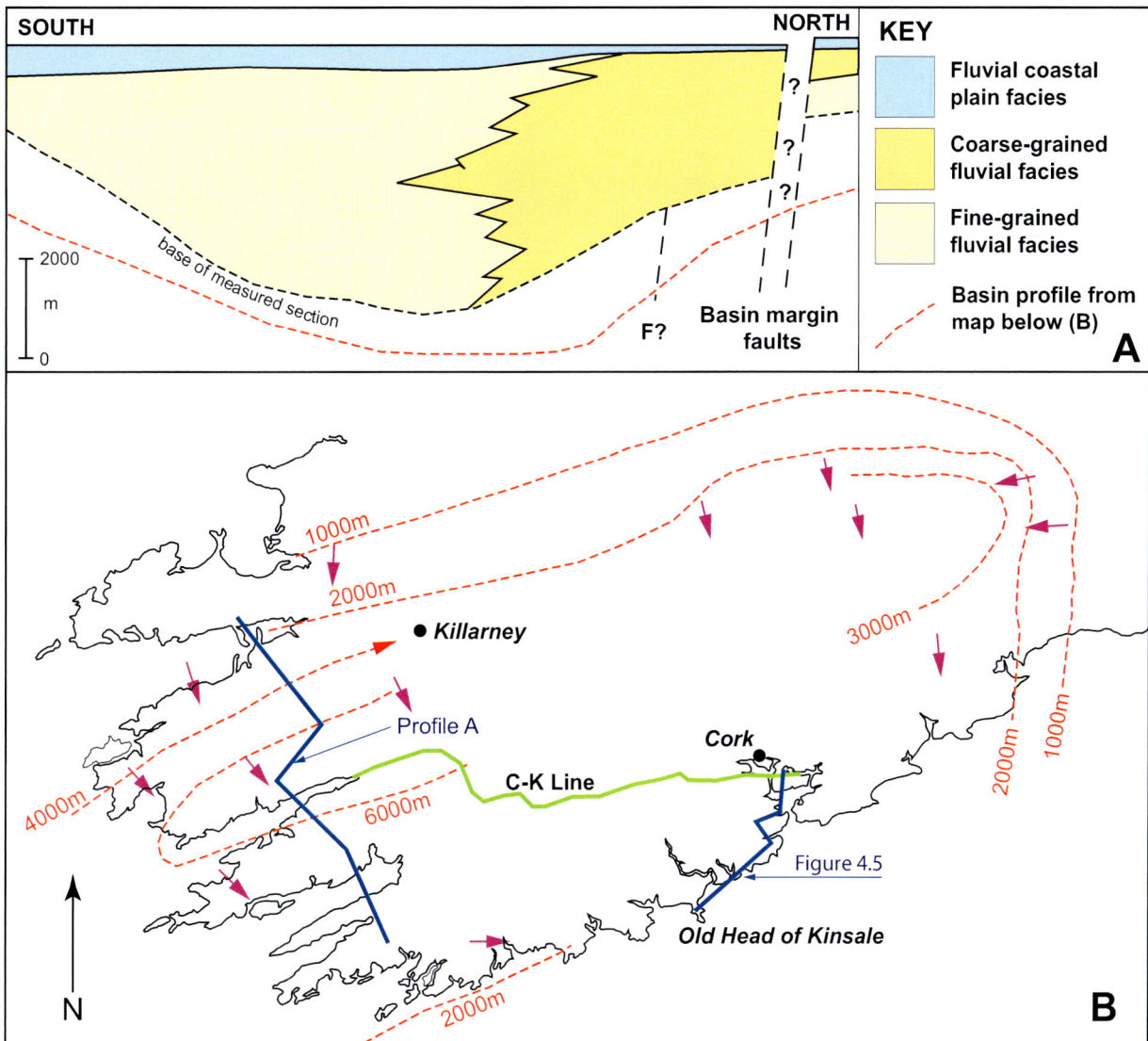

Figure 4.3. A. Simplified thickness and facies profile across the Munster Basin. B. Isopach map of the Munster Basin, with generalized palaeocurrent directions. C-K Line: Cork – Kenmare Line. Line of section in A is indicated. Both figures derived from Graham and Clayton (1988).

Central Midlands Region

The Old Red Sandstone rocks in the region north of the Munster Basin are much thinner – normally a few hundred metres in thickness. As Graham (2001) pointed out, the large volume of sediment in the Munster Basin required an extensive provenance area extending across much of the Caledonian terrain of western and northern Ireland. The marine incursion from the south was initiated during latest Devonian time in the Munster Basin (Naylor, 1969) and then progressed northwards across the country during Early Carboniferous time (Figure 4.4; Clayton and Higgs, 1979; Clayton *et al.*, 1986a). The red-bed fluvial/alluvial facies belt of the Midlands

region therefore migrated ahead of the Late Devonian–Mississippian marine transgression in response to rising base levels. It follows that the red-bed sequences in the north of the zone are mainly of Mississippian (Tournaisian) age.

Northern Basins

In the north of the island several small basins are associated with late phase Caledonian movements on NE–SW faults (Figure 4.2). The non-marine successions range up to 3 km in thickness and are largely undated. The sequences are dominated by coarse-grained sediments, with common conglomerates. A Lower Devonian (or

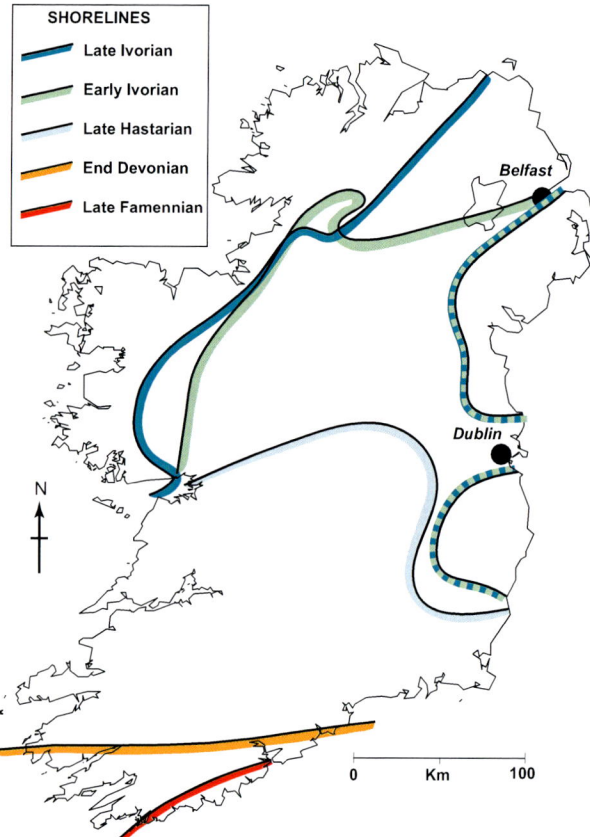

SHORELINES

— Late Ivorian
— Early Ivorian
— Late Hastarian
— End Devonian
— Late Famennian

Figure 4.4. Progressive northwards marine transgression during the Famennian and Tournaisian (based on Clayton *et al.*, 1986; Sevastopulo and Wyse Jackson, 2009).

even pre-Devonian) age is likely for most of the basin fill, although part of the Clew Bay sequence has been demonstrated to be Middle Devonian in age (Graham *et al.*, 1983).

Mississippian (Tournaisian and Viséan)

The subdivisions of the Tournaisian and Viséan shown in Table 4.1 follow the usage of Sevastopulo and Wyse Jackson (2009). The Tournaisian is divided into substages, named in Belgium, whilst the Viséan subdivisions are the stages, now substages, proposed by George *et al.* (1976), with the exception of the Chadian (replaced by the informal 'Lower Viséan'). There is conformable upward passage from the Devonian and younger red-bed sequences into marine Mississippian strata. The latter strata occupy most of the central portion of Ireland and account for approximately half the land area of the country. The sequence is dominated by carbonates, in places exceeding 1 km in thickness, with interbedded shales, and underlain by a thin basal clastic sequence. Outcrop is generally sparse, and much of the detail

Table 4.1. Carboniferous stages and substages.

	GLOBAL STAGE	REGIONAL STAGE/ SUBSTAGE	AGE (Ma)
PENNSYLVANIAN	GZHELIAN	AUTUNIAN	299.0
	KASIMOVIAN	STEPHANIAN	
	MOSCOVIAN	ASTURIAN	
		BOLSOVIAN	
		DUCKMANTIAN	311.7
		LANGSETTIAN	
		YEADONIAN	
	BASHKIRIAN	MARSDENIAN	
		KINDERSCOUTIAN	
		ALPORTIAN	
		CHOKIERIAN	318.1
MISSISSIPPIAN	SERPUKHOVIAN	ARNSBERGIAN	
		PENDLEIAN	326.4
		BRIGANTIAN	
			328
		ASBIAN	
			332
		HOLKERIAN	
			339
		ARUNDIAN	
			343
		LOWER VISÉAN	
			345
		IVORIAN	
			349
		HASTRIAN	
			360.7 ± 0.7

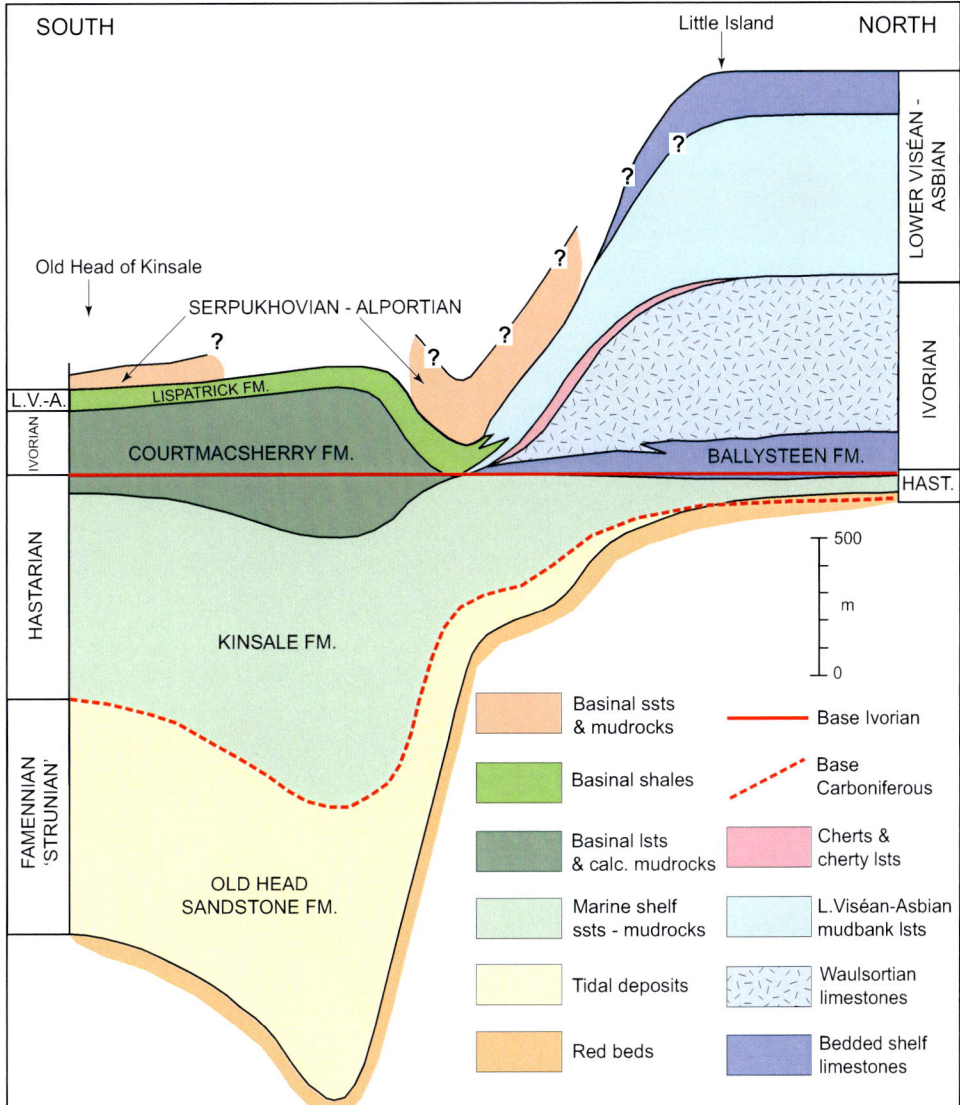

Figure 4.5. Facies and thickness relationships between the South Munster Basin and the northern shelf (modified after Naylor *et al.,* 1989). The location of the profile is shown on Figure 4.3.

of the stratigraphy is known from numerous mineral exploration boreholes and the small number of deeper oil exploration wells.

Three regions can be recognised in the Mississippian sequence:

South Munster Basin (George et al., 1976)
The basin was initiated by marine incursion late in the Fammenian ('Strunian'). Shallow-marine non-carbonate deposition was established south of a line trending between Cork Harbour and Kenmare (Figure 4.3B). Thick shallow-marine sequences (Naylor, 1966) were deposited in Strunian–earliest Tournaisian time, with thin equivalents north of the basin (Naylor *et al.,* 1989).

The main depocentre of the Old Red Sandstone Munster Basin lay to the north (Figure 4.3B) and the southward migration of the depositional axis in latest Devonian time may reflect the locking of the Munster Basin northern boundary faults (Graham, 2009). As a carbonate platform became established to the north of the South Munster Basin, the clastic input lessened, the basin deepened and deposition changed to dark mudstones with carbonate turbidites. Thickness patterns were reversed in late Tournaisian and Viséan time, with thick Waulsortian mudbank and younger limestone successions on the Midland shelf equating with thin basinal sequences in the South Munster Basin (Figure 4.5).

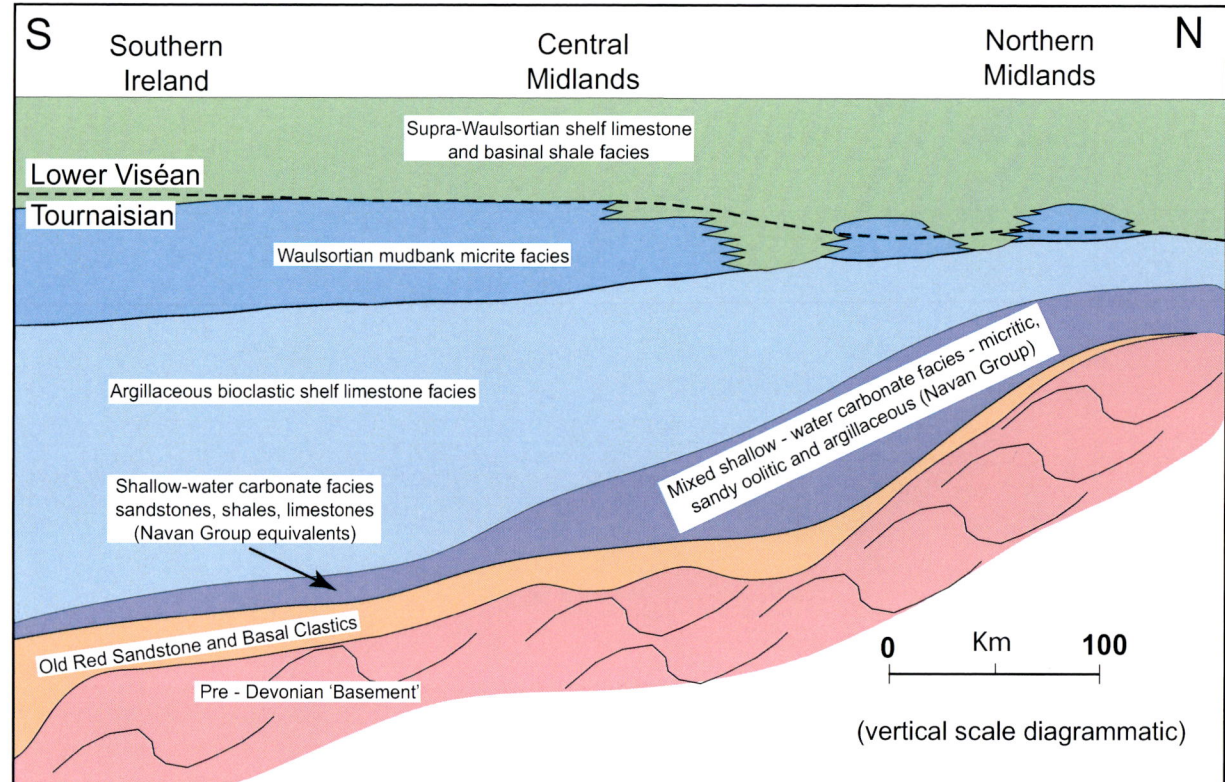

S — Southern Ireland
Central Midlands
Northern Midlands — N

Supra-Waulsortian shelf limestone and basinal shale facies

Lower Viséan
Tournaisian

Waulsortian mudbank micrite facies

Argillaceous bioclastic shelf limestone facies

Mixed shallow - water carbonate facies - micritic, sandy oolitic and argillaceous (Navan Group)

Shallow-water carbonate facies sandstones, shales, limestones (Navan Group equivalents)

Old Red Sandstone and Basal Clastics

Pre - Devonian 'Basement'

0   Km   100

(vertical scale diagrammatic)

Figure 4.6. A north-south generalized section of the Tournaisian and Lower Viséan rocks (Mississippian) between the northern Midlands and County Cork (modified after Hitzman and Large, 1986).

Central Region

A sheet of carbonate-dominated Mississippian rocks, including a widespread development of Waulsortian mudbanks (Lees, 1964), extends from the Cork–Kenmare line at the northern margin of the South Munster Basin to the Lower Palaeozoic massifs in the north of the country. A generalised view north–south across the region is shown on Figure 4.6. A feature of the Central Region is a long-lived depositional trough, approximately along the line of the River Shannon estuary, that persisted throughout the Mississippian.

Northern Region

Varied Mississippian successions are found in the structurally complex region north of the Longford-Down massif. The Carboniferous rocks in the northwest of the country were originally termed the Northwest Carboniferous Basin by the Ambassador consortium (Sheridan, 1972a, 1977: *see* Figure 3.1 showing the basins recognised and named by the Ambassador group). This was a regional name to encompass the central Lough Allen Basin, together with the contiguous and related extensions defined by Ambassador as the Omagh, Donegal,

Sligo and Slieve Beagh synclines. The main basin development is essentially grabenal in form, lying between the pre-Carboniferous inliers of the Ox Mountains and Fintona Block to the north and the Curlew Mountains and Longford-Down extension to the south (Figure 4.7).

Philcox *et al.* (1992) outlined the basinal development of much of this area, based mainly on the results of the Ambassador drilling of the early to mid-1960s, and the drilling by the Aran Energy consortia in the 1980s (*see* Chapter 3). The sequence exceeds 2500 m in thickness and encompasses much of the Mississippian (Higgs, 1984). Figure 4.8 shows the sections penetrated by petroleum exploration wells in the basin. A basal red and green clastic section of probable Tournaisian age (Kilcoo Sandstone) is succeeded by a varied clastic unit (Boyle Sandstone Formation) comprising sandstones, occasionally red, with grey and green mudstones and minor micrites, crinoidal limestones and evaporites. The overlying Kilbryan and Ballyshannon Limestone formations are predominantly argillaceous limestones passing up to shallow-water crinoidal limestones. The sequence is Ivorian to Arundian in age. The limestones are overlain by a sequence of grey calcareous mudstones

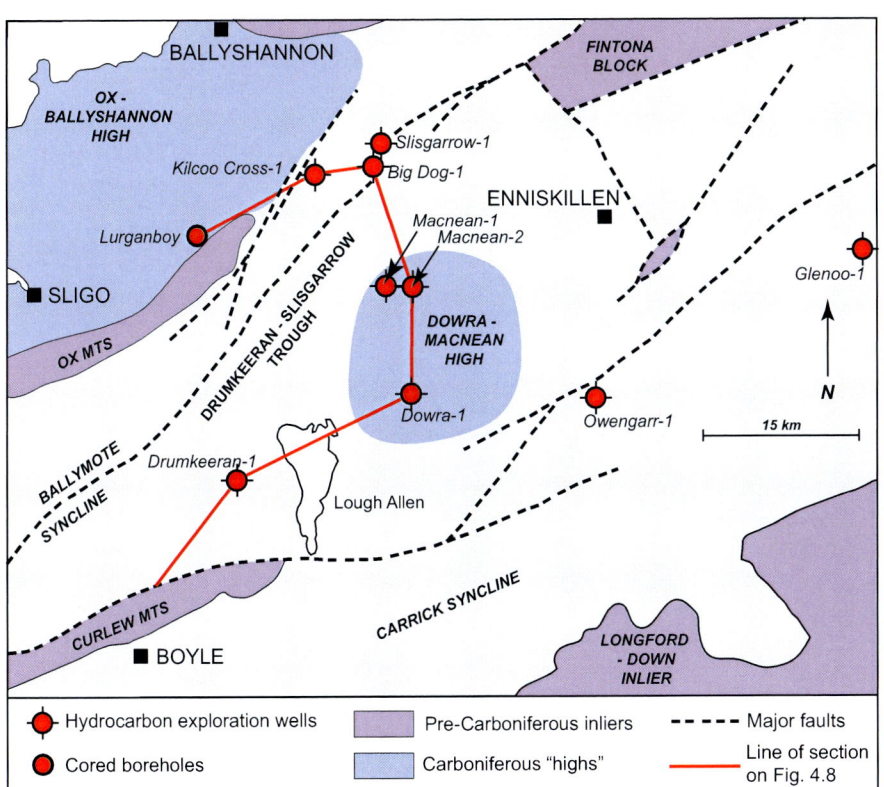

Figure 4.7. Location of exploration wells and geological elements in the Northwest Carboniferous Basin (Lough Allen Basin). Modified after Philcox *et al.*, 1992.

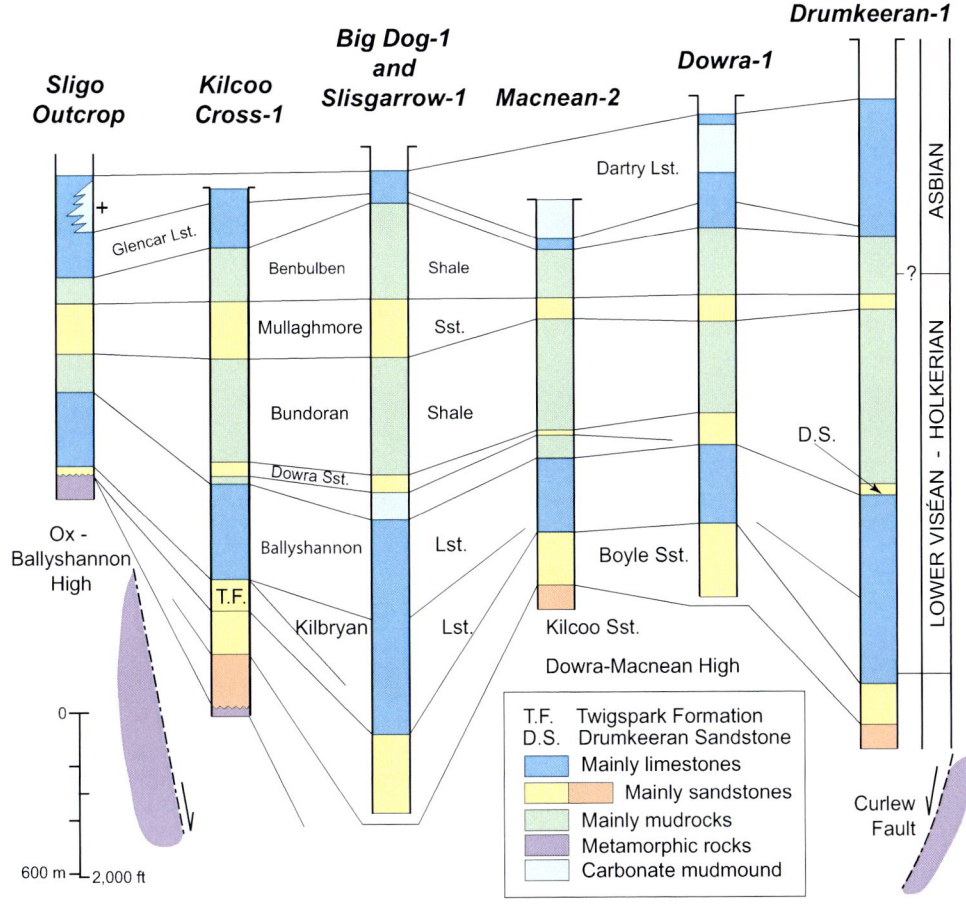

Figure 4.8. Stratigraphy of the wells and outcrop section in and around the Lough Allen Basin (modified after Philcox *et al.*, 1992). Line of section is shown on Figure 4.7.

with thin limestones of Arundian age (Bundoran Shale Formation). Two sand developments with restricted distribution have been recognised in the lower part of the formation – the Dowra and Drumkeeran Sandstone Members (Sheridan, 1977; Philcox *et al.*, 1992). The overlying deltaic Mullaghmore Sandstone Formation comprises up to 217 m of grey sandstone with intercalated mudrocks and minor limestones. As with much of the clastic section, indications are of derivation from a northerly provenance. The Mullaghmore Sandstone thins from 200 m in Big Dog-1 to less than 50 m in Drumkeeran-1, and eastwards is barely recognisable in the Owengarr and Wind Farm wells but is >200 m thick in Glenoo-1. Four coarsening-upwards cycles are recognised within the formation (Reay, 2004). The overlying calcareous mudstones of the Benbulben Shale Formation (Holkerian to Asbian) are succeeded by a carbonate sequence ranging up to late Asbian in age. Argillaceous limestones and calcareous mudstones of the transitional Glencar Limestone Formation are overlain by cherty bedded limestones with carbonate mudmound ('reef') developments (Dartry Limestone Formation). The upper part (~180 m) of the Viséan (Asbian–Brigantian) consists of goniatite-bearing mudstones and interbedded evaporites and carbonates passing up to a mudstone-dominated sequence. A southward-thinning deltaic sandstone formation is intercalated in this succession (Glenade Sandstone Formation).

Serpukhovian (Mississippian) and Pennsylvanian

We follow Sevastopulo (2009, and Table 4.1), both for the onshore and onshore sequences, in stratigraphical nomenclatural usage for the post-Viséan Carboniferous section. Specifically the stages (now reduced in rank to regional substages) of Ramsbottom *et al.* (1978) and Owens *et al.* (1985) are employed, together with the Asturian substage, whilst 'Stephanian' and 'Autunian' are retained informally as regional stages.

The Namurian subdivision of earlier usage (comprising the Serpukhovian and the lower part (Chokerian to Yeadonian) of the Bashkirian; Table 4.1) is shown as a separate unit on most published geological maps of Ireland and is moderately widespread. Younger Pennsylvanian (formerly Westphalian) rocks, on the other hand, have relatively limited preservation only in the Leinster, Slieveardagh, Crataloe, Kanturk and Coalisland coalfields, and in the Kingscourt outlier (Figure 4.9). The Serpukhovian and Pennsylvanian sequences are terrigenous in nature, with only thin

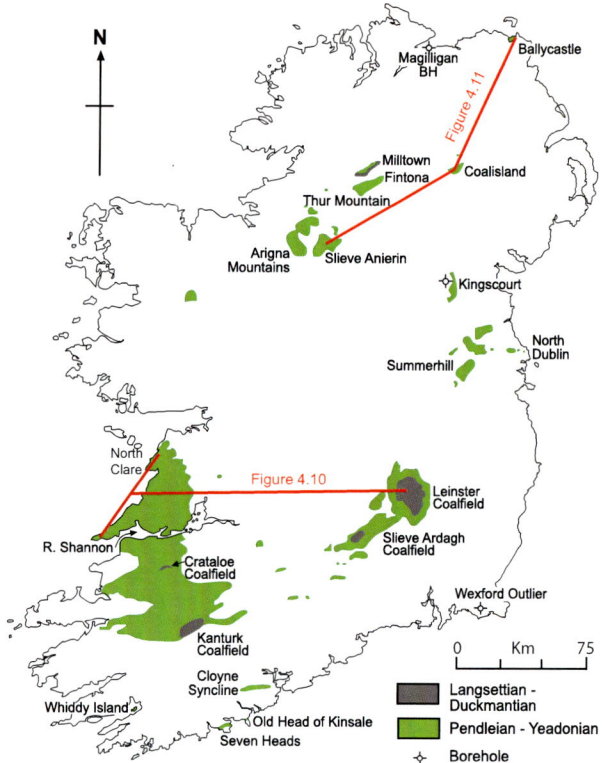

Figure 4.9. Distribution of Serpukhovian and Pennsylvanian rocks in Ireland (simplified after Sevastopulo, 2009). Lines of section illustrated in Figures 4.10 and 4.11.

limestone intercalations. The preserved sequences in the Castlecomer and Slieveardagh (Leinster) coalfields belong to the Langsettian (Westphalian A of the earlier nomenclature), and the same is probably true for the other less well documented outliers. Younger (post-Bashkirian) Pennsylvanian rocks are not known onshore. It is probable that the small remnant coalfields of Ireland were originally part of a wider and more complete area of Serpukhovian–Pennsylvanian deposition from which the younger sequences were mainly removed by pre-Permian erosion.

Serpukhovian–Bashkirian depositional environments typically vary from basinal at the base, to deltaic and fluvial at the top. Deposition took place in subsiding basins and troughs separated by positive highs. The more important highs were a buried westward extension of the Leinster massif (termed the Leinster High), a positive element north and south of Galway Bay (the Galway High), and the Longford-Down extension of the Southern Uplands High.

The Shannon Basin (in part the West Clare Basin of Gill, 1979) lay to the south of the Galway High. The most extensive Serpukhovian–Pennsylvanian outcrops are now found in a roughly north–south belt near the west coast, extending for County Clare in the north across the Shannon to Killarney in the south (Figure 4.9). In broad outline, the shallowing-upwards sequence records deep-water basin floor turbiditic sandstones and mudstones overlain by slope mudstones (Shannon Group), passing up to shallower water cyclothemic deltaic clastics (Central Clare Group: Figure 4.10).

Around the Leinster coalfields the section (Higgs, 1986; Higgs and O'Connor, 2005) comprises black shales overlain by cyclothems passing upwards into sandstones with non-marine bivalves (Figure 4.10). North of

the Leinster High there are extensive Namurian outcrops in the areas west and north of Dublin (East Leinster Basin). Sections within the basin are considerably thicker than farther south at the northern margin of the Leinster coalfield, or to the north at Kingscourt (Figure 4.9). An almost complete Serpukhovian and Bashkirian (Pennsylvanian) sandstone–shale sequence (cumulative 600 m thick) is preserved within the Kingscourt graben (Jackson, 1965; Sevastopulo, 2009). The sequence thins to 350 m and overlaps northwards across the outcrop as a result of its proximity to the northern margin of the Dublin Basin against the southern margin of the Longford-Down massif.

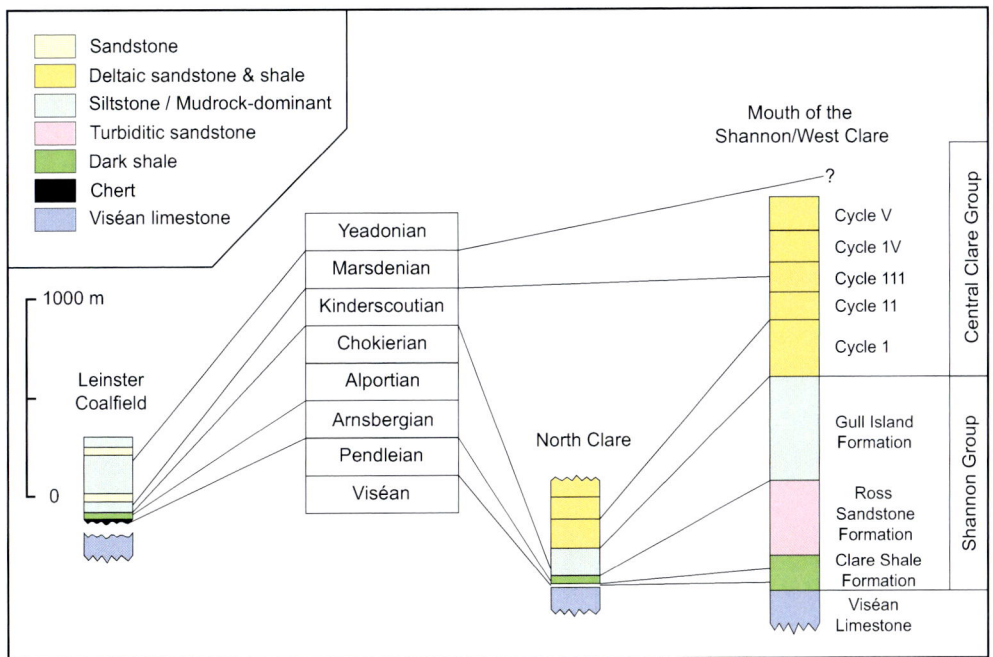

Figure 4.10. Composite Serpukhovian and Pennsylvanian stratigraphical successions in the Shannon Basin, with the Leinster coalfield succession for comparison (modified after Sevastopulo, 2009). Line of section is shown on Figure 4.9.

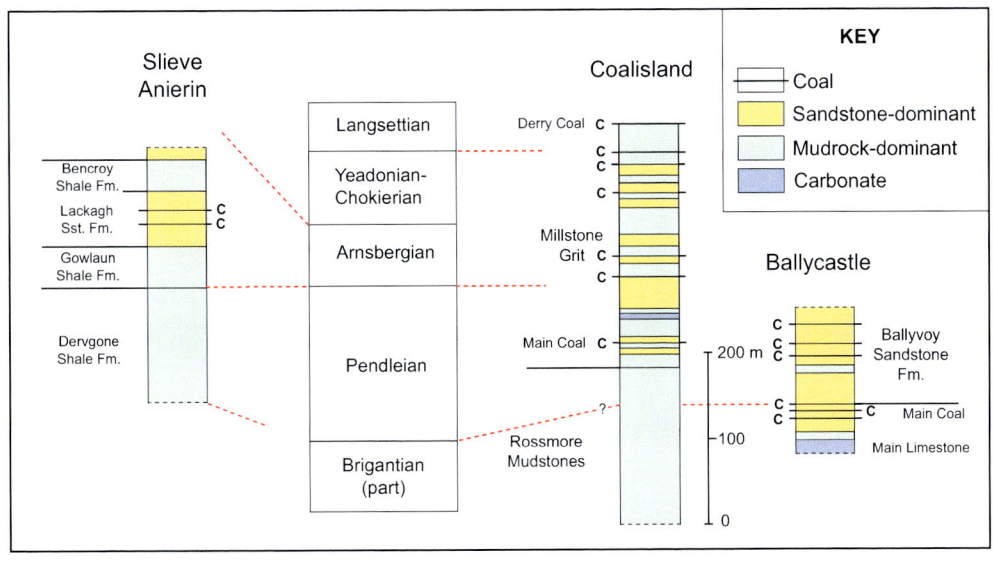

Figure 4.11. Correlation of Serpukhovian and Pennsylvanian between three areas across the north of Ireland (derived from Sevastopulo, 2009). Line of section is shown on Figure 4.9.

In the northwest, around Lough Allen, is an area of coal-bearing Serpukhovian strata, (Connaught or Arigna coalfield) that follow on from the older Mississippian section described above (Figure 4.11). There are also related outliers to the northwest and northeast. The sequence (>300 m) was probably deposited in relatively deep water and is dominated by calcareous or silty shales with rare turbiditic sandstones. The Lackagh Sandstone Formation, that contains the thin coals mined in the Lough Allen area, is interpreted as a deltaic development. Farther north, at Coalisland west of Lough Neagh, a similar shale–sandstone sequence also has intercalated coals. The sequence is poorly fossiliferous and the range of stratigraphy represented is uncertain. A sand-dominated Brigantian–Pendleian sequence is also preserved in the Ballycastle coalfield, a limited area on the north Antrim coast (Figures 4.9 and 4.11).

Permian, Mesozoic and Cenozoic

The only widespread occurrence of post-Carboniferous (Permo-Triassic to Palaeogene) strata in Ireland is in the northeast Ulster Basin. The onshore depocentres are contiguous with the offshore North Channel and Rathlin basins and therefore are described together with the offshore basins in Chapter 6. With the exception of a small outlier of Permo-Triassic rocks at Kingscourt, County Cavan, and an even smaller outlier in southern County Wexford, post-Carboniferous rocks are restricted to minor cave and karstic solution hollow fillings. The most important of these are shown on Figure 4.12. Although insignificant in area, these have some importance in terms of elucidating the post-Variscan history of the island. The most important localties are as follows:

Kingscourt Graben

With the exception of the Ulster Basin, the largest area of post-Carboniferous strata is near Kingscourt in County Cavan (Figure 4.13) where a Permo-Triassic red-bed sequence is located within a north–south trending half-graben that is faulted on its western margin (Visscher, 1971; Gardiner & Visscher, 1971). The Permo-Triassic rocks rest unconformably on synclinally folded Upper Carboniferous rocks (Namurian–Lower Westphalian: Pendleian–Langsettian; Jackson, 1965). The entire area is poorly exposed and the geology is known chiefly from borings and from the workings for gypsum. The Permo-Triassic sequence has a maximum thickness of 550 m and consists of two formations; the Kingscourt Gypsum Formation and the overlying Kingscourt

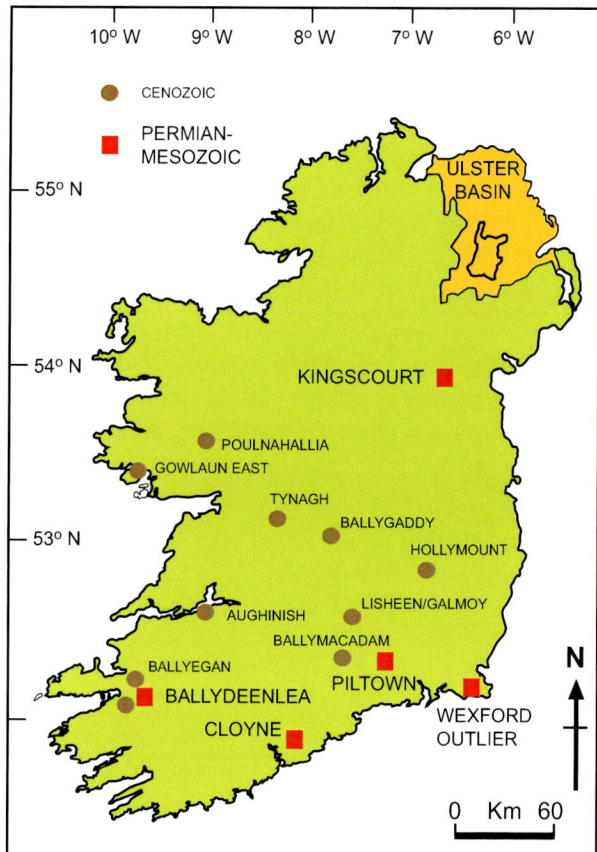

Figure 4.12. Location map showing the occurrences of post-Variscan (Permian-Mesozoic and Cenozoic) strata in Ireland (based on Naylor, 1992; Coxon and McCarron, 2009).

Sandstone Formation. The lower formation comprises a basal conglomerate overlain by red and grey mudstone and gypsum cycles. There is an apparently conformable passage upwards into the Kingscourt Sandstone Formation that consists of a basal sequence of red siltstones (80–100 m) overlain by red and grey fine-grained sandstones (400+ m). Visscher (1971) recognised distinct palynological assemblages of Thuringian and Scythian (early Triassic) age, the latter being restricted to the Kingscourt Sandstone Formation. The Permian–Triassic boundary probably coincides with the junction between the two formations.

Wexford outlier

In County Wexford, in the southeast of Ireland, more than 240 m of red siltstones, sandstones and conglomerates (Killag Formation) of probable Permo-Triassic age (Clayton et al.,1986b; Burnett et al.,1990) rest unconformably on Carboniferous rocks. However, they are poorly exposed and are known only from borehole

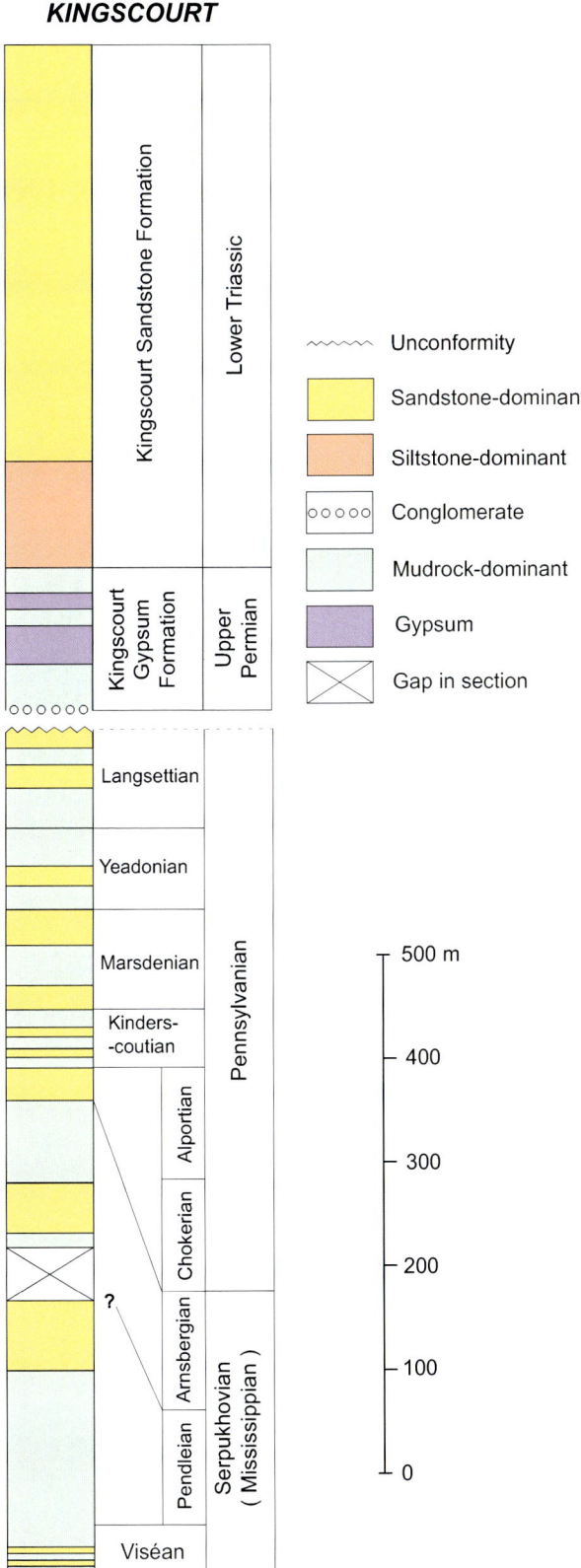

KINGSCOURT

Unconformity

Sandstone-dominant

Siltstone-dominant

Conglomerate

Mudrock-dominant

Gypsum

Gap in section

Figure 4.13. Stratigraphy of the Kingscourt area, County Cavan (after Jackson, 1965; Visscher, 1972; Naylor *et al.*, 1993, McConnell *et al.*, 2001).

information. Major NE–SW faults in the area probably controlled the deposition and preservation of this small outlier. While the rocks have not been dated, the lack of proven Permian strata in the immediate offshore area of the southern Irish Sea and Celtic Sea makes a Triassic age more likely.

Cloyne Syncline
Thin reddened clay deposits of Jurassic age (Higgs and Beese, 1986) occur within karstic solution hollows on Mississippian reef limestones near Cloyne (25 km ESE of Cork city). The silica-rich clays are 1.8 to 4.0 m thick and rest on an irregular, cavernous and altered limestone surface. Higgs & Beese (1986) recorded a microflora with a possible age range from late Lower Jurassic (Toarcian) to Middle Jurassic, with a Middle Jurassic age more likely. They suggested, on palynofacies evidence, that the clays were deposited in a lacustrine environment on carbonate lowlands. Present day steep bed attitudes suggest that there was later solution subsidence or collapse.

Carrick-on-Suir
Karstic cave deposits at Piltown, Co. Kilkenny, near Carrick-on-Suir, contain palynomorphs indicative of Late Jurassic to Early Cretacous age (Higgs and Jones, 2000). Rare dinoflagellate cysts indicate a weak marine influence, possibly a nearshore tidal location. Palaeographic considerations (*see* Chapter 7) suggest that a Late Jurassic age is more likely.

Ballydeenlea outlier, Killarney
The Ballydeenlea outlier (Walsh, 1966), north of Killarney, is the only recorded occurrence of Upper Cretaceous rocks in Ireland, other than those of the Ulster Basin. The Ballydeenlea occurrence was interpreted by Walsh (1966) as resulting from the submarine collapse of karstic caverns in Lower Carboniferous limestones. The chalk forms the matrix of a breccia containing Serpukhovian shale fragments similar to bedrock lithologies nearby immediately above the contact with the Carboniferous Limestone. However, Evans and Clayton (1998) suggested that the breccia accumulated in the chalk sea at the foot of a submarine scarp during reactivation of a Variscan fault. Preservation was thought to be due to a depositional setting within a small pull-apart structure. The surface of the bedrock at the deposit is at about 125 m O.D. and the breccia extends to a depth of at least 40 m. Foraminiferal evidence (Walsh, 1966) suggests a Campanian (*mucronata* Zone) age for the

chalk matrix. Palynofacies show the deposit to have a low terrestrial influence, supporting a picture of widespread submergence at this time.

Cenozoic deposits

Several small Cenozoic (Tertiary) deposits (Figure 4.12) have been discovered on the Carboniferous limestone surface of Ireland (Davies, 1970; Mitchell, 1980; Naylor, 1992). In addition, other undated deposits are presumed by their lithologies and geological setting to be Tertiary in age. The deposits are non-marine clays and sands associated with the karstic limestone surface, and where dated they are Oligocene or younger. The non-marine nature of the scattered deposits suggests that Ireland has been emergent throughout the period.

Exploration potential
Regional Structure

The onshore petroleum exploration programme in Ireland during the 1960s, detailed in Chapter 3, met with little success. With the benefit of hindsight, and particularly with improved understanding of maturation and source rocks, the disappointing results in the onshore can be placed in perspective.

The region was affected by the Variscan orogeny, and the putative trace of the Variscan deformation 'front', is traditionally depicted in Ireland (e.g. Cole, 1922) as a line trending between Dungarvan on the south coast to Dingle Bay on the west coast (Figure 4.14). This is often referred to as the Dingle–Dungarvan Line (*see* Chapter 1). Gill (1962) recognised this as an arcuate thrust front separating a zone of cleavage folds to the south from a zone occupying the south Irish Midlands in which concentric folds with caledonoid trend lacked strong cleavage development. The proposed line of the Variscan Front is marked by major faults in the west between Killarney and Mallow, where Mississippian limestones are displaced against Old Red Sandstone horizons possibly 3 km lower in the stratigraphy. However, as mentioned in Chapter 1, east of Mallow the line is less easy to define, becomes composite in nature (Gardiner, 1978; Naylor and Sevastopulo, 1979) and may represent the transition between zones with different structural styles. Furthermore, as pointed out by Bresser and Walter (1999) and others, there is no abrupt change in fold style at the line – a situation also noted in terms of gravity trends (Readman *et al.*, 1997). It has been suggested (Gardiner and Sheridan, 1981) that the northern margin of the Rheno-Hercynian Zone – and the Variscan

Front – be taken at the northern margin of the Cornubian massif (a line also marking the northern limit of Variscan granites (Ries, 1979), but not the northern limit of deformation. Under this model the Variscan fold belt in Ireland is recognised as a separate entity (the Southern Ireland Zone) with strong control from local basement elements. Basement influence has been put forward by various authors to explain rapid thickness and facies changes in the Upper Palaeozoic succession and changes in fold style in different areas within the southern zone (*see* Graham, 2009 for summary). An interesting concept was introduced by Murphy (1990) who suggested that the arcuate onshore fold belt represented a surge zone within northerly propagating thrust sheets. However, Graham (2009) has argued that the requirement in this model for N–S strike-slip faults at each margin of the surge zone is at variance with the field evidence. The same author also suggested that the arcuate trend of the fold belt may reflect, in part, inheritance of earlier Caledonian basement structures. There is some agreement, however, (Naylor & Sevastopulo, 1979; Gardiner and Sheridan, 1981; Ford *et al.*, 1992) that the Dingle–Dungarvan Line is a significant regional structural boundary broadly separating zones that have undergone different levels of Variscan deformation, but that it does not represent a 'thrust front' along its length. A composite lineament along its trace is discernable on Bouguer derivative maps presented by Readman *et al.* (1995, 1997), as is also the case for the sub-parallel Cork–Kenmare line to the south, discussed earlier in this Chapter (Figure 4.14). This latter line was a long-acting facies control element in Late Devonian and Mississippian time (Quin, 2008). Both the Dingle Bay–Mallow and Cork–Kenmare features were also identified on two onshore deep seismic transects (VARNET Variscan Network–1996: Figure 4.14) by Vermeulen *et al.* (1999). It appears, therefore, that lineaments exist within the basement of the southern region that are oblique to the trend of the Variscan folds. In the case of the preserved eastern section of Cork–Kenmare line, not removed by later erosion, Naylor *et al.* (1996) mapped the feature as comprising individual segments trending slightly south of west but consistently stepped northwards along NNW–SSE trending dextral faults. A number of theories have been put forward to explain the deeper structure and evolution of the Variscan fold belt – notably, a thin-skinned thrust model (Cooper *et al.*, 1984, 1986) and a thick-skinned model with steep basement faults (Sanderson, 1984; Phillips, 1985). Cooper *et al.* (1986) considered the Variscan sole thrust to lie at the base of the Old Red Sandstone in the basin centre. However, Ford *et al.* (1991) interpreted gravity data to suggest that basement was involved in the deformation and that any detachment was at mid-crustal levels. A varied and marked basement topography would be anticipated from the evidence of shallow basement blocks and basement control on Upper Palaeozoic sedimentation discussed above. The flat, undeformed basin floor interpreted on the VARNET lines by Vermeulen *et al.* (1999) may therefore represent a detachment, at least partly within basement, or may be due to poor data resolution at depth.

The strike of the Lower Palaeozoic sequence in southeast Ireland, and in the immediately adjacent Irish Sea, is generally NE–SW, but there is a marked change in the strike to ENE–WSW in the southwestern part of County Wexford and into the adjacent County Waterford (Brück *et al.*, 1979) – *see*, for instance Point D on Figure 7 of Readman *et al.* (1997). For much of the length of the coast west of Cork Harbour the onshore major folds are coast-parallel, also trending ENE–WSW. Rowell (1995) referred to this trend as a Variscan cross-trend, reflecting the orientation of Variscan thrust fronts, but it is more likely to be a reflection of the increasing influence of Caledonian basement structure on Variscan fold

trend. Although the Caledonian basement is not seen in this region, it seems likely that the strike of Variscan and Caledonian structures are here sub-parallel. It appears that the arcuate form of the Variscan fold belt in southernmost Ireland was influenced by a complex of factors – the disposition of positive elements along its northern margin (Gardiner, 1978), the location of intra-basinal basement blocks, but particularly by inherited Caledonian basement trends.

The rocks of the Munster and South Munster basins show well-developed cleavage. This is close-spaced and penetrative in the finer-grained lithologies, but a wider spaced fracture cleavage is developed in the coarser beds. There is an almost complete loss of porosity in the sandstones due to pressure solution and quartz grain overgrowths. Second and third-order folds, down to wavelength scales of a few tens of metres, normally occur on the flanks of the major WSW-trending regional folds.

In Central Ireland there are a number of large Variscan folds cored by Lower Palaeozoic rocks. As in the southern region, the orientation of the Variscan structures has been inherited from the underlying NE or ENE Caledonian trends. The Variscan phase of deformation appears to have been accommodated in the older rocks by movement on pre-existing faults. Fold style is

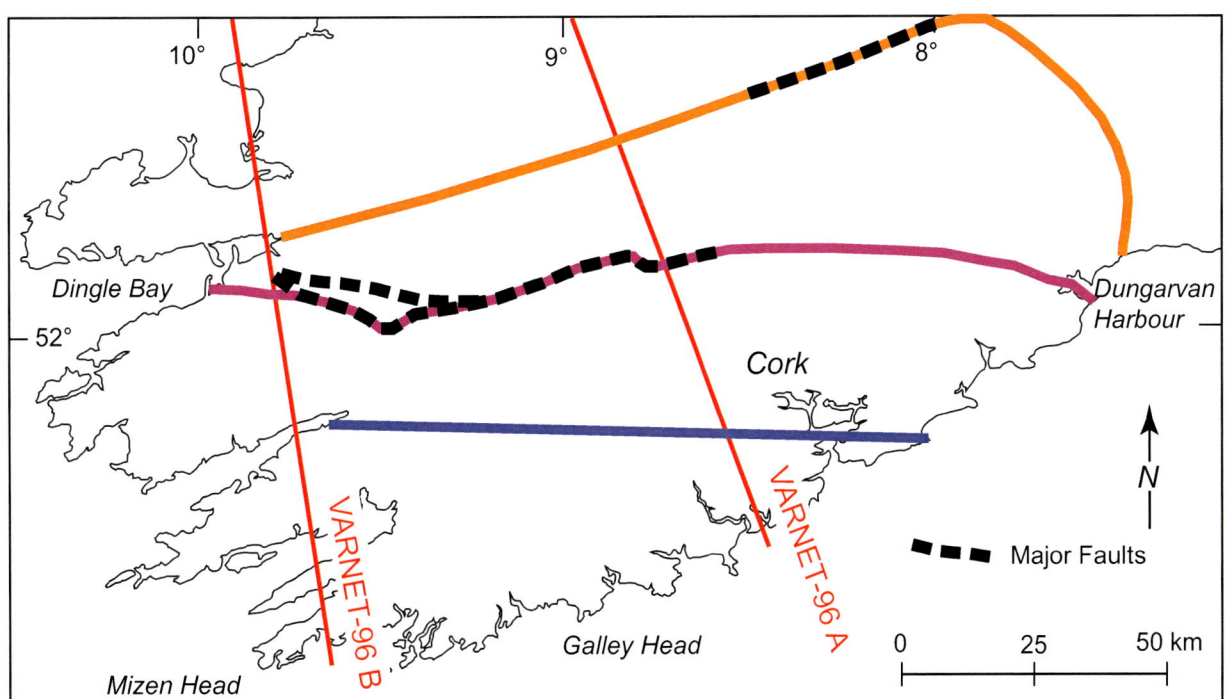

Figure 4.14. Simplified map showing three important geological boundaries in southwest Ireland. Orange: approximate limit of the Munster Basin; Purple: traditional line taken as the Variscan Front; Blue: Cork-Kenmare line. Seismic profiles: VARNET-96 Lines A and B, shown in red.

closely allied to lithology, and there is often structural disharmony between folds in the Mississippian limestone sequences and those in the overlying Namurian clastics. Severe deformation was therefore not limited to the region south of the 'Variscan Front'. Cleavage is also commonly developed in the thick fine-grained sequences of the Shannon and Dublin Basins, accompanied by small-scale folding.

Over the north and northwestern parts of the country, Variscan structure is dominated by NE–SW faults. A number of these faults are the SW prolongations of major fractures in Scotland. Synsedimentary Mississippian and Pennsylvanian movements on major faults are well documented in the Lough Allen area (Oswald, 1955; George and Oswald, 1957; Philcox et al., 1989) and in the Fintona area (Mitchell and Owens, 1990; Mitchell, 1992). The area of thickest sedimentation in the region is in the Lough Allen Basin (Northwest Carboniferous Basin), which has seen the most sustained onshore exploratory drilling activity, and this is discussed in more detail below. As Graham (2001) has pointed out, there is a northward continuation of the Variscan structural pattern 'but with gradually decreasing strain and increase in the importance of reactivation of basement structures'.

Figure 4.15. Thermal maturation levels of Devonian and Carboniferous rocks of Ireland (from Graham, 2009: based on Clayton et al., 1989).

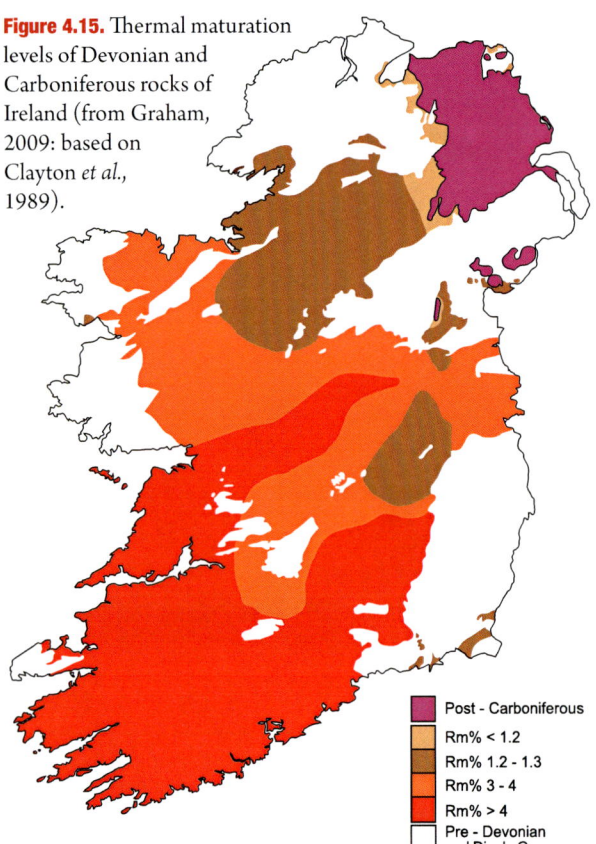

Legend:
- Post - Carboniferous
- Rm% < 1.2
- Rm% 1.2 - 1.3
- Rm% 3 - 4
- Rm% > 4
- Pre - Devonian and Dingle Group

Maturation and source rocks

Clayton et al. (1989) compiled a map of thermal maturation levels within the onshore Devonian and Carboniferous sequences (Figure 4.15). The map is based on 494 determinations from a scattering of surface or shallow borehole samples across the country, using mean random vitrinite reflectance, conodont colour, and spore colour and fluorescence characteristics. Onshore maturation values record decreasing maturation from the greenschist meta-argillite regime of the Munster Basin in the south of the country to the more promising maturation values of the northeast. Four levels or zones of maturation were recognised, numbered 1 to 4. Hydrocarbon generation is restricted to Zones 3 and 4, with only Zone 4 within the oil window. Most of the southern two-thirds of the country lies within Zones 1 and 2 (Figure 4.15) where the rocks are over-mature for hydrocarbon generation. Coals in the Leinster coalfield are of anthracitic rank (Haughey and McArdle, 1990). Vitrinite reflectance and spore coloration data from the Upper Palaeozoic rocks of the Kinsale area, on the south coast (Clayton, 1989) demonstrate two important points. The first is that the rocks are of meta-anthracite rank, but secondly that there is not a correlation between reflectance and stratigraphic position through 2 km of section across large-scale folds. This suggests that the high temperatures were maintained until cessation of the Variscan orogeny. Clayton (1989) pointed out that, even allowing for high palaeothermal gradients, a substantial thickness (5–7 km) of overlying strata was necessary in order to produce the high maturation levels.

An interesting exception to the high maturation and palaeothermal values in the southern region is found in the extreme southeast of the country, in the Wexford Outlier (Burnett et al., 1990). Organic maturation levels in the Carboniferous rocks in this area (mean Vitrinite Reflectance (Rm) values in the range 0.77% – 2.54%) are considerably lower than in the adjacent region, and can be attributed to normal burial processes. The reader is referred to Taylor et al. (1998) for details of vitrinite reflectance and their significance (e.g. Rm: mean reflectance; Ro: reflectance under oil). Present maturation levels were probably attained, together with the generation of any hydrocarbons from the section, during Late Carboniferous time. This fault-controlled outlier has much in common with the adjacent offshore St George's Channel Basin to the east (see Chapter 10). It is also interesting to note that a small karstic red clay deposit at Cloyne, near Cork, dated as (Lower-) Middle Jurassic

(Higgs and Beese, 1986), has low levels of maturation. Vitrinite reflectance determinations of 0.30–0.35% suggest only limited (<1 km) post-Middle Jurassic burial (Clayton, 1989). This, and other evidence, points to a major disconformity in the maturation profile of the region, probably coincident with the Variscan erosion surface. Maturation data are not available for Permo-Triassic rocks in this southern area. It appears likely, however, that the required additional section was Pennsylvanian (perhaps enhanced by thrust repetition), rather than a thick Permo-Triassic sequence that was subsequently eroded. For example, Naylor *et al.* (1978) recorded more than 500 m of strata in a Serpukhovian to lower Pennsylvanian section on Whiddy Island in Bantry Bay. However, Keeley (1995) has argued, using apatite fission track modelling, for the deposition of at least 1 km of Permo-Triassic strata in an area roughly coincident with the older Munster Basin. Nevertheless, maturation studies (Clayton *et al.*, 1989; Blackmore, 1995) demonstrate that the Palaeozoic rocks of the southern part of the country are over-mature (greenschist meta-argillite) for hydrocarbon sourcing and preservation. Illite crystallinity, chlorite geothermometry and fluid inclusion studies (Meere, 1995) in the same region indicate a metamorphic temperature in the range 275–325° C with geothermal gradients in excess 60° C/km. The high geothermal gradients are considered to be associated with late Palaeozoic extension, prior to Variscan deformation. From this brief description it is evident that the Munster and South Munster basins can be discounted from the viewpoint of oil and gas exploration.

Further north, the Shannon Basin lies within the zone of >Rm 4% values. However, the detailed situation is more complex than indicated on the thermal maturation map. Rm reversals have been noted at depth in both the Doonbeg-1 (Figure 3.1; Fitzgerald *et al.*, 1994) and other coreholes (Goodhue and Clayton, 1999). The section in the Doonbeg-1 well (TD 10,715 ft: 3265.8 m) shows no measurable difference in maturation level related to depth of burial. Using fluid inclusion, maturation and modelling data, Fitzgerald *et al.* (1994) concluded that peak temperatures were attained in Late Carboniferous time, prior to the onset of deformation, and that the probable cause of the anomalous maturation values was advective lateral movement of hot fluids. Corcoran and Clayton (2001) suggested that this process resulted from northwards movement of fluids from the Variscan orogenic belt, with its core of granite batholiths, and that this movement was facilitated by fracture zones. In their view, the variability and reversal of palaeogeothermal gradient in County Clare wells indicates a complex thermal regime involving advective heating and fluid circulation facilitated by fractures in the basal Clare Shale (late Brigantian to Chokerian) that rests on Viséan carbonates.

Naylor (1992) emphasised the strong contrast between the development of the onshore and offshore regions. Onshore Carboniferous sequences generally have higher maturation levels than those encountered beneath the offshore basins, probably due to higher onshore heat flows. The island of Ireland has been a positive and largely emergent feature for much of post-Variscan time. The Early Jurassic and Late Cretaceous periods were probably the main periods of post-Variscan marine submergence onshore, but evidence of these deposits has been largely removed by later erosion. A study of an extensive suite of onshore apatite fission track samples by Allen *et al.* (2002) showed a generally consistent pattern of denudation for succeeding post-Variscan time slices. The total post-Triassic denudation map, for example, shows values of <2 km, and often < 1 km, for large areas of the Midlands and Ulster, with much higher values (> 3 km) in a north–south zone along the west coast. Praeg *et al.* (2005) documented significant rapid subsidence in the offshore NW European Atlantic margin basins, coincident with kilometre-scale onshore uplift in early Palaeogene and in latest Neogene times (Figure 2.11)

Maturation levels (Figure 4.14) and reservoir quality improve northwards, and Upper Palaeozoic rocks in the northernmost areas of the Republic and in Northern Ireland have some exploration potential. Parnell (1991a) demonstrated that source rocks are present in the Carboniferous sequence of Northern Ireland and are probably gas-prone. Viséan and Sepukhovian–Bashkirian organic mudstones and coals (Figures 4.8 and 4.11) provided good quality source rocks in northeast Ireland, although studies indicate that they have generated most of their potential hydrocarbons (Reay, 2004). Vitrinite reflectance values are generally in the range Ro = 1.45–2.75%, within the gas window. Back projection of the maturity profiles indicates an uplift and erosion in the range 3.5 km to 4.5 km, suggesting that hydrocarbon generation was terminated following the Variscan tectonic episode. Hydrocarbon generation may have occurred in basinal areas in which the source rocks have been buried beneath younger Carboniferous rocks (in the west) or Permo-Triassic sequences (in the east).

Reay (2004) reports that vitrinite reflectance values in Carboniferous source sequences decrease significantly at outcrops and in boreholes around Lough Neagh, an important factor when considering the prospectivity of the Permo-Triassic basin (*see* Chapter 12). An example of this is a sample of coal from Coalisland listed by Clayton *et al.* (1989) with an Rm value of 0.57%.

Reservoir rocks

Four features of Irish geology have controlled the location and extent of reservoir quality, and especially subsurface porosity and permeability.

◆ The Republic of Ireland is comprised almost entirely of Carboniferous (Mississippian – Pennsylvanian) and older rocks.

◆ Much of the country suffered a variable amount of deformation during the Variscan (end Carboniferous) orogeny. In the southernmost part of the country this resulted in low level metamorphism and the development of a pervasive slaty cleavage.

◆ Upper Palaeozoic rocks in Ireland have undergone higher levels of thermal maturation than their counterparts in the offshore basins and in Britain. In addition to adversely affecting source rock potential, the thermal maturation levels, combined with Variscan deformation, have also increased the induration of sandstone reservoirs, resulting in low porosity and permeability.

◆ There was relatively limited deposition over the land area of Ireland during the Mesozoic, and any such deposits have been largely removed by erosion. During the Cenozoic the island was uplifted and subject to erosion. This has resulted in variable leaching and flushing of near-surface sections and the development of secondary porosity. Mesozoic rocks are well represented in the offshore basins but, with the exception of the Ulster Basin in the northeast of the island, onshore are found only in small areas at Kingscourt, County Cavan, south County Wexford, and in a number of karstic and cave deposits as outlined above.

The Lower Palaeozoic and older rocks, together with the Upper Palaeozoic rocks in the southwest of the country that lie south of the Variscan deformation front, are devoid of significant porosity or permeability and can be excluded from consideration as potential hydrocarbon reservoirs. The Mississippian carbonates that cover large tracts of central Ireland also generally lack primary porosity. There are zones of secondary dolomitisation within the carbonates. Studies have shown,

however, that although this has produced significant void space, there is only poor permeability. Seal is also likely to be a problem. The thin basal sandstones and Devonian red-beds beneath the carbonates also have limited reservoir potential. The widespread limestone terrain in Ireland also has low potential for reliable subsurface zones of porosity and permeability. Serpukhovian and Pennsylvanian sandstones with enhanced porosity due to secondary leaching may have some local reservoir potential. There are virtually no data regarding subsurface closure or seal for these strata.

As noted in Chapter 3, the Ambassador consortium (Ambassador, Conoco, Marathon) drilled nine exploration wells during the period 1962–1966, with the main emphasis on the Northwest Carboniferous Basin (Figure 3.1). The latter (Figure 4.7) is the only area in Ireland that has yielded significant shows and flows of hydrocarbons. Table 4.2 (after Sheridan, 1977) lists the shows encountered in the original drilling programme. The most encouraging result was a gas flow of 31,000 standard cubic feet per day (scfd) on test from the Dowra Sandstone Member in the Dowra-1 well. In the early 1980s a consortium led by Marinex Petroleum re-entered and stimulated the Dowra-1 well, achieving an improved a flow rate of 289,000 scfd gas. Santa Fe joined the group and, with Aran Energy now as operator, four wells were drilled during 1984–85 (Drumkeeran South-1, Macnean-2, Kilkoo Cross-1 and Slisgarrow-1: Figure 4.7). All were plugged and abandoned without encountering significant shows. More recently (2001) six wells were drilled in the basin – the Dowra-2 and Thur Mountain-1 wells in the Republic and a further four wells in County Fermanagh, Northern Ireland. These wells were hydraulically fractured and tested during 2002 and all were subsequently plugged and abandoned.

It is clear that a petroleum system has operated in the Lough Allen region, and it is probable that the source rocks attained their present high level of maturity in Late Palaeozoic time. The fact that no commercial quantity of gas has been discovered is related to reservoir quality and permeability. The Mullaghmore Sandstone Formation and the Dowra Sandstone Member are the main target reservoirs, sealed in turn by the Bundoran and Benbulaben Shale formations. Sandstones comprise up to 55% of the Mullaghmore Sandstone, but calculated and measured porosities in borehole samples range only 0% to 10% (Reay, 2004). The porosity is secondary, mainly by dissolution of grains and pore-filling cements by meteoric waters during phases of uplift. As might be

Table 4.2. Hydrocarbon shows in the Northwest Carboniferous Basin encountered by wells drilled by the Ambassador Irish Oil Company consortium.

STRATIGRAPHIC SUBDIVISIONS	DOWRA No. 1	MACNEAN No. 1	BIG DOG No. 1	OWENGARR No. 1
BENBULBEN SHALE FORMATION	————	————	Minor gas shows from silty shales at 852 and 866 ft.	———
MULLAGHMORE SANDSTONE FORMATION	Minor gas shows at 2386-2400 ft in silts.	15,000 cubic ft per day at 1,725 ft.	Four shows ranging from 10-500 units at 1530, 1796, 1842 and between 1910 and 2015 ft. Slight residue of brown oil in core at 1710 ft.	80 units in silty equivalent at 1935 ft.
BUNDORAN SHALE FORMATION	———	———	———	50 units from limestone and shales at 2120 ft.
DOWRA SANDSTONE MEMBER, BUNDORAN SHALE FORMATION	31,000 cubic ft per day from 4050-80. Traces of oil in core analysis.	———	Up to 35 units in minor shows between 3413-3470 ft.	———
BALLYSHANNON LIMESTONE FORMATION	———	———	120 units at 4013 ft. Up to 35 units between 4152 to 4180 ft.	1,000 units at 3847 ft. 400 units at 3872 ft. 20 units at 3940.
BOYLE SANDSTONE FORMATION	———	———	13 units at 5150 ft. 1,000+ units at 5,470 ft.	———

Note:
No salt water zone was encountered in the Dowra No. 1 Well.
Minor salt water zones were encountered in the lower parts of the Macnean, Big Dog and Owengarr wells.
A major salt water zone, 50 bwph of over 50,000 ppm NaCl was encountered in the Old Red Sandstone of the Glenoo No. 1 hole at 6,885 ft.

expected, outcrop samples yielded higher porosity values of 15% (Reay, 2004). The low matrix porosity means that successful production would be dependent on joints and fractures. Despite the fact that the sandstone bodies may contain some gas over a relatively wide area, it appears that these provide inadequate permeability, even with additional hydraulic fracturing. The high cost of drilling many wells makes the play non-commercial – whether horizontal drilling techniques would alter this assessment is open to question.

Four of the wells drilled by the Ambassador consortium in the first phase of exploration in the 1960s targeted the other Carboniferous basins identified by the consortium in the central and southern parts of the country (Figure 3.1). It should be noted that all four wells described below are located in areas shown on Figure 4.15 to have maturity of > Rm = 3%. However, the Trim well is close to the boundary of a zone with lower values. In the account below, the original name applied by the consortium to a basin is shown in inverted commas,

with the basin name of the current literature alongside. It is instructive to look at the results from these wells in more detail, and the following remarks are based on an examination of the completion records of the four wells, together with the account of the wells given by Sheridan (1977):

Doonbeg-1, Co. Clare ('Southwest Basin' = Shannon Basin): In terms of geological structure, the Doonbeg-1 well lies to the west of the significant NNE–SSW trending Fergus Shear Zone in west County Clare (Coller, 1984). This area is characterised by sets of ENE–WSW trending, gently plunging folds that appear to have formed in response to simple compression. The well was located on one such fold axis, with closure confirmed by a seismic survey prior to drilling. The Doonbeg well apparently drilled a closed anticlinal culmination. The succession drilled in the well consisted of Serpukhovian and Pennsylvanian) shales and sandstones to 3495 ft (1065 m), Mississippian carbonates to top Devonian at 11,000 ft (3353 m). The Namurian sequence thickens southwestwards to the mouth of the Shannon (Figure 4.10; Hodson and Lewarne, 1961; Rider, 1974). The well targeted Viséan carbonates, sealed by the overlying shales, and the basal Carboniferous clastics (Sheridan, 1977). In the event, porosity was encountered in the well only in sandstones above 2500 ft (762 m) and these yielded fresh water. The well was making about 115 barrels of fresh water per hour when conversion to mud was initiated at 2196 ft (669 m). The freshwater yields in the upper part of the hole indicate that it was flushed at these levels. Whether this took place in the Tertiary or more recently, and whether the fresh water penetrated vertically downwards or laterally down-dip from the outcrop of the sandstones, is not known. It is probable that the shallow porosity has been produced by fracturing, and leaching of silica cement by meteoric waters. Porosities up to 15%–20% were recorded, but no measurable porosity was encountered below 2500 ft (762 m). Individual sandstone beds, interbedded within siltstones and mudstones, are normally up to 3 m thick, and do not exceed 6 m.

Meelin-1, Co. Cork ('Southwest Basin' = Shannon Basin): This well was sited on a surface anticline and drilled entirely in the Mississippian carbonate sequence. The target zones were dolomitised reefal horizons and basal clastics (Sheridan, 1977). A number of zones of dolomite showed variable porosity and increased flows of fresh water. A zone between 2105 ft and 2185 ft (641–666 m), which was badly caved, had porosities of 15–20%. Abundant pyrobitumen was observed between 1312 ft (400 m) and 2250 ft (686 m), but no gas or oil shows were recorded. Porosity was confined to the dolomite zones.

Ballyragget-1, Co. Kilkenny ('Southeast Basin' = northeast Limerick Province): The well location was based on surface mapping and seismic surveys, and drilled a Mississippian carbonate section, with a thin basal shale unit that lacked sandstones. Top Devonian was at 3427 ft (1045 m) and probable Top Silurian at 3590 ft (1094 m). A dolomite section between 1565 ft (477 m) and 1735 ft (529 m) had abundant pyrobitumen. Within the dolomitic intervals calculations from the sonic and neutron logs suggest only 2% of matrix porosity, and there was no measurable porosity below 2235 ft (681 m).

Trim-1, Co. Meath ('Central Basin' = Dublin Basin): This well was located on a seismically-defined structure and drilled a typical Mississippian carbonate sequence (Sheridan, 1972b) to 5730 ft (1247 m), and then a basal section of poorly developed sandstones to Top Basement at 5960 ft (1817 m). The pre-drill targets were

the Waulsortian reef limestones and the basal clastics. Sheridan (1977) notes that 'Limestones lack porosity through recrystallisation, shales are indurated and non-hydratable, whilst the sandstones of the well-bore are semi-quartzitic'. Average calculated porosity in the limestone section was less than 1%, with a maximum in any interval of 4%. Maximum porosity in the basal sandstones was 1–2%. Bitumen was noted in several cores.

In considering possible Carboniferous sandstone reservoirs a number of additional areas of Serpukhovian–Pennsylvanian ('Namurian') outcrop are considered. The main areas of potential reservoir interest are as follows:

◆ *Leinster–Slieve Ardagh Coalfields (Figure 4.9):* A sequence in the order of 450 m (1500 ft) thick crops out around the coalfield. Although sandstones occur in the sequence, there is little reliable data concerning porosity and permeability.

◆ *'Namurian' outliers within the Dublin Basin*, west of Dublin, between Summerhill (Nevill, 1957) and north County Dublin. At outcrop, sandstones show moderate porosity, but a problem throughout these areas is the relatively shallow depth of potential target horizons.

◆ *Kingscourt, County Cavan (Figure 4.13):* The Serpukhovian and Pennsylvanian sequence contains sandstones that display some porosity at surface. In the western part of the outcrop the sequence is overlain by the Permo-Triassic succession described above (Figure 4.13). The thick Triassic sandstones lack adequate cover, but the Upper Permian gypsiferous beds rest on the Carboniferous and could provide a seal for underlying sandstones.

The rocks of the Kingscourt area are all lower than Rm = 3% in maturation, and in the central part of the graben are only Rm = 1.2%. Concerns regarding overmature source sequences, which are a problem in the Shannon Basin, therefore do not apply here. At depth in the Kingscourt graben, potential source rocks may be in the gas generation window, whilst the interbedded sandstones form potential reservoirs.

The Carboniferous and Permo-Triassic (Gardiner and Visscher, 1971) rocks of the Kingscourt area constitute a classic half-graben, with a major bounding fault (Kingscourt Fault) on the west side (Gardiner and McArdle, 1992; McConnell *et al.*, 2001). Much of the movement on this fault (*c.*2100 m) must be post-Early Triassic. The beds dip gently westwards towards the fault. However, lesser faults to the east of, and parallel to, the Kingscourt Fault have a reverse sense of movement. This produces the possibility of fault-constrained areas of closure in the Upper Carboniferous, with seal provided by the overlying Kingscourt Gypsum Formation. However, the areas under closure may be small and this, allied with modest porosities in the sandstones, could downgrade any petroleum prospectivity in the region.

◆ *Northeast Ireland*: In a study of the petrography and porosity of Carboniferous sandstones in Ulster (Figures 4.9 and 4.11) Parnell (1991b) observed an improvement of reservoir potential eastwards across the province. To the west, in areas contiguous with the Lough Allen Basin discussed above, porosity is generally < 5% and is secondary, largely as a result of cement dissolution. To the east, in the vicinity of the Permo-Triassic Ulster Basin, the sandstones attain reasonable reservoir potential and could offer targets beneath the younger cover. This aspect, together with that of Carboniferous source rock potential in this region, is discussed later when considering the prospectivity of the Permo-Triassic Ulster Basin (Chapter 12).

References

Allen, P.A., Bennett, S.D., Cunningham, M.J.M., Carter, A., Gallagher, K., Lazzaretti, E., Galewsky, J., Densmore, A.L., Phillips, W.E.A., Naylor, D., and Solla Hach, C. (2002) The post-Variscan thermal and denudational history of Ireland. *In*: Doré, A.G., Cartwright, J.A., Stoker, M.S., Turner, J.P., and White, N. (eds), *Exhumation of the North Atlantic Margin: Timing, Mechanisms and Implications for Petroleum Exploration*. Geological Society, London, Special Publications, **196**, 371–399.

Blackmore, R. (1995) Low-grade metamorphism in the Upper Palaeozoic Munster Basin, southern Ireland. *Irish Journal of Earth Sciences*, **14**, 115–133.

Bresser, G. and Walter, R. (1999) A new structural model for the SW Irish Variscides: The Variscan front of the NW European Rhenohercynian. *Tectonophysics*, **309**, 197–209.

Brück, P.M., Colthurst, J.R.J., Feely, M., Gardiner, P.R.R., Penney, S.R., Reeves, T.J., Shannon, P.M., Smith, D.G., and Vanguestaine, M. (1979) South-east Ireland: Lower Palaeozoic stratigraphy and depositional history. *In*: Harris, A.L., Holland, C.H., and Leake, B.E. (eds), *The Caledonides of the British Isles: reviewed*. Geological Society, London, Special Publications, **8**, 533–544.

Burnett, R.D., Clayton, G., Haughey, N., Sevastopulo, G.D., and Sleeman, A.G. (1990) The organic maturation levels of Carboniferous rocks in south County Wexford, Ireland. *Irish Journal of Earth Sciences*, **10**, 145–156.

Capewell, J.G. (1957) The stratigraphy and structure of the country around Sneem, County Kerry. *Proceedings of the Royal Irish Academy*, **58B**, 167–183.

Clayton, G. (1989) Vitrinite reflectance data from the Kinsale Harbour–Old Head of Kinsale area, southern Ireland, and its bearing on the interpretation of the Munster Basin. *Journal of the Geological Society of London*, **146**, 611–616.

Clayton, G. and Higgs, K. (1979) The Tournaisian marine transgression in Ireland. *Journal of Earth Sciences, Royal Dublin Society*, **2**, 1–10.

Clayton, G., Graham, J.R., Higgs, K., Holland, C.H., and Naylor, D. (1980) Devonian rocks in Ireland: a review. *Journal of Earth Sciences, Royal Dublin Society*, **2**, 161–183.

Clayton, G., Graham, J.R., Higgs, K, Sevastopulo, G.D., and Welsh, A. (1986a) Late Devonian and early Carboniferous paleogeography of southern Ireland and southwest Britain. *Annales de la Societé géologique de Belgique*, **109**, 103–111.

Clayton, G., Sevastopulo, G.D., and Sleeman, A.G. (1986b) Carboniferous (Dinantian and Silesian) and Permo-Triassic rocks in south County Wexford, Ireland. *Geological Journal*, **21**, 355–374.

Clayton, G., Haughey, N., Sevastopulo, G.D., and Burnett, R. (1989) *Thermal maturation levels in the Devonian and Carboniferous rocks in Ireland*. Geological Survey of Ireland, 36pp.

Cole, A.J. (1922) Some features of the Armorican (Hercynian) Folding in southern Ireland. *Report of the International Congress*, **XIII**, Belgium, 423–433.

Coller, D.W. (1984) Variscan structures in the Upper Palaeozoic rocks of west-central Ireland. *In*: Hutton, D.H.W. and Sanderson, D.J. (eds), *Variscan tectonics of the North Atlantic region*. Geological Society, London, Special Publications, **14**, 185–194.

Cooper, M.A., Collins, D., Ford, M., Murphy, F.X., and Trayner, P.M. (1984) Structural style, shortening estimates and the thrust front of the Irish Variscides. *In*: Hutton, D.H.W. and Sanderson, D.J. (eds), *Variscan tectonics of the North Atlantic region*. Geological Society, London, Special Publications, **14**, 167–175.

Cooper, M.A., Collins, D., Ford, M., Murphy, F.X., Trayner, P.M., and O'Sullivan, M. (1986) Structural evolution of the Irish Variscides. *Journal of the Geological Society of London*, **143**, 53–61.

Corcoran, D.V. and Clayton, G. (2001) Interpretation of vitrinite reflectance profiles in sedimentary basins, onshore and offshore Ireland. *In*: Shannon, P.M., Haughton, P.D.W., and Corcoran, D.V. (eds), *The Petroleum Exploration of Ireland's Offshore Basins*. Geological Society, London, Special Publications, **188**, 61–90.

Coxon, P. and McCarron, S.G. (2009) Cenozoic: Tertiary and Quaternary (until 11,700 years before 2000). *In*: Holland, C.H. and Sanders, I.S. (eds), *The Geology of Ireland*. Second Edition. Dunedin Academic Press, Edinburgh, 355–396.

Davies, G.L. (1970) The enigma of the Irish Tertiary. *In*: Stephens, N. and Glasscock, R.E. (eds), *Irish Geographical Studies*, Queen's University, Belfast, 1–16.

Evans, A. and Clayton, G. (1998) The geological history of the Ballydeenlea Chalk Breccia, County Kerry, Ireland. *Marine and Petroleum Geology*, **15**, 299–307.

Fitzgerald, E., Feely, M., Johnston, J.D., Clayton, Fitzgerald, L.J., and Sevastopulo, G.D. (1994) The Variscan thermal history of west Clare, Ireland. *Geological Magazine*, **131**, 545–558.

Ford, M., Brown, C., and Readman, P. (1991) Analysis and tectonic interpretation of gravity data over the Variscides of southwest Ireland. *Journal of the Geological Society of London*, **148**, 137–148.

Ford, M., Klemperer, S.L., and Ryan, P.D. (1992) Deep structure of southern Ireland: a new geological synthesis using BIRPS deep reflection profiling. *Journal of the Geological Society of London*, **149**, 915–922.

Gardiner, P.R.R. (1978) Is the Hercynian Front in Ireland a local feature? *Nature*, **271**, 538–539.

Gardiner, P.R.R. and McArdle, P. (1992) The geological setting of Permian gypsum and anhydrite deposits in the Kingscourt district, counties Cavan, Meath and Monaghan. *In*: Bowden, A.A., Earls, G., O'Connor, P.G., and Pyne, J.F. (eds), *The Irish Mineral Industry 1980–1990*. Irish Association for Economic Geology, 301–316.

Gardiner, P.R.R. and Sheridan, D.J.R. (1981) Tectonic framework of the Celtic Sea and adjacent areas with special reference to the location of the Variscan Front. *Journal of Structural Geology*, **3**, 317–331.

Gardiner, P.R.R. and Visscher, H. (1971) The Permian–Triassic Transition sequence at Kingscourt, Ireland. *Nature*, **299**, 209–210.

George, T.N. and Oswald, D.H. (1957) The Carboniferous rocks of the Donegal Syncline. *Quarterly Journal of the Geological Society of London*, **113**, 137–183.

George, T.N., Johnson, G.A.L., Mitchell, M., Prentice, J.E., Ramsbottom, W.H.C., Sevastopulo, G.D., and Wilson, R.B. (1976) *A correlation of Dinantian rocks in the British Isles*. Geological Society of London, Special Report, **7**, 87pp.

Gill, W.D. (1962) The Variscan Fold Belt in Ireland. *In*: Coe, K. (ed.), *Some Aspects of the Variscan Fold Belt*. Manchester University Press, 44–64.

Gill, W.D. (1979) *Syndepositional sliding and slumping in the west Clare Namurian basin Ireland*. Geological Survey of Ireland, Special Paper, **14**, 31pp.

Goodhue, R. and Clayton, G. (1999) Organic maturation levels, thermal history and hydrocarbon source rock potential of the Namurian rocks of the Clare Basin, Ireland. *Marine and Petroleum Geology*, **16**, 667–675.

Graham, J.R. (1983) Analysis of the Upper Devonian Munster Basin, an example of a fluvial distributary system. *In*: Collinson, J.D. and Lewin, J. (eds), *Modern and ancient fluvial systems*. Special Publication of the International Association of Sedimentologists, **6**, 473–483.

Graham, J.R. (2001) Devonian. *In*: Holland, C.H. (ed.), *The Geology of Ireland*. First Edition. Dunedin Academic Press, Edinburgh, 201–239.

Graham, J.R. (2009) Devonian. *In*: Holland, C.H. and Sanders, I.S. (eds), *The Geology of Ireland*. Second Edition. Dunedin Academic Press, Edinburgh, 175–214.

Graham, J.R. and Clayton, G. (1988) Devonian rocks in Ireland and their relation to adjacent regions. *In*: McMillan, N.J., Embry, A.F., and Glass, D.J. (eds), *Devonian of the world, Vol. 1*. Canadian Society of Petroleum Geologists, Memoir, **12**, 325–340.

Graham, J.R., James, A., and Russell, K.J. (1992) Basin history deduced from subtle changes in fluvial style: a study of distal alluvium from the Devonian of southwest Ireland. *Transactions of the Royal Society of Edinburgh: Earth Sciences*, **83**, 655–667.

Graham, J.R., Richardson, J.B., and Clayton, G. (1983) Age and significance of the Old Red Sandstone around Clew Bay, NW Ireland. *Transactions of the Royal Society of Edinburgh: Earth Sciences*, **73**, 245–249.

Haughey, N. and McArdle, P. (1990) Vitrinite reflectance data from a preliminary study on selected Irish coal seams. *Geological Survey of Ireland Bulletin*, **4**, 201–209.

Higgs, K.T. (1984) Stratigraphic palynology of the Carboniferous rocks in northwest Ireland. *Geological Survey of Ireland Bulletin*, **3**, 171–201.

Higgs, K. (1986) The stratigraphy of the Namurian rocks of the Leinster coalfield. *Geological Survey of Ireland Bulletin*, **3**, 257–276.

Higgs, K. and Beese, A.P. (1986) A Jurassic microflora from the Colbond Clay of Cloyne, County Cork. *Irish Journal of Earth Sciences*, **7**, 99–109.

Higgs, K. and Jones, G.Ll. (2000) Palynological evidence for Mesozoic karst at Piltown, County Kilkenny. *Proceedings of the Geologists' Association*, **111**, 355–362.

Higgs, K.T. and O'Connor, G. (2005) Stratigraphy and palynology of the Westphalian strata of the Leinster Coalfield. *Irish Journal of Earth Sciences*, **23**, 65–84.

Hitzman, M.W. and Large, D. (1986) A review and classification of the Irish carbonate-hosted base metal deposits. *In*: Andrew, C.J., Crowe, R.W.A., Finlay, S., Pennell, W.M., and Pyne, J.F. (eds), *Geology and genesis of mineral deposits in Ireland*. Irish Association for Economic Geology, Dublin, 341–353.

Hodson, F. and Lewarne, G. (1961) A mid-Carboniferous basin in parts of the counties Limerick and Clare, Ireland. *Quarterly Journal of the Geological Society of London*, **107**, 307–333.

Holland, C.H. and Sanders, I.S. (eds) (2009) *The Geology of Ireland*. Second Edition. Dunedin Academic Press, Edinburgh, 537pp.

Jackson, J.S. (1965) The Upper Carboniferous (Namurian and Westphalian) of Kingscourt, Ireland. *Scientific Proceedings of the Royal Dublin Society*, **A2**, 131–152.

Keeley, M.L. (1995) New evidence of Permo-Triassic rifting, onshore southern Ireland, and its implications for Variscan structural inheritance. *In*: Boldy, S.A.R. (ed.), *Permian and Triassic Rifting in Northwest Europe*. Geological Society, London, Special Publications, **91**, 239–253.

Lees, A. (1964) The structure and origin of the Waulsortian (Lower Carboniferous) 'reefs' of west central Eire. *Philosophical Transactions of the Royal Society*, Series B, **247**, 483–531.

McConnell, B., Philcox, M.E., and Geraghty, M. (2001) *Geology of Meath: a geological description to accompany the bedrock geology 1:100,000 scale map series, Sheet 13, Meath with contributions by J. Morris, W. Cox (Minerals), G. Wright (Groundwater) and R. Meehan (Quaternary)*. Geological Survey of Ireland, 78pp.

Meere, P.A. (1995) Sub-greenschist facies metamorphism from the Variscides of SW Ireland: an early syn-extensional peak thermal event. *Journal of the Geological Society of London*, **152**, 511–521.

Mitchell, G.F. (1980) The search for Tertiary Ireland. *Journal of Earth Sciences, Royal Dublin Society*, **3**, 13–33.

Mitchell, W.I. (1992) The origin of Upper Palaeozoic sedimentary basins in Northern Ireland and relationships with the Canadian Maritime Provinces. *In*: Parnell, J. (ed.), *Basins on the Atlantic Seaboard: Petroleum Geology, Sedimentology and Basin Evolution*. Geological Society, London, Special Publications, **62**, 191–202.

Mitchell, W.I. and Owens, B. (1990) The geology of the western part of the Fintona Block, Northern Ireland: evolution of Carboniferous basins. *Geological Magazine*, **127**, 407–426.

Murphy, F.X. (1990) The Irish Variscides: a fold belt developed within a major surge zone. *Journal of the Geological Society of London*, **147**, 451–460.

Naylor, D. (1966) The Upper Devonian and Carboniferous geology of the Old Head of Kinsale, Co. Cork. *Scientific Proceedings of the Royal Dublin Society*, **A2**, 229–249.

Naylor, D. (1969) Facies change in Upper Devonian and Lower Carboniferous rocks of southern Ireland. *Geological Journal*, **6**, 307–328.

Naylor, D. (1992) The post-Variscan history of Ireland. *In*: Parnell, J. (ed.), *Basins on the Atlantic Seaboard: Petroleum Geology, Sedimentology and Basin Evolution*. Geological Society, London, Special Publications, **62**, 255–275.

Naylor, D. and Jones, P.C. (1967) Sedimentation and tectonic setting of the Old Red Sandstone of Southwest Ireland. *In*: Oswald, D.H. (ed.), *International Symposium on the Devonian System*, **2**. Alberta Society of Petroleum Geologists, 1089–1099.

Naylor, D. and Sevastopulo, G.D. (1979) The Hercynian 'Front' in Ireland. *Krystallinikum*, **14**, 77–90.

Naylor, D., Haughey, N., Clayton, G., and Graham, J.R. (1993) The Kish Bank Basin, offshore Ireland. *In*: Parker, J.R. (ed.), *Petroleum Geology of North-west Europe: Proceedings of the 4th Conference*. Geological Society, London, 845–855.

Naylor, D., Jones, P.C., and Clayton, G. (1978) The Namurian stratigraphy of Whiddy Island, Bantry Bay, West Cork. *Geological Survey of Ireland Bulletin*, **2**, 235–253.

Naylor, D., Sevastopulo, G.D., and Sleeman, A.G. (1989) Subsidence history of the South Munster Basin, Ireland. *In*: Arthurton, R.S., Gutteridge, P. and Nolan, S.C. (eds), *The role of tectonics in Devonian and Carboniferous sedimentation in the British Isles*. Yorkshire Geological Society, Occasional Publication, **6**, 99–110.

Naylor, D., Sevastopulo, G.D., and Sleeman, A.G. (1996) Contemporaneous erosion and reworking within the Dinantian of the South Munster Basin. *In*: Strogen, P., Somerville, I.D., and Jones, G.Ll. (eds), *Recent advances in Lower Carboniferous geology*. Geological Society, London, Special Publications, **107**, 331–343.

Nevill, W.E. (1957) The geology of the Summerhill basin, Co. Meath, Ireland. *Proceedings of the Royal Irish Academy*, **58(B)**, 293–303.

Oswald, D.H. (1955) The Carboniferous rocks between the Ox Mountains and Donegal Bay. *Quarterly Journal of the Geological Society of London*, **111**, 167–186.

Owens, B., Riley, N.J., and Calver, M.A. (1985) Boundary stratotypes and new stage names for the lower and middle Westphalian sequences in Britain. *Compte Rendu 10ème Congrès International de Stratigraphie et de Géologie Carbonifère, Madrid*, 1983, **4**, 461–472.

Parnell, J. (1991a) Hydrocarbon potential of Northern Ireland: 1. Burial histories and source-rock potential. *Journal of Petroleum Geology*, **14**, 65–78.

Parnell, J. (1991b) Hydrocarbon potential of Northern Ireland: 2. Reservoir potential of the Carboniferous. *Journal of Petroleum Geology*, **14**, 143–160.

Philcox, M.E., Baily, H., Clayton, G., and Sevastopulo, G.D. (1992) Evolution of the Carboniferous Lough Allen Basin, Northwest Ireland. *In*: Parnell, J. (ed.), *Basins on the Atlantic Seaboard: Petroleum Geology, Sedimentology and Basin Evolution*. Geological Society Special Publications, **62**, 203–215.

Philcox, M.E., Sevastopulo, G.D., and MacDermot, C.V. (1989) Intra-Dinantian tectonic activity on the Curlew Fault, northwest Ireland. *In*: Arthurton, R.S., Gutteridge, P., and Nolan, S.C. (eds), *The role of tectonics in Devonian and Carboniferous sedimentation in the British Isles*. Yorkshire Geological Society Occasional Publication, **6**, 55–66.

Phillips, S.J.L. (1985) *The stratigraphy and structure of southwest County Cork, Ireland*. Unpublished Ph.D. thesis, Queen's University Belfast.

Praeg, D., Stoker, M.S., Shannon, P.M., Ceramicola, S., Hjelstuen, B.O., and Mathiesen, A. (2005) Episodic Cenozoic tectonism and the development of the NW European 'passive' continental margin. *Marine and Petroleum Geology*, **22**, 1007–1030.

Quin, J.G. (2008). The evolution of a thick shallow-marine succession, the South Munster Basin, Ireland. *Sedimentology*, **55**, 1052–1082.

Ramsbottom, W.H.C., Calver, M.A., Eagar, R.M.C., Hodson, F., Holliday, F., Stubblefield, C.J., and Wilson, R.B. (1978) *A correlation of Silesian rocks in the British Isles*. Geological Society, London, Special Report, **10**, 81pp.

Readman, P.W., O'Reilly, B.M., Edwards, J.W.F. and Sankey, M.J. (1995) A gravity map of Ireland and surrounding waters. *In*: Croker, P.F. and Shannon, P.M. (eds), *The Petroleum Geology of Ireland's Offshore Basins*. Geological Society Special Publication, **93**, 9–16.

Readman, P.W., O'Reilly, B.M. and Murphy, T. (1997) Gravity gradients and upper-crustal tectonic fabrics, Ireland. *Journal of the Geological Society of London*, **154**, 817–828.

Reay, D.M. (2004). Oil and Gas. *In*: Mitchell, W.I. (ed.), *The Geology of Northern Ireland: Our Natural Foundation*. Geological Survey of Northern Ireland, Belfast, 273–290.

Ries, A.C. (1979) Variscan metamorphism and K–Ar dates in the Variscan fold belt of S. Brittany and NW Spain. *Journal of the Geological Society, London*, **136**, 89–103.

Rider, M.H. (1974) The Namurian of west County Clare. *Proceedings of the Royal Irish Academy*, **74B**, 125–142.

Rowell, P. (1995) Tectono-stratigraphy of the North Celtic Sea Basin. *In*: Croker, P.F. and Shannon, P.M. (eds), *The Petroleum Geology of Ireland's Offshore Basins*. Geological Society, London, Special Publications, **93**, 101–137.

Sanderson, D.J. (1984) Structural variations across the northern margin of the Variscides in N.W. Europe. *In*: Hutton, D.H.W. and Sanderson, D.J. (eds), *Variscan Tectonics of the North Atlantic Region*. Geological Society, London, Special Publications, **14**, 149–165.

Sevastopulo, G.D. (2009) Carboniferous: Serpukhovian and Pennsylvanian. *In*: Holland, C.H. and Sanders, I.S. (eds), *The Geology of Ireland*. Second Edition. Dunedin Academic Press, Edinburgh, 269–294.

Sevastopulo, G.D. and Wyse Jackson, P.N. (2009) Carboniferous: Mississippian (Tournaisian and Viséan). *In*: Holland, C.H. and Sanders, I.S. (eds), *The Geology of Ireland*. Second Edition. Dunedin Academic Press, Edinburgh, 215–268.

Sheridan, D.J.R. (1972a) *Upper Old Red Sandstone and Lower Carboniferous of the Slieve Beagh Syncline and its setting in the northwest Carboniferous basin, Ireland*. Geological Survey of Ireland, Special Paper, **2**, 129pp.

Sheridan, D.J.R. (1972b) The stratigraphy of the Trim No. 1 well, Co. Meath and its relationship to Lower Carboniferous outcrop in east-central Ireland. *Geological Survey of Ireland Bulletin*, **1**, 311–334.

Sheridan, D.J.R. (1977) The hydrocarbons and mineralisation proved in the Carboniferous strata of deep boreholes in Ireland. *In*: Garrard, P. (ed.), *Proceedings of the Forum on Oil and Ore in Sediments*. Geology Department, Imperial College, London, 113–144.

Sleeman, A.G., Reilly, T., and Higgs, K. (1978) Preliminary stratigraphy and palynology of five sections through the Old Head Sandstone and Kinsale Formations (Upper Devonian to Lower Carboniferous) on the west side of Cork Harbour. *Geological Survey of Ireland Bulletin*, **2**, 167–187.

Taylor, G.H., Teichmüller, M., Davis, A., Diessel, C.F.K., Littke, R., and Robert, P. (eds) (1998) Organic Petrology. Gebrüder Borntraegher, Berlin Stuttgart. 740pp.

Vermeulen, N.J., Shannon, P.M., Landes, M., Masson, F., and VARNET GROUP (1999) Seismic evidence for subhorizontal crustal detachments beneath the Irish Variscides. *Irish Journal of Earth Sciences*, **17**, 1–18.

Visscher, H. (1971) *The Permian and Triassic of the Kingscourt outlier, Ireland*. Geological Survey of Ireland, Special Paper, **1**, 1–114.

Walsh, P.T. (1966) Cretaceous outliers in south-west Ireland and their implications for Cretaceous palaeogeography. *Quarterly Journal of the Geological Society of London*, **122**, 63–84.

Chapter 5

Pre-Permian stratigraphy

Deposition within the basins of what would later become the Variscan fold belt and orogenic system of west-central Europe began in Early Devonian time, following the Caledonian orogeny. The complex Rheno-Hercynian Basin extended along the southern margin of the East Avalonian plate from southwest Ireland to southeastern Poland and persisted throughout the Devonian and Carboniferous (*see* Franke, 1989 and Friend *et al.*, 2000 for discussion). The Devonian was a period of extension and transtension, alternating with periods of compression related to the continued docking of Gondwana-derived fragments. The late Viséan saw the onset of Variscan deformation, as the margins of Gondwanaland and Laurussia increasingly impinged on each other. Compressional forces became dominant in the Late Carboniferous and deposition was largely terminated in latest Carboniferous time by the northwards encroaching deformation fronts of the Variscan orogeny. As in the previous chapter (Table 4.1) we follow the stratigraphical subdivisions for the Mississippian of Sevastopulo and Wyse Jackson (2009), and of Sevastopulo (2009) for the Pennsylvanian. Specifically the stages (now reduced in rank to regional substages) of Ramsbottom *et al.* (1978) and Owens *et al.* (1985) are employed, together with the Asturian substage, whilst 'Stephanian' and 'Autunian' are retained informally as regional stages. Where possible, the agreed global series and stages are used (*see* Heckel and Clayton, 2006, for wider discussion).

Offshore distribution of Palaeozoic and older rocks

Much of the evidence regarding the offshore distribution of Palaeozoic rocks in the offshore region is derived from gravity and seismic surveys. Only a scattering of the deep petroleum exploration wells have penetrated the pre-Permian section, and many of these encountered Carboniferous strata.

The high velocity basement that underlies the Mesozoic basins in the Celtic Sea and forms the dividing platform (Pembrokeshire Ridge–Labadie Bank High)

between the parallel set of basins is probably composed of Upper Palaeozoic rocks (Figure 2.15). Within the Celtic Sea basins at least three wells (Marathon 48/30-1, Marathon 58/3-1 and Esso/Marathon 56/20-1 (Figure 3.4) bottomed in deformed Carboniferous shales (Colin *et al.*, 1981; Gardiner and Sheridan, 1981; Higgs, 1983). Palynological studies on a seabed sample from the shelf north of the North Celtic Sea Basin indicate a mid-Tournaisian age (Delanty *et al.*, 1981), while a Tournaisian age was also obtained from the Esso-Marathon 48/30-1 well (Gardiner and Sheridan, 1981). The lithology and age of these samples indicate a southwards extension of the relatively shallow marine conditions and mudrock facies that obtained onshore in southernmost Ireland during the Tournaisian.

Pennsylvanian Coal Measures are extensively developed along the southern margin of the Palaeozoic massif of St George's Land. Asturian Coal Measures are also recorded in well Texaco/HGB 103/2-1 within the St George's Channel Basin (Barr *et al.*, 1981) and might therefore be expected to occur in the eastern part of the North Celtic Sea Basin. As outlined in Chapter 4, boreholes onshore in south County Wexford (Figure 5.1) penetrated a small area of concealed thin Serpukhovian and Pennsylvanian strata (Clayton *et al.*, 1986). One borehole intersected 32 m of grey mudrock with sandstone intercalations resting on karstic Asbian limestones, and another borehole proved 38 m of Asturian interbedded mudrocks and sandstones with a thin coal. In the Fastnet Basin, pre-Permian 'basement' comprises Devonian clastics and Mississippian shelf limestones (Robinson *et al.*, 1981). The Elf Aquitaine 55/30-1 well bottomed in Devonian (possibly Frasnian) continental red-beds and tuffs, while Cities Services 63/4-1 penetrated Mississippian (late Tournaisian) shelf limestones, suggesting that a carbonate shelf existed south or southwest of the South Munster muddy starved basin tract at this time.

Upper Palaeozoic rocks have a widespread occurrence in the Atlantic basins west of Ireland (Robeson

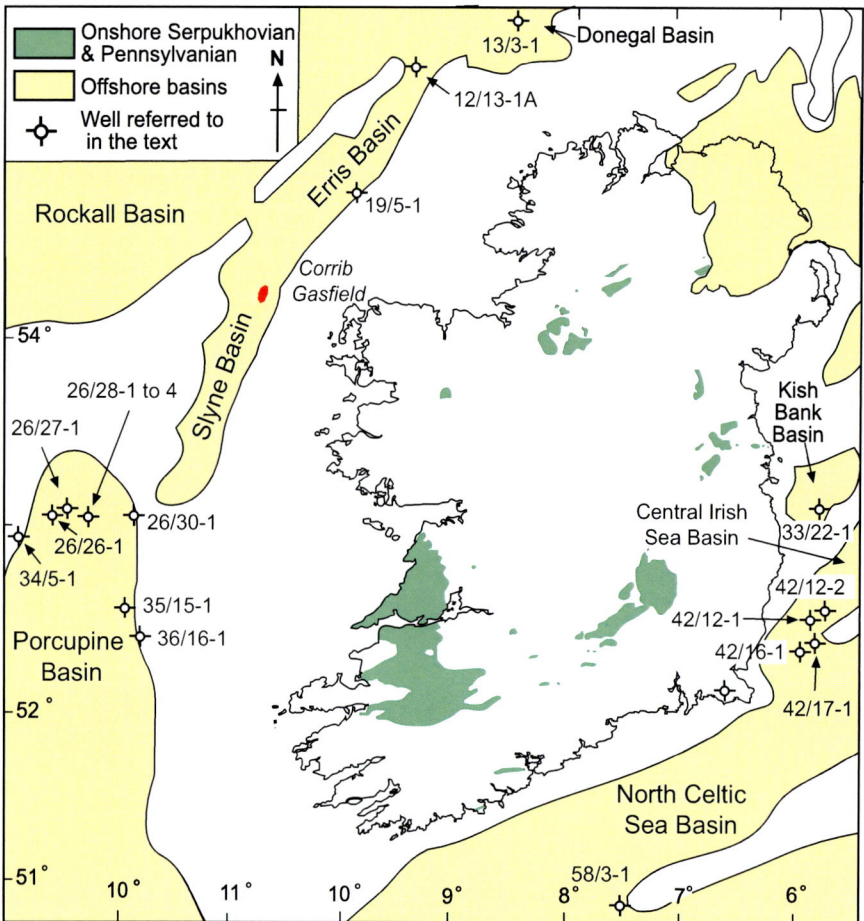

Figure 5.1. Location of wells that have intersected Serpukhovian and Pennsylvanian rocks in the Irish offshore (modified after Sevastopulo, 2009).

et al., 1988; Tate and Dobson, 1989a, 1989b). In the following descriptions the nomenclature of Naylor *et al.* (1999, 2002) is used for the major tectonic elements in the region. At the western faulted margin of the relatively shallow-water Celtic Platform, the subsurface features of the Fastnet High and Fastnet Spur separate the Fastnet Basin from the Goban Spur Basin (Figure 3.6). The Goban Spur region is composed of a set of fault-bounded basins and structural highs (Naylor *et al.*, 2002). The highs appear to be cored by Palaeozoic or older rocks, often with only thin Mesozoic and Cenozoic cover. Three NNW–SSE trending narrow horsts (the Merlin, Shackleton and Pendragon Highs) with intervening narrow basins form the outer part of the Goban Spur, developed in the footwalls of successive down-stepping faults. DSDP Site 548 (Figure 3.6) on the Merlin High (Graciansky *et al.*, 1985) at its base penetrated 20 m of quartzite and black shales, the latter containing Middle to Upper Devonian palynomorphs. On the basis of seismic interpretation, the Shackleton High is a similar feature, with only Cenozoic cover at its

crest. The seaward Pendragon High was penetrated at DSDP Site 549 (Graciansky *et al.*, 1985; Figure 3.6) that terminated in 37 m of foliated sandstones, believed to be Devonian (Old Red Sandstone) in age. It appears likely (Cook, 1987; Naylor *et al.*, 2002) that Upper Palaeozoic rocks are widespread beneath the Variscan unconformity in the Goban Spur region. Dredging on the margin of the Goban Spur (Auffret *et al.*, 1979) yielded abundant samples of granitic rocks, some clearly *in situ*, with radiometric ages of 275 Ma and 290 Ma, indicating a Variscan intrusive episode. Bioclastic limestones of possible Viséan age were also dredged on the southern margin of the Spur (Auffret *et al.*, 1979), together with fragments of high-grade metamorphic rocks and lithologies of likely Upper Palaeozoic affinity. Three basement elements (Figures 2.15 and 11.3), together with the Porcupine Fault, separate the Porcupine Basin from the Goban Spur Basin – the Balar Spur, the Tír na nÓg High and the Tóim High (Figure 3.6) – all of which, from seismic evidence, are thought to be cored by Upper Palaeozoic sequences (Naylor *et al.*, 2002). From the evidence presented above it is clear that the pre-Permian Variscan surface of the Celtic Sea and Goban Spur regions is floored mainly by Upper Palaeozoic rocks. Along the Porcupine and Rockall margin, however, positive structural elements with pre-Upper Palaeozoic cores provide a regional framework separating the younger basins. These will be described first, before considering the distribution and stratigraphy of the Upper Palaeozoic sequences.

A series of prominent highs separate the inboard basins along the west coast of Ireland from the Rockall Basin. In the south the Porcupine High (Figure 3.6) is a basement feature that underlies the bathymetric elements of the Porcupine Ridge, Porcupine Bank and part of the Slyne Ridge (Figure 2.14). The high generally has a thin Cenozoic cover, but the bathymetrically shallower parts may lack sedimentary cover. The pre-Permian geology comprises metasediments of probable Precambrian age, Caledonian granodiorites and Palaeozoic low-grade metamorphic rocks, strongly influenced in their distribution within the Porcupine Bank–Slyne Ridge area by westward extensions of the Great Glen, Leannan and Clew Bay Fault systems (Riddihough and Max, 1976; Bailey *et al.*, 1977; Max, 1978; PESGB, 2005). Cole and Crook (1910) and Auffret *et al.* (1987) recovered Lower Palaeozoic or older metasediments and gneisses by dredging on the Porcupine Ridge and Porcupine Bank. Daly *et al.* (2008) reported U–Pb zircon dating from an outcrop of granitic orthogneiss cored from the northern Porcupine High as indicating a Mesoproterozoic crystallisation age of 1.31 Ga. They suggested that this was derived from melting of pre-existing continental crust such as the Palaeoproterozoic of the Rockall Bank, and implied the existence of an important crustal boundary on the Irish continental margin, possibly aligned with the trend of the Slyne–Erris basins.

North of the Porcupine High, the positive eastern rim of the Rockall Basin, separating it from the inboard Slyne, Erris and Donegal Basins, is narrower, less continuous and crossed by major transverse faults. A narrow feature, the Slyne High, extends northwards from the Porcupine High and separates the Colm Basin from the Slyne Basin (Naylor *et al.*, 1999: Figures 2.15, 3.7 and 11.8). Further north, the Erris High (Figure 3.7) separates the Erris Basin from the main Rockall Basin (Cunningham and Shannon, 1997; Chapman *et al.,* 1999). The high probably consists of Lower Palaeozoic and/or older rocks and is overstepped by relatively thin Cenozoic sequences. The feature is a fault-bounded elongate horst that plunges southwards, where it probably has a preserved Upper

Palaeozoic cover, to form the core of a tilted Mesozoic fault block (Naylor *et al.*, 1999). Dredge samples reported by Cole and Crook (1910) along the margins of the Slyne, Erris and Donegal basins contain a wide range of metamorphic rock types (probably Dalradian) and also Carboniferous-type sandstones, shales and limestones. However, some of this material may have been transported by ice-rafting. The Amoco 12/13-1A well (Figure 5.1) in the northern Erris Basin bottomed in low-grade metasediments of uncertain (possibly Upper Palaeozoic) age. Arkosic Holkerian basal clastics and Brigantian conglomerates with granite fragments (Tate and Dobson, 1989b) in the Amoco 19/5-1 well (Erris Basin) indicate derivation from nearby granitic or metamorphic terrain.

Further west, towards the continental margin, the Rockall High (Figure 2.15) has a thin cover of sediments and Palaeogene lavas resting on acoustic basement. Dredge and drill samples have been dated from the southern part of the Rockall Bank (Miller *et al.*, 1973). It was initially thought that Lewisian (Scourian and Laxfordian), together with Grenvillian metamorphic rocks, were represented in the samples. However, more recent dating techniques (Morton and Taylor, 1991) indicate formation of the Rockall High basement rock sample suite at about 1625 Ma (Late Laxfordian), although the Rockall rocks are considered by Daly *et al.* (1995) to be comparable to those of the Annagh Gneiss Complex of north Mayo, onshore Ireland. This would suggest a somewhat older age of about 1750 Ma for the main phase of crustal growth in Rockall. The Rockall region therefore appears to comprise late (< 1.8 Ga) Palaeoproterozoic crust. Also, the southernmost of the samples on Rockall Bank yielded a possibly younger date of 987 ± 5 Ma that may represent the Grenvillian episode (Miller *et al.*, 1973). High seismic refraction velocities and reworked granulites and lavas in DSDP boreholes 403 and 404 on the southern margin of Edoras Bank suggest a positive basement element beneath the Bank. This evidence also indicates that the Bank, together with the adjacent Hatton High, is geologically similar to the Rockall High (Morton, 1984; Morton and Taylor, 1991; Roberts *et al.*, 1979). However, Hitchen (2004), interpreting seismic data acquired across Hatton Bank in UK waters, reported that the Hatton High does not comprise a simple Precambrian element, but includes discrete faulted basins with probable Mesozoic and possibly older sediments. Morton *et al.* (2009) recently reported U–Pb dating of detrital zircons from Albian sandstones drilled on Hatton Bank and suggested derivation from locally derived magmatic rocks dated at *c.*1.75 Ga (Palaeoproterozoic), while Hf isotopic data from the Hatton Bank sample suggest a large contribution from older, Archaean crust.

Data from offshore petroleum wells (Croker and Shannon, 1987; Robeson *et al.*, 1988; Tate and Dobson, 1989a, b) have added substantially to knowledge of Carboniferous stratigraphy in the Atlantic basins. Available geophysical data indicate a westwards off-shore extension of the onshore Carboniferous West Clare Basin (Gill, 1979) – herein referred to as the Shannon Basin (*see* Chapter 4) – across the offshore shelf towards the Porcupine Basin, and Croker (1995) named this the Clare Basin. In the Porcupine Basin, basement is over-lain by a thick Carboniferous succession. Duckmantian to Autunian rocks appear to be widespread, in contrast to onshore Ireland, where a thicker original sequence was probably eroded following Variscan uplift.

The well data demonstrate that positive elements at the basin margins continued to act as sediment source areas through Carboniferous time. The Shell 26/26-1 well on Finnian's Spur (separating the Porcupine Basin from the North Porcupine Basin; Figure 3.6), encoun-tered terrestrial to marginal-marine poorly dated Mississippian strata, resting unconformably on prob-able Dalradian schists and gneisses (Robeson *et al.*, 1988). Mississippian strata (with a cumulative thickness of more than 1000 m) were penetrated by only two wells – Shell 26/26-1 and Phillips 35/15-1 – in the northern part of the Porcupine Basin. The only other well on the Atlantic margin to encounter pre-Serpukhovian rocks is the Amoco 19/5-1 well in the Erris Basin (Figure 5.1). In this well, Holkerian continental sandstones (26 m pene-trated) pass upwards to a marginal to marine Asbian succession (325 m) of sandstones and shales, overlain in turn by a generally carbonate-dominated sequence. More varied sub-littoral to marine Brigantian sediments (430 m) are overlain by Serpukhovian (Pendleian-Arnsbergian) deltaic clastics with coals, terminating in an unconformity. This succession is only broadly com-parable to the same interval onshore in the northwest of Ireland, although Sevastopulo (2009) suggested that the Holkerian sandstones may equate with the Mullaghmore Sandstone Formation (Figure 4.8).

Lower Bashkirian Stage (Chokerian to Marsdenian) rocks have not been recovered in the offshore wells drilled on the Atlantic seaboard (Robeson *et al.*, 1988). Tate and Dobson (1989b) considered this to be due to a widespread hiatus or, more likely, an erosional event.

However, as Sevastopulo (2009) pointed out, evidence of this event is not seen in the west coast onshore suc-cessions. The Phillips 26/30-1 well, on the northeastern margin of the Porcupine Basin, is reported to have bot-tomed in coarse-grained biotite granodiorite (Robeson *et al.*, 1988), possibly slightly metamorphosed, uncon-formably overlain by Bolsovian sediments with a basal 'granite wash'. A Duckmantian section in the BP 26/28-1 well (Figure 5.1) contains conglomerates with a range of large clasts, including schist and granite, with a proxi-mal source from a metamorphic and granitic terrain. On the eastern margin of the Porcupine Basin the Chevron 36/16-1 recorded a thick conformable Yeadonian to Asturian section, with coals. However, the Shell 26/26-1 well on Finnian's Spur encountered Duckmantian resting on Serpukhovian. Deposition continued into Asturian time (Robeson *et al.*, 1988) and there is a gradual upward transition in the Asturian from paralic coal measures to red-beds.

Discovery of the Corrib Gasfield in the Slyne Basin (Figure 3.7) has demonstrated the presence of a working petroleum system comprising Pennsylvanian source rocks, Triassic Sherwood Sandstone reservoirs and Mercia Mudstone Group seal. Pennsylvanian (Langsettian to Bolsovian) coal measures have been penetrated in the basin (Langsettian are the oldest proven strata), and based on seismic interpretation are interpreted to be widespread (Dancer *et al.*, 1999, 2005). The sequence is dominated by siltsones and claystones, with common thin coals and sandstones. Other Carboniferous units may be present elsewhere in the basin. Langsettian beds begin the Pennsylvanian succession in the Erris Basin. In the Amoco 19/5-1 the Serpukhovian strata mentioned above are unconforma-bly overlain by paralic coal measure strata of Langsettian and Duckmantian age. Further north in the basin Upper Palaeozoic metasediments located beneath the Permian unconformity in the Amoco 12/13-1A well may indi-cate that the Serpukhovian–Pennsylvanian section is progressively truncated northwards in the basin (Chapman *et al.*, 1999). The Texaco 13/3-1 well in the Donegal Basin encountered a thick Duckmantian to Asturian coal measure sequence beneath the Jurassic, before terminating in a large intrusive dolerite body. Coal seams developed in the Porcupine region in the Langsettian, but later (Duckmantian) in the Erris and Donegal basins.

Rocks of probable Stephanian age are known from six wells in the north and west of the Porcupine Basin

(Robeson *et al.*, 1988; Ziegler, 1981; Croker and Shannon, 1987). Of these, a definite Stephanian age was established in wells Deminex 34/15-1, BP 26/28-1 and BP 26/28-2 which intersected a thin succession of grey and red-brown mudstones, calcareous sandstones and thin limestones. The Deminex 34/15-1 well (Figure 3.6) penetrated 106 m of varicoloured continental claystones and sandstones conformably underlain by interbedded sandstones, shales and thin limestones, respectively ascribed by the operator, on the basis of micropalaeontology, to the Autunian and Stephanian (Tate and Dobson, 1989b). The Autunian section has thin anhydrites and is thought to be the product of non-marine playa deposition in arid conditions. A similar section (110 m) occurs in the Gulf 26/21-1 well but it is only imprecisely dated as uppermost Carboniferous. The same is the case for 251 m of brown-grey siltstones, fine sandstones and claystones in Shell 34/19-1, which recorded minor anhydrite, limestone and coal intercalations. Unequivocal Permian strata have not been encountered by wells in the Porcupine Basin, so that the Triassic rests unconformably on the eroded Carboniferous surface, which is often reddened. In the Erris Basin, however, the uppermost Carboniferous is unconformably overlain by Upper Permian strata. Further west, on the eastern margin of the Rockall Basin, the Shell 12/2-1z well terminated in a Pennsylvanian succession of interbedded sandstones and shales, unconformably overlain by Permian (Asselian) strata (Tyrrell *et al.*, 2010).

The offshore basement geology around the north coast of Ireland comprises ridges and platforms extending southwestwards from Scotland, and bounded by major faults (Figure 6.1). The Islay–Donegal basement platform lies northwest of the major Leannan–Loch Gruinart Fault. Northwest of the Foyle Fault, a Dalradian ridge extends from Islay through Middle Bank to the north Irish coast separating the Loch Indaal Basin from the Rathlin Basin (Figure 6.1). To the southeast, the Rathlin and Larne basins are also separated by a NE–SW ridge of Dalradian metamorphic rocks, the Highland Border Ridge, that also contains younger rocks. Seismic interpretation by Shelton (1995) shows Carboniferous rocks as underlying much of the North Channel Basin, and resting on Lower Palaeozoic basement.

South of the North Channel Basin a broad syncline of Carboniferous rocks occupies much of the offshore area in the northern Irish Sea (Institute of Geological Sciences 1:250,000 Isle of Man Sheet). Erosion, following upon basin inversion or thermal doming during

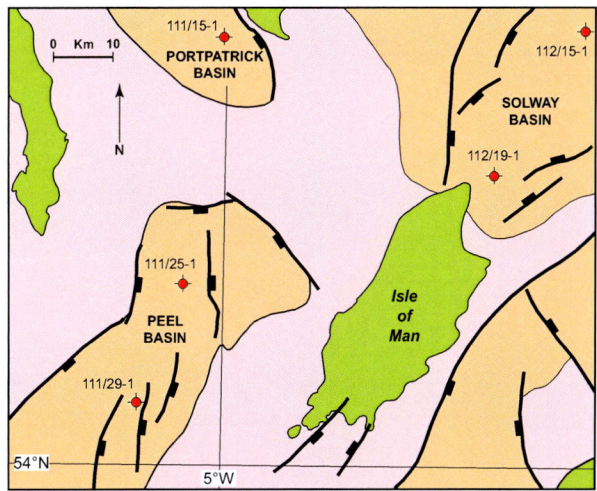

Figure 5.2. The Permo-Triassic Peel and Solway basins, north Irish Sea, showing exploration wells mentioned in the text. Based on Newman (1999).

the Variscan episode, removed the Pennsylvanian sequence beneath the Solway and Peel basins (Figure 5.2). The preserved Carboniferous sequence beneath the Variscan unconformity (Newman, 1999) comprises Late Brigantian to Pendleian shallow-marine–deltaic sandsones, shales and carbonates in the northeast (Elf 112/19-1: Solway Basin) passing to Holkerian–Asbian shallow-water carbonates in the southwest (Elf 111/29-1: Peel Basin). Pennsylvanian rocks are, however, widespread in the central and southern Irish Sea. The Amoco 33/22-1 well (Figure 5.3) on the margin of the Kish Bank Basin drilled a Duckmantian to Asturian (or possibly Stephanian) section 722 m thick, resting directly on Lower Palaeozoic strata (Jenner, 1981; Naylor *et al.*, 1993). The succession commences with a basal sandstone horizon overlain by Duckmantian cyclothemic siltstones and sandstones succeeded in turn by Bolsovian fine-grained clastics with the main development of coals. Jenner (1981) interpreted a total of 11 m of coal from wireline logs, within the Bolsovian section, with individual seams typically less than one metre, but occasionally exceeding this. The Asturian is represented by 50 m of cyclothemic deposits, topped by reddened siltstones and shales. A reworked miospore assemblage from cuttings at 338 m in the Amoco 33/22-1 well (Asturian) includes late Viséan to late Langsettian taxa (Naylor *et al.*, 1993), suggesting uplift and erosion of older Carboniferous rocks during mid-Pennsylvanian time. Reworked miospores ranging in age from Devonian to Langsettian were also recorded from an onshore Asturian section in County Wexford

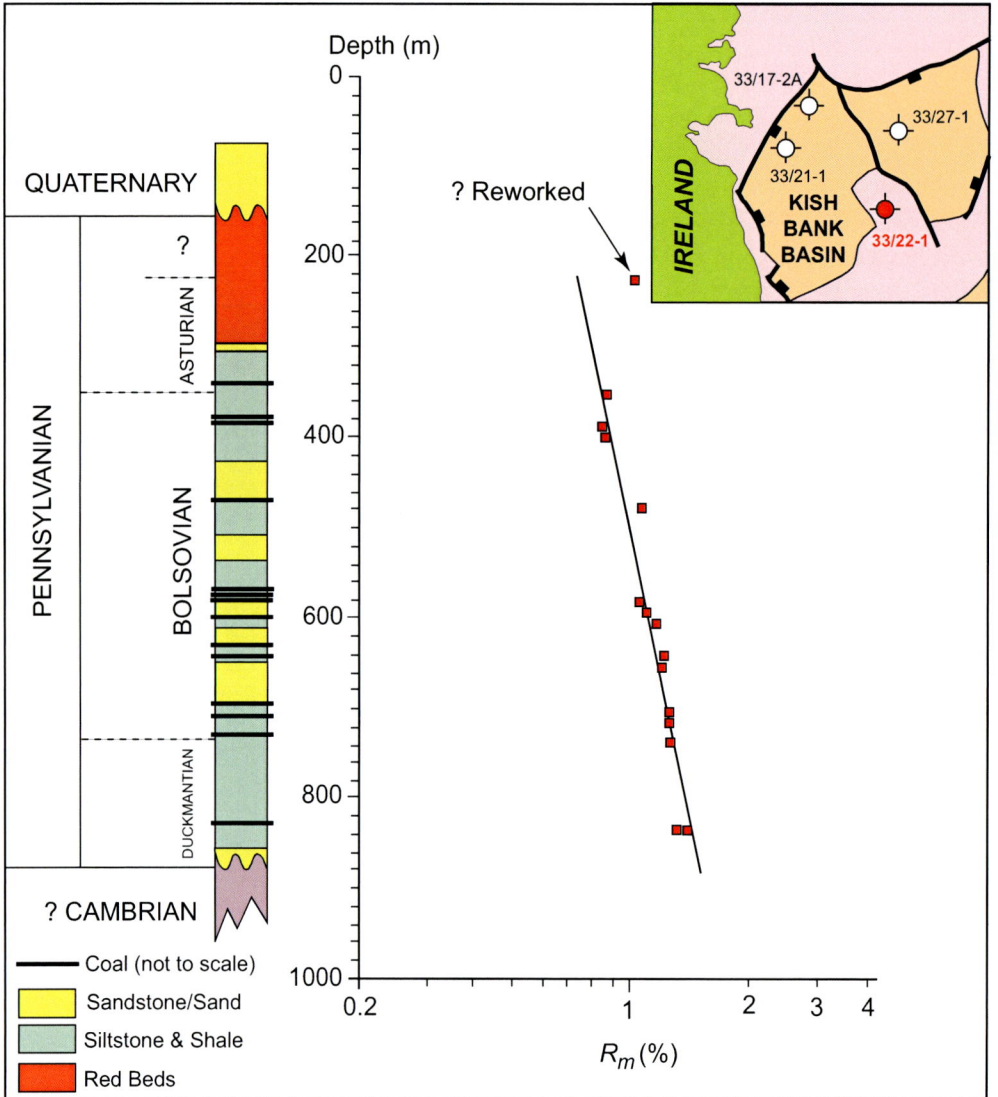

Figure 5.3. Amoco 33/22-1 well, Kish Bank Basin: simplified stratigraphy and vitrinite reflectance (Rm) data. Modified after Naylor *et al.* (1993).

(Clayton *et al.*, 1986), suggesting that the erosional episode was relatively widespread.

Wells on the southern margin of the Central Irish Sea Basin (Figure 6.7) have penetrated Duckmantian and Bolsovian deposits overlain by reddened beds of probable Asturian–Stephanian age (Maddox *et al.*, 1995). On the southwestwards extension of St Tudwal's Arch (Marathon 42/17-1) Langsettian–Duckmantian deposits are unconformable on Viséan carbonates, the intervening interval being absent through non-deposition or erosion. The Hydrocarbons Ireland Limited 42/12-1 well reached TD (Total Depth) in the Pennsylvanian, but seismic interpretation suggested that the Serpukhovian and lower Pennsylvanian are probably absent, but may be represented in the deeper undrilled parts of the basin (Maddox *et al.*, 1995).

Upper Palaeozoic palaeogeography

The data presented above provide evidence for the widespread preservation of Upper Palaeozoic, and particularly Carboniferous, strata in the Irish offshore, with intervening positive elements cored by Caledonian and older rocks. This disposition is largely the result of erosion on the Variscan landscape, although undoubtedly some upstanding elements were further modified by erosion during post-Variscan tectonic phases. A number of structural highs also acted as sediment sources during Upper Palaeozoic deposition, particularly as Variscan deformation encroached into the area. Latest Carboniferous sedimentation probably took place in basins of limited areal extent.

Terrestrial desert conditions prevailed throughout the Devonian over Ireland, and probably also over much of the offshore regions. Deposition was generally in localised alluvial fans, alluvial plains, and playa lake depocentres. Marine conditions had been established in Devon and Cornwall and a shoreline lay somewhere south of the Fastnet Basin. The thick (up to 6 km) Old Red Sandstone sequences of the onshore Munster Basin (Figure 4.3) probably range down to Givetian in age and must extend some distance westwards onto the Atlantic shelf. Much further south, in the area of the present Goban Spur, DSDP drilling (Sites 548 and 549: Figure 3.6) encountered sediments of Middle–Upper Devonian age, and it appears likely that similar rocks may be widespread in that region. Events unfolding in the southern ocean, with the northward advance of the Variscan deformation fronts, gave rise to a northward marine transgression that reached the southern shoreline of Ireland in Late Famennian time (Figure 5.4).

Old Red Sandstone facies sediments continued to be deposited across much of onshore Ireland during latest Devonian time, whilst high ground is believed to have

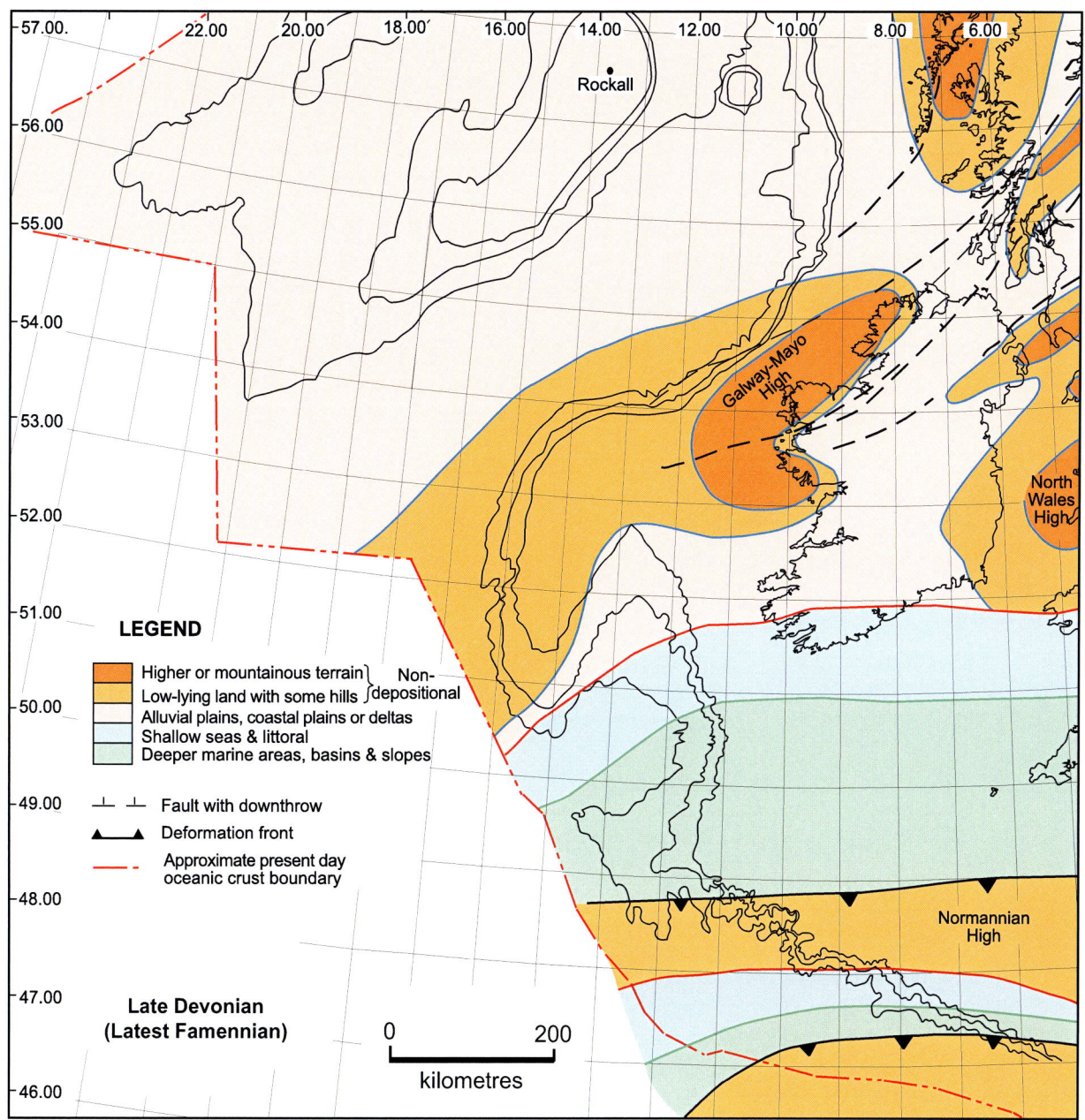

Figure 5.4. Late Devonian: latest Famennian palaeogeography. Modified after Naylor and Shannon (2009). *Principal sources*: Ziegler (1990), Bluck *et al.* (1992).

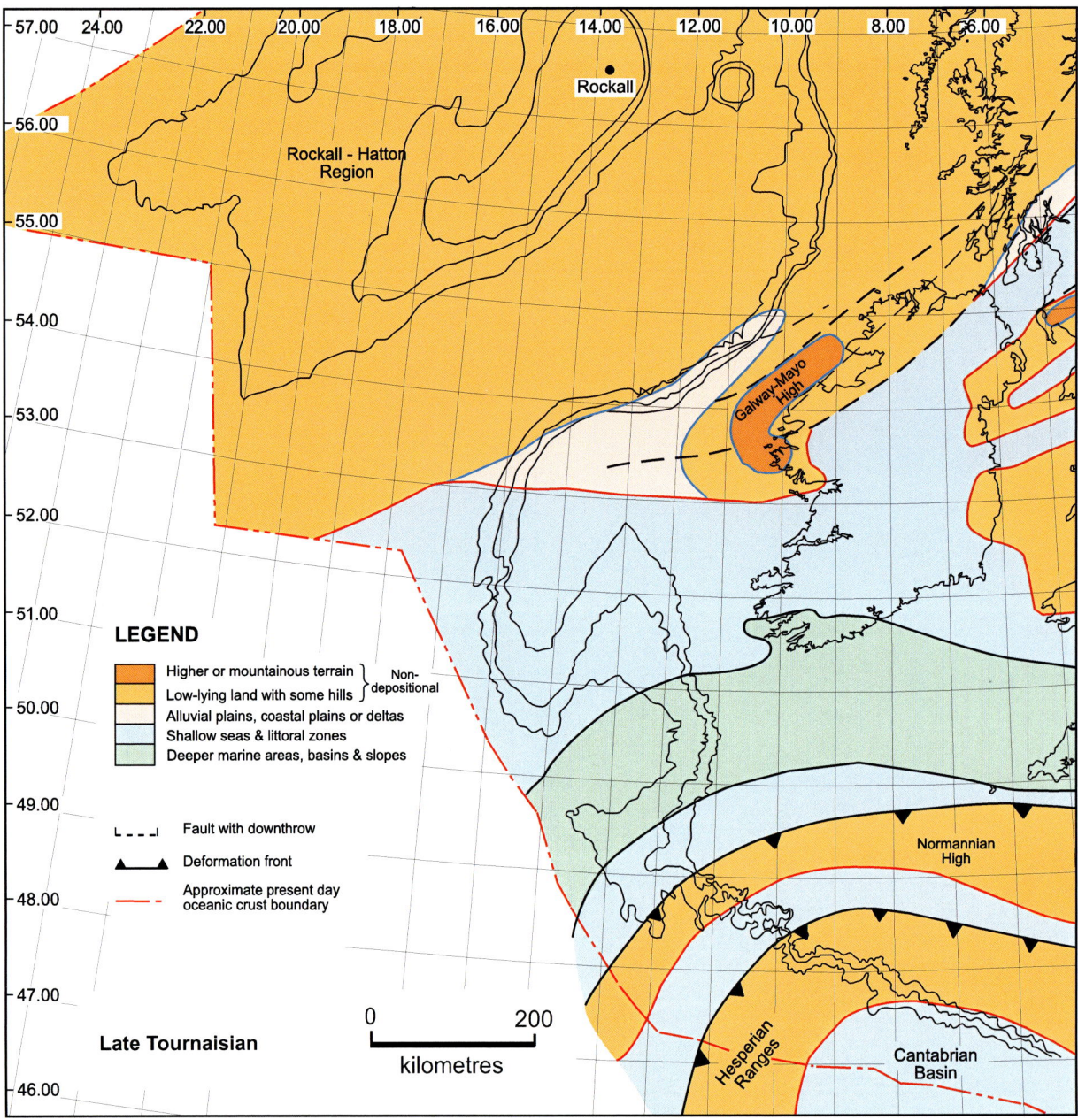

Figure 5.5. Early Carboniferous (Mississippian): latest Tournaisian palaeogeography. Modified after Naylor and Shannon (2009). *Principal sources*: Robeson *et al.* (1988), Tate and Dobson (1989b), Ziegler (1990), Cope *et al.* (1992), Naylor (1998); Sevastopulo and Wyse-Jackson (2009).

occupied the Galway–Mayo region (Bluck *et al.*, 1992). The general absence of Devonian rocks encountered in offshore drilling in the north Porcupine region (Robeson *et al.*, 1988), suggests that the Galway–Mayo High may continue as an upland extension southwestwards into this region. During the Early Carboniferous the sea continued to advance northwards across the present onshore area of Ireland (Figure 4.4), and presumably across the adjacent offshore regions. Red-bed deposition continued

north of the shoreline throughout the period (Clayton and Higgs, 1979). By latest Tournaisian time the sea had transgressed across much of the Irish mainland, although in the north the Galway–Donegal area and the southwest extension of the Southern Uplands were still emergent (Figure 5.5). The Welsh High (St George's Land) probably also extended across the present Irish Sea into Leinster. It is likely that the region from northern Scotland westwards across the Rockall–Hatton

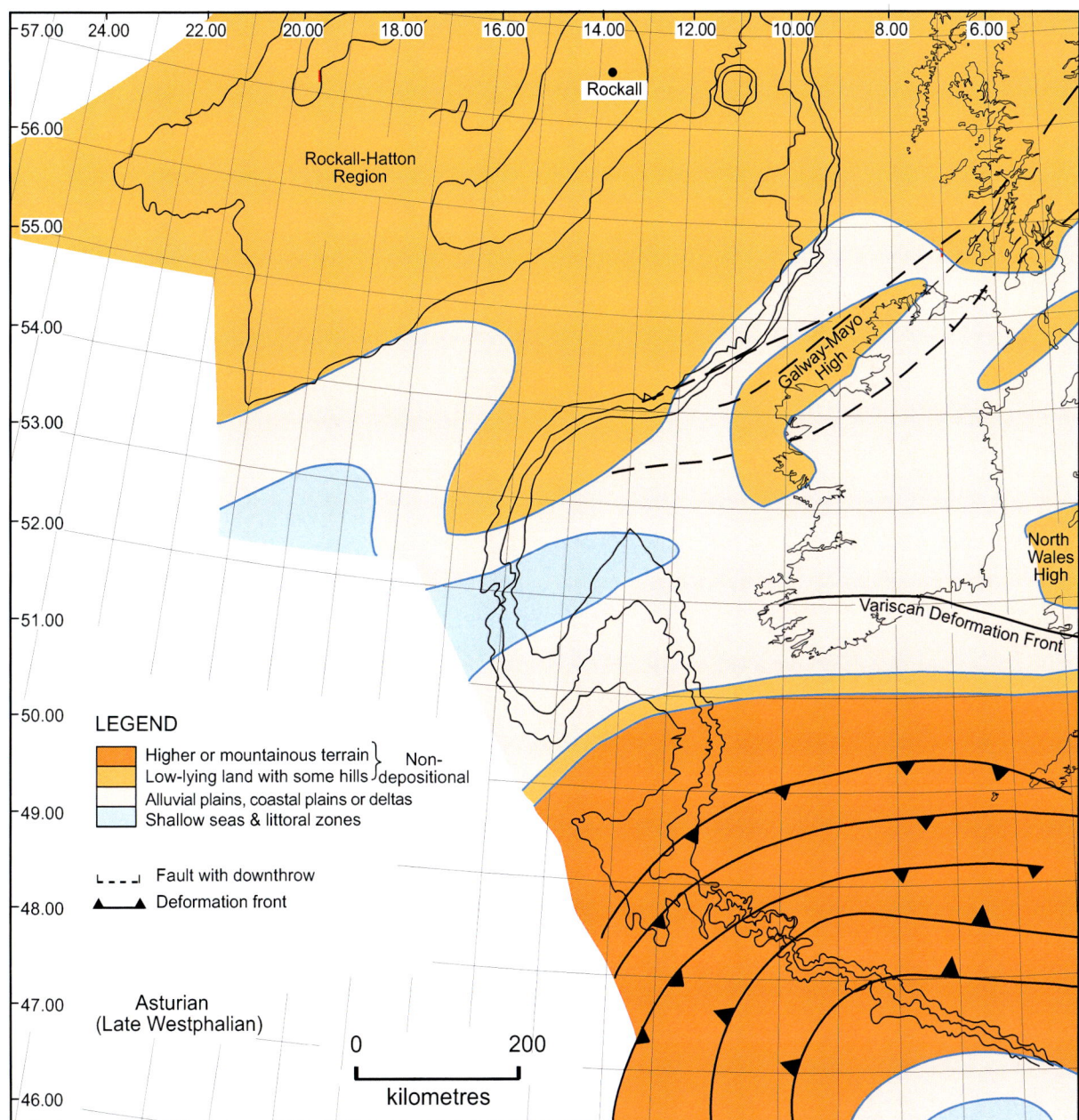

Figure 5.6. Late Carboniferous: Asturian (late Pennsylvanian) palaeogeography. After Naylor and Shannon (2009). *Principal sources*: Robeson *et al.* (1988), Tate and Dobson (1989b), Ziegler (1990), Cope *et al.* (1992).

region was largely a continental area of bare plains and low hills, with limited deposition. Carbonate sediments were deposited across much of Ireland with a transition in the southwest of the island into a southern trough with mudrock deposition (South Munster Basin) that offshore drilling suggests extended as far south as the North Celtic Sea Basin. Wells in the Fastnet Basin indicate that this area was part of a carbonate shelf south or southwest of the mudrock-dominant trough.

There was a similar disposition of land and sea areas through Viséan and Serpukhovian times, with no obvious sign of tectonic influence from the southern domain. In the north, the Galway–Mayo and Southern Uplands Highs persisted. On the north Antrim coast, thin coal seams of Brigantian age demonstrate non-marine conditions at that time. During the Serpukhovian, deltaic sandstones prograded southeastwards from the Caledonian uplands over substantial parts of Ireland. In the south,

Serpukhovian and basal Pennsylvanian sediments were probably derived from the developing orogenic areas further south. The lack of Serpukhovian strata in the southern part of the Irish Sea may indicate the presence of a landmass there at that time, or alternatively removal of the section during an intra-Carboniferous erosional episode. Evidence from reworking of Carboniferous palynomorphs, both in latest Carboniferous and Triassic strata (Naylor and Clayton, 2000), indicates that a more or less complete Carboniferous sequence was deposited in some parts of the Irish Sea area. Erosion of Carboniferous strata occurred during intra-Carboniferous episodes and during the Variscan tectonic phase. As seen earlier, drilling in the Atlantic basins has shown that Serpukhovian and Pennsylvanian deposition extended westwards into the offshore regions (Figure 5.1). Langsettian to Asturian rocks are widespread offshore, in contrast to the onshore area, and coal measure strata are known from the Porcupine Basin northwards to the Donegal Basin. Figure 5.6 shows the palaeogeography during Asturian time.

By Stephanian time the climax of the Variscan tectonic episode had produced fold mountains extending from southwest Ireland across southern Britain and into Central Europe. Stephanian and Autunian rocks are not preserved onshore in Ireland, but have been encountered in drilling off the west coast. Stephanian sediments, apparently conformable upon the Asturian, are well documented in the northern part of the Porcupine Basin (Croker and Shannon, 1987; Robeson et al., 1988). Continental sediments are interbedded with marine intercalations, and pass upwards through marginal marine deposits into anhydritic playa lake deposits. Ziegler (1982) suggested that the marine influences at this time result from marine incursions from the Tethyan domain along the developing Bay of Biscay rift system and into the Porcupine region. This marine access appears to have been closed by the end of the Autunian. Jackson et al. (1997) interpreted up to 4 km of Langsettian to Stephanian strata on seismic data south of the Isle of Man, in the Quadrant 119 Syncline. This, allied to the evidence off the Atlantic coast, suggests a pattern of limited late Carboniferous sedimentation in the axial zones of basins north of the encroaching Variscan deformation front. At the same time, earlier Carboniferous and older sediments were already being eroded at the basin margins. Along the Atlantic margin arid playa deposition persisted into latest Carboniferous (Autunian) time.

References

Auffret, G.A., Pastouret, L., Cassat, G., De Charpel, O., Gravatte, J., and Guennoc, P. (1979) Dredged rocks from the Armorican and Celtic margins. In: Montadert, L., Roberts, D. et al. (eds), Initial Reports on the Deep Sea Drilling Project, **48**, 995–1013.

Auffret, G.A., Auzende, J.M., Cousin, M., Coutelle, A., Dobson, M., Geoghegan, M., Masson, D., Rolet, J., and Valillant, P. (1987) Géologie des escarpments de Porcupine et de Goben (N.E. Atlantique). Resultats de la campagne de plongée CYAPORC. Note de groupe CYAPORC. Compte Rendu des Séances de 1 Academie des Sciences, Paris **304**, 10003–1008.

Bailey, R.J., Jackson, P.D., and Bennell, J.D. (1977) Marine geology of Slyne Ridge. Journal of the Geological Society of London, **133**, 165–172.

Barr, K.W., Colter, V.S., and Young, R. (1981) The geology of the Cardigan Bay–St. George's Channel Basin. In: Illing, L.V. and Hobson, G.D. (eds), Petroleum Geology of the Continental Shelf of North-West Europe. Heyden and Son Ltd., London, 432–443.

Bluck, B.J., Cope, J.C.W., and Scrutton, C.T. (1992) Devonian. In: Cope, J.C.W., Ingham, J.K., and Rawson, P.F. (eds), Atlas of Palaeogeography and Lithofacies. Geological Society, London, Memoir, **13**, 153pp.

Chapman, T.J., Broks, T.M., Corcoran, D.V., Duncan, L.A., and Dancer, P.N. (1999) The structural evolution of the Erris Trough, offshore northwest Ireland, and implications for hydrocarbon generation. In: Fleet, A.J. and Boldy, S.A.R. (eds), Petroleum Geology of Northwest Europe: Proceedings of the 5th Conference. Geological Society, London, 455–469.

Clayton, G. and Higgs, K. (1979) The Tournaisian marine transgression in Ireland. Journal of Earth Sciences, Royal Dublin Society, **2**, 1–10.

Clayton, G., Sevastopulo, G.D., and Sleeman, A.G. (1986) Carboniferous (Dinantian and Silesian) and Permo-Triassic rocks in south County Wexford, Ireland. Geological Journal, **21**, 355–374.

Cole, G.A.J. and Crook, T. (1910) On rock specimens dredged from the floor of the Atlantic off the coast of Ireland and their bearing on submarine geology. Memoir Geological Survey of Ireland, 84pp.

Colin, J.P., Lehmann, R.A., and Morgan, B.E. (1981) Cretaceous and Late Jurassic biostratigraphy of the North Celtic Sea Basin. In: Neale, J.W. and Brazier, M.D. (eds), Microfossils from Recent and fossil shelf seas. Ellis Horwood Ltd., Chichester, 122–155.

Cook, D.R. (1987) The Goban Spur: exploration in a deepwater frontier basin. In: Brooks, J. and Glennie, K.W. (eds), Petroleum Geology of North-West Europe. Graham and Trotman Ltd, London, 623–632.

Cope, J.C.W., Guion, P.D., Sevastopulo, G.D., and Swan, A.R.H. (1992) Carboniferous. In: Cope, J.C.W., Ingham, J.K., and Rawson, P.F. (eds), Atlas of Palaeogeography and Lithofacies. Geological Society, London, Memoir **13**, 153pp.

Croker, P.F. (1995) The Clare Basin: a geological and geophysical outline. In: Croker, P.F. and Shannon, P.M. (eds), The Petroleum Geology of Ireland's Offshore Basins. Geological Society, London, Special Publications, **93**, 327–339.

Croker, P.F. and Shannon, P.M. (1987) The evolution and hydrocarbon prospectivity of the Porcupine Basin, offshore Ireland. *In*: Brooks, J. and Glennie, K.W. (eds), *Petroleum Geology of North-West Europe*, Graham and Trotman, London, 633–642.

Cunningham, G.A. and Shannon, P.M. (1997) The Erris Ridge: a major geological feature in the NW Irish Offshore Basins. *Journal of the Geological Society of London*, **154**, 503–508.

Daly, J.S., Heaman, L.M., Fitzgerald, R.C., Menuge, J.F., Brewer, T.S., and Morton, A.C. (1995) Age and crustal evolution of crystalline basement in western Ireland and Rockall. *In*: Croker, P.F. and Shannon, P.M. (eds), *The Petroleum Geology of Ireland's Offshore Basins*. Geological Society, London, Special Publications, **93**, 433–434.

Daly, J.S., Tyrrell, S., Badenszki, E., Haughton, P.D.W., Shannon, P.M., and Whitehouse, M. (2008) Mesoproterozoic orthogneiss from the northern Porcupine High, offshore western Ireland. *Abstracts, 51st Irish Geological Research Meeting, University College Dublin*, 24.

Dancer, P.N., Algar, S.T., and Wilson, I.R. (1999) Structural Evolution of the Slyne Trough. *In*: Fleet, A.J. and Boldy, S.A.R. (eds), *Petroleum Geology of Northwest Europe: Proceedings of the 5th Conference*. Geological Society, London, 445–453.

Dancer, P.N., Kenyon-Roberts, S.M., Downey, J.W., Baillie, J.M., Meadows, N.S., and Maguire, K. (2005) The Corrib gas field, offshore west of Ireland. *In*: Doré, A.G. and Vining, B.A. (eds), *Petroleum Geology: North-West Europe and Global Perspectives. Proceedings of the 6th Petroleum Geology Conference*, Geological Society, London, 1035–1046.

Delanty, L.J., Whittington, R.J., and Dobson, M.R. (1981) The geology of the North Celtic Sea west of 7° Longitude. *Proceedings of the Royal Irish Academy*, **81B**, 37–51.

Franke, W. (1989) Tectonostratigraphic units in the Variscan belt of Central Europe. *In*: Dallmeyer, R.D. (ed.), *Terranes in the Circum-Atlantic Palaeozoic Orogens*. Geological Society of America, Special Papers, **230**, 221–228.

Friend, P.F., Williams, B.P.J., Ford, M., and Williams, E.A. (2000) Kinematics and dynamics of Old Red Sandstone Basins. *In*: Friend, P.F. and Williams, B.P.J. (eds), *New perspectives on the Old Red Sandstone*. Geological Society of London, Special Publications, **180**, 29–60.

Gardiner, P.R.R. and Sheridan, D.J.R. (1981) Tectonic framework of the Celtic Sea and adjacent areas with special reference to the location of the Variscan Front. *Journal of Structural Geology*, **3**, 317–331.

Gill, W.D. (1979) *Syndepositional sliding and slumping in the west Clare Namurian basin, Ireland*. Geological Survey of Ireland, Special Paper, **4**, 31pp.

Graciansky, P.C. de, Poag, W.C., Cunningham, R. *et al.* (1985) The Goban Spur transect: geologic evolution of a sediment-starved passive continental margin. *Bulletin of the Geological Society of America*, **96**, 58–76.

Heckel, P.H. and Clayton, G. (2006) The Carboniferous System: use of the new official names for the subsystems, series and stages. *Geologica Acta*, **4**, 403–407.

Higgs, K. (1983) Palynological evidence for the Carboniferous strata in two wells drilled in the Celtic Sea area. *Bulletin of the Geological Survey of Ireland*, **3**, 107–112.

Hitchen, K. (2004) The geology of the UK Hatton–Rockall margin. *Marine and Petroleum Geology*, **21**, 993–1012.

Institute of Geological Sciences (1979) Geological Survey of Northern Ireland, Isle of Man sheet 1:250,000 scale. HMSO.

Jackson, D.I., Johnson, H., and Smith, N.J.P. (1997) Stratigraphical relationships and a revised lithostratigraphical nomenclature for the Carboniferous, Permian and Triassic rocks of the offshore East Irish Sea Basin. *In*: Meadows, N.S., Trueblood, S.P., Hardman, M., and Cowan, G. (eds), *Petroleum Geology of the Irish Sea and Adjacent Areas*. Geological Society, London, Special Publications, **124**, 11–32.

Jenner, J. K. (1981) The structure and stratigraphy of the Kish Bank Basin. *In*: Illing, L.V. and Hobson, G.D. (eds), *Petroleum Geology of the Continental Shelf of North-West Europe*, Heyden and Son Ltd., London, 426–431.

Maddox, S.J., Blow, R., and Hardman, M. (1995) Hydrocarbon prospectivity of the Central Irish Sea Basin with reference to Block 42/12, offshore Ireland. *In*: Croker, P.F. and Shannon, P.M., (eds), *The Petroleum Geology of Ireland's Offshore Basins*. Geological Society, London, Special Publications, **93**, 59–77.

Max, M.D. (1978) Tectonic control of offshore sedimentary basins to the north and west of Ireland. *Journal of Petroleum Geology*, **1**, 103–110.

Miller, J.A., Roberts, D.A., and Dearnley, R. (1973) Precambrian rocks drilled from the Rockall Bank. *Nature*, **244**, 21–23.

Morton, A.C. (1984) Heavy minerals from Palaeogene sediments. Deep Sea Drilling Project Leg 87: their bearing on stratigraphy, sediment provenance and the evolution of the North Atlantic. *Initial Reports of the Deep Sea Drilling Project*, **81**, 653–661.

Morton, A.C. and Taylor, P.N. (1991) Geochemical and isotopic constraints on the nature and age of basement rocks from the Rockall Bank, NE Atlantic. *Journal of the Geological Society of London*, **148**, 631–634.

Morton, A.C., Hitchen, K., Fanning, C.M., Yaxley, G., Johnson, H., and Ritchie, J.D. (2009) Detrital zircon age constraints on the provenance of sandstones on Hatton Bank and Edoras Bank, NE Atlantic. *Journal of the Geological Society, London*, **166,** 137–146.

Naylor, D. (1998) Irish shorelines through geological time. Royal Dublin Society: *Occasional Papers in Irish Science and Technology*, **17**, 20pp.

Naylor, D. and Clayton, G. (2000) Palynological and maturation data and their bearing on Irish post-Variscan palaeogeography. *Irish Journal of Earth Sciences*, **18**, 33–39.

Naylor, D. and Shannon, P.M. (2009) Geology of offshore Ireland. *In*: Holland, C.H. and Sanders, I.S. (eds), *The Geology of Ireland*. Second Edition. Dunedin Academic Press, Edinburgh, 405–460.

Naylor, D., Haughey, N., Clayton, G., and Graham, J.R. (1993) The Kish Bank Basin, offshore Ireland. *In*: Parker, J.R. (ed.), *Petroleum Geology of North-west Europe: Proceedings of the 4th Conference*. Geological Society, London, 845–855.

Naylor, D., Shannon, P.M., and Murphy, N. (1999) *Irish Rockall Basin region – a standard structural nomenclature system*. Petroleum Affairs Division, Dublin, Special Publication **1/99**, 42pp.

Naylor, D., Shannon, P.M., and Murphy, N. (2002) *Porcupine–Goban region – a standard structural nomenclature system.* Petroleum Affairs Division, Dublin, Special Publication **1/02**, 65pp.

Newman, P. (1999) The geology and hydrocarbon potential of the Peel and Solway Basins, East Irish Sea. *Journal of Petroleum Geology*, **22**, 305–324.

Owens, B., Riley, N.J., and Calver, M.A. (1985) Boundary stratotypes and new stage names for the lower and middle Westphalian sequences in Britain. *Compte Rendu 10ème Congrès International de Stratigraphie et de Géologie Carbonifère, Madrid, 1983*, **4**, 461–472.

PESGB (2005) *Structural framework of the North Sea and Atlantic margin: 1:750,000 scale map, 2005 edition.* Petroleum Exploration Society of Great Britain, London.

Ramsbottom, W.H.C., Calver, M.A., Eagar, R.M.C., Hodson, F., Holliday, F., Stubblefield, C.J., and Wilson, R.B. (1978) *A correlation of Silesian rocks in the British Isles.* Geological Society, London, Special Report, **10**, 81pp.

Riddihough, R.P. and Max, M.D. (1976) A geological framework for the continental margin to the west of Ireland. *Geological Journal*, **11**, 109–120.

Roberts, D.G., Montadert, L., and Searle, R.C. (1979) The western Rockall Plateau – stratigraphy and structural evolution. *Initial Reports of the Deep Sea Drilling Project* **48**, U.S.G.P.O., Washington D.C., 1061–1088.

Robeson, D., Burnett, R.D., and Clayton, G. (1988) The Upper Palaeozoic geology of the Porcupine, Erris and Donegal Basins, offshore Ireland. *Irish Journal of Earth Sciences*, **9**, 153–175.

Robinson, K.W., Shannon, P.M., and Young, D.G.G. (1981) The Fastnet Basin: an integrated analysis. *In*: Illing, L.G. and Hobson, G.D. (eds), *Petroleum Geology of the Continental Shelf of North-West Europe.* Heyden and Son Ltd, London, 444–454.

Sevastopulo, G.D. (2009) Carboniferous: Serpukhovian and Pennsylvanian. *In*: Holland, C.H. and Sanders, I.S. (eds), *The Geology of Ireland.* Second Edition. Dunedin Academic Press, Edinburgh, 269–294.

Sevastopulo, G.D. and Wyse Jackson, P.N. (2009) Carboniferous: Mississippian (Tournaisian and Viséan). *In*: Holland, C.H. and Sanders, I.S. (eds), *The Geology of Ireland.* Second Edition. Dunedin Academic Press, Edinburgh, 215–268.

Shelton, R. (1995) Mesozoic basin evolution of the North Channel: preliminary results. *In*: Croker, P.F. and Shannon, P.M. (eds), *The Petroleum Geology of Ireland's Offshore Basins.* Geological Society, London, Special Publications, **93**, 7–20.

Tate, M.P. and Dobson, M.R. (1989a) Late Permian to early Mesozoic rifting and sedimentation offshore NW Ireland. *Marine and Petroleum Geology*, **6**, 49–59.

Tate, M.P. and Dobson, M.R. (1989b) Pre-Mesozoic geology of the western and north-western Irish continental shelf. *Journal of the Geological Society of London*, **146**, 229–240.

Tyrrell, S., Souders, A.K., Haughton, P.D.W., Daly, J.S., and Shannon, P.M. (2010) Sedimentology, sandstone provenance and palaeodrainage on the eastern Rockall Basin margin: evidence from the Pb isotopic composition of detrital K-feldspar *In*: Vining, B.A. and Pickering, S.C. (eds), *Petroleum Geology: From Mature Basins to New Frontiers – Proceedings of the 7th Petroleum Geology Conference.* Geological Society, London, 937-952.

Ziegler, P.A. (1981) Evolution of Sedimentary Basins in North-West Europe. *In*: Illing, L.V. and Hobson G.D. (eds), *Petroleum Geology of the Continental Shelf of North-West Europe.* Heyden & Son Ltd, London, 3–39.

Ziegler, P.A. (1982) *Geological Atlas of Western and Central Europe. 1st Edition.* Shell Internationale Petroleum Maatschappij, B.V., 130pp.

Ziegler, P.A. (1990) *Geological Atlas of Western and Central Europe. 2nd Edition.* Shell Internationale Petroleum Maatschappij B.V. 239pp.

Chapter 6

Permian and Triassic stratigraphy

In Early Permian time, the welding of Laurussia and Gondwanaland along the Variscan (Hercynian) megasuture was complete, resulting in the new supercontinent of Pangaea. The European region during this period was in tropical latitudes. Continued seafloor spreading in the Pacific was accompanied by rotation of the whole supercontinent (Ziegler, 1982). The resultant stresses, together with the thermal instability that built up rapidly beneath the supercontinent, gave rise to orogenic collapse depocentres adjacent to the eroding Variscan highlands, deposition in broad basins associated with wide rifting processes, reactivation of earlier fractures and then to the initiation of isolated rift systems. Although none of the rifts developed into a new spreading axis or to the production of new ocean floor, some of the incipient rift systems would later become larger and more elongate and would become important precursors to events later in the Mesozoic.

In the Irish context, deposition on the post-Variscan landscape took place in continental conditions within rift basins of various sizes, often developed along reactivated NE–SW Caledonian features (Shannon, 1991; Johnston *et al.*, 2001). Permo-Triassic rocks were deposited predominantly in non-marine red-bed facies. Seismic evidence points towards rifting in Early and Late Triassic times in different parts of the region (Shannon and MacTiernan, 1993; Musgrove *et al.*, 1995; Štolfová and Shannon, 2009).

Offshore distribution of Permian and Triassic rocks

Unequivocal Permian strata have not been encountered in offshore wells in the Celtic Sea or Fastnet basins (Figure 2.15; Naylor and Shannon, 1982; Robinson *et al.*, 1981), within which a number of wells have penetrated continental Triassic deposits resting unconformably on Palaeozoic basement. The existence of possible small pockets of Permian strata in the South Celtic Sea Basin and in other Celtic Sea basins has been suggested

by Petrie *et al.* (1989), Tappin *et al.* (1994), and Shannon (1995), largely from seismic evidence (*see* Chapter 10). Onshore along the north coast of the Bristol Channel Upper Triassic rocks rest unconformably on deformed Devonian and Carboniferous metasediments. This is also the situation in the Bristol Channel Basin (Kamerling, 1979). However, UK well 93/6-1 on the south flank of the South Celtic Sea Basin in UK waters penetrated 70 m of undated beds that may be Permian in age, resting unconformably on the Palaeozoic (Tappin *et al.*, 1994). Further west, in the small Cockburn Basin, a wedge of Permian sediments is postulated with angular contact beneath the base Triassic unconformity (Smith, 1995), while onshore in the south of England Permian aeolian and fluvial strata rest on deformed Variscan basement and underlie thick fluvial-dominant sandstones of the Triassic Sherwood Sandstone Group (McKie and Williams, 2009). However, even if undrilled pockets of Permian strata exist on the shelf south of Ireland, they are clearly of limited extent. The paucity of Permian strata in the region is interpreted as reflecting the intermontane and topographically elevated setting of the region following the Variscan orogeny, with consequent poor sediment preservation potential. In contrast, Triassic strata are well documented in the Celtic Sea region. In both the North and South Celtic Sea basins, there is the classic subdivision of the Triassic into a lower fluvial sand-dominant sequence (Sherwood Sandstone Group) and an upper red-bed mudrock and marly succession with variable amounts of evaporites (Mercia Mudstone Group), the two locally separated by a thin calcareous unit. For example, the Esso 56/20-1 well (Colin *et al.*, 1981; Figures 3.6 and 7.1) near the northern margin of the North Celtic Sea Basin penetrated a 600 m thick Triassic succession including an upper anhydritic 450 m thick Mercia Mudstone sequence. The same is also true further west in the Fastnet Basin (Robinson *et al.*, 1981), where maximum Triassic thicknesses of 600–700 m are indicated from seismic and well data. The Sherwood Sandstone Group is probably restricted to the main topographic or fault-bounded depocentres, whilst the overlying Mercia Mudstone Group extends basinwide. Thick salts are generally absent from the Fastnet and North Celtic Sea basins, except for a narrow zone along the southern margin of the latter, but are dramatically developed in the South Celtic Sea and St George's Channel basins, (Tappin *et al.*, 1994; Van Hoorn, 1987). The Mercia Mudstone Group in the Celtic Sea basins passes conformably upwards through a transitional sequence into a fully marine Jurassic succession. This marine transgression was of Rhaetian to Hettangian age. Non-marine and restricted marine facies were dominant in the Rhaetian (Millson, 1987) and these environments persisted into earliest Jurassic time in marginal and uplifted areas. In the Fastnet Basin, Triassic red-beds are succeeded by Rhaetian marine limestones, marls and mudstones broadly similar to those seen in outcrops onshore in the Bristol Channel region.

In the southern part of the Irish Sea, drilling on the southern margin of the Cardigan Bay Basin (Barr *et al.*, 1981) has revealed a Triassic succession comparable to that in the East Irish Sea Basin and the English Midlands, with the Sherwood Sandstone Group overlain by saliferous Mercia Mudstone Group strata. Northwards, the Triassic onlaps the St Tudwal's Arch, a positive fault-controlled Lower Palaeozoic and Precambrian element extending out from the Welsh Massif. The Rhaetic follow conformably on the Mercia Mudstone Group, and in the Cardigan Bay wells is about 50 m thick, consisting of red-brown claystones with occasional sandstones. The Mochras borehole on the Welsh coastline (Woodland, 1971) penetrated 48 m of dolomitic limestones above dolomitic sandstones with red-brown marl bands (Figure 6.7). The lower part of this late Triassic–Rhaetian unit is inferred to be terrestrial in origin, with increasing marine influence in the upper beds reflecting the late Triassic–early Jurassic rapid regional marine transgression.

Offshore wells in the St George's Channel Basin encountered thicknesses of up to 2000 m of Triassic sediments (Tappin *et al.*,1994). The Texaco/HGB 103/2-1 well on the southern margin of the basin penetrated 249 m of sandstones with interbedded anhydritic mud layers, resting unconformably on Carboniferous Barren Red Measures. The overlying Mercia Mudstone Group (1860 m) is dominated by red-brown mudstones, frequently anhydritic, with a saliferous middle unit. Evidence for salt diapirism is seen on seismic profiles.

Permo-Triassic strata are interpreted as extending throughout both the Goban Spur Basin and Goban Graben depocentres (Figures 2.15 and 3.6; Naylor *et al.*, 2002), and Triassic red-beds are succeeded with apparent conformity by Rhaetian marine strata illustrative of the regional marine transgression. Rifting of the Variscan basement in the Goban Spur region probably began in mid-Triassic time (Cook, 1987). The overlying succession, interpreted from seismic evidence as marly and evaporitic, is thought to have acted as a detachment zone, so that a deformed upper sequence contrasts with the relatively simple faulted horst style of the basement.

Definitive Permian strata have not been encountered by drilling in the Porcupine Basin region (Figures 2.15 and 3.6). However, two wells (Shell 34/19-1 and Deminex 34/15-1) on the western margin of the basin recorded tentatively dated Autunian mudstones with anhydrite (Croker and Shannon, 1987) that could represent earliest Permian strata. The Triassic of the Porcupine region is also little known and poorly dated, with only two wells (BP 26/22-lA and Gulf 26/21-1), both in the North Porcupine Basin, reportedly penetrating rocks of this age (Figure 7.5), and in one the sequence rests on late Pennsylvanian strata with a presumed unconformity. However, the Britoil 35/19-1 well in the Porcupine Basin also recorded salt of undated age and with unclear stratigraphic relationships. In the BP 26/22-1A well, massive shoreline sandstones are overlain by glauconitic offshore stacked bar sands, and these in turn by evaporitic shales (Croker and Shannon, 1987). These authors suggested, from seismic evidence, that Triassic sediments may be preserved in small rift basins along the southwest and possibly the southeast margins of the Porcupine Basin, and stressed the importance of reactivated Caledonian lineaments in controlling both Triassic and Lower Jurassic sedimentation. Seismic evidence suggests that more than 2 km of Triassic rocks may occur in other parts of the basin (Naylor *et al.*, 2002), while Johnston *et al.* (2001) interpreted Triassic strata as being widespread, if poorly imaged, on seismic sections in the southern part of the Porcupine Basin. Where the Triassic succession has been drilled in the Porcupine it is conformably succeeded by a limestone–shale succession of Rhaetian age.

In the Slyne Basin (Dancer *et al.*, 1999) the Enterprise 27/5-1 well drilled a massive evaporite sequence beneath the Triassic Sherwood Sandstone, that is assumed to be Zechstein equivalent. In the area of the Corrib Field in the same basin the presence of Zechstein halite is interpreted largely on the seismic sections (Dancer *et al.*, 2005; Corcoran and Mecklenburgh, 2005). Further north along the Atlantic margin (Figure 3.7), the Amoco 19/5-1 well in the Erris Basin penetrated Zechstein-equivalent marine mudstones, dolomites and anhydrites following unconformably on the Carboniferous, while the Amoco 12/13-lA well encountered similar Upper Permian rocks (Tate and Dobson, 1989a) resting on red-bed sandstones and shales of possible Lower Permian age (Murphy and Croker, 1992), although these could be Autunian in age. On seismic evidence, a base Upper Permian–Triassic sequence boundary is recognised throughout the Erris Basin (Chapman *et al.*, 1999). Little is known of the latest Palaeozoic and earliest Mesozoic history of the Rockall region (Figure 2.15). Shannon *et al.* (1995) speculated that Permo-Triassic sedimentation probably extended into the Rockall Basin. Recently, Tyrrell *et al.* (2010) reported the presence of approximately 225 m of Permo-Triassic sandstone-prone strata in the Shell 12/2-1z well on the eastern margin of the Rockall Basin. These are predominantly Lower Permian (Asselian) aeolian and fluvial sandstones and thin mudstones with a thin upper sandy unit with a poorly constrained Permo-Triassic age. Naylor *et al.* (1999), interpreting seismic data, showed a Permo-Triassic sequence within the string of perched Mesozoic basins along the western margin of the Porcupine High. Štolfová and Shannon (2009) suggested that these strata, which are generally thick and parallel bedded, are likely to have been deposited over a broader region than the current fault-preserved basins. They indicated deposition of sandstones and evaporites in a series of broad, wide-rift basins. Permo-Triassic to Lower Jurassic sequences, where locally preserved in the Rockall Basin, are therefore likely to comprise continental to shallow-marine sandstones, evaporites and limestones deposited during the initial rift phase marking the breakup of the Pangean supercontinent.

The Triassic succession within the northwest offshore basins (Figures 2.15 and 3.7) is known from a number of wells: Elf 27/13-1 (Rhaetian only: Scotchman and Thomas, 1995), Enterprise 27/5-1 and Enterprise 18/20-1 wells of the Slyne Basin (Trueblood, 1992; Dancer *et al.*, 1999), and in the Amoco 19/5-1 and Amoco 12/13-lA wells in the Erris Basin (Tate and Dobson, 1989a; Chapman *et al.*, 1999). In the Corrib Gasfield at the northern end of the Slyne Basin (Dancer *et al.*, 2005; Corcoran and Mecklenburg, 2005) a sandstone-dominant Lower Triassic Sherwood Sandstone Group equivalent (400–420 m) is overlain by a Mercia Mudstone Group comprising saliferous and anhydritic mudstones with a thick basal salt section. In general in these basins, limestones become more frequent upwards through the

succession, suggesting that nearshore sabkha-lagoonal environments were succeeded by deepening conditions as a consequence of the Rhaetian marine transgression. Volcaniclastic beds near the top of the sequence demonstrate early Mesozoic volcanism in the region (Tate and Dobson, 1988).

The Loch Indaal Basin (Figure 6.1) lies off the north coast of Ireland between the positive elements of the Islay–Donegal Platform and Middle Bank. The basin contains a Triassic continental sequence more than 2 km thick, but Permian rocks are unproven. Seabed boreholes in the area (Evans *et al.*, 1980) yielded probable Upper Triassic gypsiferous marls. In the northeastern portion of the basin there is a transition from the Mercia Mudstone Group to ammonitic black mudstones of the overlying Penarth Group (Westbury Formation; Rhaetian to Early Jurassic), reflecting the ubiquitous latest Triassic to earliest Jurassic rapid marine transgression.

Permian and Mesozoic rocks are of limited distribution onshore in Ireland, except in the northeast of the island. In this area (Figure 6.2) there is a more or less contiguous region of Mesozoic strata with intervening fault-controlled positive elements separating individual basinal depocentres. The term 'Ulster Basin' has sometimes been used for the whole region. The Mesozoic strata extend offshore into the North Channel, and Palaeogene basalts cover much of the onshore region, with the Mesozoic at subcrop beneath. The best outcrops of the sequence are along the coastline and at the margins of the plateau basalts. The sequence beneath the basalt cover is known through a number of boreholes and wells, the most important of which are listed in Table 3.1 and shown on Figure 3.2. The Permo-Triassic to Palaeogene subsurface geology of the Ulster Basin has been outlined by many authors, including Wilson, 1972; Thompson, 1979; Illing and Griffith, 1986; McCann, 1988, 1990; McCaffrey and McCann, 1992; and Naylor *et al.*, 2003. A number of slightly conflicting basin names have been used for the depositional centres in the region. To avoid confusion we use here the names recently employed by the Geological Survey of Northern Ireland (e.g. Johnston, 2004).

The major accumulations of post-Carboniferous sediments in northeast Ireland are within two fault-bounded regions that are separated by the positive Dalradian metamorphic element of the Highland Border Ridge and the Highland Boundary Fault Zone (Figure 6.2). North of this feature is the Rathlin Basin, that extends offshore. To the south, the Larne and Lough Neagh

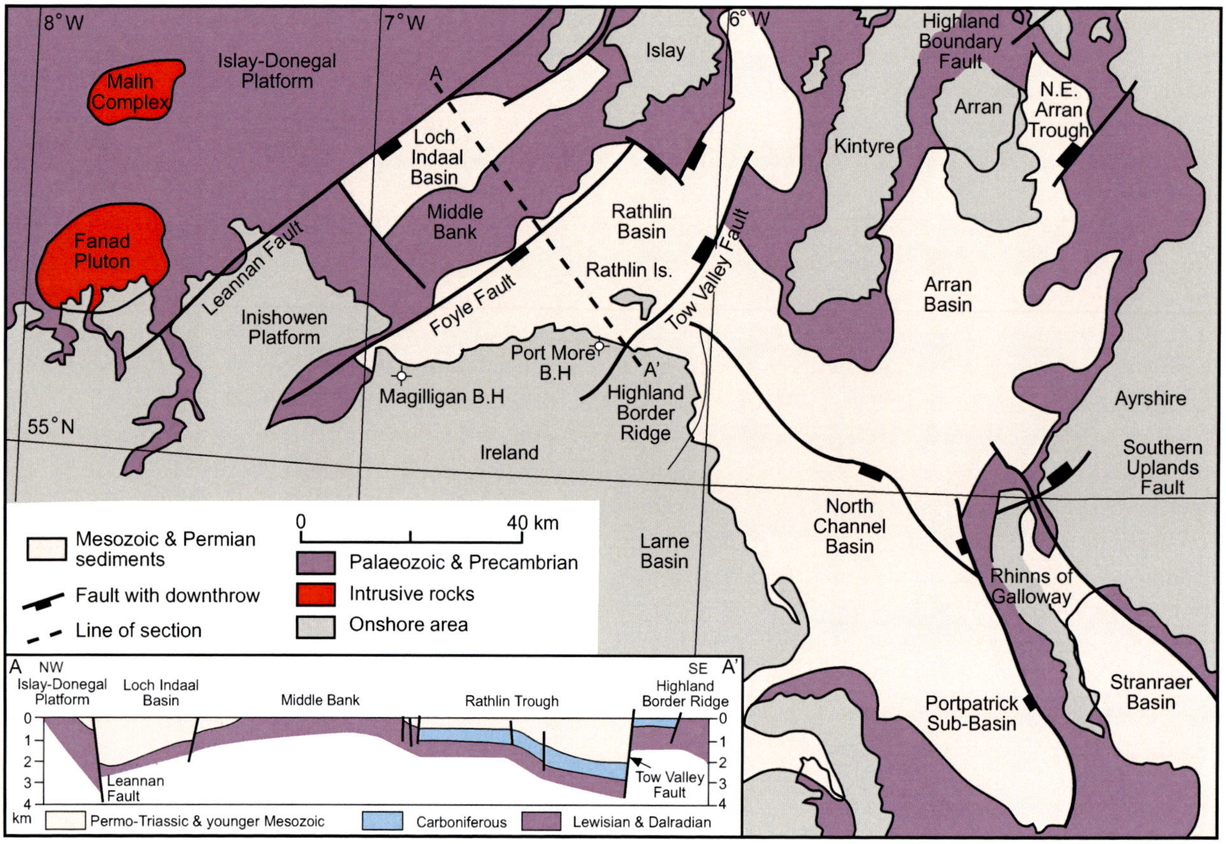

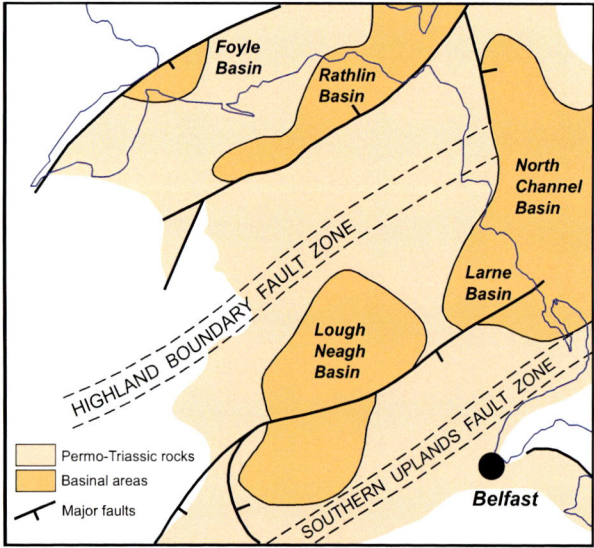

Figure 6.1. Structural elements between southwest Scotland and northeast Ireland (based on Evans et al., 1980; Naylor and Shannon, 1982; Maddox et al., 1997). Inset: Diagrammatic cross-section of the Loch Indall and Rathlin basins (adapted from Evans *et al.*, 1980).

Figure 6.2. Main structural elements and Permo-Triassic basins in northeast Ireland (modified after Johnston, 2004).

basins extend from the Highland Border Ridge southwards to the line of the Southern Upland Fault along the northern margin of the Ordovician and Silurian rocks of the Longford-Down Massif. Inland the Mesozoic

rocks extend westwards from the North Channel to the western margin of Lough Neagh. The Larne and Lough Neagh basins were formerly considered to comprise a single entity, with a northeast–southwest axial orientation (Thompson, 1979). More recent gravity modeling (Reay, 2004) has indicated two distinct depositional centres, albeit with stratigraphic continuity between the two.

On the north coast, the Rathlin Basin is contained between the Tow Valley Fault, forming the northern margin of the Highland Border Ridge, and the Foyle Fault in the northwest. In the onshore portion of the Rathlin Basin, southeast of Middle Bank and the Foyle Fault and near the western basin margin, the Magilligan borehole (Figure 6.1) penetrated a Lower Liassic to Sherwood Sandstone Group sequence (976.3 m) resting on Carboniferous rocks with thin coals. Further east in the same basin, a borehole at Port More was in probable Permian sandstones, beneath a thick Triassic sequence, when terminated at 1897 m (Table 6.1).

The onshore Permo-Triassic strata are of continental facies and extend offshore to form the bulk of the 2.4 km

Table 6.1. Stratigraphy of the Port More Borehole (Wilson and Manning, 1978)

Height above OD: 103 m	Top (m)	Thickness (m)
Superficial deposits	0.0	1.22
Tertiary		
Lower Basalts	1.22	75.41
Clay-with-flints	76.63	0.58
Cretaceous		
Ulster White Limestone Formation	77.21	91.14
Jurassic		
Lower Lias	168.35	269.45
Sill (Tertiary)	437.80	222.90
Triassic		
Penarth Group (Westbury Fmn.)	660.70	4.07
Mercia Mudstone Group	664.77	652.55
Sherwood Sandstone Group	1317.32	513.00
Permian		
Enler Group	1830.32	66.37 penetrated
Total depth	1896.69	

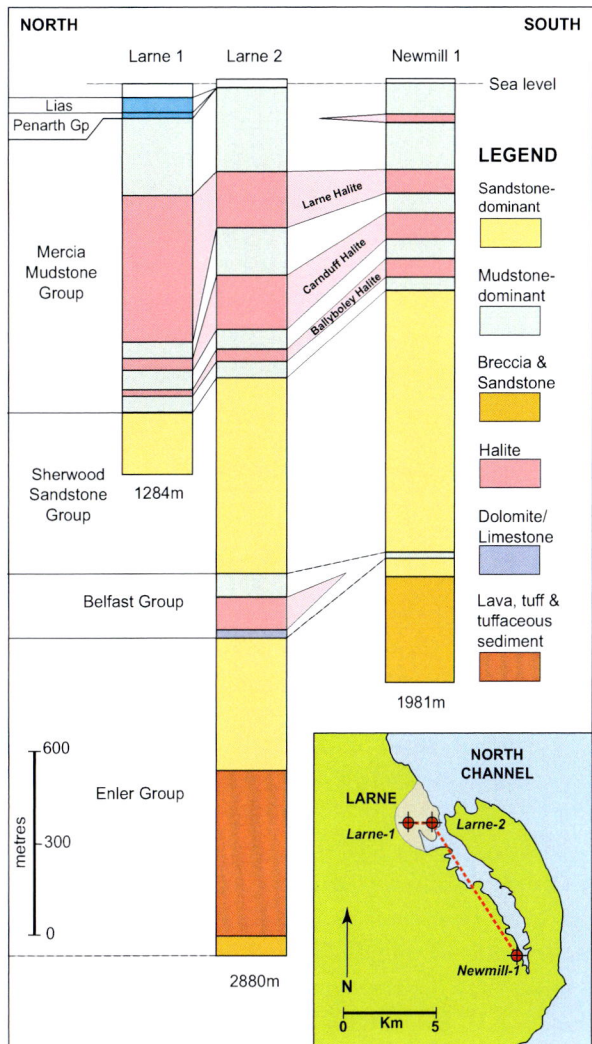

Figure 6.3. Larne Basin: stratigraphy encountered in deep tests. LH – Larne Halite; CH – Carnduff Halite; BH – Ballyboley Halite. Based on Manning and Wilson (1975), Penn (1981).

thick fill of the offshore portion of the Rathlin Basin. Seismic data indicate that the thickest development is against the Tow Valley bounding fault. Seabed boreholes in the area (Evans *et al.*, 1980) demonstrated gypsiferous marls, presumed to belong to the Mercia Mudstone Group, succeeded by Rhaetian marls and mudstones. A small, undrilled depositional centre is indicated by gravity and magnetic modeling at the southern end of the Foyle Fault at Lough Foyle (Reay, 2004) and has been named the Foyle Basin (Figure 6.1).

Drilling onshore in the Larne Basin, south of the Highland Border Ridge (*see* Wilson, 1972, Illing and Griffith, 1986 for summary), has demonstrated more than 3000 m of post-Carboniferous strata in the deeper part of the basin near the coast, thinning towards the basin margins (Smith, 1986). The Newmill-1 and Larne-2 wells in the central portion of the Larne Basin (Figure 6.3) failed to penetrate the full Permian section and terminated in Lower Permian conglomerates and breccias (Penn, 1981; Penn *et al.*, 1983; McCann, 1990). Larne-2 encountered more than 1000 m of Lower Permian strata, including a basal section (617 m thick) of volcanic tuffs,

breccias and lava flows, demonstrating an extension of the Scottish volcanic province into this region. Continental red-brown mudstones and major halite horizons more than 200 m thick characterise the Upper Permian. The overlying Triassic sequence in the borehole (1599 m) is divisible into the standard Sherwood Sandstone and Mercia Mudstone groups (Warrington *et al.*, 1980; Penn, 1981; Parnell, 1992), with three halite developments within the Mercia Mudstone Group. Northwards in the basin, Sherwood Sandstone Group strata overstep onto the long-standing positive element of the Highland Border Ridge. Mercia Mudstone Group strata, 976 m thick in the Larne-1 borehole in the basin centre, also thin markedly towards the basin margin.

The halite units demonstrated by the Larne and Newmill drilling represent the only onshore development of halite on the island of Ireland. The southern margin of the basin lies just south of Belfast Lough, where drilling shows a thickness of about 300 m of Permo-Triassic rocks, lacking salt beds. North of Belfast Lough the section thickens rapidly towards the basin centre around Larne. Immediately north of Belfast Lough the halite units crop out at the surface, and for several kilometres towards Larne Lough they are rarely deeper than 300 m, due to faulting and folding. Shallower sections in this area have been exploited for salt intermittently and at different places since 1853, by shaft, solution mining, bell mining, and currently by adit. Halite bed thicknesses of 10–25 m are reported, with shaft depths in the order of 140 m (Wilson, 1974).

The Completion Report for the Marathon/Shell Newmill-1 well on the southwest shore of Larne Lough (Marathon/Shell, 1971) records the following four salt zones within the Mercia Mudstone Group (Figure 6.3):

- ◆ Salt Zone 1 — 360–441 feet (109.7–134.4 m)
- ◆ Salt Zone 2 — 968–1219 feet (295.0–371.6 m)
- ◆ Salt Zone 3 — 1437–1720 feet (438.0–524.3 m)
- ◆ Salt Zone 4 — 1875–2120 feet (571.5–646.2 m)

Red brown silty shales are frequently interbedded with the salt, although massive unbroken salt beds up to 23 m thick also occur. Lateral variation in thickness of the halite units in the basin is demonstrated by the changes between the closely spaced Larne boreholes (Figure 6.3). The Ballytober-1 well, northeast of Larne, was drilled in 1991 on a seismically-defined structure. Details of the stratigraphy in the well have not been published, but thick halite was not encountered (Shelton, 1997), perhaps demonstrating the limited inland development of halite in the basin.

Farther west, the stratigraphy of the Lough Neagh Basin has been revealed by deep drilling (Figure 6.4) – deep boreholes at Ballymacilroy and Langford Lodge, and two closely-spaced petroleum exploration wells, Annaghmore-1 and Ballynamullan-1. Modelling of the regional gravity data (Reay, 2004) suggests that the basin is elongated north–south and centred on Lough Neagh and may contain in excess of 4 km of post-Palaeozoic sediment. The Ballymacilroy hole (2272 m), north of the Lough, drilled a normal Triassic and Permian succession and terminated in igneous rock of probable Tertiary age (54 Ma: Griffith, 1983). Naylor *et al.* (2003) reported on the stratigraphy encountered in the two exploration wells drilled on the northwest

shore of Lough Neagh. Annaghmore-1 was drilled to a depth of 5100 ft (1554.5 m) and Ballynamullan-1, which was a deviated hole, reached a vertical depth of 4500 ft (1371.6 m). Both wells drilled through the Antrim Lava Group and the thin Ulster White Limestone Formation into a Permo-Triassic red-bed sequence comprising the Mercia Mudstone, Sherwood Sandstone (Triassic), Belfast and Enler (Permian) Groups. The Annaghmore-1 well drilled, and was terminated in, an unexpectedly thick (542.5 m) red-bed section beneath the base of the Belfast Group that lacked definitive palynomorphs, but which is thought to be Permian in age. The preserved Mercia Mudstone Group sequence in the Annaghmore-1 and Ballynamullan-1 wells is similar to the lower part of the unit in Ballymacilroy-1 to the north. No halite units were encountered at either locality. Of particular interest is the 1956 Langford Lodge borehole (Figure 6.4) on the eastern shore of Lough Neagh (Manning *et al.*, 1970). This borehole penetrated directly into Lower Palaeozoic basement beneath the Sherwood Sandstone Group at 1452.7 m (TD at 1524.6 m in the same rocks). There was no Carboniferous section beneath the Triassic red-beds, although these contain derived Pennsylvanian spores. The geophysical data suggest that this well was drilled on a positive element within the basin. No halite was developed within the Mercia Mudstone Group. Thus the thick halite units penetrated by the Larne and Newmill-1 wells in the Larne Basin appear to be either substantially thinner or absent elsewhere (e.g. Ballytober-1, Ballymacilroy, Annaghmore-1, Ballynamullan-1, Langford Lodge). If halites are developed within the Lough Neagh Basin, it is likely that they underlie Lough Neagh itself.

The Mercia Mudstone Group in the Rathlin, Lough Neagh and Larne basins is succeeded by the thin Penarth Group (Rhaetian) that is variably preserved on the pre-Cretaceous erosion surface. In the Magilligan borehole the Penarth Group is 25.4 m thick, whilst 20.0 m of section are preserved in Larne-1. The basal contact on the Larne foreshore section is disconformable and the sequence comprises grey, black and brown mudrocks with thin limestones, containing bivalves and fish debris. Deposition took place in fluctuating shallow marine conditions at the onset of transgression, with periods of emergence. As water depths increased, the Penarth Group was succeeded by the Waterloo Mudstone Formation. At Larne, the basal 10 m of the Waterloo Mudstone Formation (known as the 'Pre-*planorbis* beds') are Rhaetian in age, while the upper part of the formation belongs to the Hettangian Stage (Mitchell, 2004a).

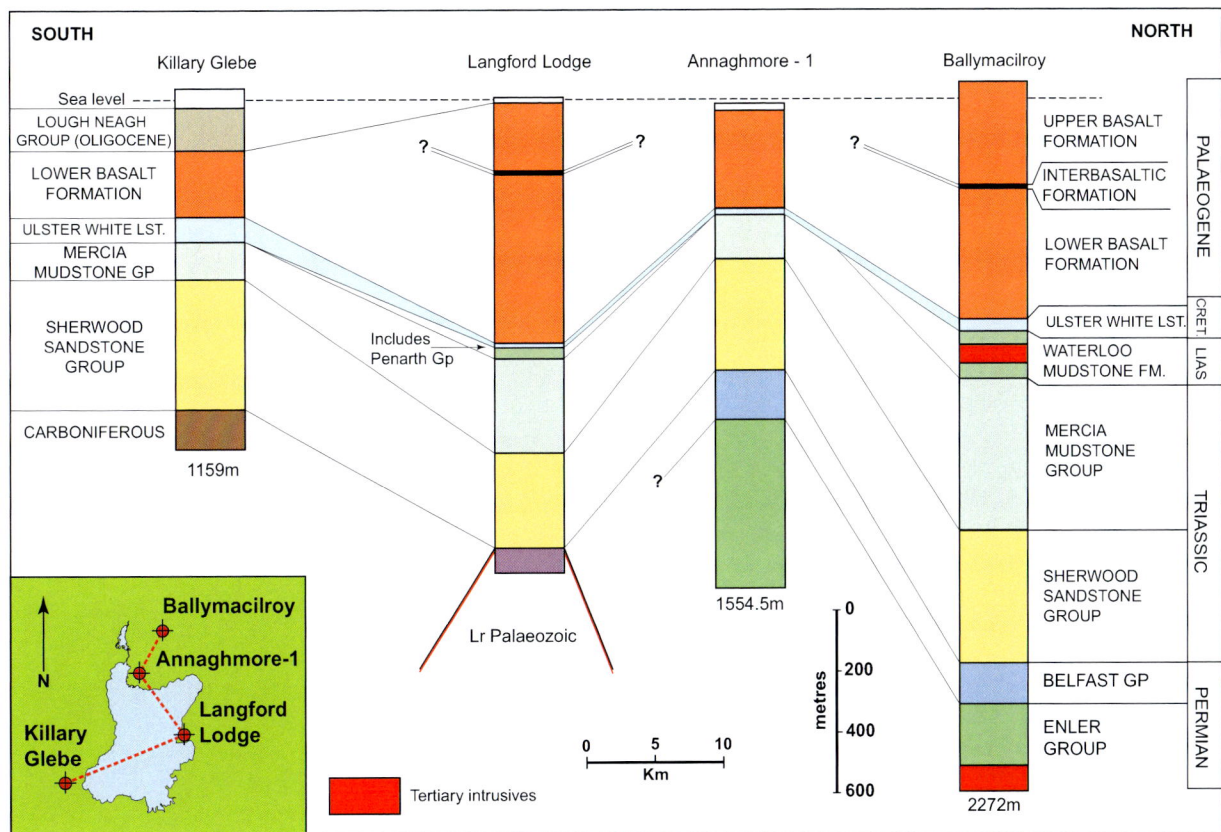

Figure 6.4. Lough Neagh Basin: correlation of stratigraphy encountered in wells and boreholes. Correlation lines are not indicative of formation depths between the wells and boreholes.

The Permo-Triassic sequence penetrated at the coast by the Larne-2 borehole is seen on seismic records to extend seawards into the North Channel (Figure 6.1), where it is as yet undrilled (Maddox *et al.*, 1997; Shelton, 1997). Seismic and gravity (Reay, 2004) data indicate that the onshore and offshore developments are part of a single basin, but to avoid confusion with the published literature, the offshore section will be referred to as the North Channel Basin. The Permo-Triassic succession in the North Channel Basin is up to 4 km thick, thinning against the Highland Border Ridge to the north and also by overlap across a series of fault blocks towards the eastern basin margin (Shelton, 1995). A succession similar to that of the Larne borehole might be anticipated. The southwards extending Portpatrick Basin is a half-graben of reversed polarity lying south of the line of the Southern Uplands Fault, with a thick Permian and Triassic section developed in the east against the boundary fault along the Lower Palaeozoic positive element of The Rhinns of Galloway. Possible halokinetic structures are evident on seismic records, which is not surprising given the thick halite units penetrated by onshore drilling.

Further south in the Irish Sea, the Permo-Triassic of the Solway and Peel basins (Figure 6.5) has been penetrated in three wells – Elf 112/19-1, Elf 111/25-1 and Elf 111/29-1 (Newman, 1999). Upper Permian mudrocks and evaporites (up to 162 m) rest on Carboniferous strata, except in well Elf 111/25-1, in which 62 m of Lower Permian sandstones (Collyhurst Sandstone Formation) intervene. The Permian is succeeded by a typical red-bed Lower Triassic Sherwood Sandstone Group sequence that thins southwestwards, from 746 m in the Solway Basin (well Elf 111/25-1) to 585 m thick in the Elf 111/29-1 well of the Peel Basin. The same southwestward thinning is also present in the Upper Triassic Mercia Mudstone Group (724 m to 174 m), in which anhydritic claystones with halites dominate.

In the Kish Bank Basin (Figure 6.5), the Carboniferous is overlain unconformably by 2000 m of Permo-Triassic rocks (Jenner, 1981; Naylor *et al.*, 1993; Dunford *et al.*, 2001). Four wells have been drilled within the basin (Figure 6.6). The Permian Collyhurst Sandstone (100 m drilled) was penetrated only at the base of the Fina 33/17-1 well where it consisted entirely of friable fine to

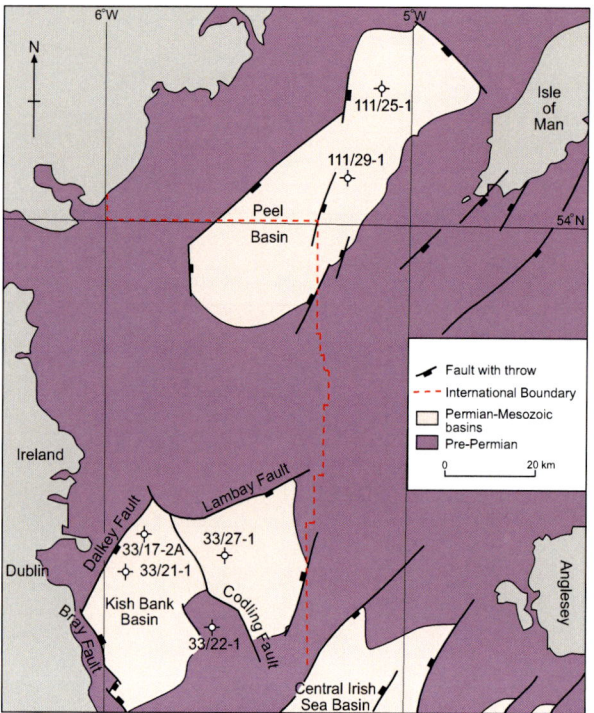

Figure 6.5. Permian–Mesozoic basins in the northern part of the Irish Sea.

coarse sandstones. It is overlain by a thin Upper Permian interval of red-brown claystone, slightly calcareous, with siltstone and sandstone interbeds. The Triassic succession penetrated by the Fina 33/17-1 well (Naylor *et al.*, 1993) consisted of the Sherwood Sandstone Group (1035 m) overlain by the Mercia Mudstone Group (535.2 m). The lowest unit of the Sherwood Sandstone Group is the St Bees Sandstone Formation, dominated by red, very fine- to fine-grained, moderately well sorted sandstones. The overlying Ormskirk Sandstone Formation is also pre-dominantly sandy, with claystone intercalations. The se-quence is comparable to that of the East Irish Sea Basin (Dunford *et al.*, 2001). The depositional environment of the overlying Mercia Mudstone Group was continental arid to semi-arid, with alternations between fluvial, la-custrine, evaporitic or aeolian conditions. The sequence is dominated by red-brown claystones, occasionally evaporitic or with minor sandstones, separated by halite developments that comprise up to 30% of the sequence in the west of the basin (Dunford *et al.*, 2001). The Mercia Mudstone Group is 535 m thick at the Fina 33/17-1

Figure 6.6. Comparative stratigraphy of the four exploration wells drilled in the Kish Bank Basin (after Jenner, 1981; Naylor *et al.*, 1993; Dunford *et al.*, 2001).

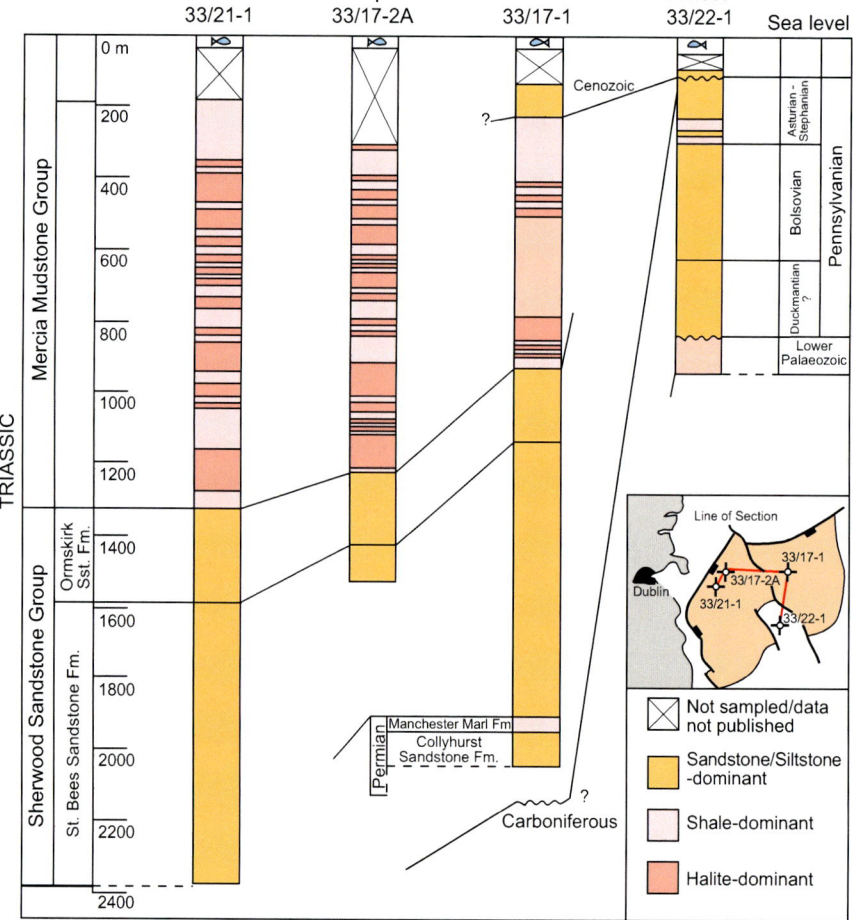

location but much thicker (1122 m) in the Shell 33/21-1 well, in part due to the presence of thicker halite units. On the basis of palynology, most of the Mercia Mudstone Group is assigned to the Scythian (Naylor *et al.*, 1993).

The Kish Bank Basin is bounded to the south by the Mid-Irish Sea High (Figure 6.7), to the south of which lies the Central Irish Sea Basin (CISB), which is strongly influenced by intra-basinal NNE–SSW and NE–SW faults. The St Tudwal's Arch, a caledonoid-trending extension from the Welsh coastline, in turn separates the Central Irish Sea Basin from the Cardigan Bay Basin, which merges southwestwards into the St George's Channel Basin (Figure 6.7). Exploration wells within the southern portion of the CISB show that Lower Triassic rocks (Sherwood Sandstone Group) rest unconformably upon deformed and reddened Carboniferous strata. Permian strata have not been encountered, but may exist in the undrilled depocentre in the northeast of the basin (Maddox *et al.*, 1995). The succession on St Tudwal's

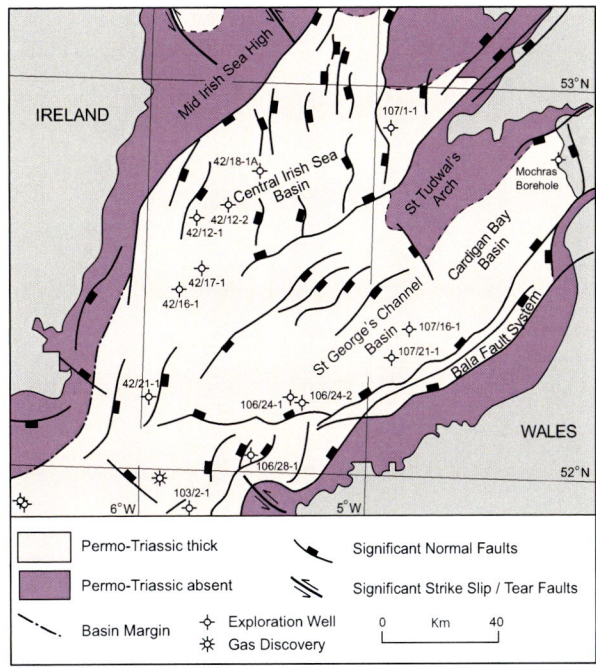

(Above) Figure 6.7. Structural elements of the southern Irish Sea with well locations mentioned in the text (modified after Maddox *et al.*, 1995).

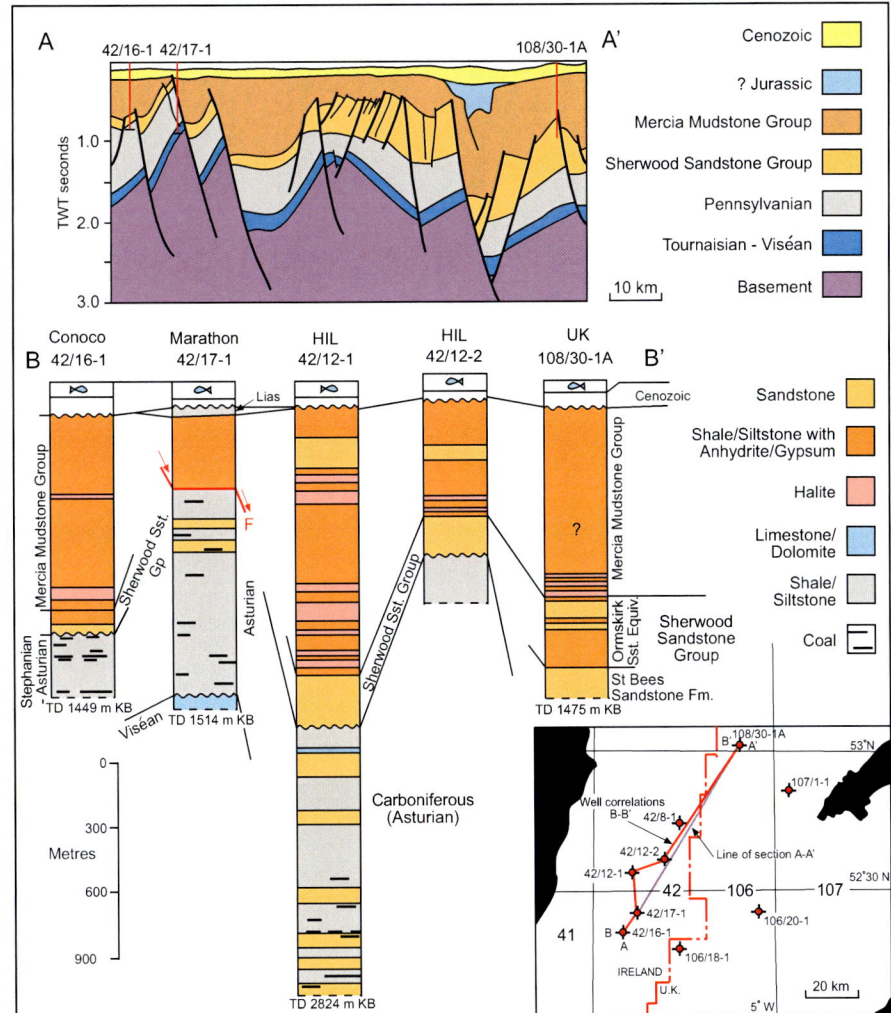

Figure 6.8. A: Geoseismic section linking wells in the Central Irish Sea Basin (after Floodpage *et al.*, 2001). B: Comparative stratigraphy of wells in the Central Irish Sea Basin (based on Green et al., 2001; Floodpage *et al.*, 2001).

Arch is thinner and shale-prone. Exploration wells within the southern part of the CISB (Figure 6.8) have penetrated Sherwood Sandstone Group sequences more than 230 m thick, overlain by in excess of 1400 m of Mercia Mudstone Group anhydritic and dolomitic mudstones interbedded with thick halites and minor sandstones (Maddox *et al.*, 1995). The Rhaetic follows conformably on the Mercia Mudstone Group in the south Irish Sea.

Permo-Triassic palaeogeography
Permian

During Permian time Ireland and much of Europe were located in low latitudes. The Variscan uplands still had considerable relief, and erosional products were deposited as fluvial and dune clastics in arid and desert conditions within limited intermontane basins. Ireland was part of a land area that extended westwards from Britain to the Porcupine Bank, with highland areas over much of the present onshore area. Permian strata are unproven on the

shelf south of Ireland and, if present, are probably of limited extent. The general absence of Permian strata in the Celtic Sea region is interpreted as reflecting the topographically elevated setting following the Variscan orogeny, with occasional fault-controlled orogenic collapse basins locally preserving remnants of the erosional products of the uplifted regions. A pattern emerges of subsidence and essentially non-marine deposition during the Permian along the northern and eastern coasts of Ireland to the north of the Variscan highlands. Volcanic activity extended from Scotland across into the Ulster Basin (Figure 6.9). Non-marine depositional conditions had persisted into latest Carboniferous (Autunian) time in the Porcupine Basin (and possibly also in the Slyne–Erris basins). It is likely that in latest Early Permian time rifting was active along the Greenland–Rockall rift and also the Biscay Rift separating Iberia from southern Europe. These are areas of possible deposition, with a Rockall–Hatton landmass of uncertain geography lying out to the west (Figure 6.9).

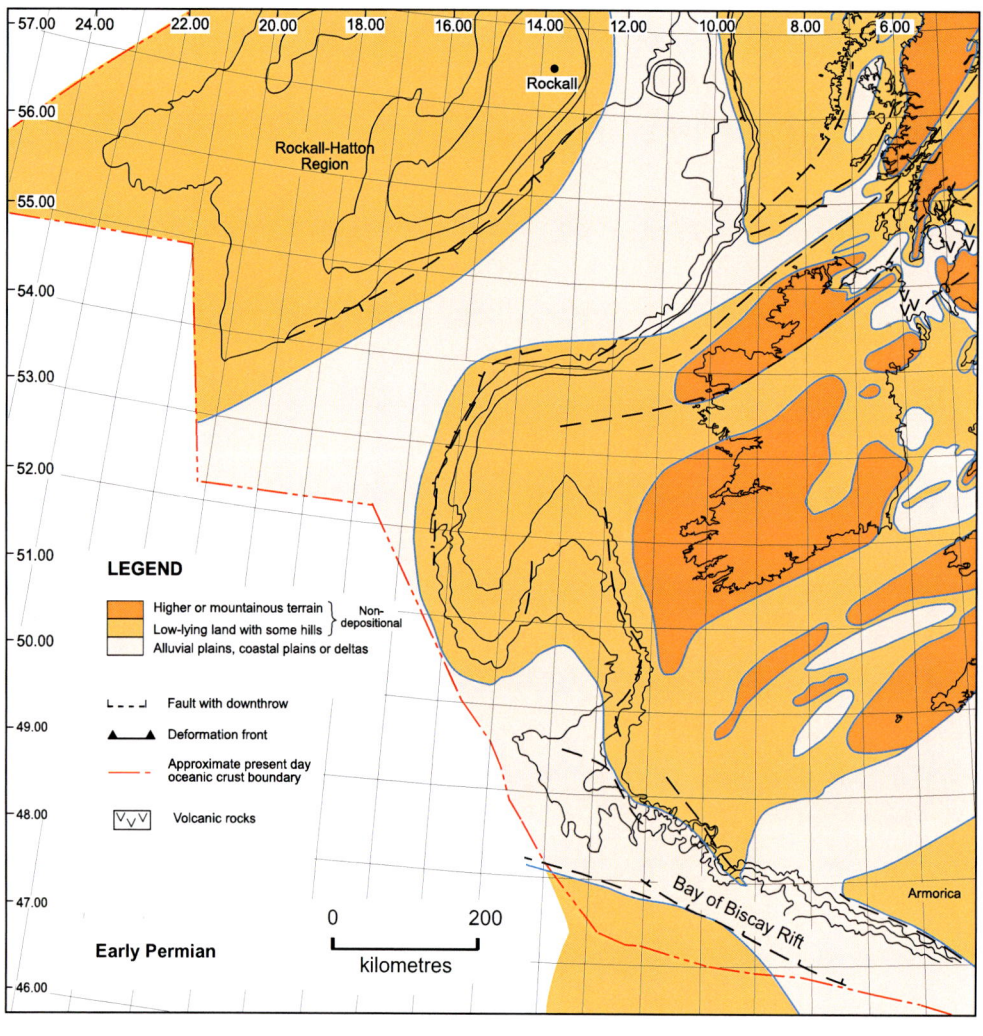

Figure 6.9. Early Permian palaeogeography (after Naylor, 1998. Principal sources: Smith, Taylor, Arthurton, Brookfield and Glennie, 1992; Ziegler, 1990).

LEGEND

Higher or mountainous terrain } Non-depositional
Low-lying land with some hills
Alluvial plains, coastal plains or deltas

Fault with downthrow
Deformation front
Approximate present day oceanic crust boundary
Volcanic rocks

Early Permian

0   200
kilometres

Rifting progressed in the Late Permian (Lopingian: Figure 6.10), and marine conditions extended from the Arctic Boreal Ocean southwards along the Norway–Greenland rift into the Rockall region, and then through the Malin Sea into the depocentres of the Irish Sea. This restricted marine area is known as the 'Bakevellia Sea'. Clastics, evaporites and minor carbonates were deposited within this basin, which had extensions into the East Irish Sea and Cheshire basins, and also into the Larne and Lough Neagh basins. Intermittent marine flooding is envisaged, and there is evidence of contemporary control on sedimentation by basin margin faults. Wells in the north Erris and Donegal basins have also encountered Zechstein-equivalent mudstone, dolomite and evaporite sequences. The extent of marine transgression southwards along the Rockall rift region is speculative, but might be expected to have reached the latitude of the Porcupine Bank. However, the presence of Late Permian fluvial deposits in the Shell 12/2-1z well (Tyrrell *et al.*,

2010) on the eastern margin of the Rockall Basin suggests that marine areas, if present, are likely to have been confined to a narrow strip in the centre of the basin. The Rockall–Hatton landmass was probably an emergent bare area with little deposition, whilst the Irish mainland also remained a positive element, despite continued erosion and deflation of the Variscan landscape.

Triassic

During the Triassic Period there was a further extension of the rift system within the Pangaean supercontinent. In the North Atlantic region this resulted in further development of the Boreal Norway–Greenland and Bay of Biscay rift systems, and inception of the Western Approaches, Celtic Sea and Porcupine rifts. The Bay of Biscay Rift represented the western extension of a rift system linking back to the Tethyan Ocean in the east. The Boreal Sea regressed in the Early Triassic, so that in Scythian time the Irish region was entirely an arid continental area with

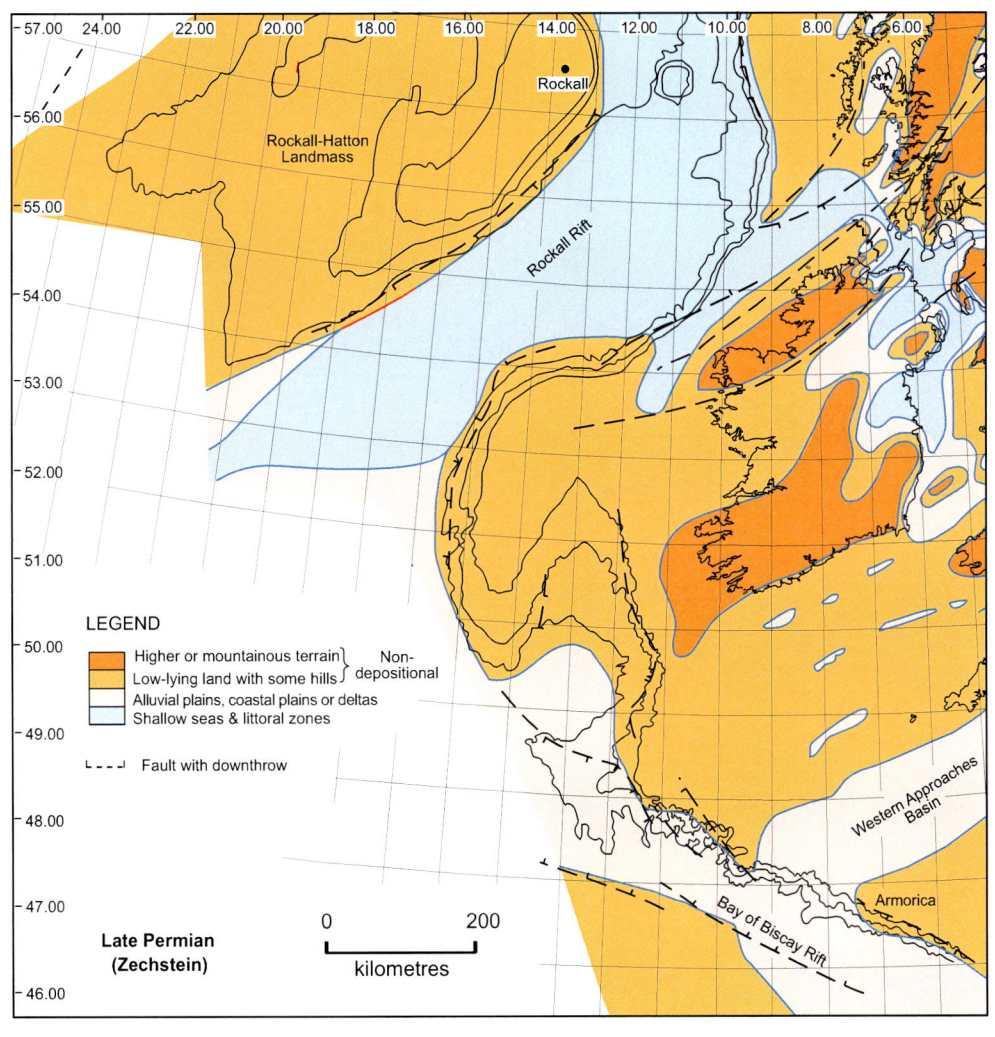

Figure 6.10. Late Permian: Thuringian (Zechstein) palaeogeography (modified after Naylor and Shannon, 2009. Principal sources: Smith *et al.*, 1992; Ziegler, 1990, Naylor, 1998, Mitchell, 2004b).

deposition of the Sherwood Sandstone Group (Figure 6.11). The Variscan mountains had suffered degradation, and bare rocky hills covered large areas. Fault-controlled sequences are seen along some margins. The lowland areas were the sites for clastic deposition by seasonal or ephemeral river systems sourced in the upland areas, with further re-deposition of sand in dune fields. Deposits of this type are found throughout the basins of the Irish Sea, Celtic Sea, Porcupine and Slyne–Erris basins. Reworked palynomorphs in the Lower Triassic of the Larne and Lough Neagh basins and the Central Irish Sea Basin (Naylor and Clayton, 2000) provide evidence of widespread erosion of Carboniferous uplands, with thick sandy deposits shed northwards from the eroding Variscan highlands into major depocentres in the Irish Sea. Dunford *et al.* (2001) envisage a main drainage system feeding northwestwards from the Cheshire Basin across the East Irish Sea Basin to the North Channel, with a tributary feeder from the Solway Basin, but bypassing a

re-entry in the area of the Central Irish Sea Basin.

The thick Lower Triassic red-bed sequences of the Loch Indaal and Rathlin troughs and the North Channel Basin extend onshore throughout the Ulster region. Deposition in the Northern Ireland basins was dominantly by fluviatile processes in floodplain and lacustrine environments, with aeolian interludes. Braided rivers carried coarser sediments at the basin margins. The Kingscourt outlier (Visscher, 1971) is the only onshore outcrop further south, and the interval there is represented by the siltstones and sandstones of the Kingscourt Sandstone Formation (Sherwood Sandstone Group equivalent). The sediments in this area may have been fed from the Irish Sea river system (Dunford *et al.*, 2001) with drainage northwards both east and west of the Longford-Down barrier and across the Northern Ireland basins.

In the Celtic Sea the Pembroke Ridge probably formed an upstanding positive element during Triassic

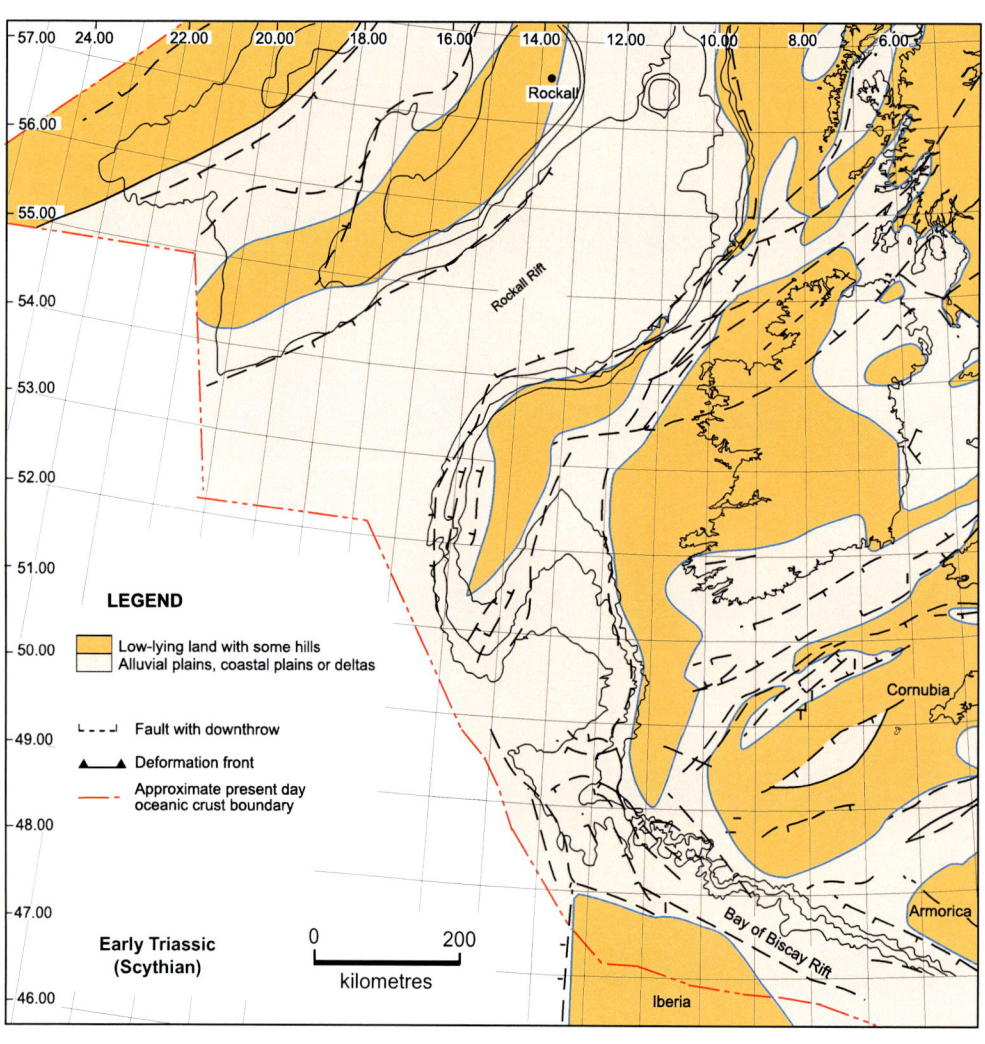

Figure 6.11. Early Triassic: Scythian palaeogeography (modified after Naylor and Shannon, 2009. Principal sources: Roberts, Montadert and Searle, 1979; Warrington and Ivimey-Cook, 1992; Naylor, 1998; Dunford *et al.*, 2001).

time. Triassic stratigraphy shows considerable differences north and south of the Ridge, suggesting that the two basinal areas were separated at that time. Sand input during the Early Triassic from the Pembrokeshire Ridge is likely to have been highly variable along strike, concentrated around fluvial systems, probably located along NW–SE cross-cutting fault zones. The Sherwood Sandstone has long been regarded as the product of a rift episode, with the overlying Mercia Mudstone developed in a thermal sag phase and extending further onto the surrounding basement elements. However, Musgrove *et al.* (1995) questioned this view, and suggested that deposition of the Sherwood Sandstone occurred within palaeogeographic lows whose positions were controlled by faults and resistant Variscan massifs. They regarded the Mercia Mudstone succession as having been deposited during a rift phase – *see also* Ruffell and Shelton (1999). Whichever model is correct, the Sherwood Sandstone is of more limited extent than the overlying

Mercia Mudstone, and it can be anticipated that sandstones would be better developed towards the source areas along the basin margin, and specifically towards fluvial input areas. Apatite fission track analysis results suggest that a relatively thick Triassic sequence extended from the Celtic Sea basins northwards over the southern part of the Irish onshore area (Keeley, 1995). Analysis of the Sherwood Sandstone Group reservoir in the Corrib Gasfield (Slyne Basin: Dancer *et al.*, 2005) indicates deposition within a braided fluvial channel system. Dancer *et al.* (2005) indicated that dipmeter and other evidence suggested that drainage flowed from SW to NE along the basin. However, Pb-isotope signatures in detrital feldspar grains (Tyrrell *et al.*, 2007) are interpreted as demonstrating derivation of sediment in Sherwood Sandstone times from the northwest (possibly Greenland), with no indications of either a southerly source or derivation of sediment from the Irish mainland. Similar arid depositional environments probably also obtained in

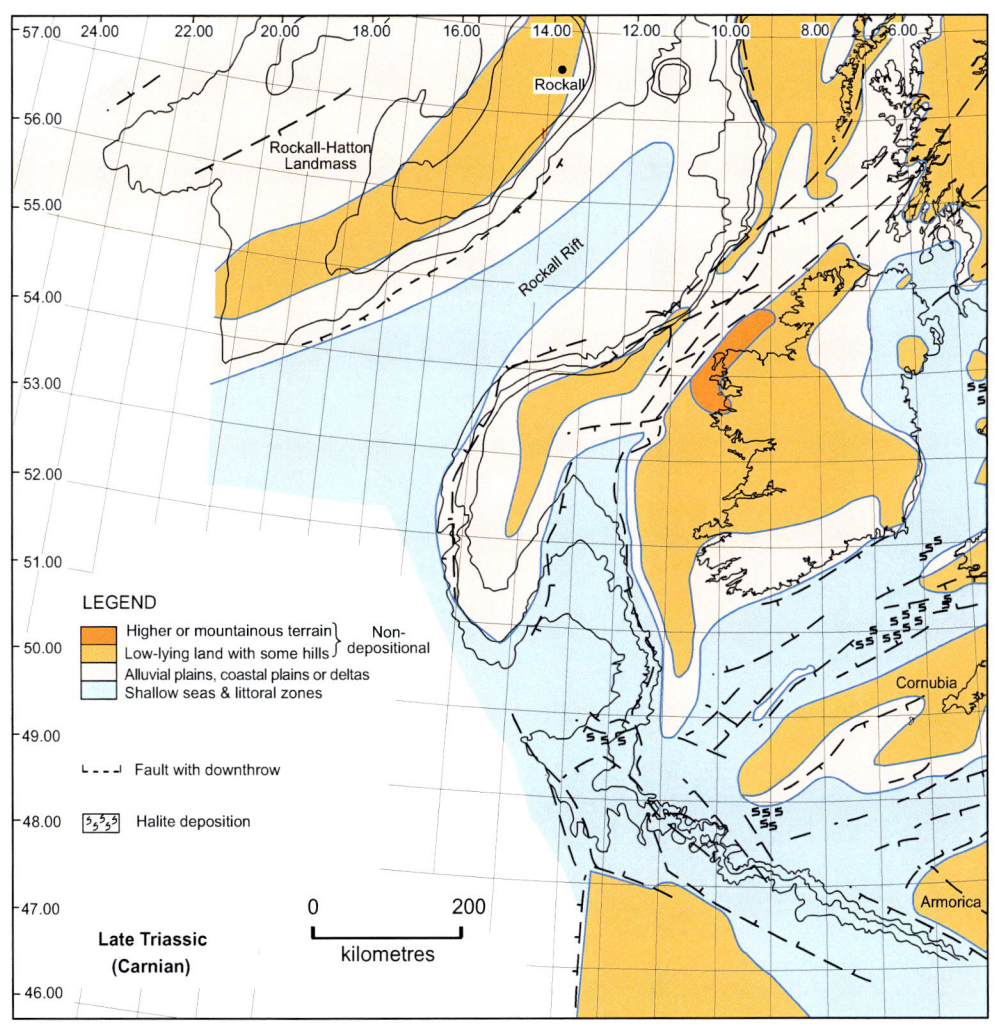

Figure 6.12. Late Triassic: Carnian palaeogeography (after Naylor and Shannon, 2009. Principal sources: Keeley, 1995; Naylor, 1992, 1998; Warrington and Ivimey-Cook, 1992; Ziegler, 1990).

LEGEND

■ Higher or mountainous terrain ⎫ Non-
■ Low-lying land with some hills ⎬ depositional
□ Alluvial plains, coastal plains or deltas
□ Shallow seas & littoral zones

⌐--⌐ Fault with downthrow

[ₛₛₛ] Halite deposition

0   200
kilometres

**Late Triassic
(Carnian)**

the Rockall and Hatton basins, with sandy non-marine strata of likely Triassic age overlying the Late Permian succession in the eastern Rockall Basin (Tyrrell *et al.*, 2010). Redfern *et al.* (2010), in a schematic Triassic palaeogeographic reconstruction, suggested that the Irish mainland was a major drainage divide between two opposing major drainage systems. The Early Triassic 'Budleighensis' drainage system (Audley-Charles, 1970), approximately 500 km in length, flowed northwards through the remnant Variscan Uplands, the south of England and the Cheshire Basin, terminating in the East Irish Sea, while another splay flowed through northeast England into the North Sea. To the west of Ireland and Scotland the predominant drainage systems flowed southeastwards from east Greenland, terminating in the Slyne Basin, and also southwestwards parallel to the axis of the Rockall Basin. McKie and Williams (2009) showed that this Atlantic Triassic river system was characterised by an overall muddying-upward trend composed of repeated progradation of terminal fans. Fluvial drainage was predominantly directed to the south, from Greenland, Fennoscandian and Scottish sources, and episodically to the north. The latter may correspond to the dipmeter evidence recorded by Dancer *et al.* (2005) in the Slyne Basin.

The Middle Triassic saw further marine transgression. Shallow marine conditions, with tidal flats, lagoons and sabkhas periodically covered large areas. However, climatic conditions in Ireland remained arid. Marine incursion progressed westwards along the Biscay rift into the Goban–Porcupine–south Rockall region. At the same time the Boreal Sea again advanced southwards towards northern Scotland, although there is no evidence that the two marine tongues became linked in Triassic time. The halite deposits within the Mercia Mudstone Group successions of the Kish and Larne basins were deposited at this time. Higher sea levels in the Late Triassic also gave rise to periodic marine incursions. In consequence there was a widespread development of playa lake and sabkha facies evaporites and marls in the Celtic Sea, Irish Sea and Ulster basins. Figure 6.12, representing the Late Triassic (Carnian), shows marine seaways established in the main basins around Ireland. Western Ireland was part of an emergent area, mostly without significant deposits, that probably extended southwards along the east side of the Porcupine Basin towards the Goban Spur. The land area also linked northwards to the Scottish landmass and through the Slyne–Erris area to the Porcupine Bank. Little is known concerning the Rockall Basin, but the likelihood is that the central area continued to subside and that marine incursion took place from the south. The Hatton region was probably emergent, with alluvial deposition in the lower-lying areas. The Triassic was brought to a close by fluctuating sea levels and eventually by a widespread marine incursion, terminating the long-lived arid and dominantly non-marine conditions that had existed through Permian and Triassic times.

References

Audley-Charles, M.G. (1970) Triassic palaeogeography of the British Isles. *Quarterly Journal of the Geological Society of London*, **126**, 49–74.

Barr, K.W., Colter, V.S., and Young, R. (1981) The geology of the Cardigan Bay–St. George's Channel Basin. *In*: Illing, L.V. and Hobson, G.D. (eds), *Petroleum Geology of the Continental Shelf of North-West Europe*. Heyden and Son Ltd., London, 432–443.

Chapman, T.J., Broks, T.M., Corcoran, D.V., Duncan, L.A., and Dancer, P.N. (1999) The structural evolution of the Erris Trough, offshore northwest Ireland, and implications for hydrocarbon generation. *In*: Fleet, A.J. and Boldy, S.A.R. (eds), *Petroleum Geology of Northwest Europe: Proceedings of the 5th Conference*. Geological Society, London, 455–469.

Colin, J.P., Lehmann, R.A., and Morgan, B. E. (1981) Cretaceous and Late Jurassic biostratigraphy of the North Celtic Sea Basin. *In*: Neale, J. W. and Brazier, M.D. (eds), *Microfossils from Recent and fossil shelf seas*. Ellis Horwood Ltd., Chichester, 122–155.

Cook, D.R. (1987) The Goban Spur: exploration in a deep-water frontier basin. *In*: Brooks, J. and Glennie, K.W. (eds), *Petroleum Geology of North-West Europe*. Graham and Trotman Ltd, London, 623–632.

Corcoran, D.V. and Mecklenburgh, R. (2005) Exhumation of the Corrib Gas Field, Slyne Basin, offshore Ireland. *Petroleum Geoscience*, **11**, 239–256.

Croker, P.F. and Shannon, P.M. (1987) The evolution and hydrocarbon prospectivity of the Porcupine Basin, offshore Ireland. *In*: Brooks, J. and Glennie, K.W. (eds), *Petroleum Geology of North West Europe*, Graham and Trotman, London, 633–642.

Dancer, P.N., Algar, S.T., and Wilson, I.R. (1999) Structural Evolution of the Slyne Trough. *In*: Fleet, A.J. and Boldy, S.A.R. (eds), *Petroleum Geology of Northwest Europe: Proceedings of the 5th Conference*. Geological Society, London, 445–453.

Dancer, P.N., Kenyon-Roberts, S.M., Downey, J.W., Baillie, J.M., Meadows, N.S., and Maguire, K. (2005) The Corrib gas field, offshore west of Ireland. *In*: Doré, A.G. and Vining, B.A. (eds), *Petroleum Geology: North-West Europe and Global Perspectives. Proceedings of the 6th Petroleum Geology Conference*. Geological Society, London, 1035–1046.

Dunford, G.M., Dancer, P.N., and Long, K.D. (2001) Hydrocarbon potential of the Kish Bank Basin: integration within a regional model for the Greater Irish Sea Basin.

In: Shannon, P.M., Haughton, P.D.W., and Corcoran, D.V. (eds), *The Petroleum Exploration of Ireland's Offshore Basins*. Geological Society, London, Special Publications, **188**, 135–154.

Evans, D., Kenolty, N., Dobson, M.R., and Whittington, R.J. (1980) *The geology of the Malin Sea*. Institute of Geological Sciences, Report **79/15**, 44pp.

Floodpage, J., Newman, P., and White, J. (2001) Hydrocarbon prospectivity in the Irish Sea area: insights from recent exploration in the Central Irish Sea, Solway and Peel basins. *In*: Shannon, P.M., Haughton, P.D.W., and Corcoran, D.V. (eds), *The Petroleum Exploration of Ireland's Offshore Basins*. Geological Society, London, Special Publications, **188**, 107–134.

Green, P.F., Duddy, I.R., Bray, R.J., Duncan, W.I., and Corcoran, D.V. (2001) The influence of thermal history on hydrocarbon prospectivity in the Central Irish Sea Basin. *In*: Shannon, P.M., Haughton, P.D.W., and Corcoran, D.V. (eds), *The Petroleum Exploration of Ireland's Offshore Basins*. Geological Society, London, Special Publications, **188**, 171–188.

Griffith, A.E. (1983) The Search for Petroleum in Northern Ireland. *In*: Brooks, J. (ed.), *Petroleum Geochemistry and Exploration in Europe*. Geological Society, London, Special Publications, **12,** 213–222.

Illing, L.V. and Griffith, A.E. (1986) Gas prospects in the 'Midland Valley' of Northern Ireland. *In*: Brooks, J., Goff, J.C., and Van Hoorn, B. (eds), *Habitat of Palaeozoic Gas in NW Europe*. Geological Society, London, Special Publications, **23**, 73–84.

Jenner, J. K. (1981) The structure and stratigraphy of the Kish Bank Basin. *In*: Illing, L.V. and Hobson, G.D. (eds), *Petroleum Geology of the Continental Shelf of North-West Europe*, Heyden and Son Ltd., London, 426–431.

Johnston, T.P. (2004) Post-Variscan Deformation and Basin Formation. *In*: Mitchell, W.I. (ed), *The Geology of Northern Ireland: Our Natural Foundation*. Geological Survey of Northern Ireland, Belfast, 205–210.

Johnston, S., Doré, A.G., and Spencer, A.M. (2001) The Mesozoic evolution of the southern North Atlantic region and its relationship to basin development in the south Porcupine Basin, offshore Ireland. *In*: Shannon P.M., Haughton, P.D.W., and Corcoran, D.V. (eds,. *The Petroleum Exploration of Ireland's Offshore Basins*. Geological Society, London, Special Publications, **188**, 237–263.

Kamerling, P. (1979) The geology and hydrocarbon habitat of the Bristol Channel Basin. *Journal of Petroleum Geology*, **2**, 75–93.

Keeley, M.L. (1995) New evidence of Permo-Triassic rifting, onshore southern Ireland, and its implications for Variscan structural inheritance. *In*: Boldy, S.A.R. (ed.), *Permian and Triassic Rifting in Northwest Europe*. Geological Society, London, Special Publications, **91**, 239–253.

Maddox, S.J., Blow, R., and Hardman, M. (1995) Hydrocarbon prospectivity of the Central Irish Sea Basin with reference to Block 42/12, offshore Ireland. *In*: Croker, P.F. and Shannon, P.M. (eds), *The Petroleum Geology of Ireland's Offshore Basins*. Geological Society, London, Special Publications, **93**, 59–77.

Maddox, S.J., Blow, R.A., and O'Brien, S.R. (1997) The geology and hydrocarbon prospectivity of the North Channel Basin. *In*: Meadows, N.S., Trueblood, S.P., Hardman, M., and Cowan, G. (eds), *Petroleum Geology of the Irish Sea and Adjacent Areas*. Geological Society, London, Special Publications, **124**, 95–111.

Manning, P.I. and Wilson, H.E. (1975) The stratigraphy of the Larne Borehole, County Antrim. *Bulletin of the Geological Survey of Great Britain*, **50**, 1–27.

Manning, P.I., Robbie, J.A. and Wilson, H.E. (1970) *Geology of Belfast and the Lagan Valley*. Memoir of the Geological Survey of Northern Ireland, Sheet 36. Her Majesty's Stationery Office, Belfast, 242pp.

Marathon Petroleum U.K. Ltd./Shell U.K. Ltd. *Newmill No.1, County Antrim, Northern Ireland. Final Well Report*, 15pp.

McCaffrey, R.J. and McCann, N. (1992) Post-Permian basin history of northeast Ireland. *In*: Parnell, J. (ed.), *Basins on the Atlantic Seaboard: Petroleum Geology, Sedimentology and Basin Evolution*. Geological Society, London, Special Publications, **62**, 277–290.

McCann, N. (1988) An assessment of the subsurface geology between Magilligan Point and Fair Head, Northern Ireland. *Irish Journal of Earth Sciences*, **9**, 71–78.

McCann, N. (1990) The subsurface geology between Belfast and Larne, Northern Ireland. *Irish Journal of Earth Sciences*, **10**, 157–173.

McKie, T. and Williams, B. (2009) Triassic palaeogeography and fluvial dispersal across the northwest European Basins. *Geological Journal*, **44**, 711–741.

Millson, J. A. (1987) The Jurassic evolution of the Celtic Sea basins. *In*: Brooks, J. and Glennie, K.W. (eds), *Petroleum Geology of North West Europe*, Graham and Trotman, London, 599–610.

Mitchell, W.I. (2004a) Triassic. *In*: Mitchell, W.I. (ed.), *The Geology of Northern Ireland: Our Natural Foundation*. Geological Survey of Northern Ireland, Belfast, 133–144.

Mitchell, W.I. (2004b) Permian. *In*: Mitchell, W.I. (ed.), *The Geology of Northern Ireland: Our Natural Foundation*. Geological Survey of Northern Ireland, Belfast, 125–132.

Murphy, N.J. and Croker, P.F. (1992) Many play concepts seen over wide area in Erris, Slyne troughs off Ireland. *Oil and Gas Journal*, Sept. 14, 92–97.

Musgrove, F.W, Murdoch, L.M. and Lenehan, T. (1995) The Variscan fold-thrust belt of southeast Ireland and its control on early Mesozoic extension and deposition: a method to predict the Sherwood Sandstone. *In*: Croker, P.F. and Shannon, P.M. (eds), *The Petroleum Geology of Ireland's Offshore Basins*. Geological Society, London, Special Publications, **93**, 81–100.

Naylor, D. (1992) The post-Variscan history of Ireland. *In*: Parnell, J. (ed.), *Basins on the Atlantic Seaboard: Petroleum Geology, Sedimentology and Basin Evolution*. Geological Society, London, Special Publications, **62**, 255–275.

Naylor, D. (1998) *Irish shorelines through geological time*. Royal Dublin Society: Occasional Papers in Irish Science and Technology, **17**, 20pp.

Naylor, D. and Clayton, G. (2000) Palynological and maturation data and their bearing on Irish post-Variscan palaeogeography. *Irish Journal of Earth Sciences*, **18**, 33–39.

Naylor, D. & Shannon, P.M. (1982) *The geology of offshore Ireland and West Britain*. Graham & Trotman, London, 161pp.

Naylor, D. and Shannon, P. M. (2009) Geology of offshore Ireland. *In*: Holland, C.H. and Sanders, I.S. (eds), *The Geology of Ireland*. Second Edition. Dunedin Academic Press, Edinburgh, 405–460.

Naylor, D., Haughey, N., Clayton, G., and Graham, J.R. (1993) The Kish Bank Basin, offshore Ireland. *In*: Parker, J.R. (ed.), *Petroleum Geology of North-West Europe: Proceedings of the 4th Conference*. Geological Society, London, 845–855.

Naylor, D., Philcox, M.E., and Clayton, G. (2003) Annaghmore-1 and Ballynamullan-1 wells, Larne–Lough Neagh Basin, Northern Ireland. *Irish Journal of Earth Sciences*, **21**, 47–69.

Naylor, D., Shannon, P.M., and Murphy, N. (1999) *Irish Rockall Basin region – a standard structural nomenclature system*. Petroleum Affairs Division, Dublin, Special Publication **1/99**, 42pp.

Naylor, D., Shannon, P.M., and Murphy, N. (2002) *Porcupine–Goban region – a standard structural nomenclature system*. Petroleum Affairs Division, Dublin, Special Publication **1/02**, 65pp.

Newman, P. (1999) The geology and hydrocarbon potential of the Peel and Solway Basins, East Irish Sea. *Journal of Petroleum Geology*, **22**, 305–324.

Parnell, J. (1992) Hydrocarbon potential of Northern Ireland: 3. Reservoir potential of the Permo-Triassic. *Journal of Petroleum Geology*, **15**, 51–70.

Penn, I.E. (1981) Larne No.2 Geological well completion report. *Report of the Institute of Geological Sciences, London*, **81/6**, 58pp.

Penn, I.E., Holliday, D.W., Kirby, G.A., Soper R.A. *et al.* (1983) The Larne borehole No.2: discovery of a new Permian volcanic centre. *Scottish Journal of Geology*, **19**, 333–46.

Petrie, S.H., Brown, J.R., Granger, P.J., and Lovell, J.P.B. (1989) Mesozoic history of the Celtic Sea Basins. *In*: Tankard, A.L. and Balkwill, H.R. (eds), *Extensional tectonics and stratigraphy of the North Atlantic Margins*. American Association of Petroleum Geologists Memoir, **46**, 433–444.

Reay, D.M. (2004) Geophysics and Concealed Geology. *In*: Mitchell, W.I. (ed.), *The Geology of Northern Ireland: Our Natural Foundation*. Geological Survey of Northern Ireland, Belfast, 227–248.

Redfern, J., Shannon, P.M., Tyrrell, S., Leleu, S., Fabuel-Perez, I., Baudon, C., Haughton, P.D.W., Daly, J.S., van Lanen, X., Stolfova, K., Hodgetts, D., Williams, B.P.J., and Speksnijder, A. (2010) An integrated study of Permo-Triassic basins along the North Atlantic passive margin: implications for future exploration. *In*: Vining, B.A. and Pickering, S.C. (eds), *Petroleum Geology: From Mature Basins to New Frontiers – Proceedings of the 7th Petroleum Geology Conference*. Geological Society, London, 921–936.

Roberts, D.G., Montadert, L., and Searle, R.C. (1979) The western Rockall Plateau – stratigraphy and structural evolution. *Initial Reports of the Deep Sea Drilling Project* **48**, U.S.G.P.O., Washington D.C., 1061–1088.

Robinson, K.W., Shannon, P.M., and Young, D.G.G. (1981) The Fastnet Basin: an integrated analysis. *In*: Illing, L.G. and Hobson, G.D. (eds), *Petroleum Geology of the Continental Shelf of North-West Europe*. Heyden and Son Ltd, London, 444–454.

Ruffell, A. and Shelton, R. (1999) The control of sedimentary facies by climate during phases of crustal extension: examples from the Triassic of onshore and offshore England and Northern Ireland. *Journal of the Geological Society, London*, **156**, 779–789.

Scotchman, I.C. and Thomas, J.R.W. (1995) Maturity and hydrocarbon generation in the Slyne Trough, northwest Ireland. *In*: Croker, P.F. and Shannon, P.M. (eds), *The Petroleum Geology of Ireland's Offshore Basins*. Geological Society, London, Special Publications, **93**, 385–411.

Shannon, P.M. (1991) The development of the Irish offshore sedimentary basins. *Journal of the Geological Society of London*, **148**, 181–189.

Shannon, P.M. (1995) Permo-Triassic development of the Celtic Sea region, offshore Ireland. *In*: Boldy, S.A.R. (ed.), *Permian and Triassic Rifting in Northwest Europe*. Geological Society, London, Special Publications, **91**, 215–237.

Shannon, P.M. and MacTiernan, B. (1993) Triassic prospectivity in the Celtic Sea, Ireland: a case history. *First Break*, 11, 47–57.

Shannon, P.M., Jacob, A.W.B., Makris, J., O'Reilly, B., Hauser, F., and Vogt, U. (1995) Basin development and petroleum prospectivity of the Rockall and Hatton region. *In*: Croker, P.F. and Shannon, P.M. (eds), *The Petroleum Geology of Ireland's Offshore Basins*. Geological Society, London, Special Publications, **93**, 435–457.

Shelton, R. (1995) Mesozoic basin evolution of the North Channel: preliminary results. *In*: Croker, P.F. and Shannon, P.M. (eds), *The Petroleum Geology of Ireland's Offshore Basins*. Geological Society Special Publications, **93**, 7–20.

Shelton, R. (1997) Tectonic evolution of the Larne Basin. *In*: Meadows, N.S., Trueblood, S.P., Hardman, M., and Cowan, G. (eds), *Petroleum Geology of the Irish Sea and Adjacent Areas*. Geological Society, London, Special Publications, **124**, 113–133.

Smith, C. (1995) Evolution of the Cockburn Basin: implications for the development of the Celtic Sea basins. *In*: Croker, P.F. and Shannon, P.M. (eds), *The Petroleum Geology of Ireland's Offshore Basins*. Geological Society, London, Special Publications, **93**, 279–295.

Smith, R.A. (1986) Permo-Triassic and Dinantian rocks of the Belfast Harbour Borehole. *Report of the British Geological Survey*, **18.** 13pp.

Smith, D.B., Taylor, J.C.M., Arthurton, R.S., Brookfield, M.E., and Glennie, K.W. (1992) Permian. *In*: Cope, J.C.W., Ingham, J.K., and Rawson, P.F. (eds), *Atlas of Palaeogeography and Lithofacies*. Geological Society, London, Memoir **13**, 153pp.

Štolfová, K. and Shannon, P.M. (2009) Permo-Triassic development from Ireland to Norway: basin architecture and regional controls. *Geological Journal*, **44**, 652–676.

Tappin, D.R., Chadwick, R.A., Jackson, A.A., Wingfield, R.T.R., and Smith, N.J.P. (1994) United Kingdom offshore regional report: *The geology of Cardigan Bay and the Bristol Channel*. London HMSO for the British Geological Survey, 107pp.

Tate, M.P. and Dobson, M. R. (1988) Syn- and post-rift igneous activity in the Porcupine Seabight basin and adjacent continental margin west of Ireland. *In*: Morton, A.C. and Parson,

L.M. (eds), *Early Tertiary volcanism and the opening of the NE Atlantic*. Geological Society, London, Special Publications, **39**, 309–334.

Tate, M.P. and Dobson, M.R. (1989a) Late Permian to early Mesozoic rifting and sedimentation offshore NW Ireland. *Marine and Petroleum Geology*, **6**, 49–59.

Thompson, S.J. (1979) Preliminary report on the Ballymacilroy No.1 borehole, Ahoghill, Co. Antrim. *Geological Survey of Northern Ireland Open File Report* **No. 63**, 17pp.

Trueblood, S. (1992) Petroleum geology of the Slyne Trough and adjacent basins. *In*: Parnell, J. (ed.), *Basins of the Atlantic Seaboard: Petroleum Geology, Sedimentology and Basin Evolution*. Geological Society, London, Special Publications, **62**, 315–326.

Tyrrell, S., Haughton, P.D.W., and Daly, J.S. (2007) Drainage reorganization during breakup of Pangea revealed by in-situ isotopic analysis of detrital K-feldspar. *Geology*, **35**, 971–974.

Tyrrell, S., Souders, A.K., Haughton, P.D.W., Daly, J.S., and Shannon, P.M. (2010) Sedimentology, sandstone provenance and palaeodrainage on the eastern Rockall Basin margin: evidence from the Pb isotopic composition of detrital K-feldspar. *In*: Vining, B.A. and Pickering, S.C. (eds), *Petroleum Geology: From Mature Basins to New Frontiers – Proceedings of the 7th Petroleum Geology Conference*. Geological Society, London, 921-936.

Van Hoorn, B. (1987) The South Celtic Sea/ Bristol Channel Basin: origin, deformation and inversion history. *Tectonophysics*, **137**, 309–334.

Visscher, H. (1971) *The Permian and Triassic of the Kingscourt outlier, Ireland*. Geological Survey of Ireland, Special Paper, **1**, 1–114.

Warrington, G. and Ivimey-Cook, H.C. (1992) Triassic. *In*: Cope, J.C.W., Ingham, J.K., and Rawson, P.F. (eds), *Atlas of Palaeogeography and Lithofacies*. Geological Society, London, Memoir **13**, 153pp.

Warrington, G., Audrey-Charles, M.G., Elliot, R.E. Evans, W.B., Ivimey- Cook, H.C., Kent, P.E., Robinson, P.L., Shotton, F.W., and Taylor, F.M. (1980) *A correlation of Triassic rocks in the British Isles*. Geological Society, London, Special Report **13**, 78pp.

Wilson, H.E. (1972) *Regional geology of Northern Ireland*. Ministry of Commerce, Geological Survey of Northern Ireland. Her Majesty's Stationery Office, Belfast, 115pp.

Wilson, H.E. (1974) The South Antrim Salt Field. *Geological Survey of Northern Ireland Open File Report* **No. 48**, 11pp.

Wilson, H.E. and Manning, P.I. (1978) *Geology of the Causeway Coast*. Memoir of the Geological Survey of Northern Ireland, Sheet 7. Her Majesty's Stationery Office, Belfast, 72pp.

Woodland, A.W (ed.), (1971) *The Llanbedr (Mochras Farm) borehole*. Institute of Geological Sciences, Report **71/18**, 115 pp.

Ziegler, P.A. (1982) *Geological Atlas of Western and Central Europe*.1st Edition. Shell Internationale Petroleum Maatschappij, B.V., 130pp.

Ziegler, P.A. (1990) *Geological Atlas of Western and Central Europe*. 2nd Edition. Shell Internationale Petroleum Maatschappij B.V., 239pp.

Chapter 7

Jurassic stratigraphy

During the Jurassic Period the chain of rifts extending from Tethys to the Central Atlantic/Caribbean region developed into the main axis along which the Pangaean supercontinent would eventually split. Shallow marine deposition became established in the Goban Spur–Celtic Sea region within basins centred over the earlier Permo-Triassic rift basins. Spreading axes developed along linked rifts that eventually gave rise to new ocean floor in the Central Atlantic region during the Late Jurassic, by which time Gondwana was almost separated from Laurasia, except for contact in the Mediterranean region. The onset of tectonism (often referred to as the Mid-Cimmerian event) in earliest Middle Jurassic time provides a natural division of the Jurassic in many of the Irish offshore basins between Lower and Middle–Upper Jurassic sequences.

Offshore distribution of Lower Jurassic rocks

Jurassic rocks are widespread in the subsurface of the North Celtic Sea and Fastnet basins (Figures 7.1 and 7.2). There is a conformable transition from Triassic red-beds to a marine Lower Jurassic succession (Robinson *et al.*, 1981; Millson, 1987). The Rhaetian transgressive sequence is succeeded by a Hettangian shallow-marine limestone-dominant sequence in the west. On the north flank of the Fastnet Basin, where carbonate deposition persisted into early Sinemurian time, the carbonate sequence attains a maximum thickness of 460 m (Deminex well 56/21-1). The carbonate-dominant sequence thins eastwards in the Celtic Sea basins, interdigitates with, and is overlain by, more argillaceous sediments. Cyclic units with an upward increase in carbonate content are typical of the Hettangian–Sinemurian mudrock sequences. A regressive phase gave rise to Upper Sinemurian deltaic sandstones in the Fastnet Basin (Robinson *et al.*, 1981), where they are thickest in the axial and southwest portions of the basin (e.g. well 56/26/-1 in Figure 7.2B). Similar sandstones were also deposited in the eastern portion of the North Celtic Sea Basin and in the St George's Channel Basin (Millson, 1987).

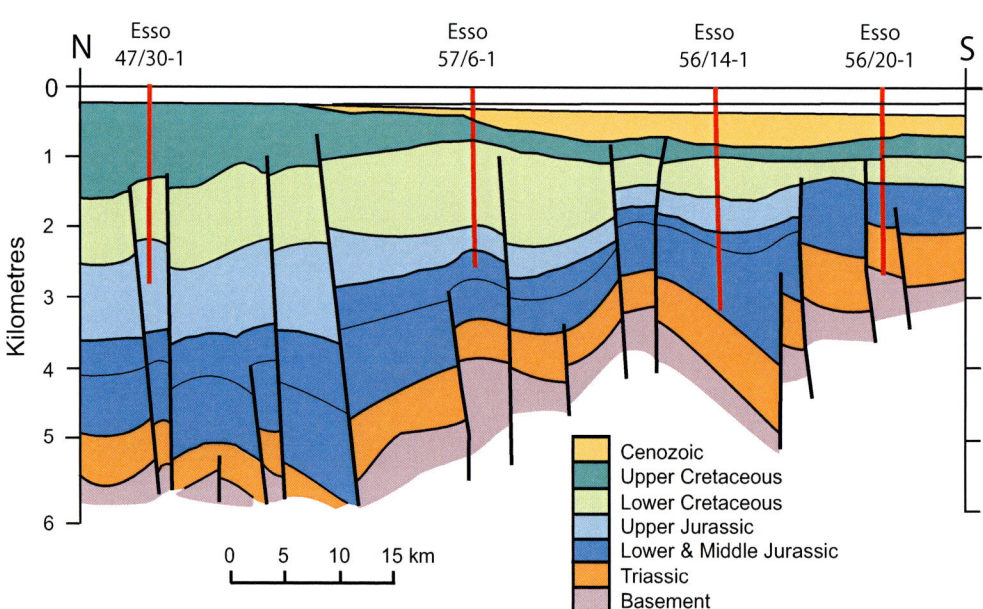

Figure 7.1. Approximate north–south cross-section through the North Celtic Sea Basin (after Colin *et al.*, 1981). The location of the section is shown in Figure 3.4.

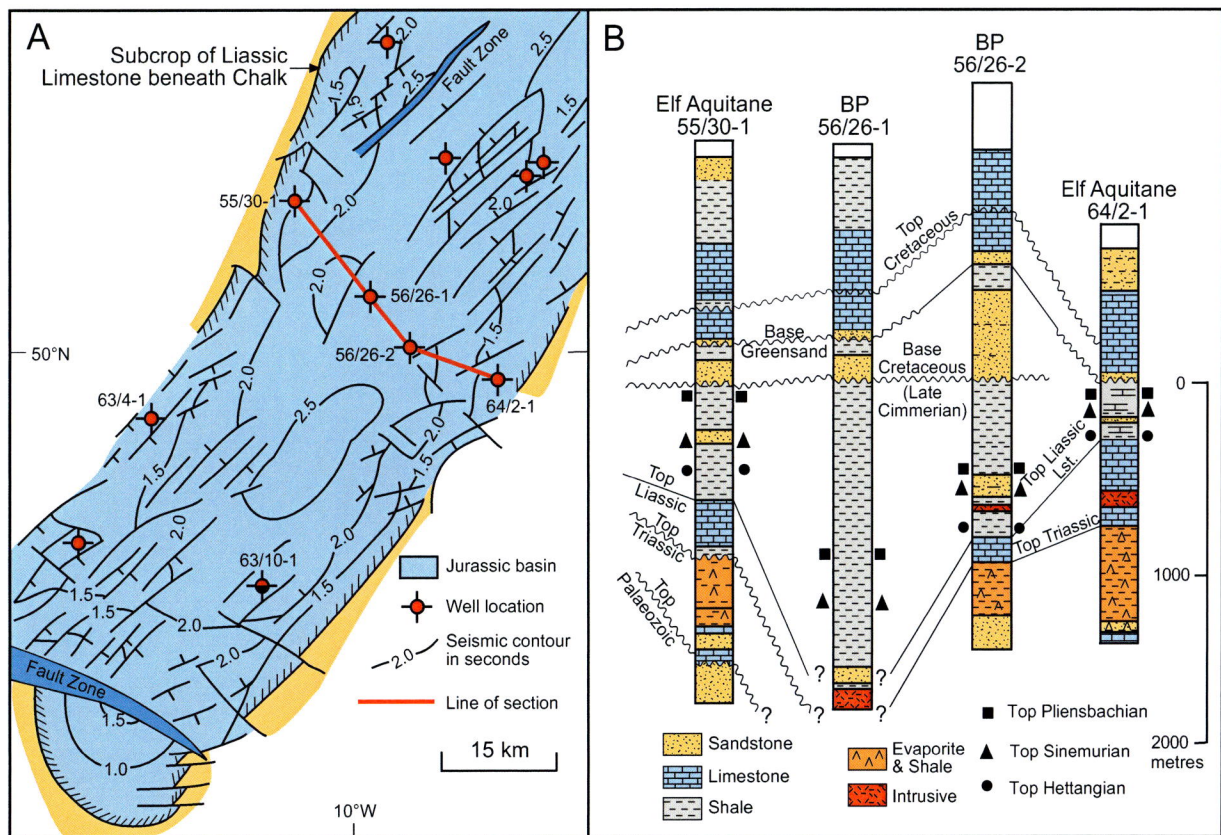

Figure 7.2. A: Time structure contour map (two-way time) of the Fastnet Basin at Top Liassic Limestone level (after Robinson *et al.*, 1981). B: Stratigraphy of exploration wells in the Fastnet Basin. The line of section is shown on the map (after Robinson *et al.*, 1981).

Petrie *et al.* (1989) regarded the sands in the Fastnet Basin as having been sourced from exposed Old Red Sandstone terrain on the Fastnet Spur north of the basin, while the coeval sands in the eastern part of the North Celtic Sea Basin came from the Leinster massif to the northeast. Kessler and Sachs (1995) saw this latter input as related to post-rift footwall readjustment at the northern margin of the North Celtic Sea and St George's Channel basins. Clastic input continued into the early Pliensbachian in the Fastnet Basin. The sandstones are everywhere succeeded by uniform argillaceous and organic-rich Pliensbachian mudstones, which in the east, around the Pembroke Ridge extension (Figure 7.3), contain coarser siliciclastics. Shallowing conditions at the end of the Pliensbachian are indicated by the development of a carbonate–sandstone member. Deltaic and shallow-marine clastics were deposited in the St George's Channel Basin, also possibly sourced from the Leinster region and from the eastern part of the Pembroke Ridge (Petrie *et al.*, 1989). There followed an abrupt return to deeper conditions in the Toarcian, and deposition of a uniform mudrock succession. The Toarcian shale sequences are

considered to have source potential, with up to 135 m of good source rocks predicted along the main Toarcian depocentre of the North Celtic Sea Basin (Murphy *et al.*, 1995: also *see* Chapter 10). Thin limestone–siltstone units, deposited in shallowing conditions at the end of the Toarcian, with deltaic and shallow-marine clastics in the St George's Channel Basin, brought Early Jurassic deposition in the Celtic Sea area to a close.

A surprisingly thick section of Lower Jurassic (Liassic) sediments within the Cardigan Bay Basin was intersected by the Mochras (Llanbedr) borehole on the Welsh coast (Woodland, 1971; Figure 6.7). The 1305 m thick marine Lower Jurassic (Hettangian to Upper Toarcian) is one of the thickest successions of this age known in Europe. The sequence comprises grey, locally silty, calcareous mudstones with some darker mudstone intercalations, particularly in the Lower Liassic. The thin-bedded mudstones alternate with bioturbated silty limestones and hard, massive calcareous siltstones. Limestone interbeds are more frequent in the lower (Hettangian to Lower Sinemurian) part of the section. The offshore wells in the Cardigan Bay and St George's Channel basins have

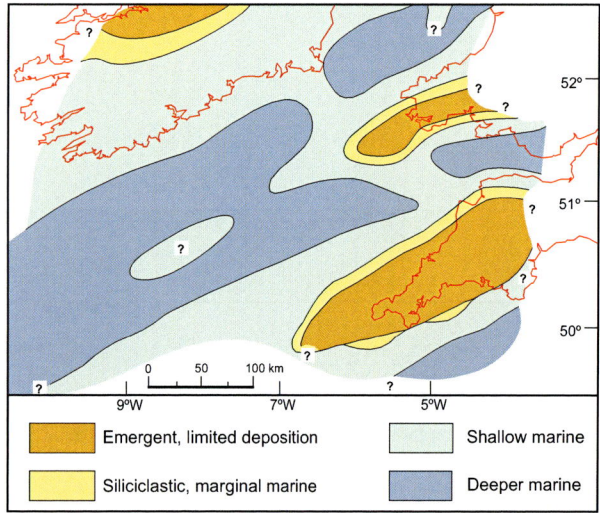

Figure 7.3. Depositional environments in the Celtic Sea region during the late Pliensbachian (modified after Millson, 1987).

all penetrated the Lower Jurassic succession. Seismic interpretation indicates thicknesses similar to those in the Mochras borehole along the axes of the basins (Barr *et al.*, 1981). However, the sequence thins towards the Pembroke Ridge extension on the south flank of the St George's Channel Basin where the Texaco/HGB well 103/2-1 (Figure 6.7) recorded 388 m of Hettangian to late Pliensbachian strata, possibly deposited in shallower water environments (Tappin *et al.*, 1994). Seabed coring on the flank of the St George's Channel Basin (Whittington *et al.*, 1981) proved Liassic grey laminated shales and limestones.

Interpretation of reflection seismic data on the Goban Spur, allied with results from the Esso 62/7-1 well (Cook, 1987), suggest that Triassic red-beds and evaporites are conformably overlain by Rhaetian to early Sinemurian marine strata. The oldest beds penetrated in the well (Figure 7.4) were Early Sinemurian cherty marine limestones with thin sandstone and siltstone interbeds. Land-derived palynomorphs and ostracods suggest inner shelf shallow-marine deposition. The overlying Upper Sinemurian to Lower Pliensbachian unit comprises marine claystones and thin limestones with thin sandstone and siltstone interbeds. An outer neritic depositional environment is suggested, indicative of continued basin subsidence. Mudstone deposition in outer shelf environments continued through Late Sinemurian to Toarcian time, with fewer limestones. On the basis of seismic interpretation Naylor *et al.* (2002) interpreted Lower–Middle Jurassic strata as extending throughout the Goban Spur Basin and Goban Graben

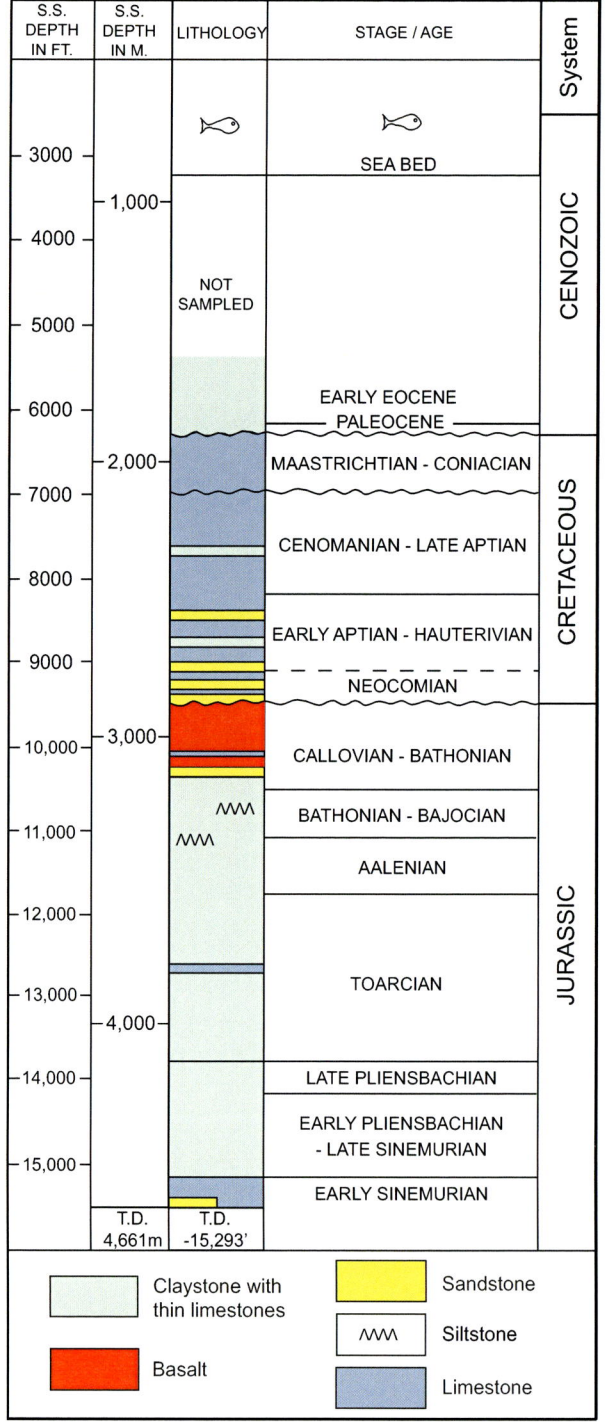

Figure 7.4. Stratigraphy of the Esso 62/7-1 well, Goban Spur Basin (modified after Cook, 1987).

(Figures 2.15 and 3.6) and into the southern part of the Porcupine Basin.

Lower Jurassic rocks in the Porcupine region have been encountered only in the North Porcupine Basin (Figure

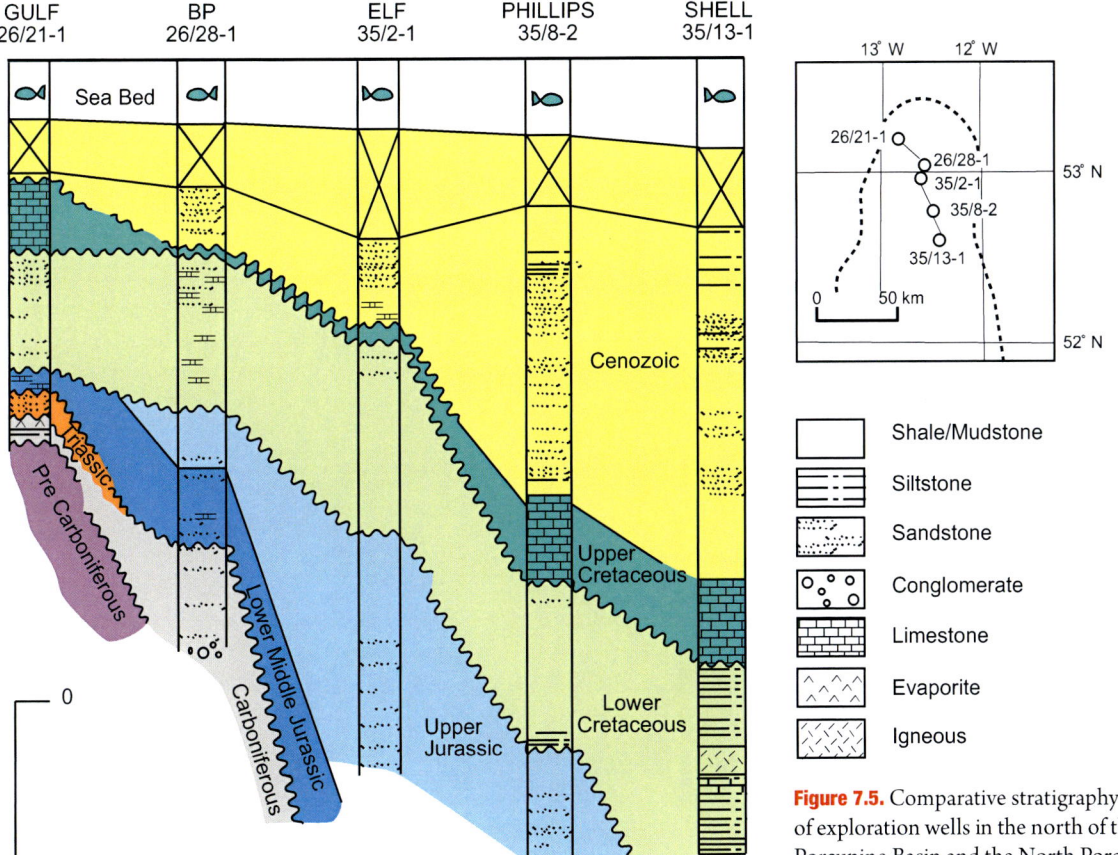

Figure 7.5. Comparative stratigraphy of exploration wells in the north of the Porcupine Basin and the North Porcupine Basin (based on Croker and Shannon, 1987).

7.5), where they appear to onlap irregular pre-Jurassic topography. Croker and Shannon (1987) suggested that Early Jurassic deposition may have been limited to small basins with Caledonian trend, as in the underlying Triassic. In Block 26/28 at the northern margin of the Porcupine Basin, for example, Bajocian strata rest unconformably on Carboniferous rocks (MacDonald *et al.*, 1987). The succession, where penetrated, is a transgressive sequence comprising limestones, shales and intercalated siltstones of Rhaetian to Hettangian age, probably deposited in brackish to freshwater conditions replaced upwards by shallow marine environments. Naylor *et al.* (1999) interpreted a seismic unit representing undifferentiated Lower– Middle Jurassic strata within the perched basins on the west flank of the Porcupine High and on the east flank of the Rockall High (Figures 2.15 and 7.6).

The Elf 27/13-1 well in Slyne Basin (Trueblood, 1992; Figure 3.7) encountered a complete Lower Jurassic sequence (Figure 7.7). Red-beds are succeeded by Sinemurian nearshore limestones and sandstones, and

then by Sinemurian to Lower Toarcian marine shales and siltstones. Toarcian shales (84 m) follow and constitute the main petroleum source rock in the basin (*see* Chapter 11). A similar sequence is recorded in the Corrib Gasfield region (Dancer *et al.*, 2005). A varied Hettangian–Sinemurian sequence of shales, sandstones, limestones, anhydrites and dolomites (Broadford Beds Formation) is conformably succeeded by grey to black shales with variable silt and organic content (Sinemurian to Pliensbachian Pabba Shale Formation, Pliensbachian Scalpa Shale Formation and Early Toarcian Portree Shale Formation). The Lower Jurassic sequence in the Slyne Basin is interpreted as a product of open marine deposition with restricted deposits in the deeper parts of the basin (Toarcian anoxic shales).

In the Amoco 19/5-1 and Amoco 12/13-1A wells in the Erris Basin (Tate and Dobson, 1989a; Figure 3.7) the Rhaetian is conformably overlain by Hettangian to early Sinemurian limestones, with claystone, siltstone and sandstone interbeds. These were probably the product of freshwater to shallow-marine sedimentation. The upper

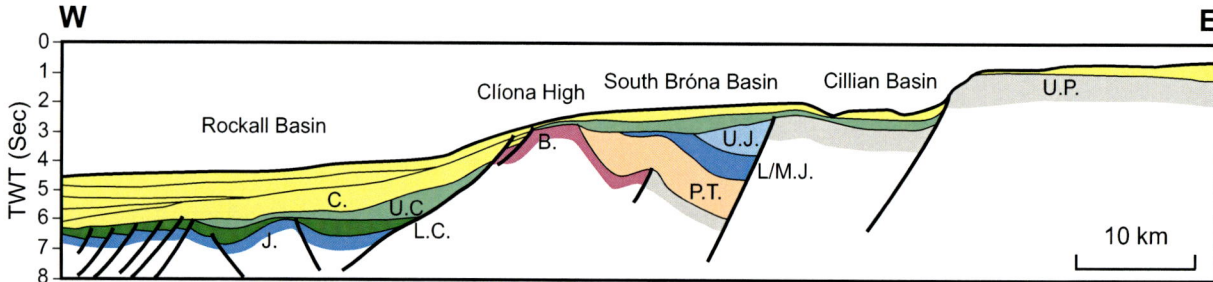

Figure 7.6. Geoseismic section across the eastern margin of the Rockall Basin and the South Bróna Basin (after Naylor *et al.*, 1999 and Naylor and Shannon, 2005). Permo-Triassic and Jurassic strata are overlain by eastward thinning Cretaceous and Cenozoic strata. B. = Basement; U.P. = Upper Palaeozoic; P.T. = Permo-Triassic; L/M.J. = Lower to Middle Jurassic; U.J. = Upper Jurassic; J. = undifferentiated Jurassic; L.C. = Lower Cretaceous; U.C. = Upper Cretaceous; C. = Cenozoic. The location of the line is shown on Figure 11.2.

zones of the Liassic are missing in these wells and the sequence is unconformably overlain by Cretaceous or younger sediments. It is likely that the Erris Basin originally had a more complete Jurassic sequence, but that this was removed by latest Jurassic and/or Early Cretaceous erosion (Morton, 1992).

All known Jurassic strata onshore in Northern Ireland belong to the Lower Liassic, although in north Antrim Wilson and Manning (1978) reported that 'derived blocks of shale containing Middle Liassic fossils have been found near Ballintoy and Upper Liassic species have been recovered from the basal conglomerate of the Cretaceous at Murlough Bay'. This suggests that a more complete Jurassic sequence may have existed onshore but was subsequently removed by pre-Cretaceous erosion, and/or that younger Jurassic strata may be present in the basins off the north Antrim coast. Jurassic rocks are found onshore intermittently at outcrop around the margins of the Antrim basalt plateau. The extant Lower Liassic deposits generally range from Hettangian to Sinemurian in age, although the sequence (248 m thick) extends up into the Lower Pliensbachian in the Port More borehole (Figures 6.1 and 12.1B). Lower Liassic rocks crop out along the Northern Ireland coastline, as at Magilligan Point, Portrush, Whitepark Bay and Larne, County Antrim, and have also been described from boreholes in the area. The Liassic sequence exposed on the east coast near Larne

Figure 7.7. Generalised stratigraphy of the Slyne and Erris basins (after Murphy and Croker, 1992; Dancer *et al.*, 1999).

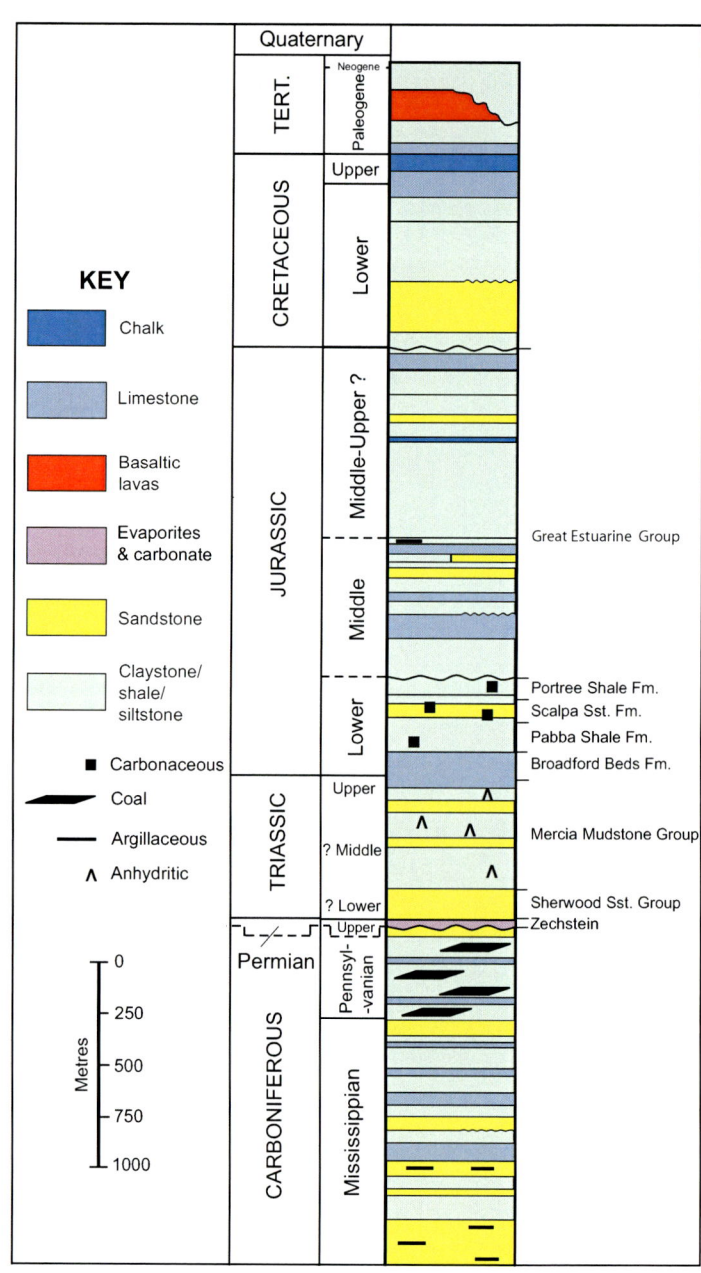

(Figure 6.3) comprises a single formation, the Waterloo Mudstone Formation. Approximately 100 m of dark grey calcareous mudstones with silt laminae and subordinate thin limestones range up to the lowest part of the Sinemurian Stage. The formation is represented in the Port More borehole by a similar sequence of calcareous mudstones and thin limestones, with siltstone beds and ironstone nodules common near the top of the sequence. Offshore, Rhaetian and Liassic strata are preserved along the Loch Gruinart–Leannan boundary fault in the Loch Indaal Basin (Fyfe *et al.*, 1993). Shallow offshore boreholes penetrated uppermost Mercia Mudstone Group red-beds conformably overlain by dark grey mudstones of the Rhaetian Westbury Formation, whilst Liassic mudstones were encountered in other boreholes.

Liassic strata are preserved locally in parts of the Solway and Peel basins (Newman, 1999; Figure 5.2), although only the Elf 111/29-1 well in the Solway Basin actually penetrated rocks of this age – 52 m of Rhaetian to Hettangian claystones. However, seismic interpretation suggests that a thin Liassic section extends over the central and southwest parts of the Solway Basin. Similarly, although the Elf 112/19-1 well on the southwest flank of the Peel Basin did not intersect Liassic strata, the well is judged on seismic evidence to be at the edge of a Liassic inlier that covers the basin centre and may be up to 800 m thick (Newman, 1999). None of the Kish Bank Basin wells (Figure 6.5) encountered Jurassic rocks. However, seismic mapping, seabed sampling and other evidence (Dobson and Whittington, 1979; Jenner, 1981; Broughan *et al.*, 1989) suggest the presence of Jurassic strata along the western and northern margins of the basin (Figure 6.5), with up to 2,700 m of possible Liassic rocks preserved against the Bray and Dalkey bounding faults (Broughan *et al.*, 1989). A sequence similar to that in Northern Ireland and Cardigan Bay is anticipated. Liassic rocks were sampled from the seabed at the northern margin of the basin (Etu-Efeotor, 1976; Dobson, 1977a, b), and Dobson and Whittington (1979) referred to dated Lower Liassic core material composed of dark grey carbonate mudstone and a light grey fine-grained limestone.

A thin succession of Hettangian to Sinemurian marine siltstones and claystones was drilled on the St Tudwal's Arch extension at the southern margin of the Central Irish Sea Basin (Marathon 42/17-1, Figure 6.7). However, seismic evidence points to the possible presence of a more complete Jurassic and Cretaceous succession in the central part of that basin (Maddox *et al.*, 1995).

Early Jurassic palaeogeography

In Western Europe a number of the main rift basins established during the Triassic continued to subside. A major marine transgression from Tethys, initiated in latest Triassic time (Rhaetian), progressed rapidly in the earliest Jurassic. Passage into the western basins was along established rift systems such as the Bay of Biscay Rift. The Early Jurassic was therefore a period of transgression on the shelf areas west of Britain. There was rapid encroachment across the subdued Rhaetian topography, on which only a few structural intra-basinal highs persisted as areas of non-marine deposition (Figure 7.8). Land area was reduced to a series of islands with Palaeozoic cores – e.g. Cornubia, part of Wales, western Ireland, Porcupine and Rockall banks. Deeper marine conditions spread along the main rift axes, and by this time there was free communication between Tethys and the Boreal Ocean.

The uniformly fine-grained and thin-bedded lithologies of the Liassic in the Irish Sea and Ulster basins is indicative of regionally extensive, relatively shallow quiet marine conditions following marine transgression across the flat late Triassic surface. The basin margin faults show little indication of movement during this period (Woodland, 1971; Cope, 1984; Broughan *et al.*, 1989). Shallow marine and non-marine environments are a feature of the Atlantic margin basins. Varied marine conditions obtained on the Celtic Sea shelf, with periodic clastic input from the exposed positive elements and the Irish mainland, whilst deeper, more uniform conditions developed towards the Goban Spur.

Offshore distribution of Middle–Upper Jurassic rocks

The variation of lithologies and facies in the Middle Jurassic of the Celtic Sea basins is a reflection of the uplift and erosion accompanying Cimmerian tectonism in the region (Naylor and Shannon, 1982; Ziegler, 1990). Shallowing conditions had developed in late Toarcian time and continued into the Middle Jurassic (Figure 7.10), producing a mixed facies regime. In the east of the basin there was thick nearshore to deltaic coarse sand input during the Aalenian to early Upper Bajocian interval whilst a more condensed argillaceous sequence, with calcareous and sand interbeds, developed in the west. Sand deposition in the east was interrupted by a transgressive marine pulse that gave rise to mud deposition in early Bajocian time, before the clastic input resumed. More widespread marine argillaceous deposition became

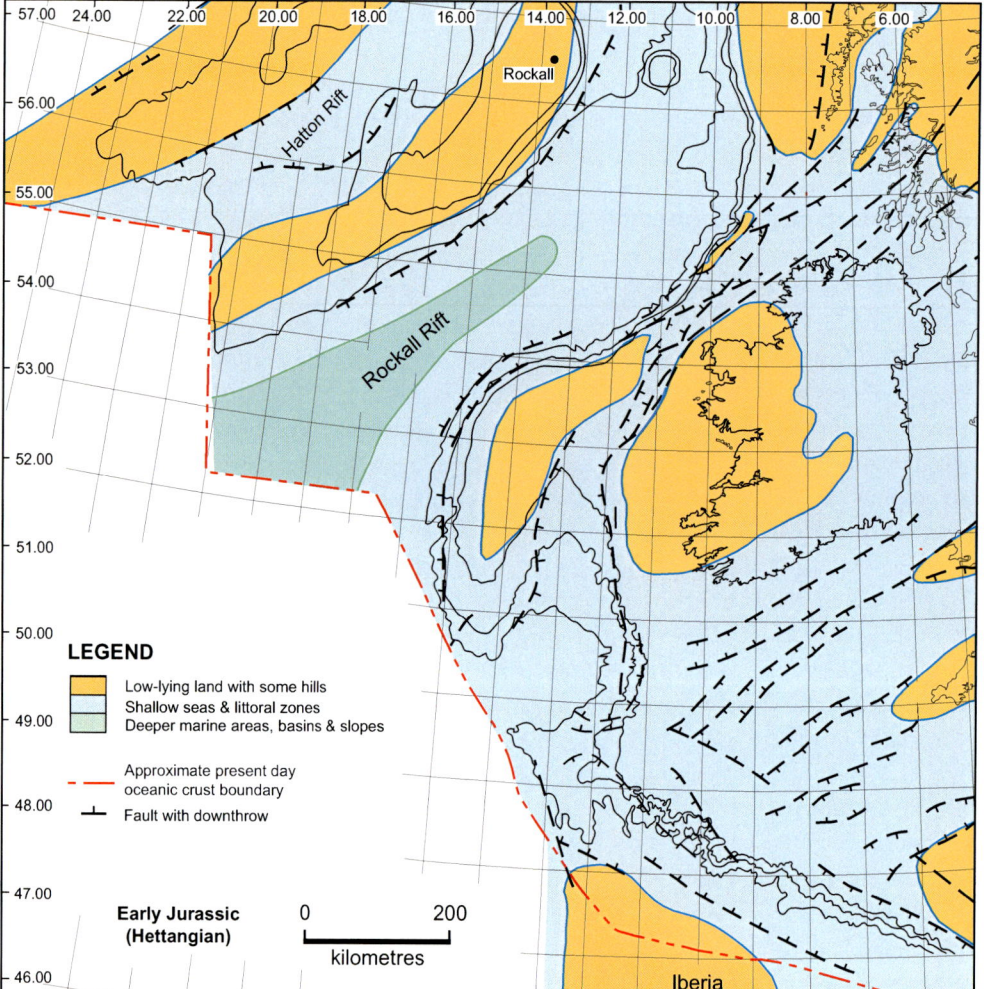

Figure 7.8. Early Jurassic (Hettangian) palaeogeography (after Naylor, 1998. *Principal sources*: Bradshaw *et al.*, 1992; Keeley, 1995; Naylor and Shannon, 1982; Naylor, 1992; Ziegler 1990).

LEGEND

Low-lying land with some hills
Shallow seas & littoral zones
Deeper marine areas, basins & slopes

— · — Approximate present day oceanic crust boundary

Fault with downthrow

Early Jurassic (Hettangian)

0　　　200

kilometres

established across the Celtic Sea region in Late Bajocian time. Basic intrusive rocks have been encountered in wells in the central part of the Fastnet Basin (Figure 7.2A), and are imaged on seismic profiles. Caston *et al.* (1981) suggested that they have a widespread occurrence throughout the central region of that basin, associated with a northeast trending fault zone. The sills are composed of olivine micro-gabbro and olivine dolerite, and K/Ar dating indicates a Late Bajocian age.

Marine conditions, with deposition of grey to grey-brown mudrocks, persisted into late Early Bathonian time, when there was a phase of regressive near-coast bioclastic sand deposition in the northeast of the North Celtic Sea Basin. This was followed by the establishment of carbonate shelf sedimentation extending from the northeast of the North Celtic Sea Basin into the St George's Channel Basin, whilst shallow-marine clastics and interbedded limestones were deposited in the Cardigan Bay Basin (Figure 7.9). Caston (1995) described the Bathonian limestone unit in Block 49/9 (Figure 7.11) as comprising up to 48 m of pelletal/oolitic limestone with thin interbedded calcareous sandstones and calcareous mudstones. These units interdigitate westwards with open-marine mudrock-prone sequence that dominated the remainder of the North Celtic Sea Basin (Millson, 1987). Further regression late in the Bathonian produced non-marine red-brown anhydritic mudrocks with local sandstones, probably across much of the basin. The latest Bathonian freshwater–?brackish strata are succeeded by possible early Callovian strata, lacking diagnostic fauna (Ainsworth *et al.*, 1987), also of non-marine continental facies. Callovian strata are lacking in the west of the region, whilst at the eastern end of the North Celtic Sea Basin (Block 49/9: Figure 7.11) there was deposition of ?Callovian to Oxfordian non-marine braided fluvial sandstones, possibly derived from outcrops along the south coast of Ireland, resting unconformably on the Bathonian (Caston, 1995). A marginal marine and non-marine facies complex was developed in the St George's Channel Basin, possibly sourced from the Leinster massif in southeast Ireland. Reddened clay deposits recorded onshore adjacent to the south coast of Ireland are probably of Middle Jurassic age (Higgs and Beese, 1986), and represent a thin non-marine deposit. This suggests that thick Jurassic sedimentation was confined to the offshore fault-controlled basins.

A major unconformity near the base of the Upper Jurassic is assigned a mid to late Oxfordian age, but could be somewhat older (Millson, 1987). This marks a further pulse of Cimmerian tectonism that gave rise to syn-rift, sand-prone strata of continental to shallow-marine origin, with sediment derivation from the re-activated basin margins. A relative rise in sea level produced more shale-prone, and locally limestone-prone, strata during the Late Jurassic throughout the North Celtic Sea Basin, although there is no evidence of widespread Upper Jurassic strata in the Fastnet and Cockburn basins. This may be the result of erosion following the Late Cimmerian tectonic phase at the end of the Jurassic. However, Shannon (1996) suggested that the basic intrusions of this age reflect the presence of a hot spot that caused local doming and prevented the deposition of a thick Upper Jurassic sequence comparable to that in the basins both to the east (North Celtic Sea Basin) and to the west (Porcupine Basin) – see the palaeogeographic reconstruction in Figure 7.12. In the eastern Celtic Sea area, grey and anhydritic mudrocks of Middle–Late Oxfordian age are mainly non-marine and rest unconformably on Middle and even Lower Jurassic strata. In Block 49/9 (Figure 7.11) a sequence of Upper Oxfordian to Lower Kimmeridgian thin sandstones, limestones and shales was drilled adjacent to the northern boundary of the basin, with marine intervals at base and top (Caston, 1995). A sharply defined marine pulse in the Late Oxfordian time is indicated from microfaunal evidence (Ainsworth *et al.*, 1987) and follows upon an unconformity in the North Celtic Sea Basin (Colin *et al.*, 1981). The Lower Kimmeridgian continues in non-marine facies and contains coarse, marginal clastics that grade basinwards into anhydritic mudrocks. Transgression then produced variable Late Kimmeridgian–Tithonian, dominantly marine, facies over much of the region, including the more persistent intra-basinal highs, with shallowing upwards at the top of the sequence. In Block 49/9 the Tithonian is represented by a thin sequence of marls deposited in a lacustrine to marginal marine environment (Caston, 1995). The South Celtic Sea Basin was probably an area of non-deposition throughout the latest Jurassic (Chapter 10). The final phase of Tithonian (to earliest Cretaceous) deposition was of argillaceous micrites, calcareous mudrocks and carbonates. Subsidence and deposition occurred during the interval along the axis of the North Celtic Sea Basin and into the St George's Channel Basin, with more varied and incomplete sequences at the basin margins.

Wells in the Cardigan Bay–St George's Channel basins (Figure 7.9) penetrated an almost complete Middle–Upper Jurassic succession (Barr *et al.*, 1981). Although sedimentation was probably continuous from the Lower Jurassic into the Middle Jurassic in the basin depocentres, there is evidence for uplift and erosion at the basin margins. Over St. Tudwal's Arch extension an estimated 650 m of Liassic strata were removed prior to ?Bajocian–Bathonian deposition, whilst on the north flank of the Pembroke Ridge sandstones of possible Bajocian age rest on the Pliensbachian succession (Tappin *et al.*, 1994). However, within the basins the Liassic is succeeded with apparent conformity by up to 760 m (HGB 107/21-1) of Middle Jurassic mudstones, sometimes calcareous, with thin limestone interbeds, probably the product of shallow marine neritic and carbonate shelf environments. As in the North Celtic Sea Basin, there was sand input in Aalenian to earliest Late Bajocian time in St George's Channel Basin wells (e.g. Texaco/HGB 103/2-1). In contrast to the Celtic Sea area, the Middle Jurassic is conformably overlain by a complete Upper Jurassic succession of shales, mudstones and calcareous mudstones (1498 m: Arco 106/24-1) ranging up to Tithonian in age. There is a distinct change late in early Oxfordian time to reddened and locally anhydritic mudrocks with a microfauna indicative of brackish to freshwater deposition (Arco 106/24-1). This was followed by a marine incursion in Late Bathonian time, before a return to continental and marginal conditions. Non-marine conditions were abruptly terminated by a widespread incursion in Early Kimmeridgian time, accompanied by deposition of grey calcareous mudrocks that persisted into the Tithonian, with increasing limestone content.

The Middle Jurassic of the Goban Spur (Cook, 1987) continues the pattern established in the Lower Jurassic and consists of Late Toarcian to Bajocian claystones and siltstones representing marine inner to outer shelf environments (Colin *et al.*, 1992), with rare paralic or

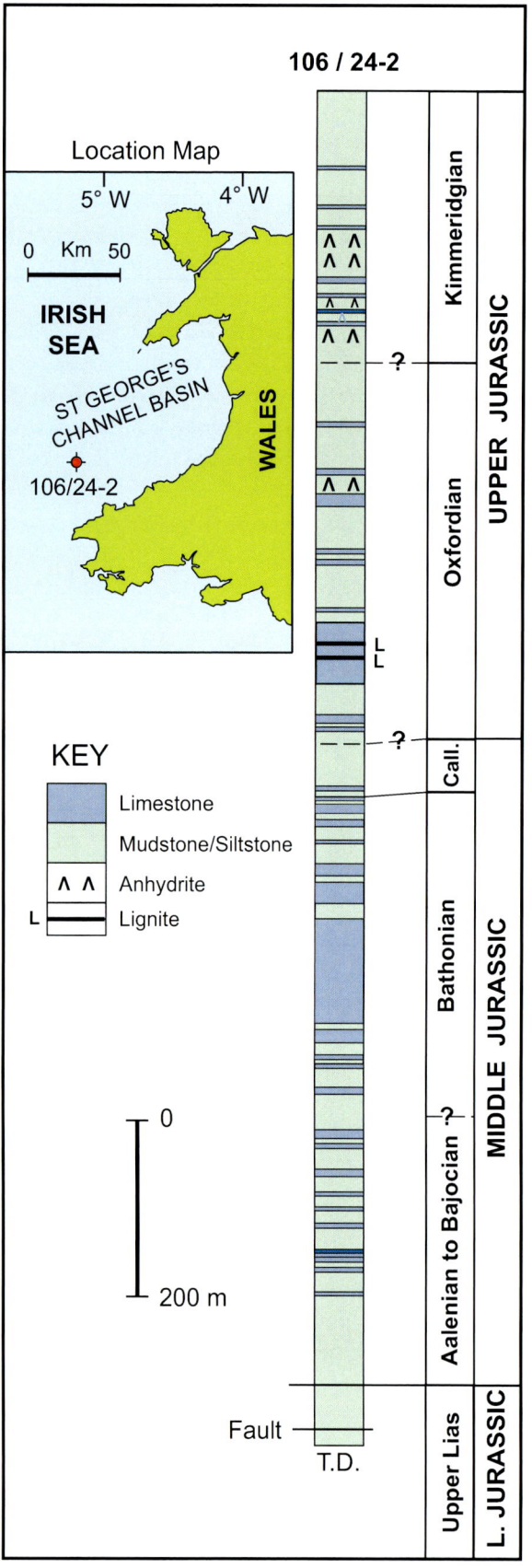

Figure 7.9. Jurassic sequence in the Arco 106/24-2 well, St George's Channel Basin (after Tappin *et al.*, 1994).

marginal sandstone intercalations (Figure 7.4). Similar marine claystones continue into the overlying Bathonian to Callovian interval, but these are then succeeded by 30 m of massive high-energy stacked beach sandstones with excellent reservoir characteristics (Cook, 1987). The subarkosic sandstones may have been sourced locally from Devonian rocks to the southeast, similar to those encountered in DSDP drilling (Site 548) in the area (Lefort *et al.*, 1985). This sequence is succeeded by 215 m of porphyritic basaltic andesite flows that K–Ar dating, reported by Tate and Dobson (1988), indicates are of Valanginian age. In the Esso 62/7-1 well on the Goban Spur Upper Jurassic strata are absent, probably due to Cimmerian tectonism (Colin *et al.*, 1992). From seismic evidence (Cook, 1987; Naylor *et al.*, 2002), Upper Jurassic rocks may occur elsewhere in the basin, possibly in marginal marine or paralic facies.

Middle Jurassic sequences are known from the northern part of the Porcupine Basin, where they rest unconformably on Liassic and older strata (Naylor and Shannon, 1982; Croker and Shannon, 1987). Continental coarse clastics in fining-up cycles were deposited in Bajocian to early Bathonian times by braided river systems issuing from a northern source and prograding southwards, sometimes as alluvial fans. Fluvial, alluvial and lacustrine deposits are recognised. This sequence was interpreted (Sinclair *et al.*, 1994) as the product of warping immediately prior to the commencement of the Late Cimmerian rifting phase. The clastics are succeeded in the north of the basin by Bathonian fine-grained sandstones and mudrocks, again dominantly non-marine, the product of meandering channels and related environments. Most of the remaining intrabasinal highs were finally inundated during the Bathonian to Callovian interval, when a sequence of thin sandstones and mudstones with rare limestones was deposited in non-marine lacustrine and fluvial conditions. The dominantly fluvial Middle Jurassic deposits in the north of the basin may pass southwards to shallow-marine sediments not yet penetrated in exploratory drilling (Croker and Shannon, 1987). The Middle Jurassic in the southern part of the Porcupine Basin is probably comparable to that of the Celtic Sea and Goban Spur. In the Rockall Basin, the Shell 12/2-1z well encountered approximately 130 m of fluvial conglomeratic sandstones and calcretes, overlain by a thin unit of marine sandstones, in turn capped by a

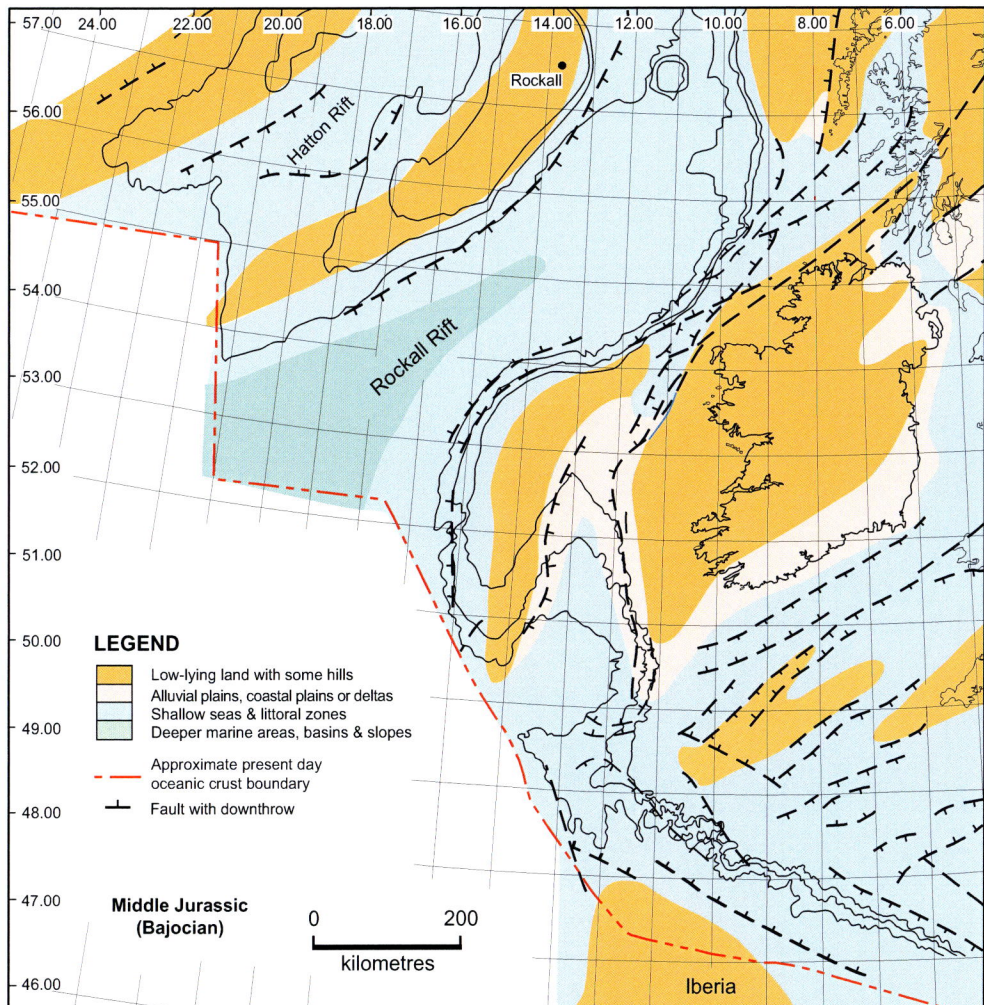

Figure 7.10. Middle Jurassic: Bajocian palaeogeography (after Naylor and Shannon, 2009. *Principal sources*: Bradshaw *et al.*, 1992; Keeley, 1995; Naylor and Shannon, 1982; Naylor, 1992, 1998; Ziegler, 1990).

LEGEND

- Low-lying land with some hills
- Alluvial plains, coastal plains or deltas
- Shallow seas & littoral zones
- Deeper marine areas, basins & slopes

--- Approximate present day oceanic crust boundary

⊥ Fault with downthrow

Middle Jurassic (Bajocian)

0 — 200 kilometres

volcanic unit (Tyrrell *et al.*, 2010). This succession, underlying Cretaceous marine mudstones and overlying Permo-Triassic sandstones, is thought to be of Middle Jurassic age. It bears similarities to the Middle Jurassic succession in the northern part of the Porcupine Basin. While a minor volcanic unit was encountered in the well, overall it appears that volcanic activity in the Atlantic basins was relatively limited in comparison with the central part of the North Sea where major plume-related volcanism resulted in extensive uplift.

Upper Jurassic (Oxfordian to Tithonian) strata have been penetrated in many of the Porcupine Basin wells. Sedimentation was apparently continuous through Middle and Late Jurassic time, with no widespread evidence of the intervening unconformity that is a feature of the Celtic Sea region. Nevertheless, late Middle and Late Jurassic time was a period of differential subsidence in the basin, reflected by varied continental sandstone

and shale sequences, interrupted by intercalations of shallow-marine strata. Oxfordian conglomerates drilled by Phillips 35/8-1, in the north part of the basin (Figure 3.6), contain quartz and fresh feldspar suggestive of an exposed nearby granite pluton, while Kimmeridgian conglomerates in Shell 34/19-1 contain schistose lithoclasts (Tate and Dobson, 1989b). The overall facies pattern suggests a northward encroachment of transgressive facies (Croker and Shannon, 1987) along the basin axis, with active movement on the marginal bounding faults. The basin was finally transgressed in the Late Oxfordian, with upward transition though a fining-upward bioclastic rich sandstone into marine shales (Sinclair *et al.*, 1994). The latest Jurassic (Tithonian) exhibits marked changes in thickness and facies, reflecting the initiation of the Late Cimmerian phase of deformation. Tithonian fan wedges have been drilled at the northern and western basin margins. The BP 26/28-1 oil discovery well at the

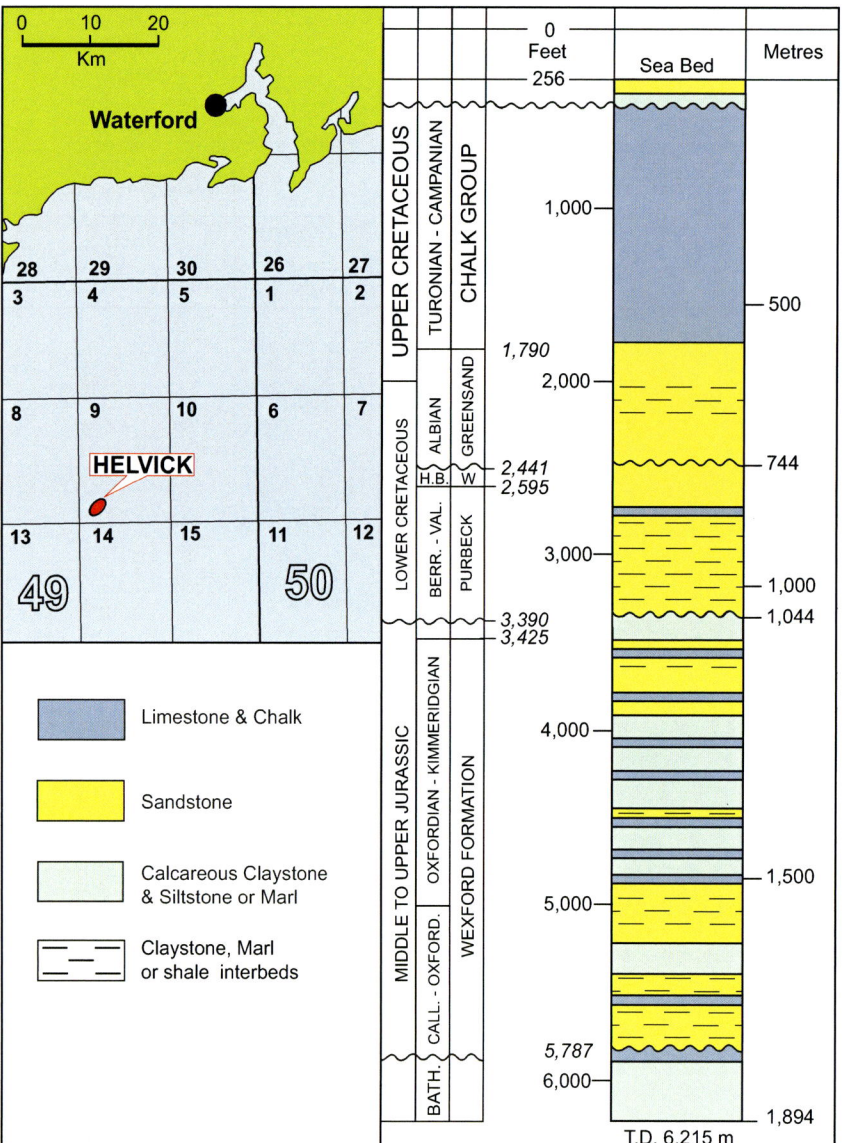

Figure 7.11. Summary stratigraphy of the Gulf 49/9-2 well (Helvick oil accumulation), North Celtic Sea Basin (after Caston, 1995). Well depths are sub-sea. W = Wealden.

northern rim of the basin (MacDonald *et al.*, 1987) demonstrated continuous subsidence from Middle Jurassic to Late Jurassic (Bajocian to Tithonian) time. The lower part of the sequence is a product of braided river or alluvial fan systems and is overlain by Bathonian strata deposited in a floodplain environment with meandering rivers. Marine influences gradually increased, and nearshore marine conditions became dominant in Kimmeridgian to early Tithonian time. Sediment derivation from Carboniferous and metamorphic provenance is indicated by petrographic studies (MacDonald *et al.*, 1987), presumably from areas north and west of the basin. Smith and Higgs (2001) reported reworked Carboniferous palynomorphs within Upper Jurassic

sandstones from wells in Quadrants 26 and 35 of the northern part of the Porcupine Basin, with probable derivation from the eastern margin of the basin. From late Kimmeridgian time onward local block faulting appears to have exercised an increasing influence on sedimentation, with variation in the thickness of stratigraphic units. The entire sequence is truncated by the Late Cimmerian (Base Cretaceous) unconformity surface (locally composite unconformities). As might be expected, the Upper Jurassic section away from the basin margin is less varied (e.g. Figure 7.5; Phillips 35/8-2 well; Croker and Shannon, 1987). Proximal to distal submarine fan sandstones and shales are overlain by marine shales (Tithonian).

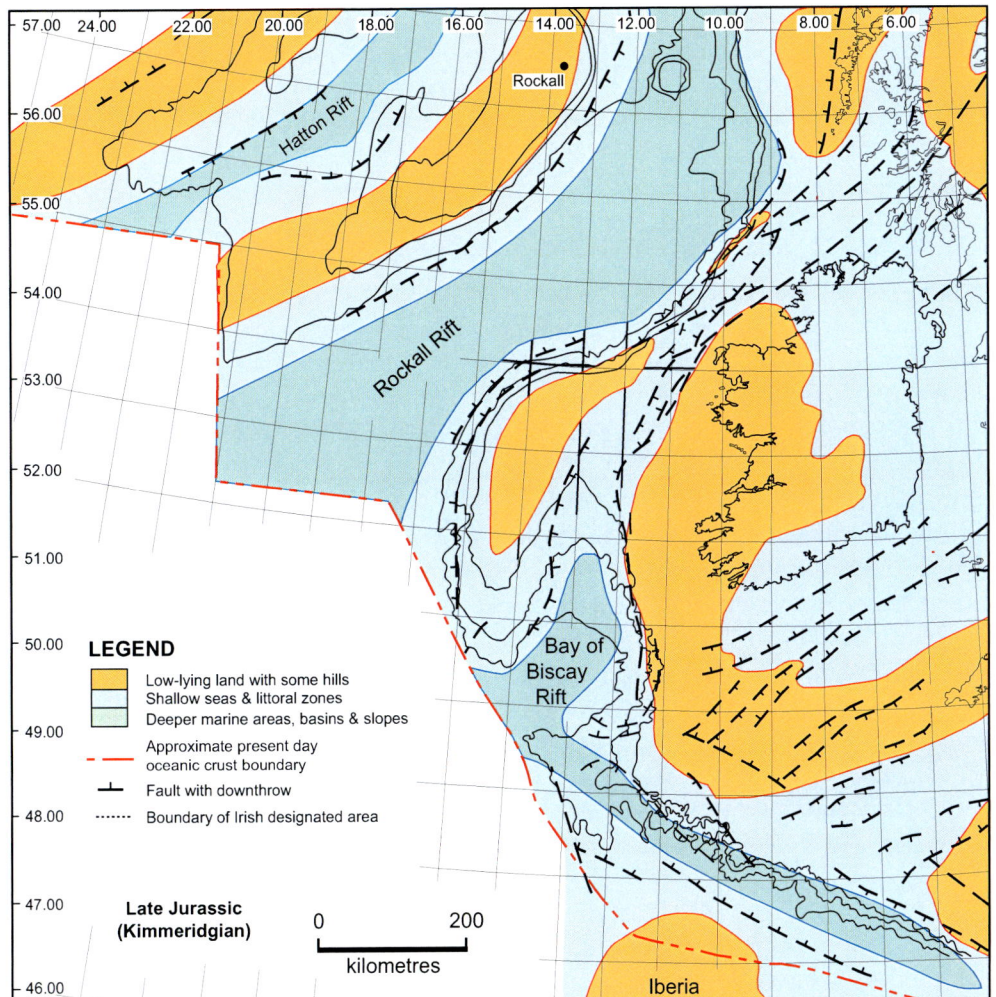

Figure 7.12. Late Jurassic (Kimmeridgian) palaeogeography (after Naylor 1998. *Principal sources*: Bradshaw *et al.*, 1992; Keeley, 1995; Naylor and Shannon, 1982; Naylor, 1992; Ziegler, 1990).

The nature of Jurassic deposition within the Rockall or Hatton basins is speculative, although Naylor and Shannon (2005) suggested that rocks of this age may be widespread (Figure 8.1). Folded and faulted pre-Cenozoic rocks (Mesozoic and possibly older) have been imaged on seismic sections on the UK portion of the Hatton High, and are thought to be widespread on the High and within the Hatton Basin (Hitchen, 2004). If a similar structural picture obtains within Rockall Basin then the disposition of Jurassic rocks may be more complex than envisaged on Figure 8.1. Middle and Upper Jurassic strata are also interpreted from seismic data (Figure 7.6) within the small perched basins along both margins of the Rockall Basin (Naylor *et al.*, 1999). Coring on the western flank of the Porcupine High (Haughton *et al.*, 2005) has confirmed the presence of Upper Jurassic strata (early Kimmeridgian to early Portlandian marine sandstones and limestones) in the North Bróna Basin,

and possibly in the South Bróna Basin (Figure 2.15).

A hiatus in sedimentation is recognised within the Slyne Basin (Figure 7.7) during the Toarcian that is not of great significance in the basin centre, but becomes more pronounced towards the basin margins. Subsidence and deposition resumed in latest Toarcian time and continued through the Middle Jurassic. The Elf 27/13-1 well in the Slyne Basin penetrated an almost complete Middle Jurassic sequence (Trueblood, 1992). The Toarcian to upper Bajocian (320 m) comprises two coarsening-upwards offshore sandbar complexes, each capped with a limestone unit. The overlying succession comprises 1320 m of upper Bajocian to upper Bathonian brackish water strata. A number of sandstone units in the lower Bathonian succession aggregate 59 m of net reservoir quality sandstone, with hydrocarbon shows. In the well, these are overlain by 358 m of strata, possibly late Bathonian in age, and these in turn by Quaternary to Recent sediments. However, seismic evidence suggests (Trueblood and Morton, 1991) that Upper Jurassic rocks are probably present in the deeper parts of the Slyne Basin (Figure 7.7). Dancer *et al.* (2005) described a Middle Jurassic (1250–1700 m thick) succession in the Corrib Gasfield area, overlying the regional hiatus, that is similar in outline to that in the Elf 27/13-1 well. A lower sequence of marine claystones with interbedded limestones that become more common near the top of the unit (Aalenian–Bathonian) is overlain by fluvio-estuarine claystones with sandstone, siltstone and coal interbeds (Bathonian). However, in the Corrib Gasfield area Upper Jurassic strata (366–506 m thick) are preserved within a collapsed graben on the crest of the field structure, and comprise silty claystones with sandstone, limestone, and coal interbeds.

As noted earlier, Middle and Upper Jurassic strata are generally absent in the Erris Basin. Middle and Upper Jurassic rocks are also unknown onshore in Northern Ireland, and in the offshore basins around the north coast. Rocks of this age could form part of the undrilled sequence in the Kish Basin (Broughan *et al.*, 1989), and in other parts of the Irish Sea, but there is no direct evidence of this. The absence of Middle and Upper Jurasssic strata around the north of Ireland and in the Irish Sea is discussed in Chapter 12.

Middle and Late Jurassic palaeogeography

Millson (1987) suggested that facies development in the Celtic Sea region was directly related to Central Atlantic rifting of Aalenian age between the African and North American plates, and that transgression in the following late Bathonian–Callovian interval coincided with the onset of seafloor spreading in the same region. The break-up of Pangaea, initiated during Middle Jurassic time, had significant impact in the whole of the Irish margin. There is evidence that the Boreal and Tethyan faunal provinces were separated during Bajocian–Bathonian times. It is likely, therefore, that the northern Rockall and Hatton areas were very shallow and emergent for periods, acting as a barrier. Connection was re-established between the two provinces later in the Middle Jurassic. The Irish mainland at this time was almost entirely emergent, with land prolongations extending southwards along both sides of the Porcupine Basin (Figure 7.10). In the Slyne Basin, stacked nearshore sand bars were succeeded by brackish water strata. Middle Jurassic sediments are not preserved in the Erris or Donegal basins, the Northern

Ireland basins or in the Irish Sea area, probably as a result of later erosional episodes (*see* Chapter 12 for discussion). A regional pattern emerges in the Middle Jurassic of intermittent subsidence around the southern part of Ireland. Relatively thick sequences accumulated within the southern basins, from Cardigan Bay out to the Goban Spur, probably with generally westwards-deepening conditions. With the separation of the European segments of Laurasia from the African portion of Gondwanaland, stresses in Europe related to the Tethyan rift system diminished. The dominant process now was the opening of the Central Atlantic Ocean and the build-up of extensional stresses in the North Atlantic domain (Ziegler, 1988). Rifting increased in the Labrador Sea and within the Greenland–Norway system, and deeper water conditions developed along the main rift axes. From an Irish viewpoint, events henceforth would be dominated by the extensional stresses related to the opening of the North Atlantic. Marine conditions persisted intermittently through Late Jurassic time along the southern seaboard of Ireland. Tectonic influences from the Atlantic domain gave rise to an important unconformity in the Celtic Sea region. The varied nature of the Upper Jurassic in the Celtic Sea contrasts with the widespread uniformity of units such as the Kimmeridge Clay in the North Sea, and probably reflects proximity to the rifting events in the proto-Atlantic (Millson, 1987). The Tithonian to early Berriasian non-marine and marginal marine sequences represent the final, areally restricted filling of the passively subsiding North Celtic Sea Basin following a Late Jurassic rifting episode (late Oxfordian to early Tithonian; Petrie *et al.*, 1989).

Upper Jurassic deposits along the Atlantic margin preceded the onset of the Main Cimmerian tectonic episode, with the development of a spectrum of facies. These ranged from basin-edge alluvial fans, braided and meandering fluvial systems to submarine fans (Croker and Shannon, 1987, MacDonald *et al.*, 1987; Shannon, 1992, 1993). Seismic evidence suggests that Upper Jurassic sequences are present in the deeper parts of the Goban Spur Basin, possibly in marginal marine facies (Auffret *et al.*, 1979; Naylor *et al.*, 2002). Elsewhere in the basin the sediments are absent due to tectonism and erosion. At the outset of Late Jurassic time in the Porcupine Basin, deltaic facies were deposited in the north of the basin, with coeval marine shales further south. Gradual transgression resulted in marine shale deposition throughout the basin. The early Tithonian marked the onset of Late Cimmerian extensional faulting, with block rotation evident on seismic sections, and this was accompanied by an influx of coarse siliciclastic sediment (Sinclair *et al*, 1994). The basin-bounding faults were active during this period of crustal extension, and submarine fans developed along the margins and in the deeper parts of the basin. Sea levels began to fall in late Kimmeridgian time and this continued during the Tithonian, with the result that shorelines gradually retreated. Marine conditions prevailed in the north of the basin during the Tithonian, first with the deposition of quiet water organic-rich shales, and then from Tithonian into earliest Cretaceous time with a more varied shallow-marine sequence. Synrift deposition was then terminated by a major angular unconformity (or sometimes a composite set of unconformities) that separates possible late Berriasian (Moore, 1992) and older strata from onlapping Valanginian and younger sediments (*see* discussion in Chapter 8).

Around Ireland, Callovian to mid-Kimmeridgian time was a period of gradually rising sea level, before regression once again set in during the final phases of the Jurassic. The suggested regional palaeogeography in Kimmeridgian time is shown on Figure 7.12. Western Ireland was probably emergent and possibly linked, through the Fastnet area, to a land extension from Cornubia. Porcupine Bank, Rockall Bank and Hatton Bank, and residual portions of the Scottish landmass were also probably emergent. Deeper water tongues occupied Rockall Trough and the Bay of Biscay Rift and extended also into the Porcupine Seabight and Hatton Trough. The region north and east of Ireland was probably a positive area of thin deposition (or even non-deposition), as in Middle Jurassic time. Thus, despite the widespread marine conditions, Upper Jurassic rocks are not preserved in the Donegal and Erris basins, the offshore Northern Ireland basins and the Central Irish Sea basins. The same is true onshore in Ireland, with the exception of the occurrence (Higgs and Jones, 2000) of karstic clay deposits with nearshore tidal influences discovered onshore at Piltown, County Kilkenny, containing palynomorphs indicative of a Late Jurassic to Early Cretaceous age (Figure 4.12). Consideration of regional palaeogeography suggests that a Late Jurassic age is the more likely – compare Figures 7.12 and 8.8.

The block faulting, tilting and erosion associated with the Late Cimmerian tectonic episode are most clearly seen in the Atlantic and Celtic Sea basins, where the time interval is relatively constrained by the beds above and below the unconformity. In the Malin, Ulster and Irish Sea basins uplift and erosion of post-Liassic strata could have taken place at any time after the Early Jurassic (*see* Chapter 12 for discussion).

References

Ainsworth, N.R., O'Neill, M., Rutherford, M.M., Clayton, G., Horton, N.F., and Penny, R.A. (1987) Biostratigraphy of the Lower Cretaceous, Jurassic and Late Triassic of the North Celtic Sea and Fastnet Basins. *In*: Brooks, J. and Glennie, K.W. (eds), *Petroleum Geology of North-West Europe, Volume 2*. Graham and Trotman, London, 611–622.

Auffret, G.A., Pastouret, L., Cassat, G., De Charpel, O., Gravatte, J., and Guennoc, P. (1979) Dredged rocks from the Armorican and Celtic margins. *In*: Montadert, L., Roberts, D. *et al*. (eds), *Initial Reports on the Deep Sea Drilling Project*, **48**, 995–1013.

Barr, K.W., Colter, V.S., and Young, R. (1981) The geology of the Cardigan Bay–St. George's Channel Basin. *In*: Illing, L.V. and Hobson, G.D. (eds), *Petroleum Geology of the Continental Shelf of North-West Europe*. Heyden and Son Ltd., London, 432–443.

Bradshaw, M.J., Cope, J.W.C., Cripps, D.T., Donovan, D.T., Howarth, M.K., Rawson, P.F., West, M., and Wimbledon, W.A. (1992) Jurassic. *In*: Cope, J.C.W., Ingham, J.K., and Rawson, P.F. (eds), *Atlas of Palaeogeography and Lithofacies*. Geological Society, London, Memoir **13**, 153pp.

Broughan, F. M., Naylor, D., and Anstey, N.A. (1989) Jurassic rocks in the Kish Bank Basin. *Irish Journal of Earth Sciences*, **10**, 99–106.

Caston, V.N.D. (1995) The Helvick oil accumulation, Block 49/9, North Celtic Sea Basin. *In*: Croker, P.F. and Shannon, P.M. (eds), *The Petroleum Geology of Ireland's Offshore Basins*. Geological Society, London, Special Publications, **93**, 209–225.

Caston, V.N.D., Dearnley, R., Harrison, R.K., Rundle, C.C., and Styles, M.T. (1981) Olivine-dolerite intrusions in the Fastnet Basin. *Journal of the Geological Society of London*, **138**, 31–46.

Colin, J.P., Ioannides, N.S., and Vining, B. (1992) Mesozoic stratigraphy of the Goban Spur, offshore south-west Ireland. *Marine and Petroleum Geology*, **9**, 527–241.

Colin, J.P., Lehmann, R.A., and Morgan, B.E. (1981) Cretaceous and Late Jurassic biostratigraphy of the North Celtic Sea Basin. *In*: Neale, J. W. and Brazier, M.D. (eds), *Microfossils from Recent and fossil shelf seas*. Ellis Horwood Ltd., Chichester, 122–155.

Cook, D.R. (1987) The Goban Spur: exploration in a deep-water frontier basin. *In*: Brooks, J. and Glennie, K.W. (eds), *Petroleum Geology of North-West Europe*. Graham and Trotman Ltd, London, 623–632.

Cope, J.C.W. (1984) The Mesozoic history of Wales. *Proceedings of the Geologists' Association*, **95**, 373–385.

Croker, P.F. and Shannon, P.M. (1987) The evolution and hydrocarbon prospectivity of the Porcupine Basin, offshore Ireland. *In*: Brooks, J. and Glennie, K.W. (eds), *Petroleum Geology of North-West Europe*, Graham and Trotman, London, 633–642.

Dancer, P.N., Algar, S.T., and Wilson, I.R. (1999) Structural

Evolution of the Slyne Trough. *In*: Fleet, A.J. and Boldy, S.A.R. (eds), *Petroleum Geology of Northwest Europe: Proceedings of the 5th Conference.* Geological Society, London, 445–453.

Dancer, P.N., Kenyon-Roberts, S.M., Downey, J.W., Baillie, J.M., Meadows, N.S., and Maguire, K. (2005) The Corrib gas field, offshore west of Ireland. *In*: Doré, A.G. and Vining, B.A. (eds), *Petroleum Geology: North-West Europe and Global Perspectives. Proceedings of the 6th Petroleum Geology Conference.* Geological Society, London, 1035–1046.

Dobson, M.R. (1977a) The geological structure of the Irish Sea. *In*: Kidson, C. and Tooley, M.J. (eds), *The Quaternary History of the Irish Sea.* Seal House Press, Liverpool, 13–26.

Dobson, M.R. (1977b) The history of the Irish Sea basins. *In*: Kidson, C. and Tooley, M.J. (eds), *The Quaternary History of the Irish Sea*, Seal House, Press, Liverpool, 93–98.

Dobson, M.R. and Whittington, R.J. (1979) Geology of the Kish Bank Basin. *Journal of Geological Society of London*, **136**, 243–249.

Etu-Efeotor, J.D. (1976) Geology of the Kish Bank Basin. *Journal of the Geological Society of London*, **132**, 708.

Fyfe, J.A., Long, D., and Evans, D. (1993) United Kingdom offshore regional report: *The geology of the Malin–Hebrides Sea area.* London HMSO for the British Geological Survey, 91pp.

Haughton, P., Praeg, D., Shannon, P., Harrington, G., Higgs, K., Amy, L., Tyrrell, S., and Morrissey, T. (2005) First results from shallow stratigraphic boreholes on the eastern flank of the Rockall Basin, offshore western Ireland. *In*: Doré, A.G. and Vining, B.A. (eds), *Petroleum Geology: North-West Europe and Global Perspectives – Proceedings of the 6th Petroleum Geology Conference.* Geological Society, London, 1077–1094.

Higgs, K. and Beese, A.P. (1986) A Jurassic microflora from the Colbond Clay of Cloyne, County Cork. *Irish Journal of Earth Sciences*, **7**, 99–109.

Higgs, K.T. and Jones, G.L. (2000) Palynological evidence for Mesozoic karst at Piltown, Co. Kilkenny. *Proceedings of the Geologists' Association*, **111**, 355–362.

Hitchen, K. (2004) The geology of the UK Hatton–Rockall margin. *Marine and Petroleum Geology*, **21**, 993–1012.

Jenner, J.K. (1981) The structure and stratigraphy of the Kish Bank Basin. *In*: Illing, L.V. and Hobson, G.D. (eds), *Petroleum Geology of the Continental Shelf of North-West Europe*, Heyden and Son Ltd., London, 426–431.

Keeley, M.L. (1995) New evidence of Permo-Triassic rifting, onshore southern Ireland, and its implications for Variscan structural inheritance. *In*: Boldy, S.A.R. (ed.), *Permian and Triassic Rifting in Northwest Europe.* Geological Society, London, Special Publications, **91**, 239–253.

Kessler, L.G. and Sachs, S.D. (1995) Depositional setting and sequence stratigraphic implications of the Upper Sinemurian (Lower Jurassic) sandstone interval, North Celtic Sea/St George's Channel Basins, offshore Ireland. *In*: Croker, P.F. and Shannon, P.M. (eds), *The Petroleum Geology of Ireland's Offshore Basins.* Geological Society, London, Special Publications, **93**, 171–192.

Lefort, J.P., Peucat, J.J., Deunff, J., and Le Herisse, A. (1985) The Goban Spur Palaeozoic basement. *In*: Bailey, M.G. (ed.), *Initial Reports of the Deep Sea Drilling Project 80.* U.S. Government Printing Office, Washington D.C., 677–679.

MacDonald, H., Allan, P.M., and Lovell, J.P.B. (1987) Geology of oil accumulation in Block 26/28, Porcupine Basin, offshore Ireland. *In*: Brooks, J. and Glennie, K.W. (eds), *Petroleum Geology of North-West Europe.* Graham and Trotman Ltd, London, 643–651.

Maddox, S.J., Blow, R., and Hardman, M. (1995) Hydrocarbon prospectivity of the Central Irish Sea Basin with reference to Block 42/12, offshore Ireland. *In*: Croker, P.F. and Shannon, P.M. (eds), *The Petroleum Geology of Ireland's Offshore Basins.* Geological Society, London, Special Publications, **93**, 59–77.

Millson, J.A. (1987) The Jurassic evolution of the Celtic Sea basins. *In*: Brooks, J. and Glennie, K.W. (eds), *Petroleum Geology of North-West Europe*, Graham and Trotman, London, 599–610.

Moore, J.G. 1992. A syn-rift to post-rift transition sequence in the Main Porcupine Basin, offshore western Ireland. *In*: Parnell, J. (ed.), *Basins on the Atlantic Seaboard: Petroleum Geology, Sedimentology and Basin Evolution.* Geological Society, London, Special Publications, **62**, 333-349.

Morton, N. (1992) Late Triassic to Middle Jurassic stratigraphy, palaeogeography and tectonics west of the British Isles. *In*: Parnell, J. (ed.), *Basins on the Atlantic Seaboard: Petroleum Geology, Sedimentology and Basin Evolution.* Geological Society, London, Special Publications, **62**, 53–70.

Murphy, N.J. and Croker, P.F. (1992) Many play concepts seen over wide area in Erris, Slyne troughs off Ireland. *Oil and Gas Journal*, Sept.14, 92–97.

Murphy, N.J., Sauer, M.J., and Armstrong, J.P. (1995) Toarcian source rock potential in the North Celtic Sea Basin, offshore Ireland. *In*: Croker, P.F. and Shannon, P.M. (eds), *The Petroleum Geology of Ireland's Offshore Basins.* Geological Society, London Special Publications, **93**, 193–207.

Naylor, D. (1992) The post-Variscan history of Ireland. *In*: Parnell, J. (ed.), *Basins on the Atlantic Seaboard: Petroleum Geology, Sedimentology and Basin Evolution.* Geological Society, London, Special Publications, **62**, 255–275.

Naylor, D. (1998) *Irish shorelines through geological time.* Royal Dublin Society: Occasional Papers in Irish Science and Technology, **17**, 20pp.

Naylor, D. and Shannon, P.M. (1982) *The Geology of Offshore Ireland and West Britain.* Graham and Trotman, London, 161pp.

Naylor, D. and Shannon, P.M. (2005) The structural framework of the Irish Atlantic Margin. *In*: Doré, A.G. and Vining, B.A. (eds), *Petroleum Geology: North-West Europe and Global Perspectives – Proceedings of the 6th Petroleum Geology Conference.* Geological Society, London, 1009–1021.

Naylor, D. and Shannon, P.M. (2009) Geology of offshore Ireland. *In*: Holland, C.H. and Sanders, I.S. (eds), *The Geology of Ireland.* Second Edition. Dunedin Academic Press, Edinburgh, 405–460.

Naylor, D., Shannon, P.M., and Murphy, N. (1999) *Irish Rockall Basin region – a standard structural nomenclature system.* Petroleum Affairs Division, Dublin, Special Publication **1/99**, 42pp.

Naylor, D., Shannon, P.M., and Murphy, N. (2002) *Porcupine–Goban region – a standard structural nomenclature system.*

Petroleum Affairs Division, Dublin, Special Publication **1/02**, 65pp.

Newman, P. (1999) The geology and hydrocarbon potential of the Peel and Solway Basins, East Irish Sea. *Journal of Petroleum Geology*, **22**, 305–324.

Petrie, S.H., Brown, J.R., Granger, P.J., and Lovell, J.P.B. (1989) Mesozoic history of the Celtic Sea Basins. *In*: Tankard, A.L. and Balkwill, H.R. (eds), *Extensional tectonics and stratigraphy of the North Atlantic Margins*. American Association of Petroleum Geologists Memoir **46**, 433–444.

Robinson, K.W., Shannon, P.M., and Young, D.G.G. (1981) The Fastnet Basin: an integrated analysis. *In*: Illing, L.G. and Hobson, G.D. (eds), *Petroleum Geology of the Continental Shelf of North-West Europe*. Heyden and Son Ltd, London, 444–454.

Shannon, P.M. (1992) Early Tertiary submarine fan deposits in the Porcupine Basin, offshore Ireland. *In*: Parnell, J. (ed.), *Basins on the Atlantic Seaboard: Petroleum Geology, Sedimentology and Basin Evolution*. Geological Society, London, Special Publications, **62**, 351–373.

Shannon, P.M. (1993) Submarine Fan Types in the Porcupine Basin, Ireland. *In*: Spencer, A.M. (ed.), *Generation, Accumulation and Production of Europe's Hydrocarbons. III*. Special Publication of the European Association of Petroleum Geoscientists **3**, Springer-Verlag, Berlin, 111–120.

Shannon, P.M. (1996) Current and future potential of oil and gas exploration in Ireland. *In*: Glennie, K. and Hurst, A. (eds.), *AD. 1995 – NW Europe's Hydrocarbon History*. Geological Society, London, 51–62.

Sinclair, I.K., Shannon, P.M., Williams, B.P.J., Harker, S.D., and Moore, J.G. (1994) Tectonic control on sedimentary evolution of three North Atlantic borderland Mesozoic basins. *Basin Research*, **6**, 193–218.

Smith, J. and Higgs, K.T. (2001) Provenance implications of reworked palynomorphs in Mesozoic successions of the Porcupine and North Porcupine basins, offshore Ireland. *In*: Shannon, P.M., Haughton, P.D.W., and Corcoran, D.V. (eds), *The Petroleum Exploration of Ireland's Offshore Basins*. Geological Society, London, Special Publications, **188**, 291–300.

Tappin, D.R., Chadwick, R.A., Jackson, A.A., Wingfield, R.T.R., and Smith, N.J.P. (1994) *United Kingdom offshore regional report: The geology of Cardigan Bay and the Bristol Channel*. London HMSO for the British Geological Survey, 107pp.

Tate, M.P. and Dobson, M. R. (1988) Syn- and post-rift igneous activity in the Porcupine Seabight basin and adjacent continental margin west of Ireland. *In*: Morton, A.C. and Parson, L.M. (eds), *Early Tertiary volcanism and the opening of the NE Atlantic*. Geological Society, London, Special Publications, **39**, 309–334.

Tate, M.P. and Dobson, M.R. (1989a) Late Permian to early Mesozoic rifting and sedimentation offshore NW Ireland. *Marine and Petroleum Geology*, **6**, 49–59.

Tate, M.P. and Dobson, M.R. (1989b) Pre-Mesozoic geology of the western and north-western Irish continental shelf. *Journal of the Geological Society of London*, **146**, 229–240.

Trueblood, S. (1992) Petroleum geology of the Slyne Trough and adjacent basins. *In*: Parnell, J. (ed.), *Basins of the Atlantic Seaboard: Petroleum Geology, Sedimentology and Basin Evolution*. Geological Society, London, Special Publications, **62**, 315–326.

Trueblood, S. and Morton, N. (1991) Comparative sequence stratigraphy and structural styles of the Slyne Trough and Hebrides Basin. *Journal of the Geological Society of London*, **148**, 197–201.

Tyrrell, S., Souders, A.K., Haughton, P.D.W., Daly, J.S., and Shannon, P.M. (2010) Sedimentology, sandstone provenance and palaeodrainage on the eastern Rockall Basin margin: evidence from the Pb isotopic composition of detrital K-feldspar *In*: Vining, B.A. and Pickering, S.C. (eds), *Petroleum Geology: From Mature Basins to New Frontiers – Proceedings of the 7th Petroleum Geology Conference*. Geological Society, London, 937-952.

Whittington, R.J., Croker, P.F., and Dobson, M.R. (1981) Aspects of the geology of the south Irish Sea. *Geological Journal*, **16**, 85–88.

Wilson, H.E. and Manning, P.I. (1978) *Geology of the Causeway Coast*. Memoir of the Geological Survey of Northern Ireland, Sheet 7. Her Majesty's Stationery Office, Belfast, 72pp.

Woodland, A.W (ed.), (1971) *The Llanbedr (Mochras Farm) borehole*. Institute of Geological Sciences, Report **71/18**, 115pp.

Ziegler, P.A. (1988) *Evolution of the Arctic–North Atlantic and the Western Tethys*. American Association of Petroleum Geologists Memoir **43**, 198pp.

Ziegler, P.A. (1990) *Geological Atlas of Western and Central Europe*. 2nd Edition. Shell Internationale Petroleum Maatschappij B.V. 239pp.

Chapter 8

Cretaceous stratigraphy

There was an increase in tectonic activity across the Jurassic–Cretaceous boundary in many of the basins. This tectonic phase, referred to as the Late Cimmerian, had a significant impact on Irish offshore geology. Crustal extension across the Arctic–North Atlantic rift system was the driving force throughout much of the Early Cretaceous. Sea levels fell from the high-stand of the Kimmeridgian and reached a low point in late Berriasian–early Valanginian time. The combination of the Late Cimmerian rifting pulse and low sea levels between latest Jurassic to early Valanginian time produced a widespread unconformity (Figure 8.1). Large areas of the shelves were exposed and subject to erosion, and deposition was continuous only along the axial zones of the main basins. There was an accompanying change from the marine and carbonate shelf regimes of the Late Jurassic to marginal and non-marine clastic wedges in the Early Cretaceous. Sea level then rose through the remainder of the Early Cretaceous, except for a period of low-stand in the Aptian. During the Early Cretaceous, rifting propagated rapidly northwards west of Greenland along the Labrador Sea into Baffin Bay. By mid-Aptian there was crustal separation between Iberia and the Goban–Ireland shelf. Seafloor spreading had commenced between Ireland and Newfoundland but was probably confined to the region south of the Charlie-Gibbs Fracture Zone as the North Atlantic began to open. As extensional stresses were focused along the North Atlantic and Labrador Sea axes, levels of rifting and tectonic activity in the surrounding regions and basins waned.

During the Late Cretaceous the North Atlantic and Arctic spreading systems continued to widen, and rifting

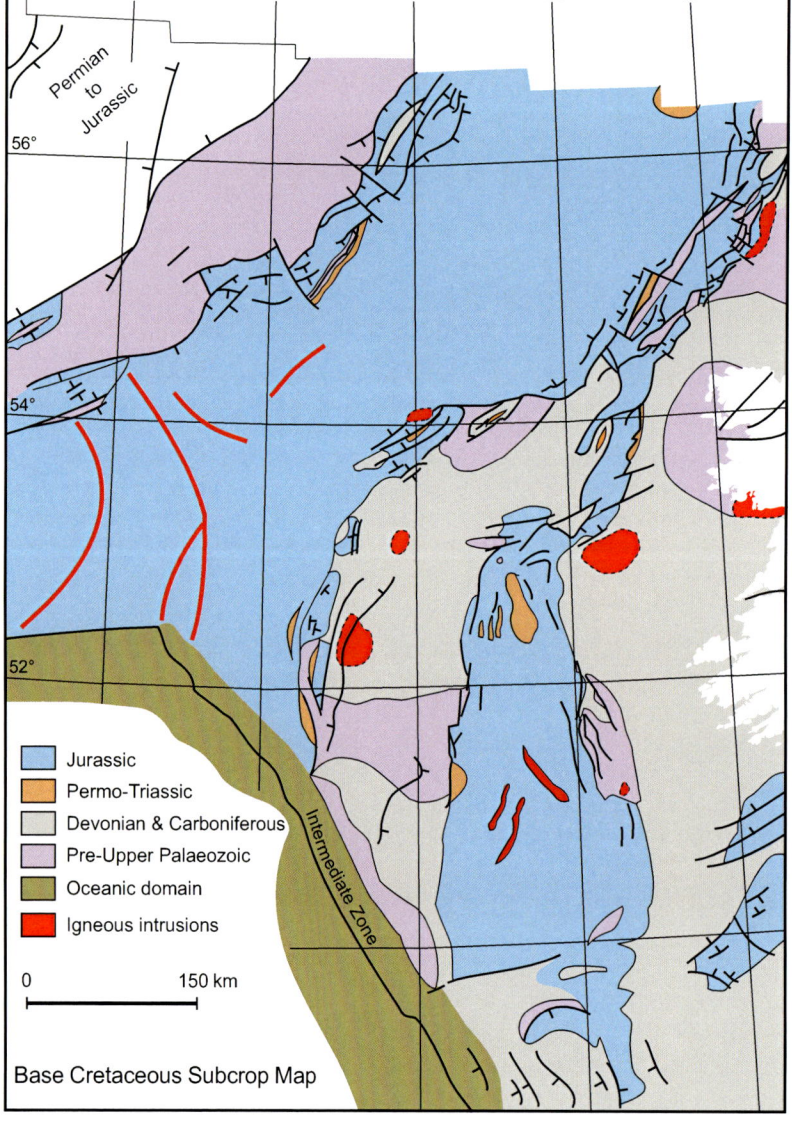

Jurassic
Permo-Triassic
Devonian & Carboniferous
Pre-Upper Palaeozoic
Oceanic domain
Igneous intrusions

0   150 km

Base Cretaceous Subcrop Map

Figure 8.1. Base Cretaceous subcrop map of the Atlantic margin basins west of Ireland (after Naylor and Shannon, 2005).

continued in the Labrador Sea. Seafloor spreading along the North Atlantic system extended north of the Charlie-Gibbs Fracture Zone in latest Cretaceous time. Tectonic activity in the other basins was at a minimum. The Late Cretaceous was a time of worldwide sea level rise (Vail *et al.*, 1977). Global sea levels began to rise during the Albian, leading to widespread marine transgression. Around Ireland the basin margins were progressively overstepped and the Late Cretaceous seas spread onto the surrounding positive massifs.

Offshore distribution of Cretaceous rocks

The Cretaceous sequence in the Celtic Sea basins begins with dominantly fluvial, alluvial, deltaic and lacustrine deposition (Robinson *et al.*, 1981; Naylor and Shannon, 1982; Shannon, 1991) comparable to the Wealden of southern England (Allen, 1981). This non-marine character is in marked contrast to coeval deep marine strata in the Porcupine Basin. Lower Cretaceous strata of pre-Aptian age are generally thin to absent in the South Celtic Sea and Cockburn basins (Smith, 1995), probably due to erosion and non-deposition. The Early Cretaceous was a quiescent period for volcanism in the Celtic Sea area, with igneous activity limited to the Cornubian platform, Western Approaches, and Goban Spur margins. In the North Celtic Sea Basin the Late Cimmerian unconformity is well developed only at the basin margins. Following upon erosion of the Late Cimmerian surface, subsidence and deposition resumed in Early Cretaceous time. Movement continued on some faults, particularly at the basin margins, whilst centrally located wells within the basin (Esso 47/30-1 and Esso 57/6-1; Figure 7.1) penetrated Berriasian non-marine clastics above Tithonian shales and carbonates, with no obvious unconformity (Colin *et al.*, 1981). Further south within the same basin (Esso-Marathon 56/14-1) Valanginian to Hauterivian beds overstep an Oxfordian–Kimmeridgian alluvial and fluvial sequence. On the flanks of the Pembrokeshire Ridge, Lower Cretaceous beds rest unconformably on Lower Jurassic strata (Figure 7.1). Despite the preservation of thick Jurassic successions in the St George's Channel and Cardigan Bay basins, the Cretaceous is not represented and Cenozoic strata rest with marked unconformity on Jurassic and older strata.

Lower Cretaceous rocks have been intersected by most of the exploration wells drilled in the Celtic Sea basins. A lower continental sandy (Wealden Group) section is overlain by a marine sand and shale sequence (Greensand Group). Here we include all the section down to the Jurassic boundary within the Wealden Group. The term 'Purbeck' (Upper Purbeck in the southern England sense) has been used informally in a lithological/facies sense for the lower beds in some cases (e.g. Caston, 1995), but this does not seem helpful in the case of the North Celtic Sea Basin, where the Lower Cretaceous facies are often diachronous. Because of the use of stratigraphic names established in England, it should be noted that Taber *et al.* (1995) take the upper boundary of the Greensand Group at the top of the Plenus Marl (near top Cenomanian: Figure 8.2), thus including the Lower Chalk of the English usage. Lower Cretaceous sedimentation in the Fastnet Basin (Figure 7.2 B: Robinson *et al.*, 1981) followed along similar lines to that in the western part of the North Celtic Sea Basin, and the two will be considered together. Ainsworth *et al.* (1987) recorded Lower Cretaceous beds resting unconformably on Aalenian to Toarcian in the centre of the Fastnet Basin, and on lower Toarcian rocks at the basin margins.

The Wealden Group section in the west of the Celtic Sea region begins with brackish to freshwater, generally alluvial, Berriasian sediments that are succeeded by Valanginian–Hauterivian continental fluvial sandstones and alluvial shales, sourced from a generally western vector (Robinson *et al.*, 1981). This western source probably foundered in Hauterivian time with brackish marine influences apparent in the lower part (late Hauterivian: Ainsworth *et al.*, 1987) of the finer-grained 'Wealden Clay' section. However, the Leinster and Welsh areas still provided coarser sediment to the eastern part of the basin during this time. Initiation of marine incursion in the west was followed by a return in the Barremian to a complex pattern of fluvial, deltaic and lagoonal facies. This facies may occur as late as Aptian to early Albian time towards the northeast of the North Celtic Sea Basin (Ainsworth *et al.*, 1987). Increasing marine transgression then continued in the southwest until the early Albian. Early Aptian shallow-marine carbonate-dominated sediments were succeeded by marine calcareous siltstones and claystones in the late Aptian. Ainsworth *et al.* (1987) noted that in general the Lower Cretaceous of the Fastnet Basin and the southwestern part of the North Celtic Sea Basin shows a stronger marine influence than further northeast, where marginal marine, deltaic and freshwater conditions prevailed. Brackish marine conditions in the early Albian were followed in mid-Albian time by a regressive phase with dominantly non-marine arenaceous sequence containing coal and lignite horizons, a facies that is strongly diachronous towards the northeastern

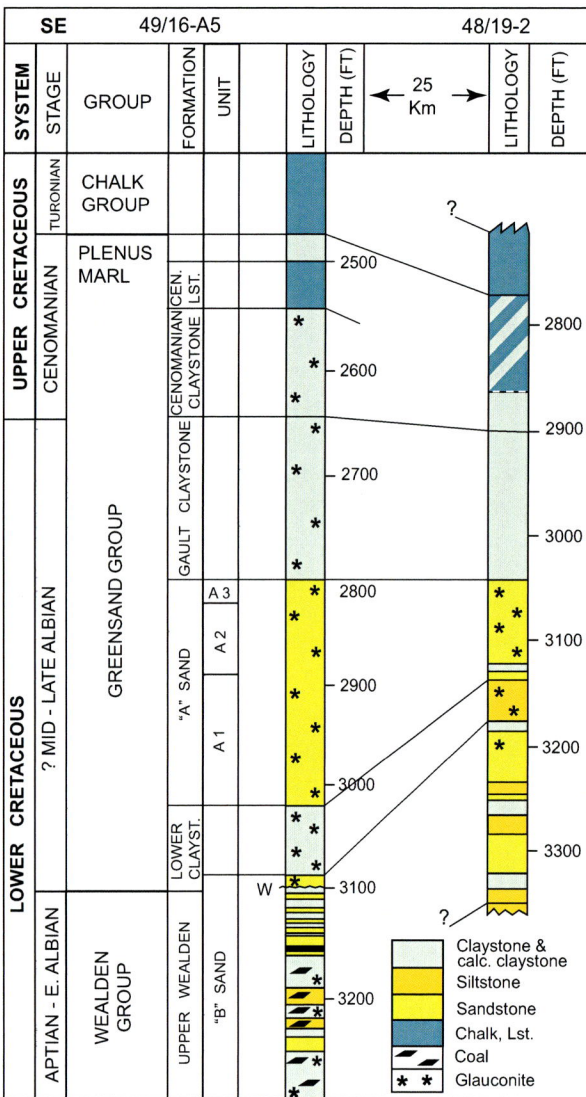

Figure 8.2. Comparative Aptian to Cenomanian stratigraphy of Kinsale Head Gasfield well 49/16-A5 (modified after Colley *et al.*, 1981; Taber *et al.*, 1995) and well 48/19-2 (modified after Howell and Griffiths,1995). Top 'A' Sand is used as datum. The well locations are shown on Figure 3.4.

sandy marls'. The Greensand Group is considered to rest with slight unconformity on the Wealden – a break that may become more marked towards the basin margins (Taber *et al.*, 1995) in a manner similar to that recorded in the Wessex Basin in southern England (Underhill and Stoneley, 1998). The Albian to Cenomanian sequence and the overlying Late Cretaceous Chalk Group represent a progressive eustatic sea level rise.

The set of varied and diachronous facies developed in the Lower Cretaceous of the north Celtic area has resulted in a marked variation in the sections encountered by exploration wells, particularly from west to east along the North Celtic Sea Basin. In the area of the Kinsale Head Gasfield, which lies centrally at the mid-point of the basin, the Wealden comprises a fine-grained delta plain interbedded sequence of sandstones, siltstones and mudstones, with intercalated coals, palaeosols containing roots, iron-rich claystones and carbonaceous shales. Coarsening-upwards cycles are common. The sandstones are normally 3–10 m thick, but major distributary sand bodies up to 25 m thick occur, although these may not be areally extensive (Colley *et al.*, 1981). There is a tripartite subdivision (Figure 8.2) of the Albian portion of the overlying Greensand Group section into a Lower Claystone, a middle 'A' Sand and an upper Gault Claystone (Taber *et al.*, 1995). This marks a progression from the estuarine/fluvio-deltaic environments of the Upper Wealden to inner shelf conditions, with sand derivation from the northern basin margin. The 'A' Sand is the main reservoir in the Kinsale Head and Ballycotton gasfields and has provided 90% of the gas production. Minor production has also come from a lower 'B' sand at the base of the Lower Claystone. The 'B' sand was a term initially restricted to the basal transgressive sand (<4 m thick) of the Greensand Group (Colley *et al.*, 1981), but Taber *et al.* (1995) later included upper sandstones within the Wealden in the unit – presumably because these also produced some gas. The 'A' sand is a bioturbated interval up to 48 m thick that comprises three coarsening-upwards units within an overall cleaning-upwards section. There is some dispute regarding the depositional environment and provenance of the 'A' sand of the Kinsale Head Gasfield. The reservoir was considered by Colley *et al.* (1981) to be the product of a stacked series of off-shore sand-bars in shallow marine conditions. Dipmeter readings and log correlations suggested sediment movement along the basin towards the southwest, with some bimodal influences. Support for the sand bar model was provided by Stonecipher *et al.* (1994), mainly based on

end of the basin. The overlying mid- to late Albian shallow-marine facies comprise an extensive transgressive succession identified in drilling or on seismic sections throughout the Celtic Sea basins (Robinson *et al.*, 1981; Taber *et al.*, 1995). In the area of the Helvick oil accumulation (Block 49/9: Figure 3.4) Caston (1995) records that the mid-Albian transgression of the Wealden floodplain was followed by deposition of the Greensand Group of Albian to Cenomanian age 'which in Helvick consists dominantly of light green to grey-green, very fine- to fine-grained glauconitic sandstones and thin, grey, soft petrographic evidence. However, Winn (1994) argued that the tabular form (80 x 50 km) of the reservoir pointed to deposition as a sand sheet below storm wave base. On the basis of wireline log, sedimentological and ichnofaunal studies Hartley (1995) viewed the sand body as being produced by the punctuated regression of a shoreface environment. The shoreface sands were regarded as prograding radially from a source lying south of the present gasfield. In this regard, it is worth noting that Robinson *et al.* (1981) suggested that a southerly source of eroding Wealden Group strata provided sediment into the Fastnet Basin during deposition of the Greensand Group. However, Taber *et al.* (1995) produced a new isopachous study of the reservoir (Figure 8.3A) which, together with an increasing amount of detrital clay in the 'A' Sand towards the south, supported the concept of a northern provenance and deposition in shallow marine conditions.

The Bula 48/19-2 well (Howell and Griffiths, 1995) is in the same sector of the basin as the Kinsale Head Gasfield, but tested a faulted anticline close to the basin margin fault northwest of the gasfield (Figure 8.2). The 'A' sand at this location was only 25 m thick and comprised intensely bioturbated brown-grey glauconitic sandstone. The 'B' sand, 21.6 m thick and ascribed entirely to the Wealden Group, consisted of friable sands with grain sizes ranging from fine sand to coarse conglomerate. Eight further reservoir quality sandstones were encountered lower in the Wealden section. In summary, the Greensand Group 'A' sand is an extensive tabular sand body with thickest development along the basin axis in the area of the Kinsale Head and Ballycotton gasfields. Sand thickness appears to have been controlled by subsidence rather than proximity to source.

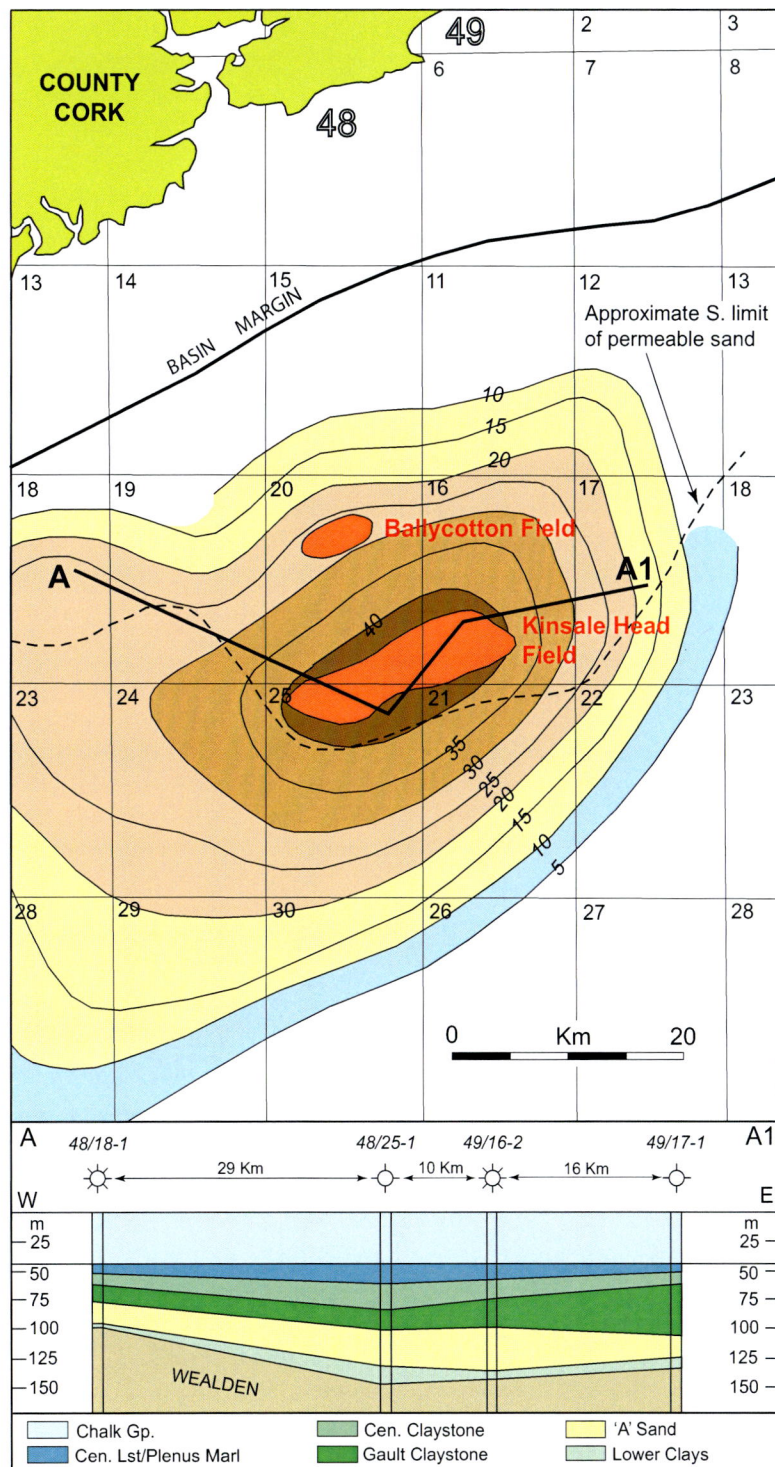

Figure 8.3. A. Kinsale Head Gasfield gross 'A' reservoir sand isopach map. Contour interval 5 m. Line A-A1 is the line of section shown in 8.3B. **B.** Simplified east-west stratigraphic cross-section through the field area showing correlations in the Greensand Group. Both modified after Taber *et al.*, 1995). Datum is Base Chalk (base Turonian).

The balance of evidence probably supports a northerly sediment provenance.

The 'A' sand is overlain by the Gault Claystone that ranges from Albian to Cenomanian in age (Figure 8.2) and acts as topseal in the gasfields. The silty claystone unit is almost 140 m thick in the Kinsale Head field area, but thins and becomes sandier towards the basin margins, and also thins along basin to the southwest. Although the unit is thicker to the northeast of the field, the facies becomes sandier and the seal potential diminishes. The uppermost units of the Greensand Group are a Cenomanian limestone that thins towards the basin margins, overlain by a distinctive thin claystone, the 'Plenus Marl'.

Chalk was deposited from Cenomanian time onwards in the depressions of the Celtic Sea, Fastnet and Western Approaches basins. Chalk of Cenomanian to Maastrichtian age crops out over a large area of the sea floor in the North Celtic Sea Basin and historically posed problems in acquiring good quality seismic images of the deeper section. The base of the Chalk Group is usually taken to be within the Cenomanian (e.g. Colin *et al.*, 1981; Murdoch *et al.*, 1995), but in the Kinsale Head field area Taber *et al.* (1995) equated the contact with the base of the Turonian. Marginal facies of Turonian to late Santonian age along the northern margin of the North Celtic Sea Basin suggest that the basin margin at that time may not have been much farther north than the present seabed outcrop limit. Considerable thicknesses of Upper Cretaceous Chalk were later eroded following Cenozoic inversion of the basin, so that the upper zones are missing over large areas of the basin. Reconstructed isopachytes of the original thickness show *c.*200 m on the basin margin and over 1300 m in the basin centre (Murdoch *et al.*, 1995 and Figure 8.4). The Chalk Group of the Fastnet Basin (Figure 7.2B) is generally only 120–160 m thick, although the unit thickens to 240 m in the Deminex 56/21-2 well (Robinson *et al.*, 1981). However, the same authors divided the chalk sequence in the basin centre into lower and upper units separated by a thin 'Plenus Marl'. The 'Lower Chalk' tends to be slightly arenaceous, being deposited in an inner shelf environment, whereas the 'Upper Chalk' is a product of deeper outer shelf conditions. It is clear from this description that the 'Lower Chalk' unit of Robinson *et al.* (1981) equates with the Cenomanian limestone of Taber *et al.* (1995) which, as noted above, together with the 'Plenus Marl' they include within the Greensand Group, rather

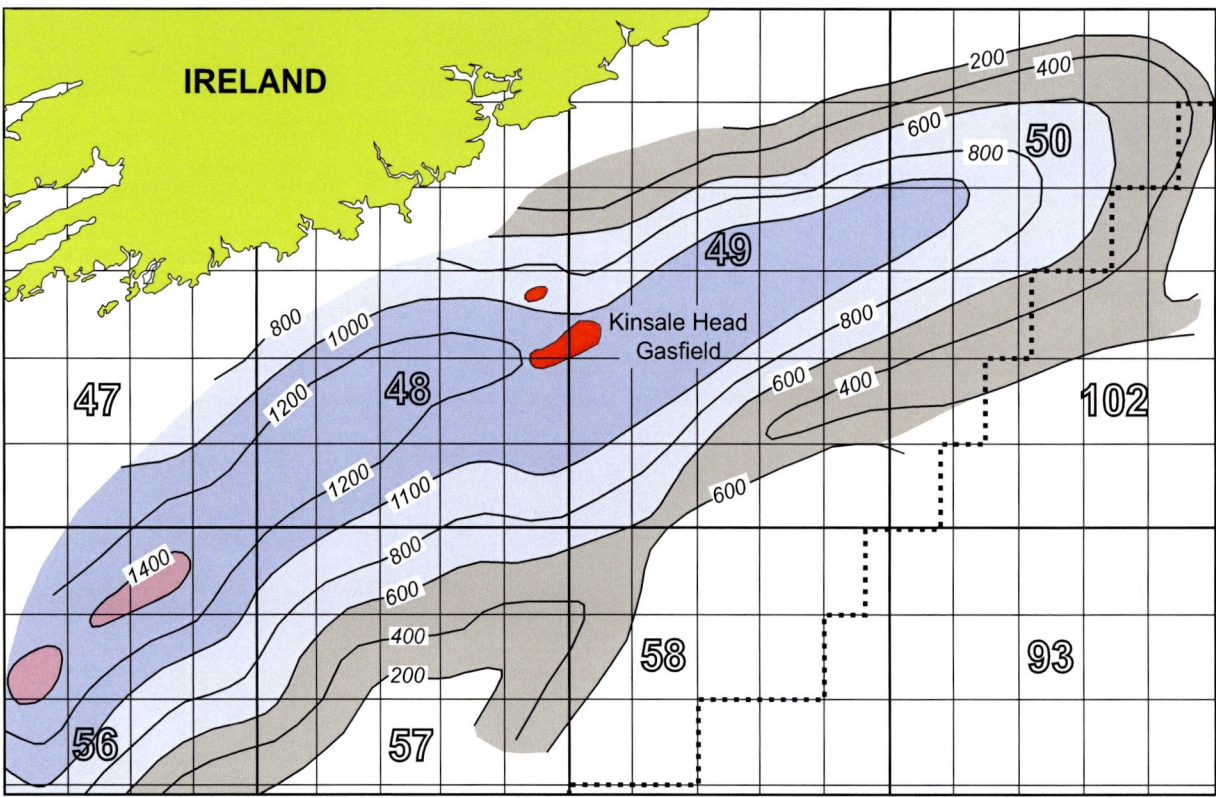

Figure 8.4. Reconstructed original Chalk isopachs in the North Celtic Sea Basin (after Murdoch *et al.*, 1995). Contour interval 200 m.

than the Chalk Group. At the southwest extremity of the Fastnet Basin (Ranger 63/8-1 well: Figure 7.2A) the lithologies of the Campanian section indicate a return to shallower inner shelf conditions.

As outlined in Chapter 4, Chalk (Upper Campanian) is preserved at only one locality onshore in southern Ireland, in a small outlier at Ballydeenlea, near Killarney, County Kerry (Walsh, 1966). This suggests that, with rising sea levels, Chalk deposition overspilled the off-shore basin margins and extended onto the Irish massif at least as far north as Killarney. Palynofacies data from the Chalk deposit show a particularly low level of ter-restrial influence, supporting a picture of widespread submergence at this time (Evans and Clayton, 1998; see Chapter 4). Marine transgression also extended to deposit a veneer of Chalk across parts of the Cornubian and Welsh landmasses.

Lower Cretaceous lithologies and facies in the Fastnet and North Celtic Sea basins contrast with those of the Goban sector. Opening of the Bay of Biscay occurred during latest Jurassic to earliest Neocomian, with sea-floor spreading beginning during the Aptian in the Meriadzek region, southeast of Goban Spur, and early Albian west of the Goban Spur (Masson et al., 1984). The Esso 62/7-1 well on the Goban Spur intersected a Lower Cretaceous section (580 m thick), resting un-conformably on the Cimmerian surface (Figures 8.1 and 7.4). The Late Cimmerian unconformity is charac-terised on the seismic data by erosion and onlap of the overlying Lower Cretaceous strata. In the well, marine sandstones and cherty limestones of possible Neocomian age, containing terrestrial palynomorphs, are overlain by shallow marine Barremian to lower Aptian limestones and claystones. Increased seafloor spreading in the Bay of Biscay then produced deeper marine conditions on the Goban Spur, with outer shelf carbonates increasingly represented in the succession. In the 62/7-1 well there is a marked facies change at the top of the lower Aptian sequence. The late Aptian to Maastrichtian strata are predominantly limestones, mainly chalk, deposited in outer-shelf to upper-bathyal environments (Cook, 1987). Seismic data in the region indicate the existence of reefs within the Lower Cretaceous carbonate sequence of the Goban Graben (Cook, 1987), and similar features are known from other localities along this section of the con-tinental margin (Masson and Roberts, 1981). Sullivan et al. (2009) pointed to the importance of a phase of litho-spheric extension and faulting in providing accommo-dation space for the carbonate-dominant sequence, and

to the possibility of intra-basinal rotated footwall ridges providing bathymetric shoals for carbonate colonisa-tion. The Turonian is not represented in the Esso 62/7-1 well, nor at DSDP Site 550, possibly due to eustatic uplift (Colin et al., 1992). Upper Maastrichtian strata are also absent from the Esso well, and are also probably absent in the Fastnet and North Celtic Sea basins. The Celtic Shelf margin proximal to the Goban Spur was intruded and covered by Lower Cretaceous (Albian–Aptian) volcanic rocks, identified from geophysical data (Cook, 1987). These are, in part, likely to be pillow lavas similar to the late Albian basaltic pillow lavas of DSDP site 550 on the southwest flank of the Goban Spur. The results of the Deep Sea Drilling Project along this portion of the Goban Spur margin are summarised in Figure 8.5. On the seaward edge of Goban Spur, at DSDP Site 549, the Upper Cretaceous sequence of Cenomanian to Maastrichtian nano-chalk, a section disrupted by hia-tuses, rests unconformably on middle Albian siltstones. Seaward of Goban Spur, Site 551 was located on a fault block and encountered Upper Cretaceous nano-chalk, mudstone and calcareous ooze resting unconformably on lower Turonian strata. The Goban Spur succession demonstrates that a major North Atlantic rifting episode began in late Hauterivian or early Barremian time (the age of the oldest syn-rift sediments) and was terminated by seafloor spreading initiation in the early Albian. The widespread post-rift unconformity here lies between Aptian and Albian strata (Graciansky et al., 1985; Hart and Crittenden, 1984; Emery & Uchupi, 1984) and extends out to water depths of 2000 m.

Further north, the Lower Cretaceous sequence throughout most of the Atlantic frontier basins consists of deepening marine shale-prone strata. Shale deposition was interrupted, especially in the Porcupine Basin and locally in the Slyne and Erris basins, by deltaic sandstones that reflect a minor rift episode of Aptian–Albian age. Moore (1992) distinguished a Tithonian to Berriasian sequence restricted to several fault-bounded sub-basins within the Porcupine Basin. This section was regarded as transitional between the main syn-rift to post-rift phases of basin development. The sub-basins occupied topo-graphic lows on the syn-rift topography, and the local controlling faults, generally oriented N–S or NE–SW, continued to be active for a time in the Early Cretaceous, particularly during the early phases of Cretaceous infill. Clastic fans prograded into the sub-basins, but were later onlapped as fault activity waned and sea levels rose. The top of the sequence is marked by an unconformity that

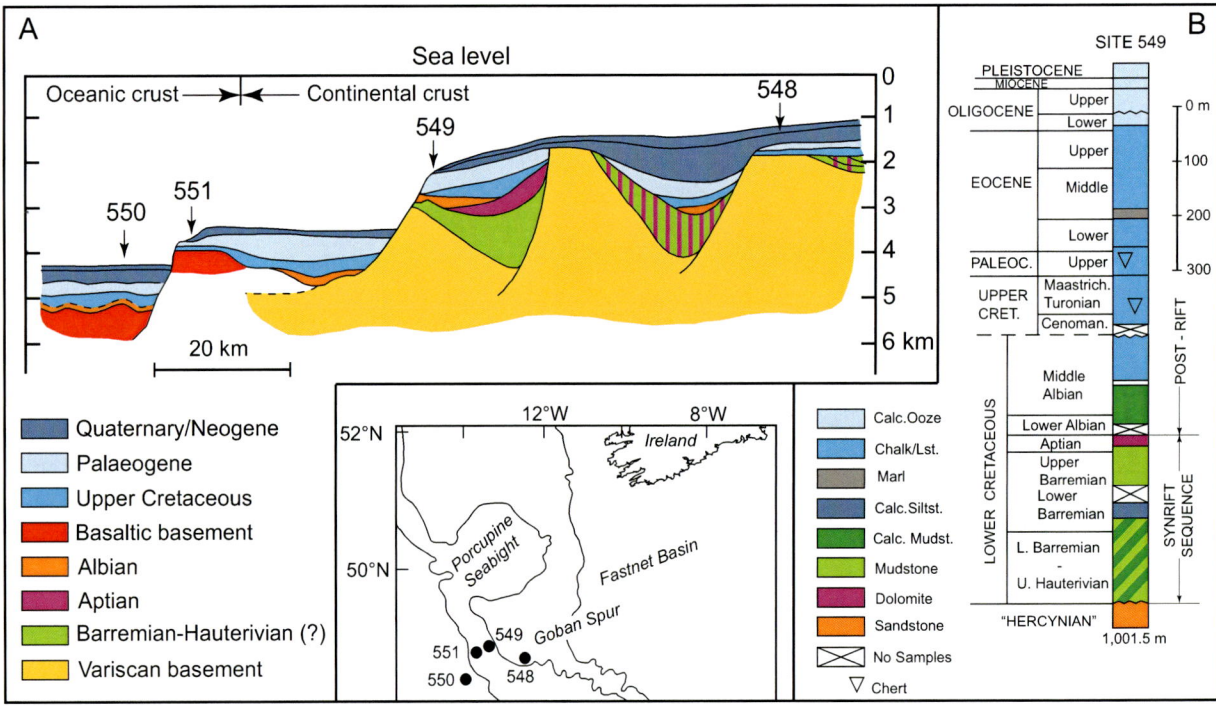

Figure 8.5. Deep Sea Drilling Project boreholes on the Goban Spur margin. A. Cross-section after Graciansky *et al.*, 1982 and Hart and Crittendon, 1985; B. Stratigraphy at Site 549, based on Graciansky and Poag, 1984; Graciansky *et al.*, 1982

represents the main Late Cimmerian surface and the onset of post-rift thermal subsidence, probably in early Valaginian time. The Elf 35/2-1 and Shell 34/19-1 wells (Figure 3.6) penetrated the transition section in two of the sub-basins and were, notably, the only wells in the basin reported to encounter Berriasian strata. The upper contact of the section is approximately end-Berriasian in age. Alluvial fan deposits in the Shell 34/19-1 well of probable Tithonian age mark the initiation of the transition sequence following upon the mid- to late Tithonian termination of the syn-rift phase (Moore, 1992).

The total Cretaceous section in the Porcupine Basin is up to 4 km thick. The thickest development of Lower Cretaceous strata above the Late Cimmerian surface is generally seen along the eastern margin, where the basin margin faults remained active. The Base Cretaceous unconformity (sometimes several close-spaced local unconformities) has an irregular topography reflecting the burial of a topography created by partly eroded footwall crestal highs and intervening partly-filled Late Jurassic half-graben basins and depocentres. Parts of the western margin of the Porcupine Basin are characterised by progressive onlap across older tilted fault blocks. At the centre of the basin the Lower Cretaceous is a thick and uniform sequence, but towards the basin margins it separates into several seismo-stratigraphical units bounded

by unconformities and overlain by Cenomanian to Maastrichtian chalk. The Shell 35/13-1 well (Figure 3.6) drilled 658 m of Albian to Cenomanian calcareous siltstones with sandstones, and this succession thins westwards to the Deminex 34/15-1 well where there are only 60 m of Lower Cretaceous (Hauterivian to Albian) marine claystones. Lower Cretaceous fan deposits related to active fault scarps along the margins of the Porcupine Basin are overlain by younger Lower Cretaceous units with less marked internal structure, indicating lower energy environments. Reworked Carboniferous palynomorphs in Barremian and Berriasian sand-prone horizons in Quadrant 35 wells are considered to have been derived from the east flank of the basin (Smith and Higgs, 2001). Masson and Roberts (1981) interpreted reefal structures, similar to those identified on the Goban margin, extending northwards into the southeastern Porcupine Basin. The structures are considered as being within the Late Jurassic to Early Cretaceous (pre-Aptian) stratigraphic section. It is worth noting in this regard, that the Esso 62/7-1 well in the Goban Spur Basin penetrated a carbonate-dominant Cretaceous section but did not encounter significant carbonates in the Jurassic. O'Sullivan *et al.* (2009) reported that seismic interpretation in the southern Porcupine Basin–Goban province demonstrates a range of facies from carbonate shelf through shelf margin to isolated offshore carbonate build-ups within the Lower Cretaceous section.

In the northern part of the Porcupine Basin (Figure 7.5), the Late Cimmerian unconformity surface is overlain by a northwards-onlapping sequence. Moore and Shannon (1995) divided the Cretaceous succession into three chronostratigraphic sequences. The lowermost sequence (PK1) is Valanginian to Late Aptian in age and comprises shallow-marine shelf facies sediments around the basin margins and deeper basinal sediments in the basin centre. The lower sequence boundary either rests on the transition sequence described above, or unconformably overlies older eroded Jurassic strata. The overlying sequence (PK2) is Late Aptian–Cenomanian in age and is dominated by marine shelf mudstones and sandstones, with deep-water basinal facies only further south in the basin (Figure 8.6). According to Croker and Shannon (1987) Aptian/Albian rift-generated deltaics prograded from the northern and southeastern basin margins, and a series of glauconitic offshore bar sandstones rimmed the basin during the Albian. Further basinwards the clastic pulses may have produced sandy and silty turbidites in basin floor fans such as those recorded in the Early Cretaceous of the Phillips 35/8-1 well (Shannon, 1993). MacDonald *et al.* (1987) recorded a Barremian to early Albian sequence at the northern margin of the basin (Block 26/28) comprising mudstones with thin dolomites. Shallowing in the upper part of the interval was followed by a marked change in sedimentation pattern and the deposition of a widespread 200 m thick blanket of glauconitic calcareous sandstones that pass upwards with apparent conformity into Upper Cretaceous chalk deposition.

Evidence of Lower Cretaceous volcanic activity has been encountered in a number of wells in the Porcupine Basin (Croker and Shannon, 1987; Tate and Dobson, 1988; Tate, 1993). The Phillips 35/8-1 well (Figure 3.6) penetrated 244 m of Hauterivian–Barremian claystones with multiple interlaminated tuff horizons, and 70 m of scoriaceous waterlain airfall deposits, pumice and vitric volcanogenic sediment within the Aptian–Albian interval. Pyroclastics were also recognised at the Barremian–Aptian boundary in the Gulf 26/21-1 well, and at the top of the Albian in the Shell 35/13-1 well. The presence of a Median Volcanic Ridge within the Lower Cretaceous sequence of the Porcupine Basin has been postulated from geophysical evidence by a number of authors (Young and Bailey, 1974; Roberts *et al.*, 1981; Masson and Miles, 1986; Ziegler, 1990;). This broadly non-penetrative

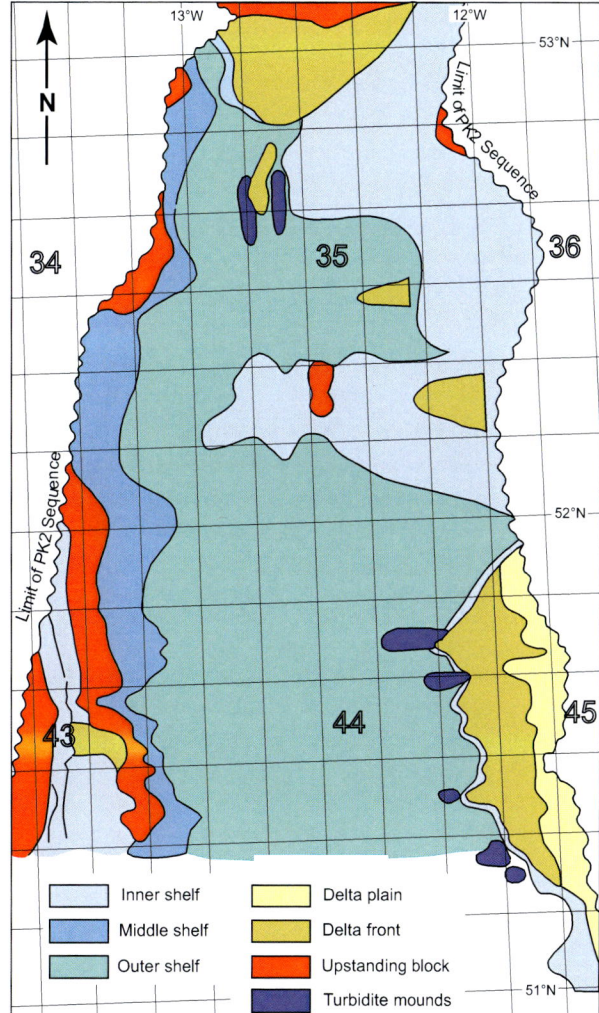

Figure 8.6. Depositional environments in the northern Porcupine Basin during the late Aptian to Cenomanian period (PK2 interval: after Moore and Shannon, 1995).

feature (discussed in more detail in Chapter 11; Figure 11.6) was further analysed, using multichannel seismic data, by Naylor and Anstey (1987), Tate and Dobson (1988), and Tate (1993). The body (now sometimes referred to as the Dunquin Prospect, Figure 3.6) appears to comprise two or three discrete units or extrusions and is thought (Masson and Miles, 1986) to be early Aptian or older. Naylor and Anstey (1987) mapped the Late Cimmerian unconformity surface as passing beneath the Ridge and considered the extrusion, together with related sill intrusion, as related to continued fault activity in Early Cretaceous time. A further phase of Palaeogene sill and dyke intrusion may also be involved. Naylor *et al.* (2002) identified additional similar ridges within the basin and named the whole group as the Porcupine

Volcanic Ridge System (Figure 11.6). A different model was proposed by Reston *et al.* (2001) who suggested serpentinisation of thinning crust producing uplift of crustal blocks to form the northern part of the Ridge. This was then followed by sepentinite diapirism at the sea floor to form the southern portion of the feature. However, new deep seismic data have been interpreted by Davison *et al.* (2009) as indicating that the structure is not rooted, since continuous Late Jurassic–Early Cretaceous reflectors are visible below the feature, and is thus unlikely to be a serpentinised mantle intrusion. The same authors suggested that it is a high level (4–9 km depth) intrusion. Further igneous activity in the region is indicated by a marked magnetic and gravity anomaly at the southeastern margin of the Porcupine Basin (49°50' N; 11°30' W), in the region of Balar Spur, which has been interpreted as an igneous centre (Roberts *et al.*, 1981; Naylor and Shannon, 1982). Later work (Cook, 1987; Tate and Dobson, 1988) showed this to be a penetrative igneous body of possible Aptian–Albian age, which was named the Seabight Igneous Centre (Figure 11.3).

A Cenomanian to Danian chalk sequence is found in the Porcupine Basin (PK3 sequence of Moore and Shannon, 1995). About 400 m of Upper Cretaceous is preserved near the margins of the basin, while in excess of 1000 m occurs towards the central axis (Figure 11.6). The succession rests unconformably on an erosional sequence boundary at the basin margins (Moore and Shannon, 1995). The Upper Cretaceous Chalk interval in Block 26/28 at the northern rim of the basin is less than 10 m thick (MacDonald *et al.*, 1987) and is dated as possibly Cenomanian to Maastrichtian. In contrast, possible high-energy calciturbidites are present on the southeastern and southwestern basin margins, probably related to movement on the bounding faults. Reefal and biohermal build-ups are also interpreted in these areas, both at the top of, and within, the chalk sequence (Moore and Shannon, 1995). Within the main Porcupine Basin the lower part of the chalk sequence is interpreted as marly, with a poor impedance contrast with the underlying section (confirmed in wells 35/9-2 and 34/15-1; Figure 3.6). In contrast, the North Porcupine Basin sequence is homogeneous chalk.

Relatively little is known of the Cretaceous succession in the Rockall Basin (Figure 2.15) due to the paucity of wells in the basin. The BP 132/15-1 well, on the eastern margin of the basin in UK waters, drilled a thin section of Aptian/Albian mudstones with interbedded limestones. On seismic profiles (Musgrove and Mitchener, 1996) this

interval thickens basinwards and onlaps a thin wedge of undated, possibly reworked granite resting on crystalline basement. This is overlain by sandstones with interbedded basalts and tuffs of Paleocene to mid-Eocene age. Various authors (e.g. Shannon and Naylor, 1998; Corfield *et al.*, 1999; Walsh *et al.*, 1999; Shannon *et al.*, 1999) have speculated that marine-dominant Lower Cretaceous strata were deposited in the Rockall Basin. The Shell 12/2-1z well on the eastern margin of the Rockall Basin drilled a thin (c.40 m) section of Lower Cretaceous deepwater mudstones overlain by a thick Upper Cretaceous succession (Tyrrell *et al.*, 2010). Further south along the margins of the Irish sector of the Rockall Basin Haughton *et al.* (2005) documented a thin, Early Cretaceous condensed ('brownsand'), succession in a series of shallow boreholes. Further west, in the Hatton region, little or no drilling has taken place and the Mesozoic succession is very poorly constrained. However, seismic evidence from the Rockall Basin and the eastern margin of the Rockall Bank indicates the presence of tilted fault blocks. These have been interpreted as being of pre-Cenozoic age (e.g. Keser Neish, 1993; Shannon *et al.*, 1994, 1995, 1999). Recent BGS boreholes (99/1 and 99/2A) on the Hatton Bank drilled up to 30 m of Albian freshwater or lacustrine mudstones and paralic medium- to coarsegrained sandstones and siltstones lying beneath a veneer of Quaternary and Plio-Pleistocene sands and muds (Hitchen, 2004). Within the southern Rockall Trough a number of large curvilinear volcanic ridges (Figure 11.2) – the Barra Volcanic Ridge System – were identified by Scrutton and Bentley (1988). The ridges are 2 km high and 20 km wide and are considered to be of possible mid-Early Cretaceous age, with construction continuing through Late Cretaceous time. Burial beneath sediment occurred in Eocene time (Naylor *et al.*, 1999). More recently, Gernigon *et al.* (2009), using new seismic data, have interpreted the Barra Volcanic Ridge as a structural horst cut by Early Cretaceous extrusives and intrusives, and later by Cenozoic dykes. The Anton Dohrn, Hebrides Terrace and Helens Reef igneous centres in the northern part of the Rockall Trough are all of probable Late Cretaceous age and are the harbingers of the major volcanism that characterised the early Palaeogene of the Atlantic margin basins.

Upper Cretaceous strata are inferred from geophysical evidence in the Rockall Basin (Naylor *et al.*, 1999). Cretaceous sequences are also interpreted in the perched basins along the eastern margin of the Rockall Basin (Figure 7.6). In addition, Haughton *et al.* (2005)

recorded a generally thin mixed clastic-carbonate (glauconitic 'greensand' and chalky micrite) shallow-marine Cretaceous succession in a series of shallow boreholes in this region. The same authors also suggested a minor phase of intra-Cretaceous (probably Cenomanian–Turonian) extension and faulting. Intra-Cretaceous extrusive volcanism was also suggested, probably linked to the nearby Drol Igneous Centre (Naylor *et al.*, 1999). Marine conditions extended from the Rockall Basin eastwards across parts of Ireland and the Celtic Sea. Upper Cretaceous strata in the basins of the Rockall region may be marls rather than pure chalk (Shannon *et al.*, 1993). In some of the frontier basins (e.g. the Hatton Basin) much of the Cretaceous succession may be absent owing to basin inversion linked to thermal and lower crustal underplating effects (Shannon *et al.*, 1995). However, two shallow boreholes on the UK sector of the Hatton Bank penetrated Albian shales and sandstones beneath thin Cenozoic cover. Seismic imaging in the same area revealed folded and faulted Mesozoic (and possibly older) strata beneath the base Cenozoic unconformity (Hitchen, 2004). Similar features may be assumed to extend southwards into the Irish region. This is in contrast to Rockall High, on which basement rocks are at shallow depth.

Lower Cretaceous rocks were identified in parts of the Slyne and Erris basins on geoseismic sections by Tate and Dobson (1989a) and Murphy and Croker (1992). It was known that the picture was not simple, since the Elf 27/13-1 well in the south Slyne Basin (Figure 3.7) penetrated possible Quaternary sediments resting on Bathonian strata (Trueblood, 1992). The sequence in the Corrib Gasfield region at the northern end of the Slyne Basin (Dancer *et al.*, 2005) comprises 50–96 m of Albian claystones and glauconitic sandstones passing upwards to sandy claystones with occasional limestones, and resting unconformably on Middle to Upper Jurassic strata. Lower Cretaceous rocks are apparently restricted to a relatively thin layer in the northern part of the Slyne Basin (Dancer *et al.*, 1999; *see also* Chapman *et al.*, 1999 Fig. 3H). In contrast, the Lower Cretaceous sequence thickens into the Erris Basin and is present throughout the basin. In the Amoco 12/13-1A Erris well, Jurassic strata are succeeded by 1.2 km of Cretaceous to Cenozoic rocks (Cunningham and Shannon, 1997). Lower Cretaceous (late Berriasian–Valanginian) marine claystones and sandstones are overlain by thick submarine-fan sandstones (late Valanginian–Hauterivian: >120 m) and then by marine claystones grading up to

Albian argillaceous limestones (Murphy and Croker, 1992). Seismic interpretation in the area of the 12/13-1A well (Chapman *et al.*, 1999) indicates several intra-Lower Cretaceous unconformities, including a major Aptian–Albian angular unconformity related to a phase of movement on the basin bounding faults. The control exercised on the thickness and extent of the Cretaceous section by the positive element of the Erris Ridge along the western boundary is clearly illustrated by a series of geoseismic sections presented by Chapman *et al.* (1999). Cunningham and Shannon (1997) showed the Lower Cretaceous section as wedging out entirely against the northern part of the high, whilst in the central sector Naylor *et al.* (1999) interpret the entire Cretaceous section as abutting the feature.

A reduced Upper Cretaceous succession is present in the northern part of the Slyne Basin (Dancer *et al.*, 1999), but absent elsewhere due to subsequent erosion. In the area of the Corrib Gasfield (north Slyne Basin) the Upper Cretaceous Chalk Group, with occasional claystone interbeds, ranges from 79 to 293 m thick (Dancer *et al.*, 2005) and rests unconformably on the Lower Cretaceous. In common with the Lower Cretaceous, the Upper Cretaceous Chalk Group is thicker and more continuous in the Erris Basin and may range up to the Danian in age (Chapman *et al.*, 1999). Murphy and Croker (1992) reported a typical Upper Cretaceous section comprising micritic limestones and chalks in the Amoco 12/13-1A well in the Erris Basin. A total Cretaceous section up to 1000 ms (TWT: suggested conversion 200 ms = 271 m) is interpreted for the axial portion of the Erris Basin (Chapman *et al.*, 1999).

Lower Cretaceous strata are absent from the basins lying north and east of Ireland – that is from the Malin Sea Basin (Evans *et al.*, 1980), the onshore basins of northeast Ireland, and the basins of the Irish Sea. By the beginning of the Late Cretaceous a shallow marine embayment extended eastwards from the Rockall Basin, eventually submerging much of the Inner Hebrides, Malin Shelf, Firth of Clyde and Ulster regions. A thin layer of basal sand and chalk was laid down unconformably across the denuded surface of Jurassic, Triassic and older rocks. Much of this Upper Cretaceous Chalk layer has subsequently been removed by erosion. However, it is preserved over parts of the basins of western Scotland, but in the Firth of Clyde it survives only as blocks within a volcanic centre on Arran. These younger Mesozoic sequences are not preserved in the basins of the Malin Shelf, but are widely preserved beneath, and exposed around the rim of, the plateau basalt sheet that covers most of the northeast Irish onshore basins and in the immediate offshore of the Rathlin Basin. Basal Greensands (Hibernian Greensand Formation) of early Cenomanian to early Santonian age are followed by widespread Chalk (Ulster White Limestone Formation: up to 120 m) of late Santonian to early Maastrichtian age (Fletcher, 1977; Figure 8.7).

Upper Cretaceous sedimentation onshore in the Ulster Basin began with a marine invasion in the southeast, and glauconitic marls followed by yellow and green weathering sandstones and siltstones were deposited. The total thickness of the Hibernian Greensand Formation is only about 30 m and four members are recognised. The uppermost member (Coniacian–Santonian) rests with significant unconformity on the eroded surface of the lower members (Cenomanian–Turonian), with a basal conglomerate (Mitchell, 2004). The formation is mainly restricted to southeast County Antrim and is not found in the Rathlin Basin. After a period of non-deposition in the Santonian renewed transgression eventually extended deposition throughout the region. Chalk sedimentation began at the same time in the Larne, Lough Neagh and Rathlin basins. The Ulster White Limestone Formation is subdivided into fourteen members, on the basis of lithology or flint and fossil content. The pore spaces of the chalk lithology contain secondary calcite cementation, with the result that the rock is both harder and denser than its counterpart in the English Chalk Group. The basal member of the formation is glauconitic in the Larne and Lough Neagh basins due to re-working of the underlying greensands, but is glauconite-free in the Rathlin Basin where the Hibernian Greensand Formation was not deposited. There was only progressive transgression of the Highland Border Ridge, which remained upstanding during deposition of the first six members of the formation (Mitchell, 2004), and the whole structure was not submerged until the mid-Santonian. There was further inundation during the late Campanian, extending chalk deposition to its present outcrop limits. Maastrichtian chalk is known and was probably originally extensive, but now has a restricted distribution due to pre-basalt erosion. The maximum exposed thickness of the Ulster White Limestone is more than 120 m on the coast between Ballycastle and Portrush in the Rathlin Trough, where the succession is almost complete.

Outcrop of the Ulster White Limestone Formation is interpreted as extending up to 6 km offshore from the north coast between Castlerock in the west to Port More

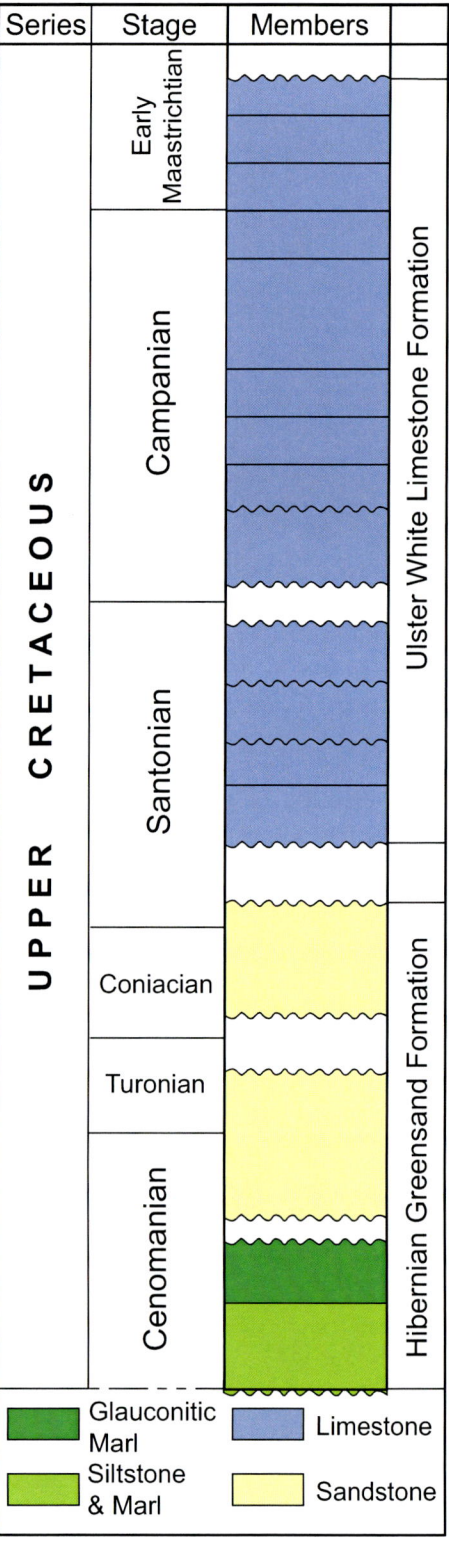

Figure 8.7. Generalised Cretaceous stratigraphy in Northern Ireland (after Mitchell, 2004 and Simms 2009). The Ulster White Limestone Formation is divided into 14 local 'Chalk' members, and the Hibernian Greensand Formation into four members. Wavy contacts indicate erosion surfaces.

in the east, and to extend just to the north of Rathlin Island (Fyfe *et al.*, 1993). Only one offshore borehole has pennetrated chalk – BGS 75/39 about 3 km offshore from Castlerock – penetrated 2.5 m of ?Santonian chalk with flints. The offshore outcrop terminates eastwards against the Tow Valley Fault extension, and there is no evidence for Cretaceous strata east of the fault in the offshore. The lack of Cretaceous sediments in the North Channel and throughout the Irish Sea is discussed in Chapter 12.

Cretaceous palaeogeography

During the Berriasian low-stand episode (Figure 8.8) the Irish massif formed part of a larger landmass which incorporated the Hebridean Platform, Scotland, much of western England, and extended southwards across the Celtic Sea and Western Approaches areas to Armorica. In the Celtic Sea basins the Early Cretaceous deposition began with alluvial and fluvial sediments that gradually

extended westwards into the Fastnet Basin. This kind of complex marginal deposition may also have occurred in the extreme northern part of the Porcupine Basin. Marine conditions obtained throughout most of the Porcupine– Goban Spur region with no evidence of non-marine exposure or deposition seen in any of the wells, and also through most of the Rockall Basin. Non-marine deposition took place in the Hatton Bank region to the west. The continental sandy (Wealden) section in the southern regions is overlain by a marine sand and shale sequence. Marine transgression advanced during the Aptian and extended across much of the region in Albian time. In the Bristol Channel–South Celtic Sea basins, Aptian marine sands transgress over eroded Jurassic and older rocks.

The Late Cretaceous was a time of worldwide sea level rise (Vail *et al.*, 1977) in response to major seafloor spreading and plate re-organisation. Global sea levels began to rise during the Albian, leading to widespread transgression. Sea levels in the Campanian may have

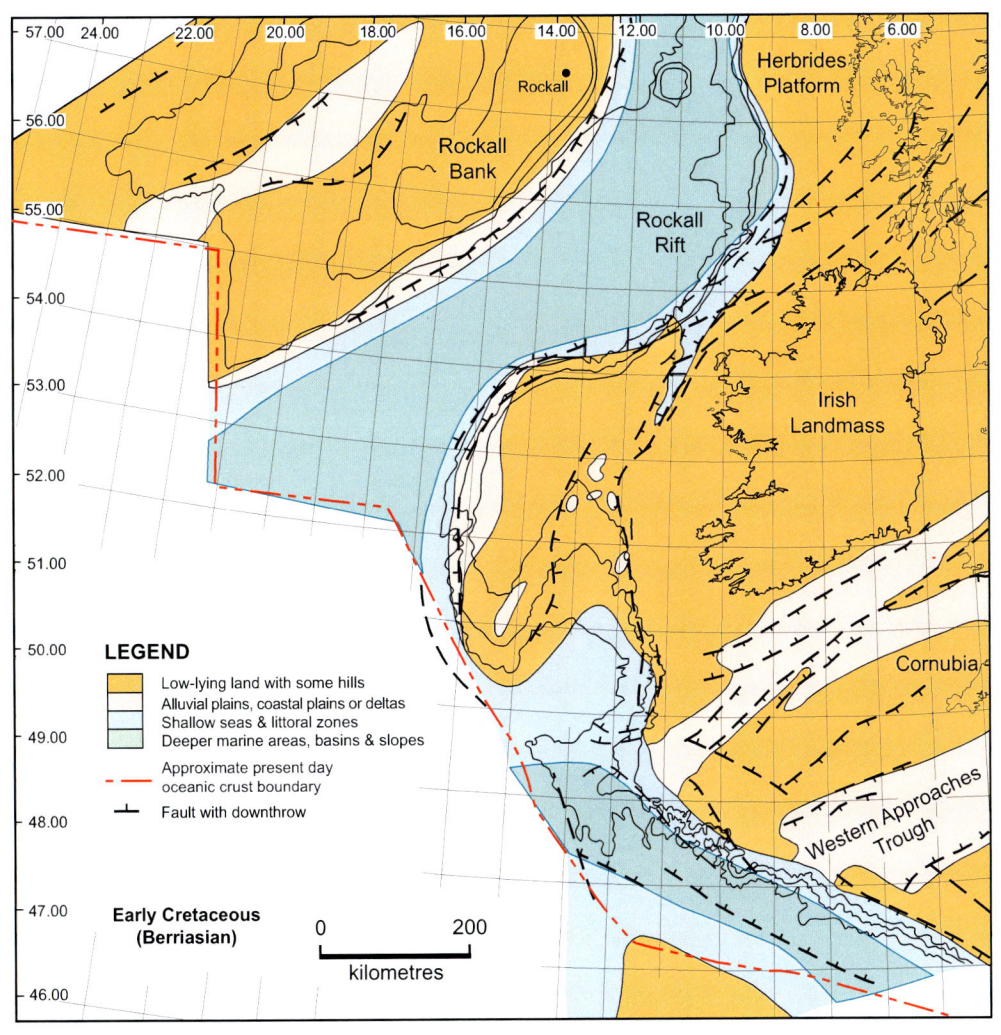

Figure 8.8. Early Cretaceous: Berriasian palaeogeography (after Naylor and Shannon, 2009. *Principal sources*: Hancock and Rawson, 1992; Naylor and Shannon, 1982; Naylor, 1992, 1998; Ziegler, 1990).

LEGEND

- Low-lying land with some hills
- Alluvial plains, coastal plains or deltas
- Shallow seas & littoral zones
- Deeper marine areas, basins & slopes
- ----- Approximate present day oceanic crust boundary
- ⊥ Fault with downthrow

Early Cretaceous (Berriasian)

0   200

kilometres

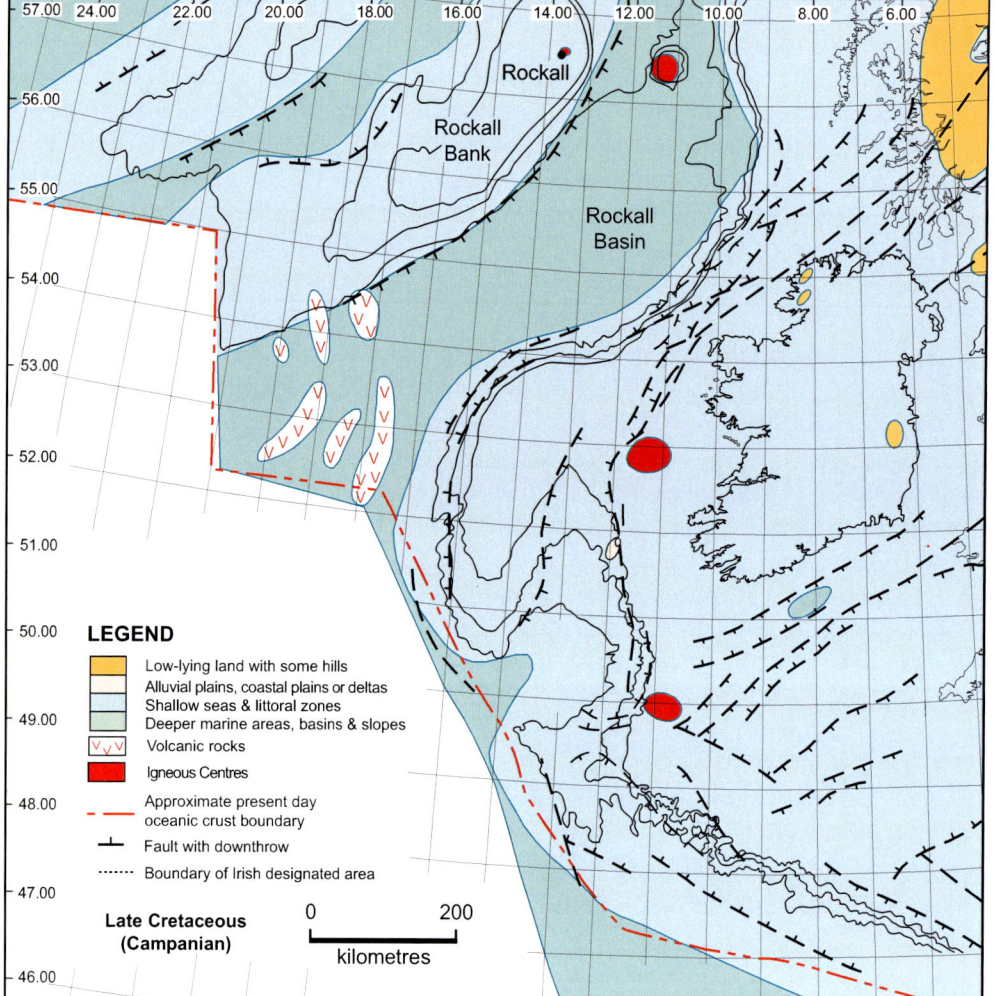

Figure 8.9.
Late Cretaceous:
Campanian palaeogeo-
graphy (after Naylor,
1998. *Principal sources*:
Hancock and Rawson,
1992; Naylor and
Shannon, 1982; Naylor,
1992; Ziegler, 1990).

LEGEND

	Low-lying land with some hills
	Alluvial plains, coastal plains or deltas
	Shallow seas & littoral zones
	Deeper marine areas, basins & slopes
V V	Volcanic rocks
	Igneous Centres
— · —	Approximate present day oceanic crust boundary
⊥	Fault with downthrow
· · · · · ·	Boundary of Irish designated area

Late Cretaceous
(Campanian)

0   200
kilometres

been 200–300 m above present levels, and chalk deposition was widespread. Around Ireland the basin margins
were progressively overstepped and the Late Cretaceous
seas spread onto the surrounding positive massifs. By
Campanian time (Figure 8.9) Ireland was probably
totally immersed, with the possible exception of small
islands. Marine transgression also extended to deposit a
veneer of chalk across parts of the Cornubian and Welsh
landmasses. Towards the end of the Cretaceous the first
major pulses (Laramide) of the Alpine orogeny began to
affect the region, causing uplift and erosion. The Chalk
Sea retreated and Ireland became emergent, together
with large tracts of the surrounding continental shelf.

References

Ainsworth, N.R., O'Neill, M., Rutherford, M.M., Clayton, G.,
Horton, N.F., and Penny, R.A. (1987) Biostratigraphy of the
Lower Cretaceous, Jurassic and Late Triassic of the North
Celtic Sea and Fastnet Basins. *In*: Brooks, J. and Glennie,
K.W. (eds), *Petroleum Geology of North West Europe, Volume
2*. Graham and Trotman, London, 611–622.

Allen, P. (1981) Pursuit of Wealden models. *Journal of the
Geological Society of London*, **138**, 375–405.

Caston, V.N.D. (1995) The Helvick oil accumulation, Block
49/9, North Celtic Sea Basin. *In*: Croker, P.F. and Shannon,
P.M. (eds), *The Petroleum Geology of Ireland's Offshore Basins*.
Geological Society, London, Special Publications, **93**,
209–225.

Chapman, T.J., Broks, T.M., Corcoran, D.V., Duncan, L.A., and
Dancer, P.N. (1999) The structural evolution of the Erris
Trough, offshore northwest Ireland, and implications for hy-
drocarbon generation. *In*: Fleet, A.J. and Boldy, S.A.R. (eds),
*Petroleum Geology of Northwest Europe: Proceedings of the 5th
Conference*. Geological Society, London, 455–469.

Colin, J.P., Ioannides, N.S., and Vining, B. (1992) Mesozoic stratigraphy of the Goban Spur, offshore south-west Ireland. *Marine and Petroleum Geology*, **9**, 527–241.

Colin, J.P., Lehmann, R.A., and Morgan, B. E. (1981) Cretaceous and Late Jurassic biostratigraphy of the North Celtic Sea Basin. *In*: Neale, J. W. and Brazier, M.D. (eds), *Microfossils from Recent and fossil shelf seas*. Ellis Horwood Ltd., Chichester, 122–155.

Colley, M.G., McWilliams, A.S.F., and Myers, R.C. (1981) Geology of the Kinsale Head gas field, Celtic Sea, Ireland. *In*: Illing, L.V. and Hobson, G.D. (eds), *Petroleum Geology of the Continental Shelf of North-West Europe*. Heyden and Son Ltd., London, 504–510.

Cook, D.R. (1987) The Goban Spur: exploration in a deep-water frontier basin. *In*: Brooks, J. and Glennie, K.W. (eds), *Petroleum Geology of North-West Europe*. Graham and Trotman Ltd, London, 623–632.

Corfield, S., Murphy, N., and Parker, S. (1999) The structural and stratigraphic framework of the Rockall Trough. *In*: Fleet, A.J. and Boldy, S.A.R. (eds), *Petroleum Geology of Northwest Europe: Proceedings of the 5th Conference*. Geological Society, London, 407–420.

Croker, P.F. and Shannon, P.M. (1987) The evolution and hydrocarbon prospectivity of the Porcupine Basin, offshore Ireland. *In*: Brooks, J. and Glennie, K.W. (eds), *Petroleum Geology of North West Europe*, Graham and Trotman, London, 633–642.

Cunningham, G.A. and Shannon, P.M. (1997) The Erris Ridge: a major geological feature in the NW Irish Offshore Basins. *Journal of the Geological Society of London*, **154**, 503–508.

Dancer, P.N., Algar, S.T., and Wilson, I.R. (1999) Structural Evolution of the Slyne Trough. *In*: Fleet, A.J. and Boldy, S.A.R. (eds), *Petroleum Geology of Northwest Europe: Proceedings of the 5th Conference*. Geological Society, London, 445–453.

Dancer, P.N., Kenyon-Roberts, S.M., Downey, J.W., Baillie, J.M., Meadows, N.S., and Maguire, K. (2005) The Corrib gas field, offshore west of Ireland. *In*: Doré, A.G. and Vining, B.A. (eds), *Petroleum Geology: North-West Europe and Global Perspectives. Proceedings of the 6th Petroleum Geology Conference*. Geological Society, London, 1035–1046.

Davison, I., Steel, I., Nutall, P., Keane P., Taylor, M., and Baptista, P. (2009) Deep structure of the Porcupine and Goban Spur basins, SW Ireland. *Atlantic Ireland 2009 Conference, Dublin. Abstracts Volume*, 29.

Emery, K.O. and Uchupi, E. (1984) *The Geology of the Atlantic Ocean*. Springer-Verlag, New York, 1050pp.

Evans, A. and Clayton, G. (1998) The geological history of the Ballydeenlea Chalk Breccia, County Kerry, Ireland. *Marine and Petroleum Geology*, **15**, 299–307.

Evans, D., Kenolty, N., Dobson, M. R., and Whittington, R. J. (1980) *The geology of the Malin Sea*. Institute of Geological Sciences, Report **79/15**, 44pp.

Fletcher, T.P. (1977) *Lithostratigraphy of the Chalk (Ulster White Limestone Formation) in Northern Ireland*. Report of the Institute of Geological Sciences, No. **77/24**, 33pp.

Fyfe, J.A., Long, D., and Evans, D. (1993) *United Kingdom offshore regional report: The geology of the Malin–Hebrides Sea area*. London HMSO for the British Geological Survey, 45pp.

Gernigon, L., Ravaut, C., Shannon, P.M., Chabert, A., O'Reilly, B.M., and Readman, P.W. (2009) The evolution of Irish passive margins: implications for locating the transition between continental and oceanic crust. *Atlantic Ireland 2009 Conference, Dublin. Abstracts Volume*, 31–32.

Graciansky, P.C. de and Poag, W.C. (1984) Geologic history of Goban Spur, Northwest Europe Continental Margin. *Initial reports of the Deep Sea Drilling Project*, **80**, 1187–1216.

Graciansky, P.C. de, and 14 others. (1982) La croisière 80 du vaisseau de reserches *Glomar Challenger* au pied de la marge celtique (Atlantique Nord). *Comptes rendus de l'Academie des Sciences (Paris)*, Serie 2, **294**, 793–798.

Graciansky, P.C. de, Poag, W.C., Cunningham, R. *et al.* (1985) The Goban Spur transect: geologic evolution of a sediment-starved passive continental margin. *Bulletin of the Geological Society of America*, **96**, 58–76.

Hancock, J.M. and Rawson, P.F. (1992) Cretaceous. *In*: Cope, J.C.W., Ingham, J.K., and Rawson, P.F. (eds), *Atlas of Palaeogeography and Lithofacies*. Geological Society, London, Memoir **13**, 153pp.

Hart, M.B. and Crittenden, S. (1984) Early Cretaceous Ostracoda from the Goban Spur, D.S.D.P. Leg 80, Site 549. *Cretaceous Research*, **6**, 219–233.

Hartley, A. (1995) Sedimentology of the Cretaceous Greensand, Quadrants 48 and 49 North Celtic Sea Basin: a progradational shoreface deposit. *In*: Croker, P.F. and Shannon, P.M. (eds), *The Petroleum Geology of Ireland's Offshore Basins*. Geological Society, London, Special Publications, **93**, 245–257.

Haughton, P., Praeg, D., Shannon, P.M., Harrington, G., Higgs, K., Amy, L., Tyrrell, S., and Morrissey, T. (2005) First results from shallow stratigraphic boreholes on the eastern flank of the Rockall Basin, offshore western Ireland. *In*: Doré, A.G. and Vining, B.A. (eds), *Petroleum Geology: North-West Europe and Global Perspectives – Proceedings of the 6th Petroleum Geology Conference*. Geological Society, London, 1077–1094.

Hitchen, K. (2004) The geology of the UK Hatton–Rockall margin. *Marine and Petroleum Geology*, **21**, 993–1012.

Howell, T.J. and Griffiths, P. (1995) A study of the hydrocarbon distribution and Lower Cretaceous Greensand prospectivity in Blocks 48/15, 48/17, 48/18 and 48/19, North Celtic Sea Basin. *In*: Croker, P.F. and Shannon, P.M. (eds), *The Petroleum Geology of Ireland's Offshore Basins*. Geological Society, London, Special Publications, **93**, 261–275.

Keser Neish, J. (1993) Seismic structure of the Hatton–Rockall area: an integrated seismic/modelling study from composite datasets. *In*: Parker, J.R. (ed.), *Petroleum Geology of Northwest Europe: Proceedings of the 4th Conference*. Geological Society, London, 1047–1056.

MacDonald, H., Allan, P.M., and Lovell, J.P.B. (1987) Geology of oil accumulation in Block 26/28, Porcupine Basin, offshore Ireland. *In*: Brooks, J. and Glennie, K.W. (eds), *Petroleum Geology of North-West Europe*. Graham and Trotman Ltd, London, 643–651.

Masson, D.G. and Miles, P.R. (1986) Structure and development of Porcupine Seabight sedimentary basin, offshore Southwest Ireland. *Bulletin of the American Association of Petroleum Geologists*, **70**, 563–584.

Masson, D.G. and Roberts, D. (1981) Late Jurassic–Early

Cretaceous reef trends on the Continental Margin SW of the British Isles. *Journal of the Geological Society of London*, **138**, 437–443.

Masson, D.G., Montadert, L., and Scrutton, R.A. (1984). Regional geology of the Goban Spur margin. *Initial reports of the Deep Sea Drilling Project*, **80**, 1115–1139.

Mitchell, W.I. (2004) Cretaceous. *In*: Mitchell, W.I. (ed.), *The Geology of Northern Ireland: Our Natural Foundation*. Geological Survey of Northern Ireland, Belfast, 149–160.

Moore, J.G. (1992) A syn-rift to post-rift transition sequence in the Main Porcupine Basin, offshore western Ireland. *In*: Parnell, J. (ed.), *Basins on the Atlantic Seaboard: Petroleum Geology, Sedimentology and Basin Evolution*. Geological Society, London, Special Publications, **62**, 333–349.

Moore, J.G. and Shannon, P.M. (1995) The Cretaceous succession in the Porcupine Basin, Offshore Ireland: facies distribution and hydrocarbon potential. *In*: Croker, P.F. and Shannon, P.M. (eds), *The Petroleum Geology of Ireland's Offshore Basins*. Geological Society, London, Special Publications, **93**, 345–370.

Murdoch, L.M., Musgrove, F.W., and Perry, J.S. (1995) Tertiary uplift and inversion history in the North Celtic Sea Basin and its influence on source rock maturity. *In*: Croker, P.F. and Shannon, P.M. (eds), *The Petroleum Geology of Ireland's Offshore Basins*. Geological Society, London, Special Publications, **93**, 297–319.

Murphy, N.J. and Croker, P.F. (1992) Many play concepts seen over wide area in Erris, Slyne troughs off Ireland. *Oil and Gas Journal*, Sept. 14, 92–97.

Musgrove, F.W. and Mitchener, B. (1996) Analysis of the pre-Tertiary rifting history of the Rockall Trough. *Petroleum Geoscience*, **2**, 353–360.

Naylor, D. (1992) The post-Variscan history of Ireland. *In*: Parnell, J. (ed.), *Basins on the Atlantic Seaboard: Petroleum Geology, Sedimentology and Basin Evolution*. Geological Society, London, Special Publications, **62**, 255–275.

Naylor, D. (1998) *Irish shorelines through geological time*. Royal Dublin Society: Occasional Papers in Irish Science and Technology, **17**, 20pp.

Naylor, D. and Anstey, N.A. (1987) A reflection seismic study of the Porcupine Basin, offshore West Ireland. *Irish Journal of Earth Sciences*, **8**, 187–210.

Naylor, D. and Shannon, P.M. (1982) *The Geology of Offshore Ireland and West Britain*, Graham and Trotman, London, 161pp.

Naylor, D. and Shannon, P.M. (2005) The structural framework of the Irish Atlantic Margin. *In*: Doré, A.G. and Vining, B.A. (eds), *Petroleum Geology: North-West Europe and Global Perspectives – Proceedings of the 6th Petroleum Geology Conference*. Geological Society, London, 1009–1021.

Naylor, D., Shannon, P.M., and Murphy, N. (1999) *Irish Rockall Basin region – a standard structural nomenclature system*. Petroleum Affairs Division, Dublin, Special Publication **1/99**, 42pp.

Naylor, D., Shannon, P.M., and Murphy, N. (2002) *Porcupine–Goban region – a standard structural nomenclature system*. Petroleum Affairs Division, Dublin, Special Publication **1/02**, 65pp.

Sullivan, J.M., Jones, S.M., and Hardy R.J. (2009) The Lower Cretaceous carbonate potential of the South Porcupine and Goban Spur basins, offshore Ireland. *Atlantic Ireland 2009 Conference, Dublin. Abstracts Volume*, 11.

Reston, T.J., Pennell, J., Stubenrauch, A., Walker, I., and Perez-Gussinye, M. (2001) Detachment faulting, mantle serpentinization, and serpentinite-mud volcanism beneath the Porcupine Basin, southwest of Ireland. *Geology*, **29**, 587–590.

Roberts, D.G., Masson, D.G., Montadert, L., and De Charpal, O. (1981) Continental margin from the Porcupine Seabight to the Armorican marginal basin. *In*: Illing, L.G. and Hobson, G.D. (eds), *Petroleum Geology of the Continental Shelf of North-West Europe*. Heyden and Son Ltd, London, 455–473.

Robinson, K.W., Shannon, P.M., and Young, D.G.G. (1981) The Fastnet Basin: an integrated analysis. *In*: Illing, L.G. and Hobson, G.D. (eds), *Petroleum Geology of the Continental Shelf of North-West Europe*. Heyden and Son Ltd, London, 444–454.

Scrutton, R.A. and Bentley, P.A.D. (1988) Major Cretaceous volcanic province in the southern Rockall Trough. *Earth and Planetary Science Letters*, **91**, 198–204.

Shannon, P.M. (1991) Tectonic framework and petroleum potential of the Celtic Sea, Ireland. *First Break*, **9**, 107–122.

Shannon, P.M. (1993) Submarine Fan Types in the Porcupine Basin, Ireland. *In*: Spencer, A.M. (ed.), *Generation, Accumulation and Production of Europe's Hydrocarbons. III*. Special Publication of the European Association of Petroleum Geoscientists **3**, Springer-Verlag, Berlin, 111–120.

Shannon, P.M. and Naylor, D. (1998) An assessment of Irish Offshore Basins and petroleum plays. *Journal of Petroleum Geology*, **21**, 125–152.

Shannon, P.M., Jacob, A.W.B., Makris, J, O'Reilly, B., Hauser, F., Readman, P.W., and Makris, J. (1999) Structural setting, geological development and basin modelling in the Rockall Trough. *In*: Fleet, A.G. and Boldy, S.A.R (eds), *Petroleum Geology of Northwest Europe: Proceedings of the 5th Conference*. Geological Society, London, 421–431.

Shannon, P.M., Jacob, A.W.B., Makris, J., O'Reilly, B., Hauser, F., and Vogt, U. (1994) Basin evolution in the Rockall region, North Atlantic. *First Break*, **12**, 515–522.

Shannon, P.M., Jacob, A.W.B., Makris, J., O'Reilly, B., Hauser, F., and Vogt, U. (1995) Basin development and petroleum prospectivity of the Rockall and Hatton region. *In*: Croker, P.F. and Shannon, P.M. (eds), *The Petroleum Geology of Ireland's Offshore Basins*. Geological Society, London, Special Publications, **93**, 435–457.

Shannon, P.M., Moore, J.G., Jacob, A.W.B., and Makris, J. (1993) Cretaceous and Tertiary basin development west of Ireland. *In*: Parker, J.R. (ed.), *Petroleum Geology of Northwest Europe: Proceedings of the 4th Conference*. Geological Society, London, 1057–1066.

Simms, M.J. (2009) Permian and Mesozoic. *In*: Holland, C.H. and Sanders, I.S. (eds), *The Geology of Ireland*. Second Edition. Dunedin Academic Press, Edinburgh, 311–332.

Smith, C. (1995) Evolution of the Cockburn Basin: implications for the structural development of the Celtic Sea basins. *In*: Croker, P.F. and Shannon, P.M. (eds), *The Petroleum Geology of Ireland's Offshore Basins*. Geological Society, London,

Special Publications, **93**, 279–295.

Smith, J. and Higgs, K.T. (2001) Provenance implications or reworked palynomorphs in Mesozoic successions of the Porcupine and North Porcupine basins, offshore Ireland. *In*: Shannon, P.M., Haughton, P.D.W., and Corcoran, D.V. (eds), *The Petroleum Exploration of Ireland's Offshore Basins*. Geological Society, London, Special Publications, **188**, 291–300.

Stonecipher, S.A., Spaw, J.M., and Hammes, U. (1994) Diagenetic modeling as an exploration/exploitation tool: the search for an elusive unconformity in the North Celtic Sea. *Annual American Association of Petroleum Geologists Convention, Denver, Colorado, June 12-15*. Abstracts of Papers, 265.

Taber, D.R., Vickers, M.K., and Winn, R.D. Jr. (1995) The definition of the Albian 'A' Sand reservoir fairway and aspects of associated gas accumulations in the North Celtic Sea Basin. *In*: Croker, P.F. and Shannon, P.M. (eds), *The Petroleum Geology of Ireland's Offshore Basins*. Geological Society, London, Special Publications, **93**, 227–244.

Tate, M.P. (1993) Structural framework and tectono-stratigraphic evolution of the Porcupine Seabight Basin, offshore western Ireland. *Marine and Petroleum Geology*, **10**, 95–123.

Tate, M.P. and Dobson, M. R. (1988) Syn- and post-rift igneous activity in the Porcupine Seabight basin and adjacent continental margin west of Ireland. *In*: Morton, A.C. and Parson, L.M. (eds), *Early Tertiary volcanism and the opening of the NE Atlantic*. Geological Society, London, Special Publications, **39**, 309–334.

Tate, M.P. and Dobson, M.R. (1989a) Late Permian to early Mesozoic rifting and sedimentation offshore NW Ireland. *Marine and Petroleum Geology*, **6**, 49–59.

Trueblood, S. (1992) Petroleum geology of the Slyne Trough and adjacent basins. *In*: Parnell, J. (ed.), *Basins of the Atlantic Seaboard: Petroleum Geology, Sedimentology and Basin Evolution*. Geological Society, London, Special Publications,

62, 315–326.

Tyrrell, S., Souders, A.K., Haughton, P.D.W., Daly, J.S., and Shannon, P.M. (2010) Sedimentology, sandstone provenance and palaeodrainage on the eastern Rockall Basin margin: evidence from the Pb isotopic composition of detrital K-feldspar. *In*: Vining, B.A. and Pickering, S.C. (eds), *Petroleum Geology: From Mature Basins to New Frontiers – Proceedings of the 7th Petroleum Geology Conference*. Geological Society, London, 937-952.

Underhill, J.R. and Stoneley, R (1998) Introduction to the development, evolution and petroleum geology of the Wessex Basin. *In*: Underhill, J.R. (ed.), *Development, Evolution and Petroleum Geology of the Wessex Basin*. Geological Society, London, Special Publications, **133**, 1–18.

Vail, P.R., Mitchum, Jr., R.M., Todd, R.G., Widmier, J M., Thompson, S., Sangree, J.B., Bubb, J.N., and Hatfield, W.G. (1977*) Seismic stratigraphy: application to hydrocarbon exploration*. American Association of Petroleum Geologists Memoir **26**, 42–212.

Walsh, A., Knag, G., Morris, H., Quinquis, H., Tricker, P., Bird, C., and Bower, S. (1999) Petroleum geology of the Irish Rockall Trough. *In*: Fleet, A.J. and Boldy, S.A.R. (eds), *Petroleum Geology of Northwest Europe: Proceedings of the 5th Conference*. Geological Society, London, 433–444.

Walsh, P.T. (1966) Cretaceous outliers in south-west Ireland and their implications for Cretaceous palaeogeography. *Quarterly Journal of the Geological Society of London*, **122**, 63–84.

Winn, R.D. Jnr. (1994) Shelf sand-sheet reservoir of the Lower Cretaceous Greensand, North Celtic Sea Basin, offshore Ireland. *AAPG Bulletin*, **78**, 1775–1789.

Young, D.G.G. and Bailey, R.J. (1974) An interpretation of some magnetic data off the west coast of Ireland. *Geological Journal*, **9**, 137–146.

Ziegler, P.A. (1990) *Geological Atlas of Western and Central Europe. 2nd Edition*. Shell Internationale Petroleum Maatschappij B.V. 239pp.

Chapter 9

Cenozoic stratigraphy

By earliest Cenozoic (Tertiary) time continental breakup had been completed and seafloor spreading had commenced along the Atlantic margin west of Ireland, north of the Charlie-Gibbs Fracture Zone. The early stages of the seafloor spreading in the region was coincident with voluminous volcanism and other igneous activity of the North Atlantic Igneous Province, spectacularly displayed in the north of Ireland and Scotland. There was a change from the marine chalk deposition of Late Cretaceous to pulsed clastic deposition in the Palaeogene, the response to a major eustatic regression. The base Cenozoic subcrop map of the Atlantic region is shown in Figure 9.1. A number of episodes of Cenozoic epeirogenesis coincided with plate reorganisation, as outlined in Chapter 2 (Figure 2.10). Seafloor spreading in the Labrador Sea ceased in latest Eocene to earliest Oligocene times and resulted in a period of rapid sagging, differential subsidence and a decrease in coarse clastic deposition, accompanied by a regionally correlatable major unconformity. This marked a major change from predominantly downslope sediment transport to a system of largely alongslope sediment transport that has persisted in the deep-water basins up to the present day. This change was also broadly coincident with the culmination of the first Alpine (Pyrenean) orogenic events. Early Cenozoic sedimentation patterns in the basins were therefore controlled by a series of major, coeval thermal and tectonic events, some emanating from the Atlantic domain, others from further south in the Alpine domain.

The early to mid-Miocene period saw the separation of Jan Mayen and Greenland, and this was accompanied by localised compressional inversion and the formation of a regional set of unconformities. It was also coincident with the Late Alpine orogenic effects. The expansion of the deep-water oceanic system of the North Atlantic resulted in the accelerated buildup of a series of large and extensive deep-water contourite drift deposits in most of the basins along the Atlantic margin, including the Rockall, Porcupine and Hatton basins. Contourite development was initiated with the development of deep-water conditions following the rapid sagging and differential subsidence of the large basins in earliest Oligocene time. However, the opening of the Fram Strait between NE Greenland and Svalbard in the Miocene established a deep Arctic–NE Atlantic connection, facilitated geostrophic current sediment transport and contourite deposition, and triggered a progressive Northern Hemisphere climatic cooling (Stoker et al., 2005b).

Early Pliocene tectonic movements, associated with late Neogene global plate reorganisation, affected much of the NE Atlantic region and resulted in kilometre-scale domal uplift of onshore areas coeval with offshore subsidence and tilting. Closure of the Central American Seaway by the formation of the Isthmus of Panama resulted in the redirection of warm water masses to northern latitudes. The Miocene–Pliocene reorganisation is characterised by a pronounced regional unconformity. The late Pliocene marked the onset of the widespread northern hemisphere glaciation that prevailed through Pleistocene times. This produced pulsed sediment discharge that prograded across the shelf/slope break and resulted in a series of major sediment wedges, particularly pronounced west of Scotland and Norway.

The Holocene development of the region has been characterised by relatively little sediment discharge from the shelf into the deep-water basins, with the predominant sedimentation patterns resulting from redistribution of sediment by geostrophic currents, together with pelagic sedimentation. However, current sweeping of the basin margins has facilitated a combination of canyon development (some inherited from Plio-Pleistocene times) and scouring of the lower slopes, resulting in periodic dramatic and substantial slope failure. This in turn has led to slump, slide, turbidite and debrite deposition interspersed with the pelagic sedimentation.

Offshore distribution of Cenozoic rocks

Cenozoic sedimentary strata are generally thin throughout the basins in the Irish Sea and northwards into the NE onshore basins. Typically a few hundred metres at most

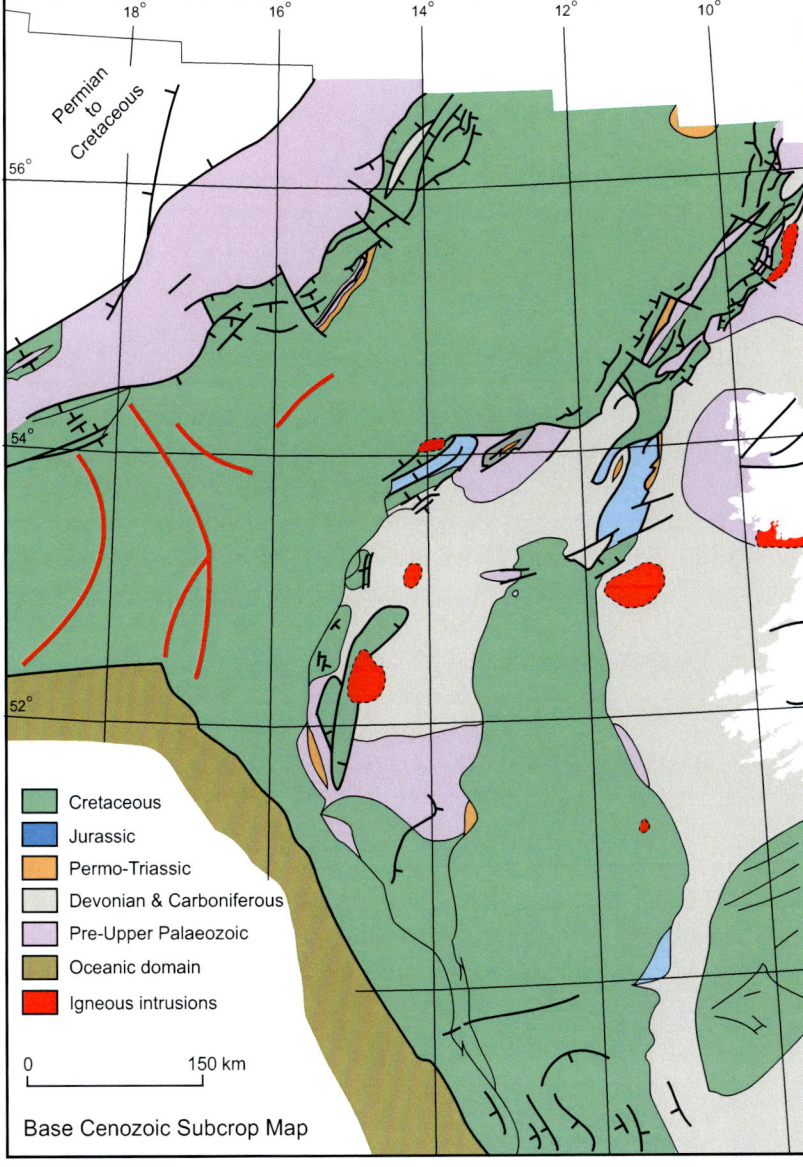

Figure 9.1. Base Cenozoic subcrop map of the Atlantic margin basins west of Ireland (*after* Naylor and Shannon, 2005, 2009).

Base Cenozoic Subcrop Map

Legend:
- Cretaceous
- Jurassic
- Permo-Triassic
- Devonian & Carboniferous
- Pre-Upper Palaeozoic
- Oceanic domain
- Igneous intrusions

0 — 150 km

are preserved. In northeast Ireland and the Firth of Clyde, the Cenozoic succession is dominated by the Eocene Plateau basalts, which are often in excess of 400 m thick. However, in the Lough Neagh area of Northern Ireland a relatively thick succession of Chattian (late Oligocene) age rests upon older Tertiary strata. Weathered basalts are overlain by a succession of interbedded ?Eocene lithomarge, pyroclastics and lacustrine deposits up to 300 m thick in boreholes (Parnell *et al.*, 1989). These are overlain by the Lough Neagh Clays, which extend over about 500 km², of which 300 km² underlie the Lough while the remainder is covered by Quaternary boulder clay. The Oligocene succession consists mainly of clays and silts, which reach a maximum known thickness of 363 m (Wilkinson *et al.*, 1980) and represents the deposits of a large shallow lacustrine complex, extending beyond the limits of the present Lough Neagh. This was fed by rivers, mostly flowing from the west and occasionally from the northeast, and building a series of river deltas into the tectonically-controlled freshwater lacustrine system. The clays have been used for more than a century as pottery clay. However, they also contain significant quantities of lignites, deposited in a swampy environment at the lake margin with the thickest bands in the lower part of the succession. Boreholes have intersected a cumulative lignite thickness of up to 140 m and the material is reported to have a high calorific value (Parnell *et al.*, 1989). In the Kish Bank Basin the upper part of the Fina 33/17-1 well encountered approximately 212 m of presumed Cenozoic to Recent strata. This comprised variable red-brown to grey sandstone with increasing amounts of claystone with depth. Lignite beds were common in the upper part of the section. On the basis of very poor miospore recoveries, the uppermost 60 m or so was interpreted as Pleistocene to Recent in age, while the lower section was regarded as being Pliocene or older (Naylor *et al.*, 1993). Further south in the Irish Sea, a thin Cenozoic succession overlies St Tudwal's Arch, while an unusually thick succession of probable Neogene sediments was encountered in the Mochras borehole at the eastern margin of the Cardigan Bay Basin (Figure 6.7). Here, beneath a thin veneer of Pleistocene sediments, a thick (524 m) sequence of Oligo-Miocene clays and sands unconformably overlies the thickest known (1305 m) succession of Lower Jurassic strata in the British Isles (Woodland, 1971). Further west, up to 800 m of latest Paleocene–earliest Eocene to possibly Neogene claystones, thin sandstones and lignites are documented in the St George's Channel Basin.

Palaeothermal and compaction studies in the Mochras borehole by Holford *et al.* (2005) suggest that early Cretaceous exhumation removed up to 2.5 km of mid-Jurassic–early Cretaceous section while a Neogene exhumation episode led to the erosion of *c.*1.5 km of mid-Miocene–?Pliocene sediments. Similar major inversion and erosional events have been postulated throughout the Irish Sea region (further discussion on this subject can be found in Chapter 12). A thin (less than a couple of hundred metres) succession of poorly dated Cenozoic clastic strata was recorded in the Central Irish Sea Basin (e.g. Floodpage *et al.*, 2001). The thin nature of the succession is also attributed, as with other parts of the region, to mid-Cenozoic inversion and erosion events so that the preserved strata are typically of Miocene or younger age. This is the general pattern within the broad Irish Sea region, where typically only a veneer of such sediments rests unconformably upon Jurassic and older strata.

The Cenozoic succession in the general Celtic Sea region is also relatively thin and incomplete, largely due to similar Cenozoic tectonism and erosional events to those recorded in the Irish Sea. However, the overall post-Mesozoic succession thickens and becomes more complete westwards towards the Fastnet Basin and beyond into the Atlantic domain. Cenozoic strata are generally thin to absent through much of the central part of the North Celtic Sea Basin, where a veneer of probable Neogene strata rest unconformably upon inverted

and eroded Cretaceous Chalk. However, towards the northern margin of the basin, the Gulf 49/9-1 well recorded approximately 300 m of Eocene–Oligocene strata resting unconformably on Upper Cretaceous Chalk. Paleocene sediments may have been deposited conformably on the Cretaceous succession in places, but any such strata have since been removed by erosion. The biostratigraphy in all the wells with thick Cenozoic strata shows the Paleocene to be missing (Murdoch *et al.*, 1995). An Eocene to Oligocene sequence, of fluvial and later marine affinities, dated with confidence on the basis of angiosperm pollen, generally forms the greater part of the preserved section. The Cenozoic strata recorded in wells towards the northern part of the South Celtic Sea Basin (e.g. Marathon 49/29-1 and Marathon 58/3-1) are also dated as Eocene to Oligocene. This section rests with slight angular unconformity on the Upper Cretaceous Chalk section with only the Paleocene section missing. A Pliocene to Pleistocene succession was recorded in Marathon 49/29-1 and in other wells. It is likely that a thin veneer of sediments of this age is widely developed throughout the basin, unconformably overlying the inverted Eocene–Oligocene and Chalk strata. Murdoch *et al.* (1995) recognised two major Cenozoic erosion events. The first was a regional uplift in the Paleocene, with erosion of up to 1100 m taking place in the centre of the North Celtic Sea Basin. The second was a phase of compressive inversion, probably associated with an Oligo-Miocene phase of the Alpine Orogeny. The two tectonic events were separated by a period of basinal deposition in the Eocene to Oligocene.

Further west, a thicker Cenozoic succession has been encountered by wells in the Fastnet Basin, with more than 1 km of strata recorded (Robinson *et al.*, 1981). Again the succession is attenuated by the two erosional events. The succession typically commences with a mid-Eocene to Oligocene shallow-marine limestone, of the order of 350 m thick throughout much of the basin but thin or absent in some areas to the southwest. Marine Miocene and Pliocene clays, with some sandy intercalations, overlie the limestones and the overall succession shows a broad pattern of marine deepening though time.

In the large deep-water basins west of Ireland a thick (up to 4 km) relatively complete Cenozoic succession is preserved (Figure 9.2). A combination of well and seismic data from the Porcupine and Rockall basins allow the piecing together of a comprehensive picture of the sedimentary and oceanographic development of the European passive margin during the period of plate

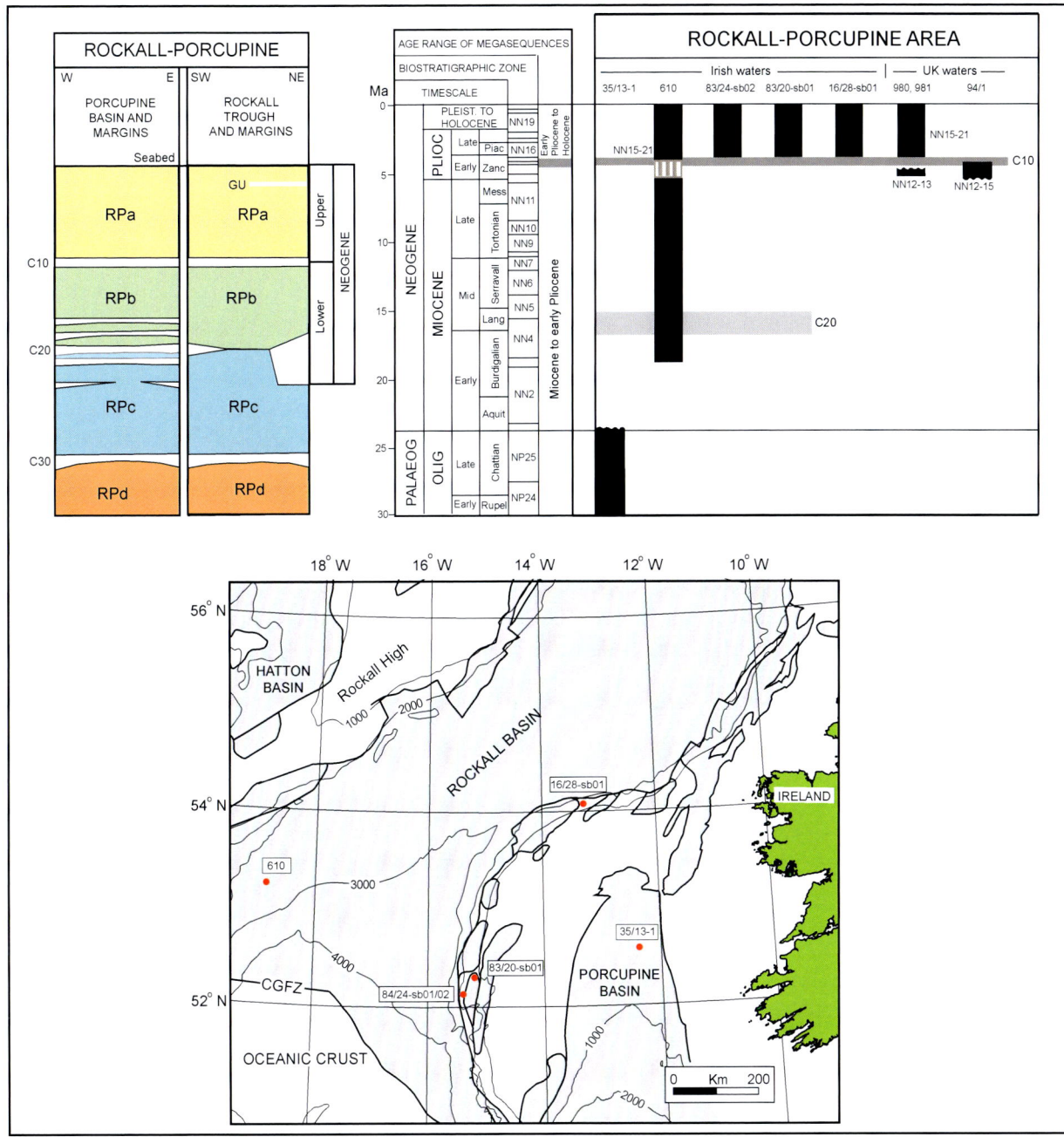

Figure 9.2. Stratigraphic nomenclature (megasequences and key unconformities) for the Cenozoic of the Porcupine and Rockall region and stratigraphic range chart key of exploration wells and boreholes in the Rockall and Porcupine area (based on Stoker *et al.*, 2005b, and Haughton *et al.*, 2005). Timescale is after Berggren *et al.* (1995).

reorganisation that followed the breakup and separation of the European and North American continents. Regional seismic mapping, tied to key wells, has shown that the Cenozoic succession is punctuated by a series of regional unconformities that reflect the effects of major configurational changes within the basins in response to lithospheric causes (Stoker *et al.*, 2005a). These unconformities (*see* detailed descriptions in McDonnell and

Shannon, 2001; Stoker *et al.*, 2001, 2005b) are the C40 of early Paleocene age, C30 of latest Eocene to earliest Oligocene age, C20 of mid-Miocene age and C10 of early Pliocene age (Figure 9.2). They bound a set of distinct megasequences that reflect the regional changes within the basins. The unconformities can be mapped in both the Porcupine and Rockall basins, with the major ones (C10, C30 and C40) also identified in the Hatton Basin.

Unfortunately, the amount of well control in the Hatton and Rockall basins is sparse, with most of the main well information coming from exploration drilling in the northern part of the Porcupine Basin. However, valuable information is also provided by a set of deep stratigraphic boreholes along the eastern margin of the Rockall Basin (Haughton *et al.*, 2005).

Where drilled in the northern part of the Porcupine Basin, the earliest Cenozoic strata, of approximately mid-Paleocene age, rest disconformably on Upper Cretaceous strata, with some of the uppermost Cretaceous succession missing. In the northern part of the Porcupine Basin a number of wells have penetrated the succession above the C40 unconformity. Here the earliest Cenozoic strata are typically of marly siltstones or limestones, deposited in a marine setting. These give way to a late Paleocene to early Eocene succession of sandstone and siltstone packages that represent the deposits of a series of pulsed deltaic progradational episodes (Figure 9.3).

The C40 unconformity is defined by a high-amplitude, continuous reflection marking the Cretaceous to Cenozoic boundary (Figure 9.4) and representing a major change from carbonate to clastic deposition. Well data (e.g. BP 26/28-5, Shell 34/19-1, Shell 35/13-1, Deminex 35/15-1) indicate that the Paleocene succession in the Porcupine Basin is mudstone dominated, in contrast to the major sandstone-dominated succession of mid-Eocene to earliest late Eocene times. The earliest Cenozoic seismic strata along the western margin of the Porcupine Basin comprise a number of wedge-shaped seismic packages, each extending for up to 8 km in the dip direction. The structural setting of these packages, together with their seismic configuration, is consistent with the development of a series of marginal alluvial fans.

In the northern part of the Porcupine Basin, Moore and Shannon (1992) show that on seismic data a succession of five pulsed clinoform packages, of probably predominantly Eocene age, prograde southwards and overstep a mounded sequence of Paleocene age (Figure 9.4). The latter is interpreted as a basinfloor fan sequence resting unconformably upon Upper Cretaceous Chalk. Each progradational package is separated by a continuous high-amplitude reflection consistent with a flooding surface (Croker and Shannon, 1987; Moore and Shannon, 1992; Shannon *et al.*, 2005). Each package records sediment input from a northerly quadrant of the basin margin. The preservation of topsets in the deltaic clinoforms points to a relative sea-level rise coincident with sediment progradation. These packages are coeval

with localised basin floor submarine fan deposition in the deeper parts of the Porcupine Basin (Shannon *et al.*, 1993), sourced by sediment shed from the Porcupine Bank to the west. On seismic profiles a set of elongate mounded seismic facies, interpreted as being largely of Eocene age, was described by Shannon (1992) and McDonnell and Shannon (2001) from the southwest of the basin. These mounds are approximately 0.2–0.3 s TWT thick, corresponding to c. 200–300 m. They extend up to 50 km into the basin and are characterised by bi-directional downlap at the mound edges. Internally the mounds comprise moderate- to high-amplitude, laterally continuous reflections of moderate frequency. These mounds were interpreted by Shannon (1992) as submarine fan deposits, with the lateral seismic character change towards the mound edges representing a shale-prone fan fringe to a more sandy, channelised inner fan zone. The mound margins frequently show evidence of incision and erosional modification. The stacking pattern, with the thickest parts of successive mounds offset from thickest parts of the underlying mounds, indicates topographic control on fan progradation. The individual fan sequence may be tentatively correlated with individual deltaic progradational events in the northern part of the basin, suggesting a regional control on the pulsed nature of sediment supply into the basin. The overall pattern of pulsed clastic input in latest Paleocene to Eocene times has been explained by shelf/flank uplift and basin centre subsidence controlled by regional tectonics (Shannon *et al.*, 2005). It was previously interpreted as being due to ridge-push (Shannon *et al.*, 1993), while in the North Sea a similar pattern has been attributed to mantle plume tectonics (White and Lovell, 1997).

The latest Eocene succession is dominated by moderate to high-amplitude continuous reflections that drape the mounds and onlap the southerly-prograding deltaic clinoforms in the north of the basin. Wells Phillips 35/8-2 and Shell 35/29-1 record a mud-prone probable Upper Eocene succession. Northerly flooding of the Eocene deltaic sequences at the end of Eocene time is interpreted to signify an end to significant coarse clastic input into the basin and culminates in the development of the regionally extensive C30 unconformity.

The interpreted Paleocene to Eocene succession in the Rockall Basin is also bounded at the base by the C40 unconformity, a basinwide, high-amplitude reflection which is locally onlapped by lower amplitude, less continuous reflections. The overlying succession thickens into the basin centre and is characterised by moderate- to

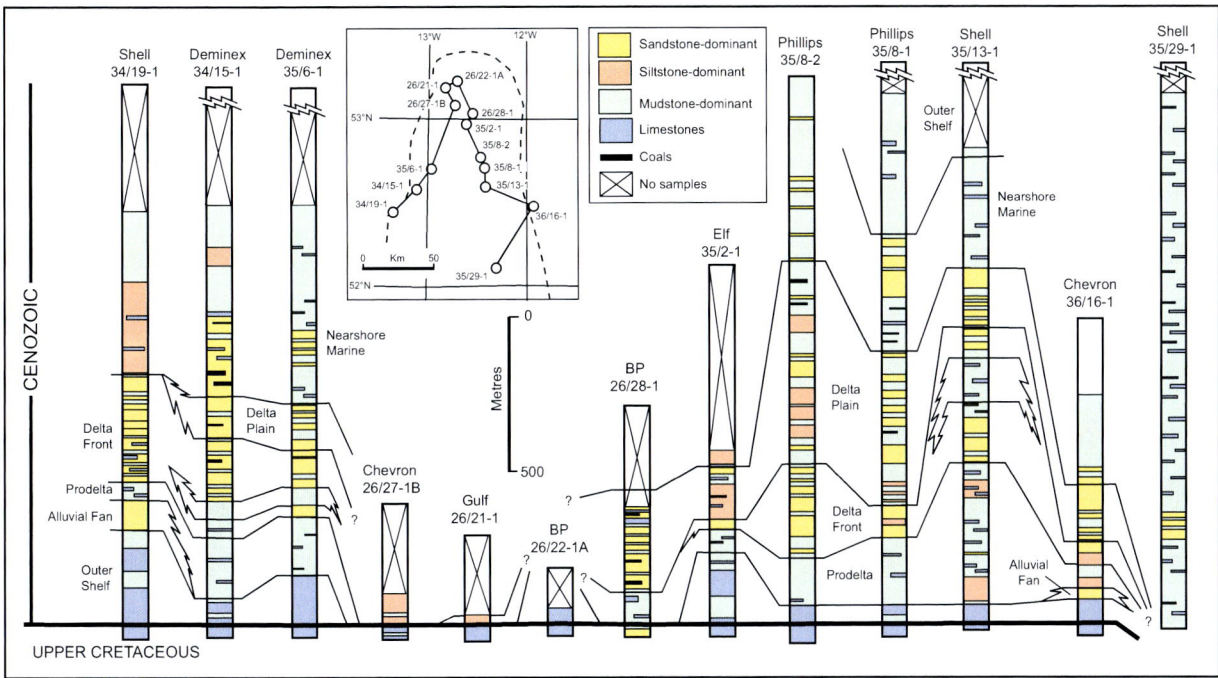

Figure 9.3. Correlation of wells drilled through the Cenozoic succession in the northern part of the Porcupine Basin (adapted from Croker and Shannon, 1987).

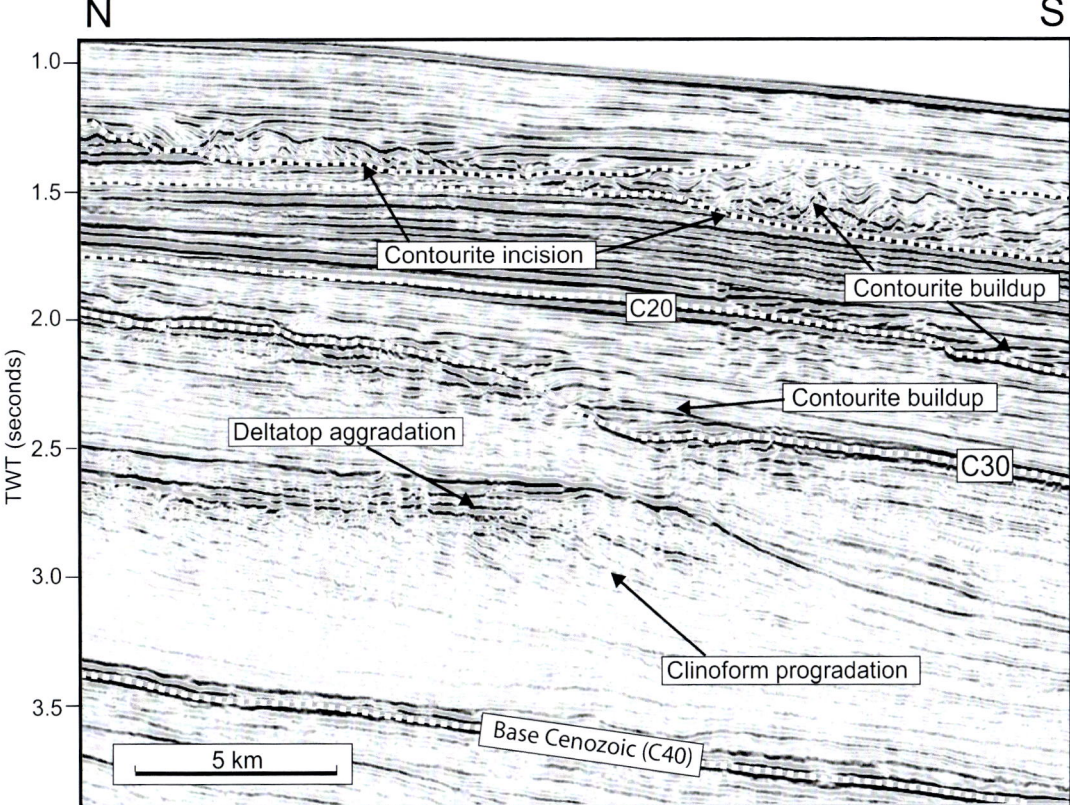

Figure 9.4. Seismic character of the Cenozoic succession in the northern part of the Porcupine Basin (from Shannon *et al.*, 2005). The C40, C30 and C20 major unconformities are indicated. The C40-C30 interval consists of a sand-prone prograding succession, while the post-C30 succession represents a mud-prone succession characterized by contourite erosional and depositional features.

high-amplitude reflections with reflection frequency decreasing basinwards (McDonnell and Shannon, 2001). Seismic reflection amplitude tends to increase towards the top of the sequence, similar to the pattern observed in the Porcupine Basin, although reflection continuity is poorer in the Rockall Basin. At the eastern margin of the basin a wedge of low- to moderate-amplitude seismic reflections near the base of the sequence suggests that the Porcupine High may have acted as a sediment supply provenance for the Rockall Basin, as well as the Porcupine Basin, during early Palaeogene time. This wedge is onlapped by a seismic package of inferred mid-Eocene age. Well BP 132/15-1, on the eastern flank of the Rockall Basin in UK waters, recorded a highly tuffaceous, mud-dominated Paleocene succession with thin sandstones. The main sandy interval encountered in this well was of Early Eocene (Ypresian) age. This is somewhat earlier than the main sandy succession in the Porcupine Basin.

Some broad similarities in the stratal geometries of the Paleocene to Eocene sequences are observed between the Porcupine and Rockall basins. In particular, the onlapping relationship of the lowest Paleocene strata onto the high-amplitude basal reflection marking the top of the Cretaceous Chalk succession is markedly similar. Wedge-shaped basal units are recorded in both basins, while the overlying seismic packages in both basins show a general upwards increase in reflection amplitude. However, there is no evidence for the development of regionally extensive, large-scale progradational clinoforms in the Eocene succession of the Rockall Basin comparable to those in the Porcupine Basin.

In addition to the well-known Palaeogene igneous province of Scotland and Northern Ireland, a significant igneous history in the Atlantic offshore basins has been documented by well and seismic data in the Irish Atlantic offshore basins. Major igneous centres have been identified on the basis of seismic, gravity and magnetic mapping (Naylor *et al.*, 1999, 2002). They include the Brendan Igneous Centre in the northeast of the Porcupine Basin, which is accompanied by large gravity and magnetic anomalies (Riddihough, 1975; Mohr, 1982; Tate and Dobson, 1988). A number of volcanic cones have been recorded along the eastern margin of the Rockall Basin (Haughton *et al.*, 2001). Although few of the major centres have been dated, their setting is strongly suggestive of an early Palaeogene age. High-amplitude discontinuous seismic reflectors are prevalent through much of the Rockall Basin and the northeastern part of

the Porcupine Basin and have been interpreted as extensive transgressive sills. Seeman (1984) reported tholeiitic sills in the Cenozoic succession of the Shell 35/13-1 well and suggested a late Oligocene age (25.8 ± 2.6 Ma) on the basis of K–Ar dating. Undated dolerites and other intrusives in other wells are also likely to be of Palaeogene age. Possible feeder dykes to the transgressive sills have been identified on some seismic profiles in the Porcupine Basin. In the north of the basin, Tate and Dobson (1988) identified a series of parallel small igneous plugs, named the Slyne Fissures, and inferred their age to be Palaeogene. Wells drilled on the Corrib Gasfield (Dancer *et al.*, 2005), in the Slyne Basin, encountered a Cenozoic succession typically less than 300 m thick, including a succession of lavas that impede seismic data quality across the field. Dancer and Pillar (2001) recorded approximately 40 m of Late Palaeogene (possibly Oligocene) to Recent clays and sandstones resting unconformably on 70–120 m of Early Cenozoic basaltic lavas and tuffs. Dancer *et al.* (2005) provided K/Ar ages for volcanics in the area within the range 40–54.3 Ma, suggesting that the succession is of Eocene age. The upper surface of the volcanic succession is highly rugose, interpreted as the result of subaerial exposure and weathering. Several distinct periods of eruption are inferred from magnetic phase reversal recorded in the lavas. By analogy with outcrops from County Antrim in the north of Ireland, discrete lave flows are likely to be separated by interbedded weathered layers. These, in turn, rest unconformably on Upper Cretaceous chalks, interbedded stringers of limestones and calcareous claystones.

A major regional tectonic and facies change in latest Eocene time is coincident with a major unconformity (C30 of McDonnell and Shannon, 2001; Stoker *et al.*, 2001, 2005b,c). This unconformity (Figure 9.4) locally cuts through several hundred metres of strata in the Porcupine Basin and is also mapped in the Rockall Basin (Shannon *et al.*, 2005). It resulted from a rapid tilting (*see* Chapter 2) causing deepening and the onset of vigorous large-scale ocean current circulation. The unconformity also marks a major change from downslope sediment transportation to mainly alongslope sediment movement. From latest Eocene time basin subsidence outstripped sedimentation in the Atlantic margin basins. This, in turn, has led to the development of undersupplied deep-water basins with steep sides.

Praeg *et al.* (2005) suggested a minimum post-Eocene deepening of 700 m in the vicinity of the Anton Dohrn seamount, where mid- to late Eocene nearshore conglomerates crop out on the crest of the seamount. The same authors also illustrated the strongly differential nature of the deepening further north, where subaerial Paleocene basalts between Bill Bailey and Lousy banks are overlain by post-Eocene strata and lie in water depths ranging from 500 m to more than 2 km. A similar Oligocene basinward tilt of the Eocene and older strata by several degrees was documented along the northeast Rockall margin by other authors (e.g. Cunningham and Shannon, 1997). The early Oligocene succession in the Porcupine and Rockall basins typically shows very continuous, flat-lying parallel seismic reflections, overlying the major C30 unconformity. Where drilled they are typically mudstones with occasional silt and marl layers.

Unlike the Norwegian and parts of the UK Atlantic basins further north, where the base of the Neogene is marked by a pronounced unconformity, the base of the Neogene in the central and southern Rockall Basin lies within the post C30 megasequence (the RPc megasequence of McDonnell and Shannon, 2001; Stoker *et al.*, 2005a). In the Porcupine Basin, the base of the Neogene corresponds to a mainly conformable seismic reflector (McDonnell &Shannon, 2001), recognised in wells as an early Miocene hiatus, dated as intra-NN2 biozone (Dobson *et al.*, 1991).

The early to mid-Miocene successions of the Atlantic margin region, like those of the Oligocene, are characterised by flat-lying parallel reflections within the basins. Where drilled, these are typically mudstones with occasional silt and marl layers. Towards the basin margins the succession typically displays an onlapping, sometimes mounded and locally significant upslope accretion (Stoker *et al.*, 2005c). This represents the first development of muddy and silty contourites within the Atlantic margin basin.

In the Rockall Basin, a pre-C20 high-amplitude Early Miocene seismic marker was identified by McDonnell and Shannon (2001) as a diagenetic horizon, related to the deposition of reworked volcanic ash, which was subsequently altered to a smectite-rich horizon. Dolan (1986) attributed this to volcanic ash derived from volcanic eruptions in the Norwegian Basin. While there is no evidence for a comparable coeval seismic horizon within the Porcupine Basin, numerous wells (e.g. Chevron 36/16-1) in the basin recorded a number of glauconite-rich horizons in the Upper Eocene to Lower Miocene interval. Volcanogenic glauconite in Early and Late Cretaceous clays age in southern England and Northern Ireland was suggested by Jeans *et al.* (1982) to have developed from volcanic debris of mafic composition. McDonnell and Shannon (2001) suggested that the glauconite-rich horizons in the Porcupine Basin are derived from volcanic debris of comparable mafic origin, indicating a different source provenance to the altered (acidic or alkaline) volcanic glass of the Rockall Basin.

Along the eastern margin of the Rockall Basin, a series of boreholes penetrated a generally thin Cenozoic succession, with Plio-Pleistocene strata resting above the C10 unconformity, and underlain by variable Eocene to mid Miocene strata bounded beneath by the C30 unconformity (Haughton *et al.*, 2005). The base Cenozoic surface in the 83/20-sb01 well is reported to be characterised by a phosphatised crust and is a composite unconformity with early to early mid-Miocene well-oxygenated bathyal or slope clays and silts resting upon Late Cretaceous chalks. However, in the other two boreholes (84/24-sb02 and 16/28-sb01) the C30 unconformity is overlain by early Eocene outer shelfal strata. The Eocene deposits vary from clastic to carbonate-prone from north to south, possibly reflecting preferential sequestration of clastic sediment by the inboard adjacent Porcupine Basin. Early Eocene smectitic clays are common in the 16/28 borehole consistent with a source in earlier volcanic deposits. These weak layers may have played a role in promoting the widespread slope failure features along the basin margin.

A major intra-Miocene unconformity occurs throughout the Porcupine and Rockall basins. In the southern Rockall Basin, the C20 reflector is a regional submarine surface that onlaps the basin margins and has been dated as late early–early mid-Miocene (NN4–5 biozone) at DSDP site 610, where it is diagenetically enhanced (*see* Stoker *et al.*, 2001 and references therein). In the Porcupine Basin, C20 corresponds to one of several angular unconformities that grade into conformity in the basin centre (McDonnell & Shannon, 2001).

The mid- to late Miocene megasequence is seen in both the Rockall Basin and Porcupine Basin. From hereon in the shallow stratigraphy we will refer to the Rockall Trough and the Porcupine Seabight, which are the bathymetric features overlying the sedimentary basins of the Rockall Basin and Porcupine Basin respectively, as their sedimentary composition and architecture are more appropriately regarded as being controlled by the modern bathymetry. The thickness of these deposits remains unknown in the Trough, but up to 0.5 s TWT of lower Miocene strata are inferred from the Porcupine Seabight. The succession displays an onlapping, ponded, basin-floor-fill geometry throughout the region, commonly mounded, and with locally significant upslope accretion onto the basin margin (Figure 9.5). In the Rockall Trough, this geometry reflects three major styles of contourite drift accumulation: (1) flat-lying to gently mounded, sheeted drifts that occur in the axial region of the basin floor; (2) elongate mounded drifts onlapping onto, and prograding up, the margins of the basin; and, (3) the giant, elongate, Feni Drift (Figure 9.6) in the southern Rockall Trough (Stoker *et al.*, 1998; 2001). A smaller drift also developed adjacent to Porcupine Bank. In the Porcupine Seabight the succession shows a broad catenary profile, although a north–south subsidiary depocentre lying to the west of the eastern basin margin is superimposed on the broader infill. This corresponds to the Porcupine Drift (McDonnell and Shannon, 2001), an elongate drift that is coeval with the Feni Ridge but is more topographically subdued than the latter. The contourite drifts are up to 40 m high with a wavelength of 2–3 km (Stoker *et al.*, 2001). In areas of upslope progradation, sediment drifts have migrated from the basin onto the slope to a point close to the contemporary shelf break. Internal reflections display both onlapping and downlapping terminations, the latter generally associated with moat development on the flanks of the mounded drifts.

In the southern Rockall Trough, Deep Sea Drilling Project (DSDP) site 610 sampled the Feni Ridge and proved lower–middle Miocene to lower Pliocene nanofossil chalk and ooze, deposited in a lower bathyal setting (Hill, 1987). Lower Pliocene nanofossil ooze was recovered from a basinal, sheeted drift at Ocean Drilling Program (ODP) site 981, onlapping onto the western flank of the Trough (Jansen *et al.*, 1996). On the eastern flank of the Rockall Trough, borehole 83/20-sb01 cored middle–upper Miocene muds onlapping onto the slope of Porcupine Bank (Haughton *et al.*, 2005). Further north, the British Geological Survey (BGS) borehole 94/1 proved upper Miocene–lower Pliocene bioclastic sand and gravel from the moat of an elongate mounded drift on the slope of Rockall Bank (Stoker *et al.*, 2001). In the NE Rockall Trough, a sand-dominated succession was proved in well BP 164/25-2 that penetrated the feather-edge of a flat-lying basinal succession (Stoker *et al.*, 2005a). In the axis of the Rockall Trough, BGS borehole 90/15 cored middle Miocene bioclastic sands and muds from the top of Anton Dohrn Seamount, whereas 90/18 recovered middle to upper Miocene nanofossil ooze from the top of Rosemary Bank (Stoker *et al.*, 1993). The tops of the seamounts were probably located

at upper to middle bathyal depths (Stoker, 1997). At shallower depths, middle–upper Miocene onlapping, transgressive, glauconitic sandstones, deposited in a deep neritic to upper bathyal environment, have been proved in BGS boreholes 88/7,7A, 90/12,12A and 90/13 on the upper Hebrides Slope and Shelf, and in the Texaco 13/3-1 and Amoco 19/5-1 wells in the Donegal and Erris basins respectively on the Malin Shelf (Stoker *et al.*, 1993). In the Porcupine Seabight, the Shell 35/13-1 well sampled the upper part of the RPc megasequence and proved a sequence of upper Oligocene–lowest Miocene silty and calcareous claystones, interbedded siltstones and argillaceous limestones unconformably overlain by a lower Miocene conglomerate (Dobson *et al.*, 1991). Slump deposits and mass failures became a feature of the basin margins from early Miocene time. However, the most extensive and largest mass failures occurred in Pliocene to Recent times (Evans *et al.*, 2005) and are discussed further below. A large buried slide complex on the western margin of the Porcupine Seabight has been described by Moore and Shannon (1991). This measures 80 km in length and runs parallel to the basin margin. It is 20 km wide and reaches a maximum thickness of some 400 m at the centre of the body. Slide movement, suggested as being of mid-Miocene age, was facilitated by a basal detachment interpreted to be an overpressured muddy horizon, and the maximum amount of lateral movement is estimated to be approximately 1 km in the central part of the complex.

An early Pliocene (C10) unconformity has been identified along the entire length of the NW European Atlantic margin (Stoker *et al.*, 2001, 2005a). It heralded the onset of global cooling and Northern European glaciation. The C10 unconformity is mapped as a regionally extensive seismic reflector through the Porcupine and Rockall regions. In the latter basin it has been dated as intra-early Pliocene in age, between 3.85 and 4.5 Ma based on biostratigraphic data from ODP site 981 (Stoker, 2002). This age is consistent with a marked change in sedimentation style across C10 at DSDP site 610 (Stoker *et al.*, 2001) and an early Pliocene break in deposition reported from UK well Mobil North Sea 214/4-1, in the Faroe–Shetland Channel (Davies and Cartwright, 2002), close to, but slightly below, the level of the C10 equivalent (Stoker *et al.*, 2005b). An intra-early Pliocene age has also been reported on the basis of biostratigraphic data from BGS boreholes 88/7,7a and 77/9 from the shelf-margin off NW Britain, where lower Pliocene strata immediately overlie both C10 and the intra-Neogene unconformity

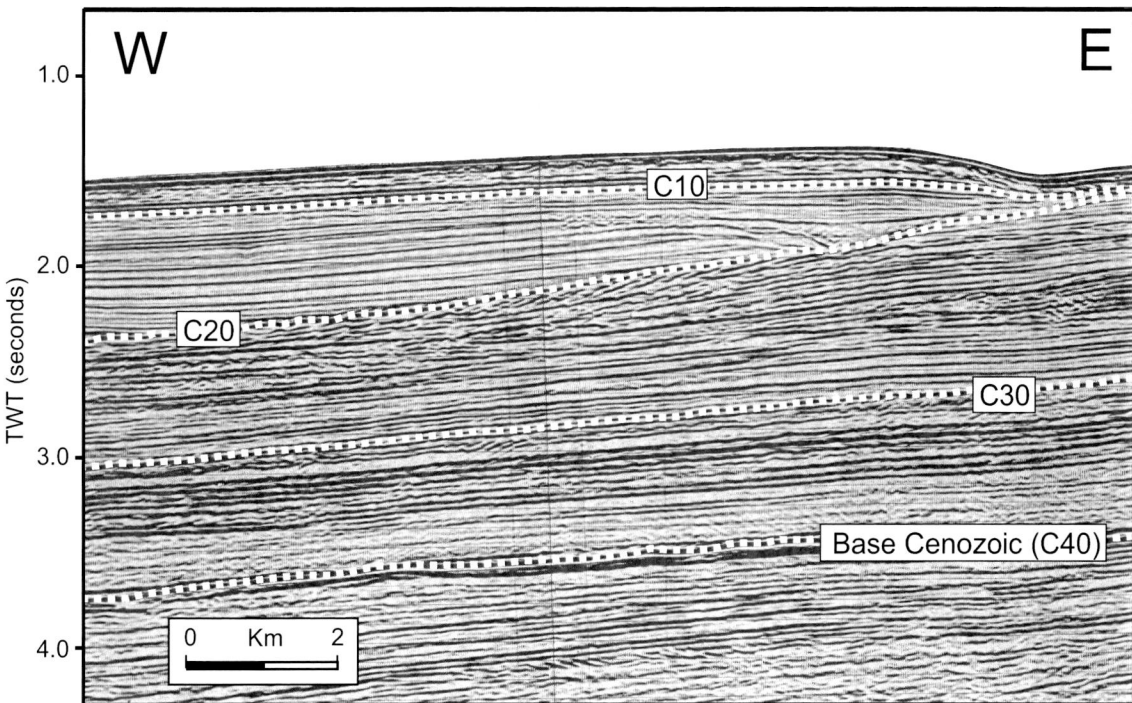

Figure 9.5. Seismic profile from the eastern margin of the Porcupine Basin (from Shannon *et al.,* 2005). The C40, C30, C20 and C10 unconformites are shown. Contourite incision and aggradational onlap onto the C20 unconformity is seen. The base-Pliocene C10 unconformity marks a change in seismic character corresponding to the onset of major cooling and glacial conditions.

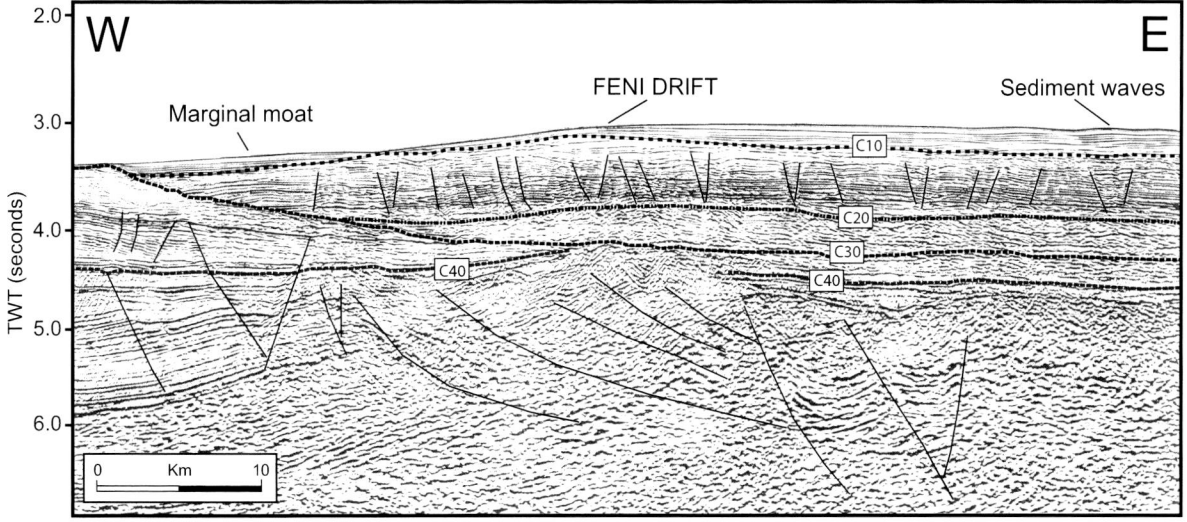

Figure 9.6. Seismic profile from the western flank of the Rockall Basin illustrating the sedimentary architecture of the Miocene Feni Drift and highlighting the four regional unconformities (C40, C30, C20 and C10) in the basin. Note the abundance of synsedimentary polygonal faulting in the C20-C10 interval (from McDonnell and Shannon, 2001).

in the proximal part of the prograding upper Neogene successions (Stoker, 2002).

The Pliocene to Holocene succession in the Irish Atlantic offshore is generally thin (typically less than 150 m) and lies above the C10 unconformity (Figures 9.5 and 9.6). The succession is widely distributed as a thin draping sheet of sediment within the Porcupine, Rockall and Hatton regions. On the basin floor the strata are typically pelagic foraminiferal oozes. Occasional Pleistocene sediment wedges, such as the Barra-Donegal Fan, prograde from the glaciated margins of the northern part of Ireland and Scotland. This is in contrast with the succession in the mid-Norway region where the mega-sequence is dominated by thick glaciomarine clastic wedges (Dahlgren *et al.*, 2005).

The deep-water basins and their flanking highs west of Ireland contain clusters of carbonate mounds, some of which are relatively well documented using geophysical and video data and also by direct sampling (Shannon *et al.*, 2007; Bailey *et al.*, 2003) and drilling (Williams *et al.*, 2006). They occur on or above the C10 unconformity (Figure 9.7). More than a thousand mounds, some on the seabed while others are buried, have been identified by geophysical means in the Porcupine Seabight and on the flanks of the Rockall Trough (Croker and O'Loughlin, 1998; Huvenne *et al.*, 2007). Most commonly recorded in the northern part of the Porcupine Seabight and on the shoulders of the Porcupine and Rockall banks, they generally lie in water depths ranging from 500 m to 900 m. They typically occur in clusters, generally referred to in the literature as 'mound provinces', with individual mounds sometimes more than 100 m high and up to a few kilometres in diameter. The carbonate mounds typically contain large numbers of deep-water corals, most notably *Lophelia pertusa*, whilst foraminifera and ostracods are other important biogenic components (Ainsworth *et al.*, 1996).

The mounds along the eastern margin of the Porcupine Seabight are barrier-like buildups and are situated in a region of relatively steep seabed slopes associated with Cenozoic mass wastage and margin-parallel erosional channels (De Mol *et al.*, 2002). The mounds along the northern margin of the Porcupine Seabight are divided into large seabed mounds in the southwest and smaller, mainly buried, mound clusters in the northwest (Henriet *et al.*, 1998, 2001). In contrast to the mounds along the eastern margin, these mound populations are located in an area of low seabed dips and the mounds, including the buried ones, have pronounced north–south erosional moats. The mound cluster on the eastern margin of the Rockall Trough/western margin of the Porcupine Bank also lies in deep water (700–900 m). More than 140 individual mounds have been identified on sidescan sonar data in an elongate band parallel to the eastern margin of the Rockall Trough (O'Reilly *et al.*, 2003). The smaller mounds, in particular, are frequently associated with recent slope failure escarpments, while evidence for recent slope failure escarpments is lacking around the clusters of large mounds.

Significantly, with very few exceptions, the carbonate mounds in the Porcupine and Rockall region all appear to root on this same regionally continuous sub-horizontal C10 seismic reflector. This suggests a geologically instantaneous or very rapid growth event of early Pliocene age. In 2005 a number of the mounds in the Porcupine Seabight were drilled on IODP expedition 307. The partly buried Challenger Mound was drilled to the base of the mound (Williams *et al.*, 2006). It consists of a 155 m sequence of cold-water coral-bearing Pleistocene sediments (floatstone, rudstone, and wackestone), characterised by at least 10 distinct 10 m scale layers alternation of light grey and dark green intervals. The carbonate-rich and light-colored layers are partially lithified. The mound is rooted on an erosive unconformity that has been identified in all three sites drilled. Directly below the mound a thin layer of early Pliocene sediments overlies a thick early Miocene package of green-grey calcareous siltstones. However, the triggering mechanism and origin of the mounds remain conjectural, with both geological and/or oceanographic controls suggested for their development. There is no strong geophysical or geochemical evidence to support a regional origin by thermal hydrocarbon seepage along faults. However, the overall sedimentary basin architecture of the basins could have facilitated up-dip fluid migration towards the margins and shallower parts of the basin, where the mounds are developed. Potential fluid migration pathways are available through a variety of lithofacies, stratal surfaces and unconformities, faults and detachments, and volcanic centres (Shannon *et al.*, 2007). These fluids may have created diagenetic hardgrounds at escape areas along the C10 unconformity. Alternatively such fluids may have played a role in triggering slope failures which exhumed lithified horizons on which the deep-water corals nucleated (Figure 9.7). However, the fact that the major initiation of mound growth coincides with the C10 unconformity that heralded a major oceanographic and climate change, together with the similarity in the bathymetric setting of the main mound provinces, is highly suggestive of a major component of oceanographic control on their development.

The present day bathymetry of the deep-water basins of the Irish Atlantic margin reveals a wealth of features, ranging from the contourite drifts outlined above, to dramatic slope failure and canyon features (Figures 1.1 and 9.8). The contourite buildups are most pronounced along the western margin of the Rockall Trough. In contrast, the complex canyon systems are best seen along the eastern, sediment-starved steep (3–7°) slopes of the eastern margin of the Rockall Trough and were

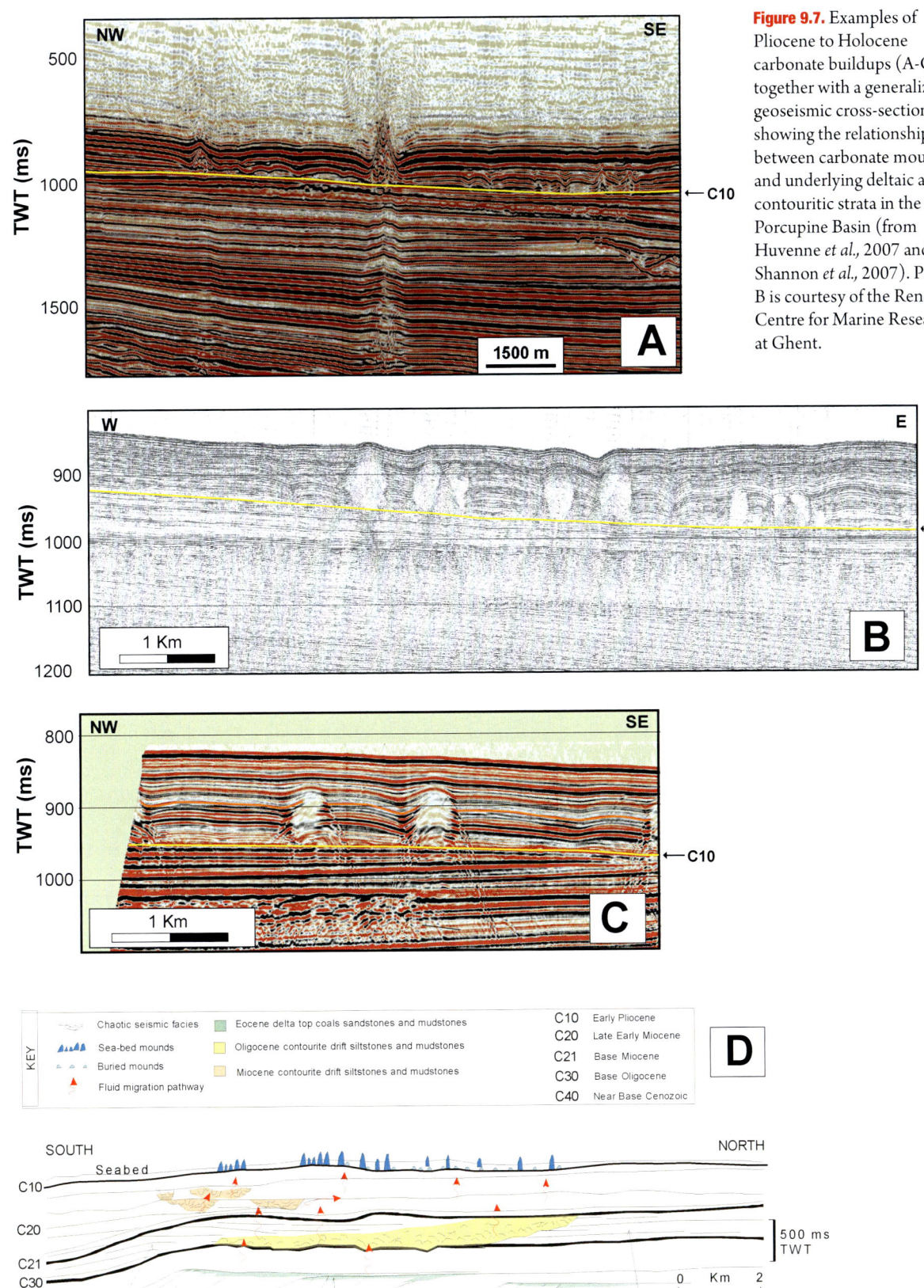

Figure 9.7. Examples of Pliocene to Holocene carbonate buildups (A-C), together with a generalized geoseismic cross-section (D) showing the relationship between carbonate mounds and underlying deltaic and contouritic strata in the Porcupine Basin (from Huvenne *et al.*, 2007 and Shannon *et al.*, 2007). Profile B is courtesy of the Renard Centre for Marine Research at Ghent.

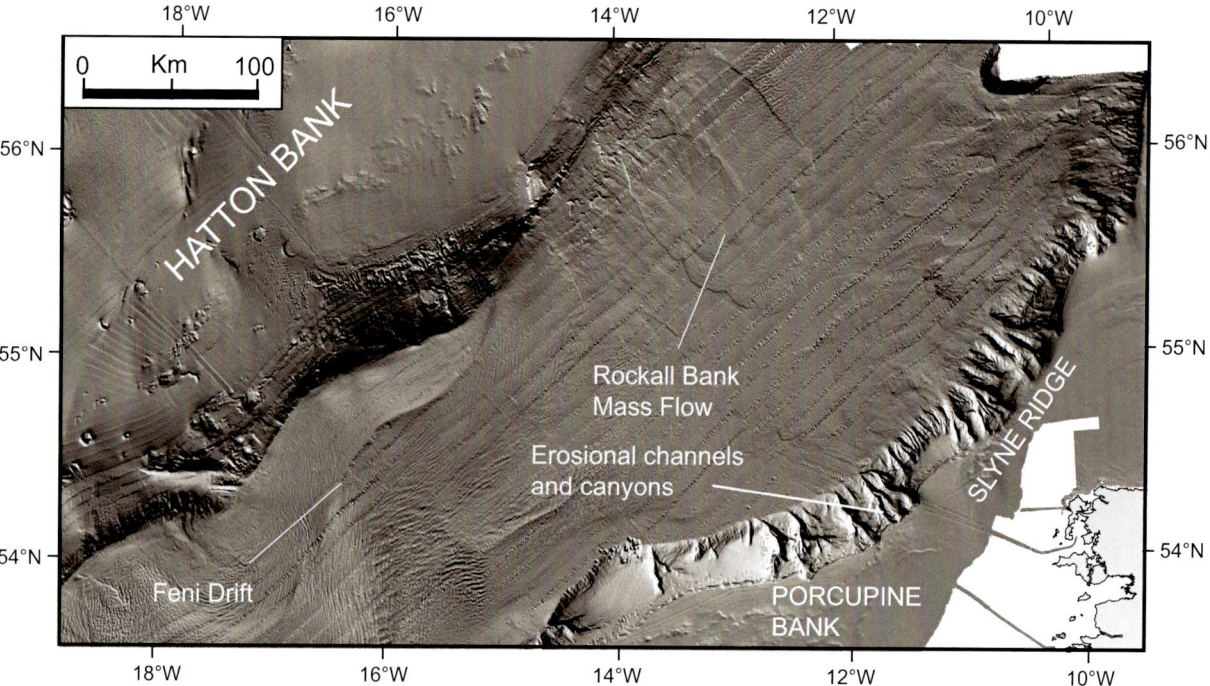

Figure 9.8. Swath bathymetry map of the Rockall Trough showing erosional and depositional features (courtesy of the Geological Survey of Ireland). Erosional channels and canyons dominate the eastern flank of the trough while the Feni Drift and the Rockall Bank Mass Flow are the predominant features seen on the western flank.

described by Elliott *et al.* (2006). While the canyons have a fresh appearance, they are thought to have developed in immediate post-C30 times following the major episode of basin sagging and differential subsidence. Elliott *et al.* (2006) suggested that rapid deepening of the basin at this time led to the formation of steep slopes, triggering mass failure that localised and evolved into canyon incision. The canyons cut the C30 unconformity but are themselves cut by the C10 unconformity. They probably formed in latest Eocene to Oligocene times and are clearly pre-Pliocene in age. The canyons are therefore likely to be relatively old features that persist due to the undersupplied nature of the basin margin. Present day canyon activity is limited to occasional sidewall collapse, but during the Plio-Pleistocene the canyons in the northeast Rockall were active sediment conduits for shelf-derived sediment feeding onto the Barra–Donegal Fan. Further south, the continental slope along the southern margin of the Goban Spur is deeply incised along its entire length by submarine canyons (Day, 1959; Naylor *et al.*, 2002). There are more than 20 canyons along this part of the margin, and they are typically steep-walled and V-shaped in transverse section.

The post-C10 succession on the shelf and slope regions of the Porcupine Seabight and especially the Rockall

Trough is also characterised by a number of dramatic slope failure features (Shannon *et al.*, 2001; Unnithan *et al.*, 2001; Elliott *et al.*, 2006). A complex system of canyons, channels and slope failure features is developed on the steep margins of the Rockall Trough, especially in the sediment-starved regions south of the Barra–Donegal Fan region (O'Reilly *et al.*, 2007). However, one of the largest and most dramatic slope failure complexes is the Rockall Bank Mass Flow '(Figure 9.8), emanating from the western margin of the Rockall Trough, recently re-evaluated by Elliott *et al.* (2010).

A 6100 km^2 region of slope failure scarps, with individual scarps reaching up to 22 km long and 150 m high, lies upslope of a series of massflow lobes that cover at least 18000 km^2 of the base of slope and floor of the Rockall Trough. Several phases of movement resulted in large sediment lobes that can be followed up to 120 km downdip across the floor of the Rockall Trough. The lobe region is unusual in comparison with the many slumps of broadly similar Plio-Pleistocene age documented along the NW European offshore margin (Evans *et al.*, 2005) in having a generally negative, inset topography relative to the adjacent undisturbed contourite deposits along much of its length. Following the major phases of slope collapse, smaller scale slab failures generated turbidity currents that are recorded in the gravity cores, as the slope continually readjusted. Although the short term trigger for failure is not known, the long-term instability control is suggested to be differential sedimentation and erosion on the slope associated with contourite drift development. This produced an erosional moat system along the mid-slope which may have undermined the upper slope, promoting failure and overloading the lower slope through deposition. While the age of movement is poorly constrained, some evidence of the timing of slope failure comes from the cored turbidites and slides which were suggested to be of last glacial age (10.2–21.7 ka). Flood *et al.* (1979) dated the slumping as occurring at approximately 15–16 ka, which is in broad agreement with the ^{14}C dates of 10.2–21.7 ky recorded by Øvrebø *et al.* (2005).

Seabed gravity cores have penetrated glacial and interglacial successions with characteristic sedimentological patterns (Øvrebø *et al.*, 2006). During full glacial conditions, bottom current activity was weak and multiple episodes of mass wasting took place. During interstadials and interglacials, mud-prone deposition took place along the western margin of the Rockall Trough while the eastern margin of the basin was characterised by strong bottom currents, giving rise to periods of erosion and winnowing and the deposition of sandy biogenic (interglacial) and mixed clastic-biogenic (interstadial) contourites.

Cenozoic Palaeogeography

Shortly after the commencement of the Cenozoic Era, crustal separation between the European and Greenland plates was complete and new oceanic crust extended northwards through the North Atlantic. The final breakup was accompanied by a change from extensional rifting to passive margin tectonics, and by an increase in igneous activity that was centred around Scotland and Northern Ireland and adjacent parts of the trailing

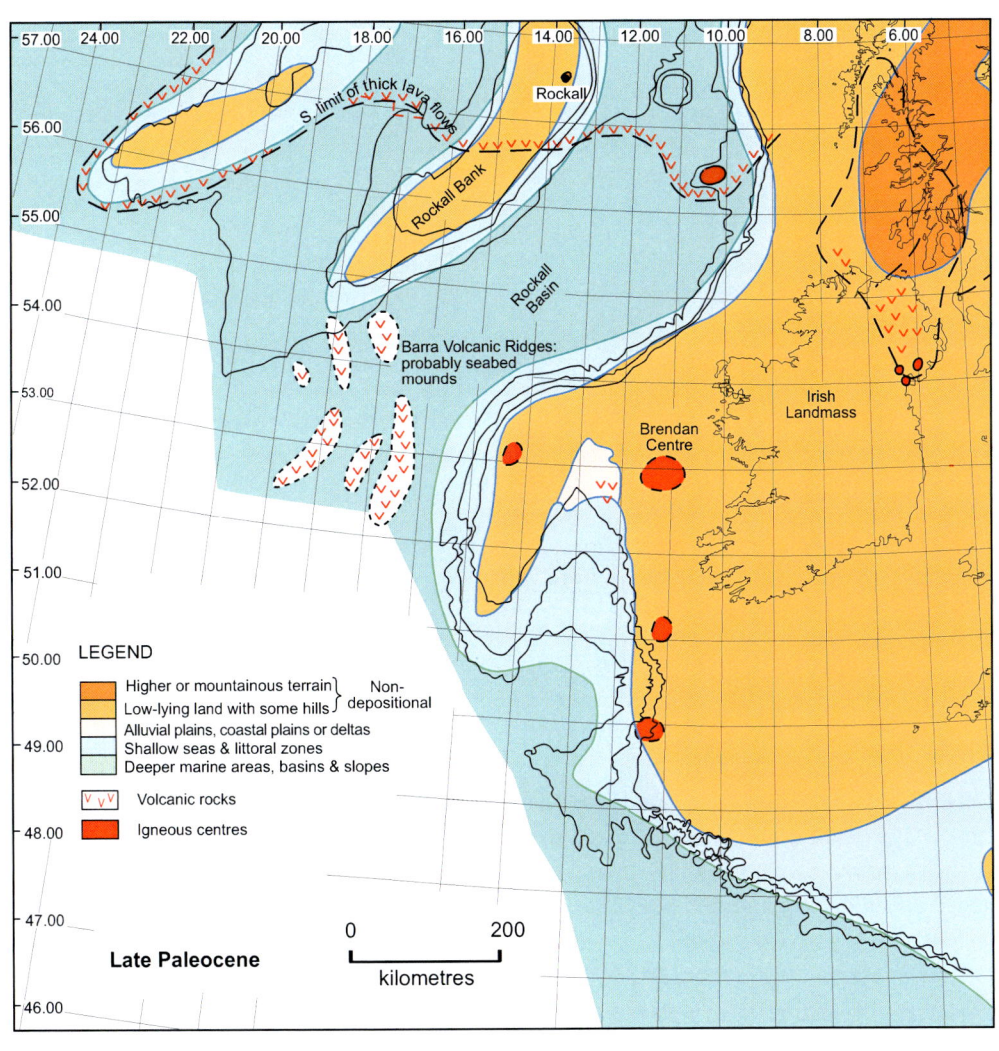

Figure 9.9. Paleocene palaeogeography (from Naylor and Shannon, 2009).

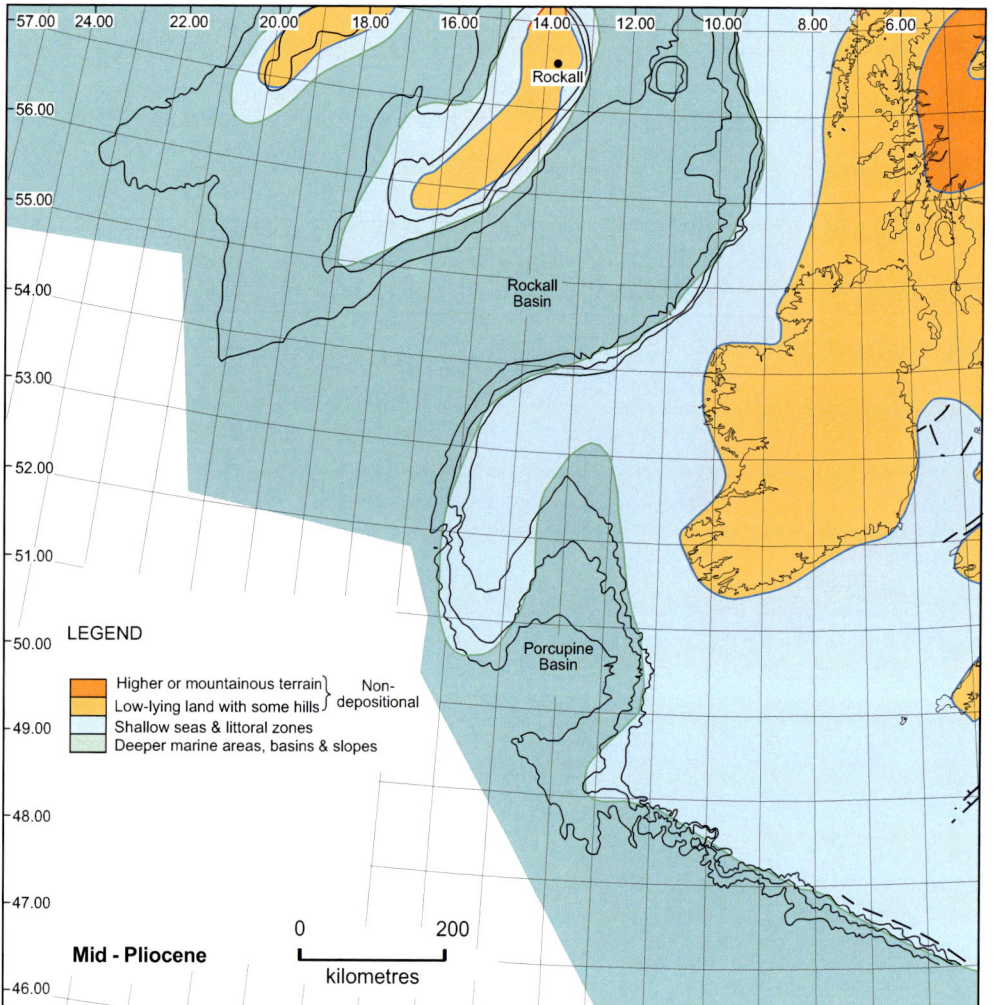

Figure 9.10. Pliocene palaeogeography (from Naylor and Shannon, 2009).

passive margin (Figure 9.9). This igneous activity possibly reflected mantle circulation cells whose location was influenced by large crustal thickness differences between the major basins and their flanking basement highs (Praeg *et al.*, 2005; Stoker *et al.*, 2005b, c). The overall setting of epeirogenic tilting (coeval subsidence and uplift) caused the basinward-progradation of major sediment wedges deposited as deltas in shallow parts of basins such as the northern part of the Porcupine, and as basin floor fans in the more rapidly subsiding deep-water parts of the basins. This kilometre-scale epeirogeny saw onshore uplift of up to 1.5 km, coincident with offshore subsidence of up to 600 m in the major sedimentary basins (Praeg *et al.*, 2005). Major domal uplift has also been suggested for the Irish Sea region (Cope, 1994), and the attenuated Palaeogene succession in the Celtic Sea region may be consistent with a period of regional uplift.

The major basin readjustment in latest Eocene–earliest Oligocene times reflects a phase of sagging, which was accompanied by the cessation of major coarse clastic deposition, the differential deepening of major basin depocentres, and the onset of large-scale deep-water current systems. The basins became under-supplied and the resulting strata were largely deep-water and fine-grained, with alongslope reworking of sediment to form shelfward accreting silty contourite deposits. The largest of these deposits lies along the western margin of the Rockall Trough, with a smaller contourite drift deposit in the smaller Porcupine Seabight. Typified by the pattern observed in the Rockall Trough, currents were largely anticlockwise, with erosion along the eastern margin and deposition on the western margin. This erosion led to the destabilisation of the scoured margin with the formation of canyons, slides and slope failures that provided sediment for reworking on the lower parts of the slope. Local domal inversion in the Miocene resulted in localised topography, erosion and deposition. Such inversion domes (Stoker *et al.*, 2005a) are relatively rare in the Irish offshore, being more prevalent further north in the UK and Norwegian parts of the Rockall and other large basins.

The early Pliocene (Figure 9.10) witnessed the third Cenozoic epeirogenic episode in the Atlantic region. Tilting, resulting from plate reorganisation, resulted in a regional unconformity. The Pliocene to Holocene succession was relatively thin throughout most of the Irish offshore, thickening rapidly towards the Irish/UK boundary in the Rockall Trough, where the Barra Donegal Fan provided a thick succession of glaciogenic and debris flow clastics shed from the adjacent highlands.

References

Ainsworth, N.R., Whatley, R.C., and Jones, R.W. (1996) Foraminifera and Ostracoda from Quaternary carbonate mounds associated with gas seepage in the Porcupine Basin, offshore Western Ireland. *Revista Española de Micropaleontologia,* **38**, 113–151.

Bailey, W., Shannon, P.M., Walsh, J.J., and Unnithan, V. (2003) Spatial distribution of faults and deep sea carbonate mounds in the Porcupine Basin. *Marine and Petroleum Geology,* **20**, 509–522.

Berggren, W.A., Kent, D.V., Swisher III, C.C., and Aubry, M.-P. (1995) A revised Cenozoic geochronology and chronostratigraphy. *In*: Berggren, W.A., Kent, D.V., Aubry, M.-P., Hardenbol, J. (eds), *Geochronology, Time Scales and Stratigraphic Correlation: Framework For an Historical Geology.* Society for Economic Palaeontologists and Mineralogists, Special Publication, **54**, 129–212.

Cope, J.C.W. (1994) A latest Cretaceous hotspot and the south-easterly tilt of Britain. *Journal of the Geological Society, London,* **151**, 729–731.

Croker, P.F. and O'Loughlin, O. (1998) A catalogue of Irish carbonate mud mounds. *TTR-7 post cruise conference on carbonate mud mounds and cold water reefs.* University of Ghent, Belgium, 1–13.

Croker, P.F. and Shannon, P.M. (1987) The evolution and hydrocarbon prospectivity of the Porcupine Basin, offshore Ireland. *In*: Brooks, J. and Glennie, K.W. (eds), *Petroleum Geology of North West Europe,* Graham and Trotman, London, 633–642.

Cunningham, G.A. and Shannon, P.M. (1997) The Erris Ridge: a major geological feature in the NW Irish Offshore Basins. *Journal of the Geological Society, London,* **154**, 503–508.

Dahlgren, K.I.T., Vorren, T.O., Stoker, M.S., Nielsen, T., Nygård, A., Sejrup, H.P., and De Santis, L. (2005). Late Cenozoic prograding wedges on the NW European Atlantic margin: their formation and relationships to tectonics and climate. *Marine and Petroleum Geology,* **22**, 1089–1110.

Dancer, P.N. and Pillar, N.W. (2001) Exploring in the Slyne Basin: a geophysical challenge. *In*: Shannon, P.M., Haughton, P.D.W., and Corocoran, D.V. (eds), *The Petroleum Exploration of Ireland's Offshore Basins.* Geological Society, London, Special Publications, **188**, 209–222.

Dancer, P.N., Kenyon-Roberts, S.M., Downey, J.W., Baillie, J.M., Meadows, N.S., and Maguire, K. (2005) The Corrib gas field, offshore west of Ireland. *In*: Doré, A.G. and Vining, B.A. (eds), *Petroleum Geology: North-West Europe and Global Perspectives: Proceedings of the 6th Petroleum Geology Conference.* Geological Society, London, 1035–1046.

Davies, R.J. and Cartwright, J. (2002) A fossilized Opal A to Opal C/T transformation on the northeast Atlantic margin: support for a significantly elevated palaeogeothermal gradient during the Neogene? *Basin Research,* **14**, 467–485.

Day, A.A. (1959) The continental margin between Brittany and Ireland. *Deep-Sea Research,* **5**, 249–265.

De Mol, B., Van Rensbergen, P., Pillen, S., Van Herreweghe, K., Van Rooij, D., McDonnell, A., Huvenne, V., Ivanov, M., Swennen, R., and Henriet, J.P. (2002) Large deep-water coral banks in the Porcupine Basin, southwest of Ireland. *Marine Geology,* **188**, 193–231.

Dobson, M.R., Haynes, J.R., Bannister, A.D., Levene, D.G., Petrie, H.S., and Woodbridge, R.A., (1991). Early Tertiary palaeoenvironments and sedimentation in the NE Main Porcupine Basin (well 35/13-1), offshore western Ireland – evidence for global change in the Tertiary. *Basin Research,* **3**, 99–117.

Dolan, J.F. (1986) The relationship between the 'R2' seismic reflector and a zone of abundant detrital and authigenic smectites, Deep Sea Drilling Project Hole 610, Rockall Plateau region, North Atlantic. *In*: Ruddiman, W.F., Kidd, R.B., Thomas, E. *et al.* (eds), *Initial Reports of the Deep Sea Drilling Project,* **94**. US Government Printing Office, Washington, DC, 1105–1109.

Elliott, G.M., Shannon, P.M., Haughton, P.D.W., and Øvrebø, L.K. (2010) The Rockall Bank Mass Flow: Collapse of a moated contourite drift onlapping the eastern flank of Rockall Bank, west of Ireland. *Marine and Petroleum Geology,* **27**, 92–107.

Elliott, G.M., Shannon, P.M., Haughton, P.D.W., Praeg, D., and O'Reilly, B. (2006) Mid- to late Cenozoic canyon development on the eastern margin of the Rockall Trough, offshore Ireland. *Marine Geology,* **229**, 113–132.

Evans, D., Harrison, Z., Shannon, P.M., Laberg, J.S., Nielsen, T., Ayers, S., Holmes, R., Hoult, R.J., Lindberg, B., Haflidason, H., Long, D., Kuijpers, A., Andersen, E.S., and Bryn, P. (2005) Palaeoslides and other mass failures of Pliocene to Pleistocene age along the Atlantic continental margin of NW Europe. *Marine and Petroleum Geology,* **22**, 1131–1148.

Flood, R.D., Hollister, C.D., and Lonsdale, P. (1979) Disruption of the Feni sediment drift by debris flows from Rockall Bank. *Marine Geology,* **32**, 311–334.

Floodpage, J., Newman, P., and White, J. (2001) Hydrocarbon prospectivity in the Irish Sea area: insights from recent exploration in the Central Irish Sea, Solway and Peel basins. *In*: Shannon, P.M., Haughton, P.D.W., and Corcoran, D.V. (eds), *The Petroleum Exploration of Ireland's Offshore Basins.* Geological Society, London, Special Publications, **188**, 107–134.

Haughton, P., Morrissey, T., and Praeg, D. (2001) Volcanoes on the edge: new insights from shallow drilling on the margins of the Rockall Trough. *Ireland's Deepwater Frontier Conference, Killiney, Ireland, Extended Abstracts,* 86–88.

Haughton, P., Praeg, D., Shannon, P.M., Harrington, G., Higgs, K., Amy, L., Tyrrell, S., and Morrissey, T. (2005) First results from shallow stratigraphic boreholes on the eastern flank of the Rockall Basin, offshore western Ireland. *In:* Doré, A.G. and Vining, B.A. (eds), *Petroleum Geology: North-West Europe and Global Perspectives – Proceedings of the 6th Petroleum Geology Conference.* Geological Society, London, 1077–1094.

Henriet, J.P., De Mol, B., Pillen, S., Vanneste, M., Van Rooij, D., Versteeg, W., Croker, P.F., Shannon, P.M., Unnithan, V., Bouriak, S., Chachkine, P., and The Porcupine BELGICA 97 Shipboard Party. (1998) Gas hydrate crystals may help build reefs. *Nature,* **391,** 648–649.

Henriet, J.P., De Mol, B., Vanneste, M., Huvenne, V., Van Rooij, D., and the 'Porcupine-Belgica' 97, 98 and 99 Shipboard Parties. (2001) Carbonate mounds and slope failures in the Porcupine Basin: a development model involving fluid venting. *In:* Shannon, P.M., Haughton, P.D.W., and Corcoran, D.V. (eds), *The Petroleum Exploration of Ireland's Offshore Basins.* Geological Society, London, Special Publications, **188,** 375–383.

Hill, P.R. (1987) Characteristics of sediments from Feni and Gardar drifts, sites 610 and 611, Deep Sea Drilling Project Leg 96. *In:* Ruddiman, W.F., Kidd, R.B., Thomas, E., *et al.* (eds), *Initial Reports of the Deep Sea Drilling Project,* **94,** U.S. Government Printing Office, Washington D.C., 1075–1082.

Holford, S.P., Green, P.F., and Turner, J.P. (2005). Palaeothermal and compaction studies in the Mochras borehole (NW Wales) reveal early Cretaceous and Neogene exhumation and argue against regional Palaeogene uplift in the southern Irish Sea. *Journal of the Geological Society, London,* **162,** 829–840.

Huvenne, V.A.I., Bailey, W.R., Shannon, P.M., Naeth, J., di Primio, R., Henriet, J.P., Horsfield, B., de Haas, H., Wheeler, A., and Olu-Le Roy, K. (2007) The Magellan mound province in the Porcupine Basin. *International Journal of Earth Sciences,* **96,** 85–101.

Jansen, E., Raymo, M.E., Blum, P. *et al.* (1996). 3. Sites 980/981. *In:* Jansen, E., Raymo, M.E., Blum, P. *et al.* (eds.), *Proceedings ODP, Initial reports,* **162:** College Station, TX (Ocean Drilling Program), 49–90.

Jeans, C.V., Merriman, R.J., Mitchell, J.G., and Bland, D.J. (1982) Volcanic clays in the Cretaceous of southern England and Northern Ireland. *Clay Minerals,* **17,** 105–156.

McDonnell, A. and Shannon, P.M. (2001) Comparative Tertiary basin development in the Porcupine and Rockall basins. *In:* Shannon, P.M., Haughton, P.D.W., and Corcoran, D.V. (eds), *The Petroleum Exploration of Ireland's Offshore Basins.* Geological Society, London, Special Publications, **188,** 323–344.

Mohr, P. (1982) Tertiary dolerite intrusions of west-central Ireland. *Proceedings of the Royal Irish Academy,* **82B,** 53–82.

Moore, J.G. and Shannon, P.M. (1991) Slump structures in the Late Tertiary of the Porcupine Basin, offshore Ireland. *Marine and Petroleum Geology,* **8,** 184–197.

Moore, J.G. and Shannon, P.M. (1992) Palaeocene–Eocene deltaic sedimentation, Porcupine Basin, offshore Ireland: a sequence stratigraphic approach. *First Break,* **10,** 461–469.

Murdoch, L.M., Musgrove, F.W., and Perry, J.S. (1995) Tertiary uplift and inversion history in the North Celtic Sea Basin and its influence on source rock maturity. *In:* Croker, P.F. and Shannon, P.M. (eds), *The Petroleum Geology of Ireland's Offshore Basins.* Geological Society, London, Special Publications, **93,** 297–319.

Naylor, D. and Shannon, P. M. (2005) The structural framework of the Irish Atlantic Margin. *In:* Doré, A.G. and Vining, B.A. (eds), *Petroleum Geology: North-West Europe and Global Perspectives – Proceedings of the 6th Petroleum Geology Conference.* Geological Society, London, 1009–1021.

Naylor, D. and Shannon, P. M. (2009) Geology of offshore Ireland. *In:* Holland, C.H. and Sanders, I.S. (eds), *The Geology of Ireland.* Second Edition. Dunedin Academic Press, Edinburgh, 405–460.

Naylor, D., Haughey, N., Clayton, G., and Graham, J.R. (1993) The Kish Bank Basin, offshore Ireland. *In:* Parker, J.R. (ed.), *Petroleum Geology of North-west Europe: Proceedings of the 4th Conference.* Geological Society, London, 845–855.

Naylor, D., Shannon, P.M., and Murphy, N. (1999) *Irish Rockall Basin region – a standard structural nomenclature system.* Petroleum Affairs Division, Dublin, Special Publication **1/99,** 42pp.

Naylor, D., Shannon, P.M., and Murphy, N. (2002) *Porcupine-Goban region – a standard structural nomenclature system.* Petroleum Affairs Division, Dublin, Special Publication **1/02,** 65pp.

O'Reilly, B.M., Readman, P.W., Shannon, P.M., and Jacob, A.W.B. (2003) A model for the development of a carbonate mound population in the Rockall Trough based on deep-towed sidescan sonar data. *Marine Geology,* **198,** 55–66.

O'Reilly, B.M., Shannon, P.M., and Readman, P.W. (2007) Shelf to slope sedimentation processes and the impact of Plio-Pleistocene glaciations in the northeast Atlantic, west of Ireland. *Marine Geology,* **238,** 21–44.

Øvrebø, L.K., Haughton, P.D.W., and Shannon, P.M. (2005) Temporal and spatial variation in Late Quaternary slope sedimentation along the margins of the Rockall Trough, offshore west Ireland. *Norwegian Journal of Geology,* **85,** 279–294.

Øvrebø, L.K., Haughton, P.D.W., and Shannon, P.M. (2006) A record of fluctuating bottom currents on the slopes west of the Porcupine Bank, offshore Ireland – implications for Late Quaternary climate forcing. *Marine Geology,* **225,** 279–309.

Parnell, J., Shukla, B., and Meighan, I.G. (1989) The lignite and associated sediments of the Tertiary Lough Neagh Basin. *Irish Journal of Earth Sciences,* **10,** 67–88.

Praeg, D., Stoker, M.S., Shannon, P.M., Ceramicola, S., Hjelstuen, B.O., and Mathiesen, A. (2005) Episodic Cenozoic tectonism and the development of the NW European 'passive' continental margin. *Marine and Petroleum Geology,* **22,** 1007–1030.

Riddihough, R.P. (1975) *A magnetic map of the continental margin of west of Ireland involving part of the Rockall Trough and the Faeroe Plateau.* Dublin Institute for Advanced Studies, Geophysical Bulletin **33,** 1p. and Enclosure.

Robinson, K.W., Shannon, P.M., and Young, D.G.G. (1981) The Fastnet Basin: an integrated analysis. *In:* Illing, L.G. and Hobson, G.D. (eds), *Petroleum Geology of the Continental Shelf of North-West Europe.* Heyden and Son Ltd, London, 444–454.

Seeman, U. (1984) Tertiary intrusives on the Atlantic continental margin off southwest Ireland. *Irish Journal of Earth Sciences*, **6**, 229–236.

Shannon, P.M. (1992) Early Tertiary submarine fan deposits in the Porcupine Basin, offshore Ireland. *In:* Parnell, J. (ed.), *Basins on the Atlantic Seaboard: Petroleum Geology, Sedimentology and Basin Evolution.* Geological Society, London, Special Publications, **62**, 351–373.

Shannon, P.M., McDonnell, A., and Bailey, W.R. (2007) The evolution of the Porcupine and Rockall basins, offshore Ireland: the geological template for carbonate mound development. *International Journal of Earth Sciences*, **96**, 21–35.

Shannon, P.M., Moore, J.G., Jacob, A.W.B., and Makris, J., (1993) Cretaceous and Tertiary basin development west of Ireland. *In*: Parker, J.R. (ed.), *Petroleum Geology of Northwest Europe: Proceedings of the 4th Conference.* Geological Society, London, 1057–1066.

Shannon, P.M., O'Reilly, B.M., Readman, P.W., Jacob, A.W.B., and Kenyon, N. (2001) Slope failure features on the margin of the Rockall Trough. *In:* Shannon, P.M., Haughton, P.D.W., and Corcoran, D.V. (eds), *The Petroleum Exploration of Ireland's Offshore Basins.* Geological Society, London, Special Publications, **188**, 455–464.

Shannon, P.M., Stoker, M.S., Praeg, D., van Weering, T.C.E., de Haas, H., Nielsen, T., Dahlgren, K.I.T. and Hjelstuen, B.O. (2005) Sequence stratigraphic analysis in deep-water, underfilled NW European passive margins basins. *Marine and Petroleum Geology*, **22**, 1185–1200.

Stoker, M.S. (1997) Mid- to Late Cenozoic sedimentation on the continental margin off NW Britain. *Journal of the Geological Society, London*, **154**, 509–515.

Stoker, M.S. (2002) Late Neogene development of the UK Atlantic margin. *In*: Dore, A.G., Cartwright, J.A., Stoker, M.S., Turner, J.P., and White, N. (eds.), *Exhumation of the North Atlantic Margin: Timing, Mechanisms and Implications for Petroleum Exploration.* Geological Society, London, Special Publications, **196**, 313–329.

Stoker, M.S., Hitchen, K., and Graham, C.C. (1993) *United Kingdom offshore regional report: the geology of the Hebrides and west Shetland shelves and adjacent deep-water areas.* HMSO for the British Geological Survey, London. 149pp.

Stoker, M.S., Akhurst, M.C., Howe, J.A., Stowe, D.A.V. (1998) Sediment drifts and contourites on the continental margin off northwest Britain. *Sedimentary Geology*, **115**, 33–51.

Stoker, M.S., Hoult, R.J., Nielsen, T., Hjelstuen, B.O., Laberg, J.S., Shannon, P.M., Praeg, D., Mathiesen, A., van Weering, T.C.E., and McDonnell, A. (2005c) Sedimentary and oceanographic responses to early Neogene compression on the NW European margin. *Marine and Petroleum Geology*, **22**, 1031–1044.

Stoker, M.S., Praeg, D., Hjelstuen, B.O., Laberg, J.S., Nielsen, T., and Shannon, P.M. (2005b) Neogene stratigraphy and the sedimentary and oceanographic development of the NW European Atlantic margin. *Marine and Petroleum Geology*, **22**, 977–1005.

Stoker, M.S., Praeg, D., Shannon, P.M., Hjelstuen, B.O., Laberg, J.S., van Weering, T.C.E., Sejrup, H.P., and Evans, D. (2005a) Neogene evolution of the Atlantic continental margin of NW Europe (Lofoten Islands to SW Ireland): anything but passive. *In*: Doré, A.G. and Vining, B. (eds), *Petroleum Geology: North-West Europe and Global Perspectives – Proceedings of the 6th Petroleum Geology Conference.* Geological Society, London, 1057–1076.

Stoker, M.S., van Weering, T.C.E., and Svaerdborg, T. (2001) A Mid–Late Cenozoic tectonostratigraphic framework for the Rockall Trough. *In:* Shannon, P.M., Haughton, P.D.W., and Corcoran, D. (eds), *The Petroleum Exploration of Ireland's Offshore Basins.* Geological Society, London, Special Publications, **188**, 411–438.

Tate, M.P. and Dobson, M. R. (1988) Syn- and post-rift igneous activity in the Porcupine Seabight basin and adjacent continental margin west of Ireland. *In*: Morton, A.C. and Parson, L.M. (eds), *Early Tertiary volcanism and the opening of the NE Atlantic.* Geological Society, London, Special Publications, **39**, 309–334.

Unnithan, V., Shannon, P.M., McGrane, K., Readman, P.W., Jacob, A.W.B., Keary, R., and Kenyon, N.H. (2001) Slope instability and sediment re-distribution in the Rockall Trough: constraints from GLORIA. *In*: Shannon, P.M., Haughton, P.D.W., and Corcoran, D.V. (eds), *The Petroleum Exploration of Ireland's Offshore Basins.* Geological Society, London, Special Publications, **188**, 439–454.

White, N. and Lovell, B. (1997) Measuring the pulse of a plume with the sedimentary record. *Nature*, **387**, 888–891.

Wilkinson, G.C., Bazley, R.A.B., and Boulter, M.C. (1980) The geology and palynology of the Oligocene Lough Neagh Clays, Northern Ireland. *Journal of the Geological Society, London*, **137**, 65–75.

Williams, T., Kano, A., Ferdelman, T., Henriet, J.-P. *et al.* (2006) Cold-water coral mounds revealed. *Eos*, **87**, 525–527.

Woodland, A.W (ed.), (1971) *The Llanbedr (Mochras Farm) borehole.* Institute of Geological Sciences, Report **71/18**, 115pp.

Chapter 10

Celtic Sea basins

Regional setting

The basins of the Celtic Sea region, south of Ireland, lie in relatively shallow waters (100–200 m) and are noticeably elongate (Figure 10.1; Shannon, 1991a). The main orientation is ENE–WSW, with some strike swings at the western and eastern ends of the basin system where the basins take on a more NE–SW orientation. Two main parallel sets of basins occur, separated by an intermittent basement high (Labadie Bank–Pembrokeshire Ridge). The northern set of basins comprises the St George's Channel, North Celtic Sea and Fastnet basins, while the southern set includes the South Celtic Sea and the Cockburn basins. The largest basins, with the thickest sediments, are those to the north of the Labadie Bank–Pembrokeshire Ridge basement high, containing up to 9 km of Triassic to Cenozoic strata, resting upon deformed Variscan basement (Shannon and Naylor, 1998). The basins are characterised by having a thick Triassic and Jurassic succession, with a variable thickness of younger strata. A thicker salt succession, with impressive halokinetic structures, occurs in the South Celtic Sea Basin, suggesting that the Labadie Bank–Pembrokeshire Ridge is a long-standing feature that controlled the location, orientation and sedimentary history of the basins. The Cretaceous and Cenozoic succession is thinner than in the Atlantic basins while, in contrast to the Irish Sea basins, little or no Permian strata have been encountered, whereas a considerable Jurassic succession is preserved. A number of inversion episodes, notably in the Cenozoic, also characterise the geological development of these basins.

The development of the Celtic Sea basins needs to be considered within their wider structural setting. The onshore geology of southernmost Ireland is dominated by major Variscan structures, as described and discussed in Chapter 4. The Variscan folds and thrust structures have an arcuate trend, changing from ESE–WNW in the east to ENE–WSW in the west of the belt. The strike of the Lower Palaeozoic sequence in southeast Ireland and in Irish Sea is generally NE–SW, but there is a marked strike change to ENE–WSW in the southwestern part of

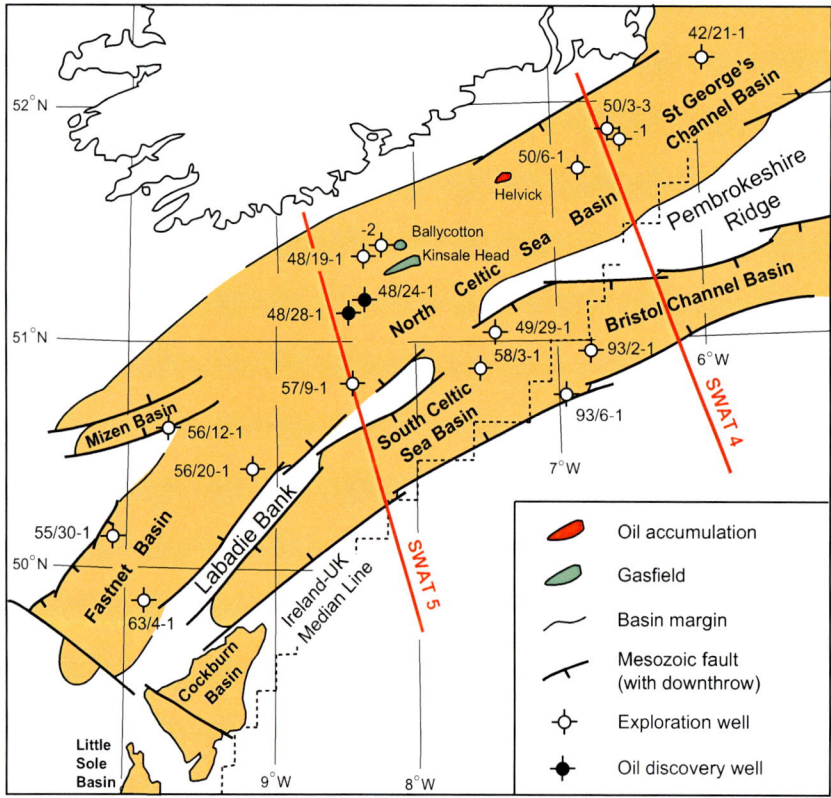

Figure 10.1. Generalised map of the Celtic Sea basins showing the positions of two SWAT deep-profile seismic lines and the locations of wells mentioned in the text.

County Wexford and westwards into County Waterford (Brück *et al.*, 1979; Readman *et al.*, 1997). Along the coast west of Cork Harbour the onshore major folds also trend ENE–WSW. Although the Caledonian basement is not seen in this region, it seems likely that the strike of Variscan and Caledonian structures are here sub-parallel – a point that is pertinent when considering the offshore Celtic Sea region.

Published seismic profiles have illustrated an underlying structural control on the geometry of the Celtic Sea basins, although the details of that control are still somewhat unclear. Two NNW–SSE deep seismic profiles – SWAT 4 and 5 (Figure 10.1) – were acquired across the Celtic Sea basins under the BIRPS Deep Seismic Profiling programme (McGeary *et al.*, 1987, Coward and Trudgill, 1989). The SWAT 4 profile across the eastern part of the North Celtic Sea Basin shows the basin lying in the hanging-wall of a low-angle deep crustal structure interpreted by McGeary *et al.* (1987) as the Variscan Front. The upper crust beneath the Mesozoic basins on the SWAT profiles (to about 7 seconds TWT) is largely seismically transparent, whilst the lower crust is typified by strong, short and relatively flat reflectors. The Moho is taken to be the base of the zone of prominent reflectors at 10 to 11 seconds TWT, overlying seismically transparent upper mantle. The upper crust on both SWAT profiles displays a number of other southerly-dipping reflectors, similar to the SWAT 4 feature identified as the Variscan Front. This latter feature lies southeast of Dungarvan Harbour and approximately on the line shown on many maps as depicting the Variscan Front, trending from South Wales across the east Celtic Sea to Dungarvan. Murphy (1990) suggested that the SWAT 4 feature linked onshore to the tip line of the Variscan sole thrust. However, it is likely that the position of the structure on the traditional line of the Variscan Front is simply fortuitous. As Ford *et al.* (1992) have pointed out, since there is no thrust feature onshore, for instance at Dungarvan, there is no reason to associate the offshore feature with the putative trend of the Variscan Front onshore. Indeed, the balance of evidence suggests that the dipping reflectors originated as Caledonian structures, some of which were re-activated during the Variscan episode and later as controlling elements of the Mesozoic basins. This situation is seen on other BIRPS lines around the British Isles (BIRPS and ECORS, 1986; McGeary *et al.*, 1987) where basin-controlling structures can be traced to depth and merge into sub-horizontal reflectors in the lower crust. The strong control of NE–SW caledonoid elements on basement

development in the south Irish Sea–St George's Channel area is discussed below, and the change westwards of Lower Palaeozoic trend to ENE–WSW is noted above. It is reasonable to suppose that this basement control on basin development continued westwards along the Celtic Sea. Thus, on SWAT 4 one such structure is utilised as the northern margin of the North Celtic Sea Basin (McGeary *et al.*, 1987). There are insufficient data to say whether this situation obtains along the whole length of the northern margin of the basin, and it is not clear in the case of the SWAT 5 profile. However, industry seismic surveys have shown the northern basin margin to be strongly fault-controlled along much of its length and, given the caledonoid trend of the margin, it would be surprising if older structures had not exercised strong control. Support for this view is found in the Bouguer gravity data for the region presented by Readman *et al.* (1995) which shows a well-developed Caledonian fabric. It is clear, also on some seismic sections, that structures such as Variscan backthrusts have undergone phases of re-activation (e.g. Shannon, 1991c: Figure 7.5).

The northwestern segment of the North Celtic Sea Basin, sometimes referred to as the Mizen Basin, has a pronounced half-graben morphology, and it has been argued (McCann and Shannon, 1993) that the basin shape was controlled by extensional re-activation of a deep-seated Variscan backthrust structure. However, these authors noted that the orientation is somewhat oblique to the dominant Variscan trend and also suggested the possibility that the reactivated structure is an older Caledonian structural lineament. While the effects of the Variscan compression and orogeny on the underlying Caledonian and older structures is uncertain, it seems more plausible to presume that the major controls are reactivated Caledonian structures and that the strike-swing of the Caledonian structures to the south of Ireland was reinforced by the Variscan deformation in this region.

The two groups of Celtic Sea basins are separated by the Labadie Bank–Pembrokeshire Ridge complex. Although the Pembrokeshire Ridge extends from the Lower Palaeozoic Welsh massif, in the Celtic Sea region it is probably composed of Upper Palaeozoic rocks. The indications are that it formed an upstanding positive element during Triassic time, since Triassic stratigraphy shows considerable differences north and south of the Ridge, suggesting that the two basinal areas were separated at that time. Shannon and MacTiernan (1993) report a gravity low on the feature in the south of Quadrant 50 that can be modelled in terms of a large, high-level granite. This may have been unroofed by Early Triassic time and could have acted as a local sand source at that time.

While dating of the extensive early Mesozoic red-bed facies in the region is generally poor, regional lithostratigraphic correlations are relatively robust and suggest the regional presence of a Triassic succession, with a lower Sherwood Sandstone and an upper Mercia Mudstone succession. However, it is notable that there is little evidence for Permian strata in the region. Exceptions to this are the half-graben lying beneath interpreted Triassic strata in the undrilled Cockburn Basin (Smith, 1995) and probable Permian intersected in drilling in the eastern part of the South Celtic Sea Basin (van Hoorn, 1987). This suggests that the Celtic Sea remained an elevated region, shedding sediment into the subsiding Permian basins of the Irish Sea, such as the Kish Bank Basin and the Irish Sea Basin, which lay to the north of the orogenic zone and were generally unaffected by the Variscan orogenesis. Regional early Mesozoic deposition commenced in Triassic times, following crustal cooling, erosion and Variscan orogenic collapse. Štolfová and Shannon (2009) suggested that the generally uniform shape of the Triassic succession, with the exception of the halokinetic structures, in the Celtic Sea region, was the result of wide rifting following the orogenic collapse of the Variscan fold belt.

With the exception of a narrow belt at the southern margin of the North Celtic Sea Basin, thick Triassic salt in the region is confined to the belt of basins lying south of the Labadie Bank–Pembrokeshire Ridge and also farther east in the St George's Channel Basin region. The basement ridge system played an important role in basin development from early Mesozoic time onwards – a view supported by the general lack of thick Upper Jurassic sediments from the basins south of the ridge, while the effects of Cenozoic inversion are more pronounced north of the ridge system.

Following thermal subsidence, with a regional marine transgression in earliest Jurassic time, minor rift events or localised faulting were manifest by sandy deltaic and shallow-marine strata in the western Fastnet and the eastern North Celtic Sea basins. The main rifting in the basins took place in Late Jurassic time, but the deposits of this phase are largely confined to the North Celtic Sea Basin. It is unclear if this reflects lack of deposition elsewhere, or removal through later, earliest Cretaceous, erosion. The general consensus in the published literature

appears to favour the former. The presence of Middle Jurassic volcanics in the Fastnet Basin was interpreted by Shannon (1995) as the possible effect of a thermal anomaly that caused local doming and prevented the deposition of a thick Upper Jurassic sequence in the basin. Little evidence is seen in the South Celtic Sea Basin to suggest deposition and later erosion of a significant thickness of Upper Jurassic strata.

The Lower Cretaceous succession in the Celtic Sea basins is of variable thickness. The underlying Upper Jurassic, where present, is dominated by syn-rift deposition. However, the Cretaceous shows little evidence of major syn-depositional faulting and exhibits a broad thermal subsidence geometry. In contrast to the marine Lower Cretaceous sediments of the Atlantic basins, the depositional environment in the Celtic Sea was non-marine, dominated by fluvial deposits. Deltaic to marine conditions in the mid-Cretaceous and open marine terrigenous-free chalk in the Upper Cretaceous reflected a sea level rise. The non-marine deposition is likely to be due to the effects of the rotation of the Iberian peninsula (Robinson *et al.*, 1981). This imposed compressional stresses on the Celtic Sea region, resulting in uplift, erosion and a marked unconformity. However, as seafloor spreading ceased in the Iberian region the Celtic Sea region experienced delayed thermal subsidence in response to the Late Jurassic rifting, resulting in the widespread sag-like geometry of the Cretaceous. The variations in thickness are due partly to the residual rift topography, the effects of the Iberian compression, and the effects of later uplift and erosional events.

Evidence is seen for several uplift and erosional episodes in the Celtic Sea region. As mentioned above, the base of the Cretaceous is generally marked by a pronounced unconformity through much of the region, although the amount of Jurassic strata removed is uncertain. Two Cenozoic inversion and erosional events are recorded in the seismic and well records in the region (Murdoch *et al.*, 1995), and vitrinite reflectance, apatite fission track and seismic velocity analyses provide constraints on the magnitude of erosion. A regional uplift occurred in the Paleocene, but this was not associated with significant inversion or compression. Up to 500 m of strata were removed during this event. Compressive inversion with doming and reverse fault movement occurred in the late Eocene to early Oligocene. The later event is most pronounced in the northeast of the region, with up to 1.1 km of strata eroded (*see* Chapters 8 and 9 for further discussion). This event is significant in many regards in the region. It resulted in the exhumation of Upper Cretaceous chalk to the seabed, which is the main cause of the poor seismic data quality in the North Celtic Sea Basin. It also created a number of elongate inversion structures, some of which provide traps for hydrocarbons. The most obvious of these are the Kinsale Head and the Ballycotton gasfield structures (Figure 10.1).

Key observations regarding the geological development of the Celtic Sea region are as follows:

a) the Mesozoic sediments rest upon folded and indurated Devonian and Carboniferous (largely Mississippian) metasediments;

b) the basins appear to have an underlying Variscan and older structural control on their location and orientation;

c) although the pre-Jurassic strata are poorly dated, there is little evidence for any significant occurrence of Permian strata in the major Celtic Sea basins;

d) thick Upper Jurassic sediments have only been encountered in the North Celtic Sea Basin; and

e) the region experienced several phases of basin inversion, most notably in the Cenozoic.

Regional Exploration History

Following the Petroleum and Other Minerals Development Act of 1960 and the Continental Shelf Act of 1968, acreage was designated in the Irish offshore, as outlined in Chapter 3 (*see also* Croker and Shannon, 1995). It is probably pertinent here to provide a brief summary of the subsequent exploration history in the Celtic Sea region. Marathon Petroleum received exclusive rights over large tranches of acreage in the Celtic Sea (and also to the west and northwest of Ireland). Seismic exploration was followed by drilling of Ireland's first offshore well (Marathon 48/25-1), which spudded in 1970. This well, targeted on a shallow Palaeogene inversion structure, encountered gas shows. Marathon returned to Block 48/25 the following year and, with only the third well to be drilled in the Irish offshore (Marathon 48/25-2), located the Kinsale Head Gasfield (Colley *et al.*, 1981), with reserves of approximately 1.5 Tcf. Esso farmed into the acreage in the west of the basin, and during the 1970s these two companies drilled a number of wells, most of which located on shallow Kinsale Head look-alike inversion anticlinal structures. These were four-way dip-closed structures that could be mapped at Aptian to Upper Cretaceous levels. Most of these were dry, probably due to a combination of the poor seismic data quality, the late structuring and the shallow nature of many of the domal features.

There were occasional technical successes, one of which was the Seven Heads gas accumulation. The discovery well (Esso 48/24-1), drilled in 1973, flowed 780 barrels of oil per day (bopd) and 10 million (MM) standard cubic feet of gas per day (scfd). Reserves were estimated at about 100 billion cubic feet of gas, and 1–2 MM barrels of oil (Naylor and Shannon, 1982). As mentioned in Chapter 3, Ramco was later awarded the Seven Heads acreage and, following appraisal, brought the gasfield on stream in 2003, with a subsea tie back to the Kinsale Head Alpha platform. While estimates for both reserves (reportedly up to 390 bcf) and production rates (up to 60 MMscfd) at the time were very optimistic, the geological complexities of the reservoir resulted in a poor production performance. Operatorship returned in 2006 to Marathon (now Star Energy Group Plc). The latest reported estimates for the original recoverable reserves are somewhat less than 100 bcf and daily production is approximately 20 MMscfd. Another minor success in the early phase of exploration was the Ardmore gas discovery (Marathon 49/13-1 and offset well Marathon 49/14-1 which flowed at a rate of 8.81 MMscfd) in 1974. Providence Resources have recently completed a 3D seismic survey over the field and have indicated the likelihood of a recoverable reserve potential of *c.* 30 bcf of gas within the uppermost Lower Cretaceous reservoir interval. The original discovery wells by Marathon also indicated the presence of a heavy oil accumulation (16° API) beneath the gas. Analysis of the recent Providence Resources 3D seismic survey have suggested that this could be a significant accumulation in the region, with an in-place resource potential of up to *c.* 230 MMBO being reported in the industry.

It was not until 1976 that licences were granted in the Celtic Sea and Fastnet basins to companies other than Marathon and Esso, as part of the First Round of Licensing. The late 1970s and early 1980s saw exploration move westwards to the Fastnet Basin. Seismic data quality was better than in the North Celtic Sea Basin, largely due to the absence of major inversion which exposed the hard Upper Cretaceous chalk on the seabed in the latter basin. Approximately 10 wells were drilled, largely on tilted Jurassic fault-block structures. However, although some oil shows and occasional flows were encountered from Lower Jurassic sandstones, no commercial successes resulted. Minor renewal of exploration interest and activity saw a few additional wells drilled in the mid 1980s, but overall the results were disappointing.

In 1982, a further 18 blocks were awarded in the Celtic Sea basins under the Second Round of Licensing. The Round was followed by an intensive phase of seismic acquisition, and then in 1983 by the drilling of further wells in the North Celtic Basin. Improved seismic techniques allowed better imaging of the deeper plays, particularly tilted fault blocks in the Jurassic, which were judged to have considerable potential. Hydrocarbons were discovered in 1983 by a Gulf-operated consortium within the Middle–Upper Jurassic section in the discovery well for the Helvick oil accumulation (Figure 10.1). This well (Gulf 49/9-2) tested hydrocarbons from four intervals flowing a cumulative total of 9911 bopd and 7.44 MM scfd, and demonstrated the potential for light oil accumulations in the Celtic Sea area. Unfortunately, follow-up drilling showed that the accumulation was small, with oil-in-place estimates in the range 8.4–12.3 MMbo reported (Caston, 1995). In 2000 Providence Resources drilled the 49/9-6z well which flowed at a rate of approximately 5200 bopd. However, subsequent analysis confirmed that the reservoir is compartmentalised and complex. In 1985, Gulf drilled a further success when well 50/6-1 flowed 2074 bopd and 1.3 MM scfd from an Upper Jurassic sandstone reservoir. In 2008 Providence Resources drilled an appraisal well on this structure (Dunmore discovery). The well failed to prove the lateral extent of the original reservoir sandstones but encountered a new oil-bearing potential Jurassic carbonate reservoir. By the mid-1980s the pace of Irish offshore exploration in the Celtic Sea region had slowed considerably. Most of the obvious large four-way dip-closed anticlinal structures in the region had been drilled, with generally disappointing results and little encouragement for almost a decade. One exception was the discovery by Marathon in 1989 of the Ballycotton Gasfield, a small (89.9 bcf in place) accumulation fortuitously located close to the producing Kinsale Head gasfield. A number of Jurassic tilted fault-block structures had also been drilled, with generally disappointing results.

The Bula 48/19-2 well (Figure 10.1), drilled in 1992 west of the Ballycotton Gasfield, encountered an accumulation of heavy biodegraded oil in Greensand and Wealden sandstones (Howell and Griffiths, 1995), similar to the Seven Heads discovery. Relatively few exploration or appraisal wells were drilled in the Celtic Sea region from the mid-1990s for five or six years. Marathon's Southwest Kinsale Head accumulation (reserves estimated to be approximately 28 bcf) was appraised by wells in 2001 and 2003 and was brought on stream in 2003 through the Kinsale Head Bravo platform. By this time, seismic data quality had improved significantly in the area, with newer acquisition and processing methods and the thrust of renewed exploration moved to re-examining some structures that showed indications of hydrocarbons in earlier phases of drilling, with detailed seismic analysis allowing identification of stratigraphic aspects of trapping in structures.

Drilling activity in the Celtic Sea area underwent a revival from 2006 to 2008. Island Oil and Gas drilled a successful gas appraisal well (48/23-3) at the southwestern end of the Seven Heads Gasfield, and also drilled a successful exploration well (49/23-1) on the Old Head prospect, east of Kinsale. In 2007, Island Oil and Gas drilled an exploration/appraisal well (57/2-3) on the Schull gas accumulation and an appraisal/development well on the Old Head of Kinsale discovery (Block 49/23). Providence Resources drilled an appraisal well on the Hook Head structure (Block 57/11) and encountered good quality oil. In 2008, hydrocarbon shows were encountered in two wells drilled by Providence Resources in the North Celtic Sea Basin: on the Dunmore (Block 50/6) and Hook Head (Block 50/11) accumulations.

A summary of the structural and stratigraphic framework of each of the major basins, and a discussion of their exploration potential, follows.

St George's Channel and Cardigan Bay basins
Basin framework

There is, unfortunately, confusion in the literature over the names for the southern Irish Sea basins. The 'Cardigan Bay Basin' of Maddox *et al.* (1995), Naylor and Shannon (1999), Izzat *et al.* (2001) and Dunford *et al.* (2001) amongst others, equates to the 'St. George's Channel Basin' of Tappin *et al.* (1994), Corcoran and Clayton (1999), Green *et al.* (2001), Williams *et al.* (2005) and others. The second group of authors also generally used 'Cardigan Bay Basin' for the inshore part of the basin trend, termed the 'Tremadoc Bay Basin' by others. As a result, care is needed when reading the literature. Here we use the nomenclature of Tappin *et al.* (1994), which is also close to that used in a seminal paper on the area by Barr *et al.* (1981; *see* Figure 6.7).

The St George's Channel Basin is a northeast-trending extension into United Kingdom waters of the North Celtic Sea Basin. A further northeast prolongation of this basin trend – the Cardigan Bay Basin – extends to the Welsh coast. These basins are separated from the Central Irish Sea Basin by the St Tudwal's Arch and its

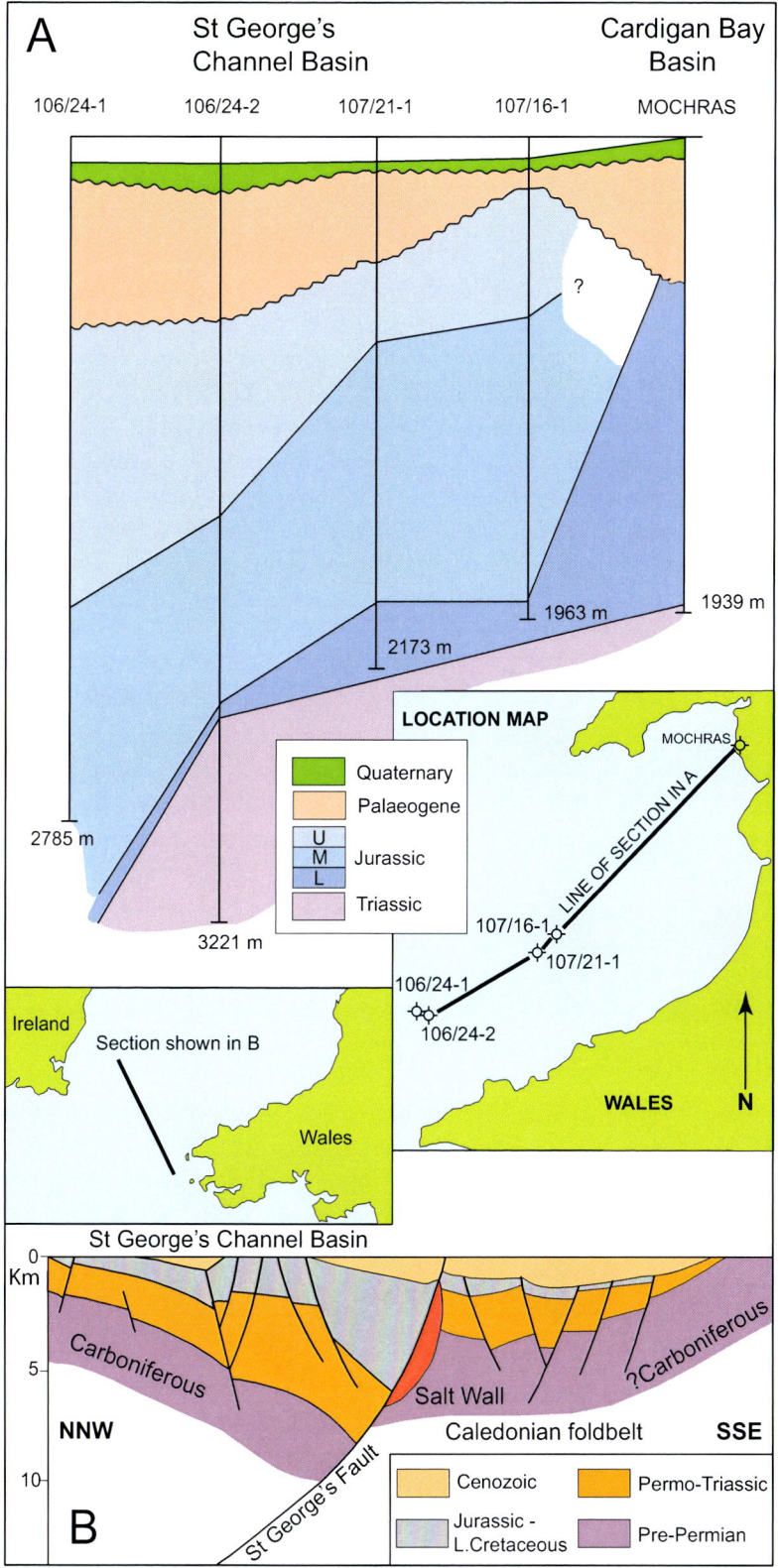

Figure 10.2. A. Stratigraphic units in the inner St George's Channel and Cardigan Bay basins. **B.** Geological cross-section NNW-SSE across the St George's Channel Basin, based on seismic data. Both adapted after Tappin *et al.* (1994).

NE–SW trending offshore extension (Figure 6.7). A southeastern margin of Mesozoic sediments in the basins is defined by the offshore prolongation of the NE–SW Bala Fault zone. The depocentre of the St George's Channel Basin takes the form of an asymmetric deep graben whose southeastern margin is formed by the St George's Fault (Figure 10.2B). Unlike the basin-bounding structures, the St George's Fault is arcuate in trace, being WSW trending in the east, possibly as a splay from the Bala Fault, and almost E–W at its western end.

The basins are crossed by a number of NW–SE trending faults, most notably by extensions of the Sticklepath–Lustleigh Fault System (Tappin *et al.*, 1994) and the Codling Fault, both of which underwent sinistral lateral phases of movement in the Cenozoic. Maddox *et al.* (1995) took the northwestward offshore extension of the Sticklepath–Lustleigh System as marking the southwestern limit of the St George's Channel Basin. However, this is less evident on the detailed isopachyte maps of Tappin *et al.* (1994) and Welch and Turner (2000), on which, towards the western end of the St George's Fault, the basin merges southwards into the North Celtic Sea Basin. The St George's Basin is separated from the Cardigan Bay Basin to the northeast by a shallow structural saddle, on trend with the Codling Fault further north.

The post-Variscan basin fill is very thick. Tappin *et al.* (1994) suggest that it exceeds 7000 m, while PESGB (2007) maps indicate thicknesses in excess of 8000 m, with synsedimentary movement on the controlling faults. Units of the stratigraphy are seen to thicken into the Bala, and particularly the St George's, faults (Welch and Turner, 2000). Thick halite-bearing Permo-Triassic beds overlain by some 4000 m of Jurassic strata occur in the hanging-wall of the St George's Fault, and up to 1500 m of Cenozoic sediments. The basin fill sequence is not well known, since exploration wells have not drilled the full section. Accommodation was achieved during a protracted period of Triassic–Jurassic extension, with discrete increments of movement during the period (Welch and Turner, 2000). The same authors conclude that there was only minor modification of Mesozoic extensional fault geometry during Miocene basin inversion, although this did include transpressional reactivation of NW-trending faults, such as the Sticklepath and Codling systems. Turner (1997) proposed a dextral shear model during the Cenozoic with clockwise rotation between the Codling Fault to the north and the Variscan Front and Sticklepath–Lustleigh Fault System in the south. In this model pre-existing NE-trending faults underwent sinistral displacement. There is evidence of salt migration in the St George's Channel Basin, and a massive salt wall (55 km long, 3 km wide) has developed along the St George's Fault. Salt migration may have begun in Early Cretaceous time (Tappin et al., 1994), or in the Late Cretaceous, with a further phase subsequently in the Miocene (Welch and Turner, 2000).

The St George's Channel Basin is considered by Tappin et al. (1994) to be underlain by up to 3 km of relatively undeformed Carboniferous strata, with the overlying sequence unconformable on the Variscan surface. Seismic interpretation suggests that pockets of Permian strata may exist in the basin. The Texaco/HGB 103/2-1 well on the southern margin of the basin penetrated 76 m of undated beds unconformable on basement that may be Permian in age (Tappin et al., 1994). Seabed samples and wells, together with drilling on the southern margin of the Cardigan Bay Basin (Barr et al., 1981), have revealed a Triassic succession comparable to that in the East Irish Sea Basin and the English Midlands, with the Sherwood Sandstone Group overlain by saliferous Mercia Mudstone Group strata. To the north, the Triassic onlaps the St Tudwal's Arch, a positive fault-controlled Lower Palaeozoic and Precambrian element extending out from the Welsh Massif. Offshore wells in the St George's Channel Basin have encountered thicknesses of up to 2000 m of Triassic sediments. The Texaco/HGB 103/2-1 well (Figure 6.7) on the southern margin of the basin encountered 249 m of sandstones containing anhydritic mud partings, overlain by the Mercia Mudstone Group (1860 m), which is dominated by red-brown mudstones, frequently anhydritic, with a saliferous middle unit.

Exploration drilling in Cardigan Bay and St George's Channel basins, south of St Tudwal's Arch, has revealed an almost complete Lower Jurassic to lower Purbeckian succession (Barr et al., 1981 and Figures 7.9 and 10.2A). The Mochras borehole on the Welsh coast penetrated 1305 m of Lower Jurassic (Woodland, 1971), whilst in excess of 3 km of Middle and Upper Jurassic sediments were possibly deposited in the centre of the St George's Channel Basin (Maddox et al., 1995). In the Cardigan Bay Basin, a half-graben that deepens to the southeast, there is evidence for ongoing extensional fault movement and subsidence through Early Jurassic time that was terminated by an interval of mid-Cimmerian (mid-Jurassic) uplift. This was followed by renewed subsidence and the deposition of Middle–Upper Jurassic sediments, in turn halted by widespread Late Cimmerian uplift in the Cretaceous. This episode was particularly marked at the basin margins, with movement on the basin margin faults (Tappin et al., 1994). The phase of Palaeogene uplift and erosion that removed considerable thicknesses of Cretaceous strata from the North Celtic Sea Basin resulted in the erosion of the entire Cretaceous sequence in the St George's Channel and Cardigan Bay basins, if indeed such sediments were originally deposited. The Palaeogene phase was accompanied by erosion and movement along the Sticklepath and Lundy faults, and on the marginal faults of the Cardigan Bay Basin, possibly also with related uplift of the Welsh and Cornubian massifs (Dobson and Whittington, 1987). Renewed subsidence in the Cardigan Bay Basin was accompanied by deposition of a Palaeogene sequence of claystones, lignites and sandstones. Mild mid-Cenozoic (Oligo-Miocene) inversion then produced some reversal on faults, tightening of earlier inversion structures, and localised upwarping. Dextral movement probably took place along the Sticklepath Fault Zone at this time, as on the Codling Fault further north. Up to 300 m of Quaternary to Recent deposits completed sedimentation in the Cardigan Bay Basin.

Figure 10.3A is a geoseismic section at the western margin of the St George's Channel Basin based on seismic line BU86-12. The seismic line extends ESE from a point close to the Wexford coast across the southern part of Block 41/19 and the southwest corner of Block 41/20. By reference to the nearest wells (Maddox et al., 1995) the post-Palaeozoic section is thought to be mainly Triassic in age. A strong reflector beneath the base Triassic horizon is identified as near Top Lower Carboniferous. The thick sedimentary sequence above this reflector in

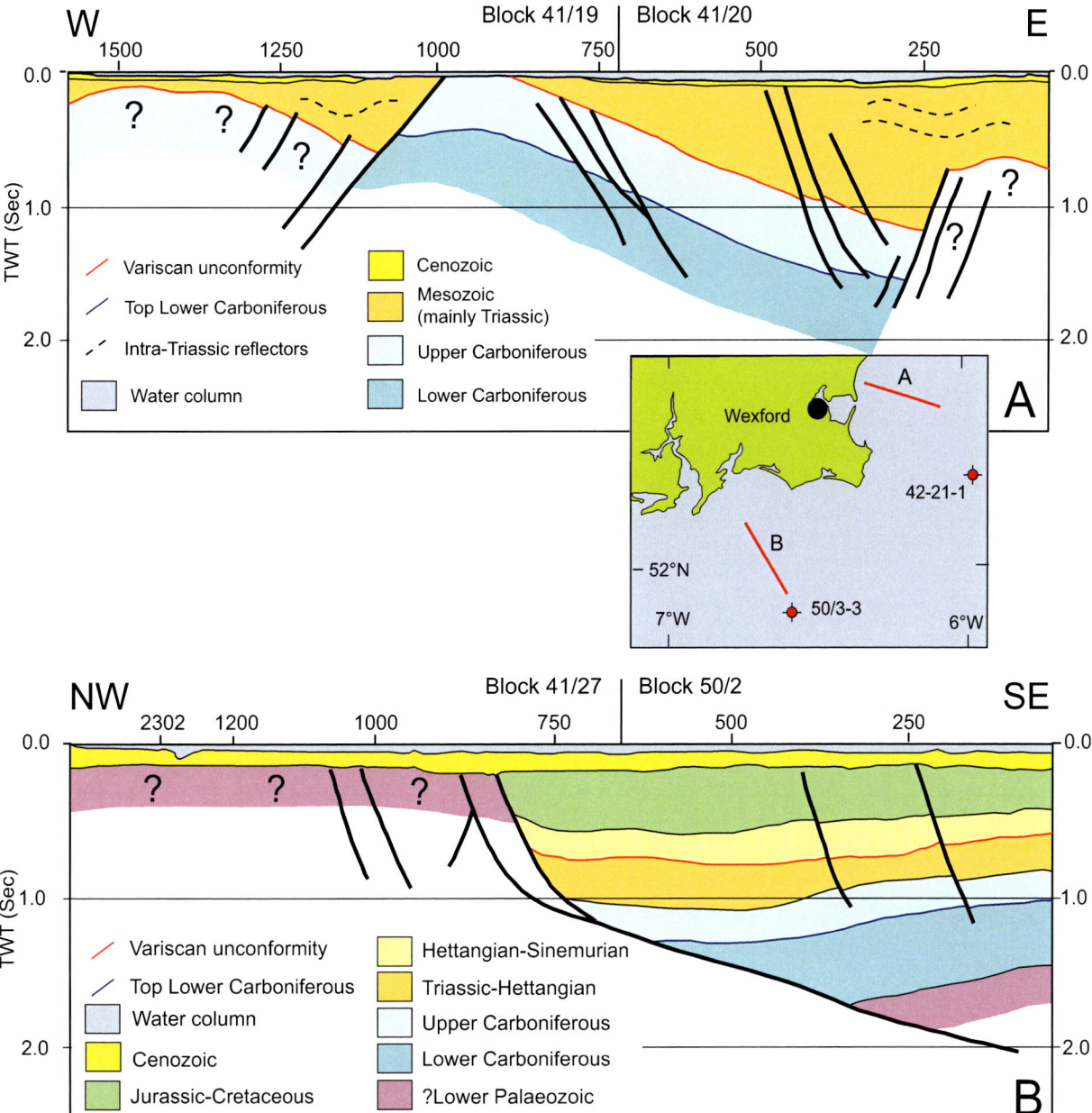

Figure 10.3. A. Geoseismic profile, offshore Wexford, illustrating the nature of the western margin of the St George's Channel Basin. Based on the interpretation of seismic line BU86-12. **B.** Geoseismic profile, offshore south County Wexford, crossing the northern margin of the North Celtic Sea Basin. Based on the interpretation of seismic line BP87–50-32. Shotpoint numbers are indicated on each line.

Block 41/20 is thinned westwards by erosion and the Triassic is entirely eroded in the southwestern part of Block 41/19. Faulting appears to have played only a minor role in this thinning and there is little sign of westward stratigraphic thinning of the Triassic or Lower Carboniferous units. At the eastern end of the seismic line, within Block 41/20, a positive structural element with little internal seismic character may represent a basement high. This feature, the lateral extent of which is unknown, is on approximate trend with faults at the northern margin onshore of the Precambrian–Lower Palaeozoic Carnsore structural element. Structure within the overlying Triassic on Line BU86-12 may result from re-activation of this feature, but the possibility of salt movement also exists. At ~ S.P.1000 (Figure 10.3A) a significant down-to-the-coast fault, and related minor faults to the west, produce a half-graben, probably containing mainly Triassic strata. Up to 0.5 sec Two-Way-Travel time of fill is present in the half-graben, and variation of dip within the Triassic section suggests fault re-activation or, possibly, halokinesis. The pre-Triassic section in the western part of the line lacks coherent reflectors. Dinantian limestones occur onshore immediately west of Line BU86-12. Cenozoic sediments form only a thin veneer extending across the eroded surface of the older structures.

In Figure 10.3A, the Triassic rests without noticeable angular unconformity on the Variscan erosion surface. The Sherwood Sandstone Group is only 120 m thick and mud-prone in the Conoco 42/16-1 well. However, reservoir quality may improve westwards towards the sediment source areas – a concept supported by the indications of offlapping reflectors at the base of the Triassic succession. The Sherwood Sandstone unit is not clearly definable on the seismic line, and may be thin and represented by the group of strong reflections associated with the base Triassic event. Permian strata are unlikely to be present. Much of the Triassic section is anticipated to belong to the Mercia Mudstone Group and to comprise anhydritic and dolomitic mudstones with interbedded halites and minor sandstones. Maddox *et al.* (1995) referred to minor halokinetic deformation in this succession, observed on seismic data.

In a recent study of the St George's Channel Basin, Williams *et al.* (2005) proposed at least two major inversion episodes – Late Cretaceous and Neogene, with a minor shortening event in the Eocene. The same authors also presented evidence from the basin margin supporting earlier cooling phases in the Permo-Triassic, following rifting, and also in the mid-Late Jurassic. Sonic velocity profiles were used to constrain the minimum thickness of cumulative Mesozoic and Cenozoic exhumation and erosion, and this produced estimates of 1 km in the basin centre and 2.24 km at its margins. In a review of the same basin Holford *et al.* (2005) concluded that Early Cretaceous exhumation was restricted to basin margins, whereas both the depocentre and margins were exhumed during the early Palaeogene (~1–2 km) and late Palaeogene to Neogene (1–1.5 km). The smaller Cardigan Bay Basin is seen as having major phases of cooling and exhumation during the Early Cretaceous (2–3 km) and Neogene (1–1.5 km), together with limited evidence for an early Palaeogene episode.

Exploration potential

Source rocks

The St George's Channel Basin lies north of the putative trace of the Variscan Thrust Front as it crosses from South Wales to Ireland. For this reason Tappin *et al.* (1994) considered the potential of Carboniferous source rocks beneath the basin to be higher than in the basins south of the line, where the Carboniferous sequences have suffered deformation and possible low grade (greenschist facies) metamorphism. Based on vitrinite reflectance data from wells drilled in the Central Irish Sea Basin, Corcoran and Clayton (1999, 2001) concluded that the Carboniferous (Duckmantian–Stephanian) source sequence was exposed to maximum palaeotemperatures prior to the Early Jurassic, and probably during Stephanian to Early Permian time. The rocks are reported as having modest gas source potential. This effectively means that hydrocarbon generation from this source was 'switched off' during Variscan uplift. A similar situation is envisaged for the St George's Channel Basin.

Liassic (Lower Jurassic) rocks are organic-rich and are oil-prone where penetrated in the offshore wells (Barr *et al.*, 1981). In the Marathon 42/17-1 well, on the St Tudwal's Arch extension, they are immature, but in the Gulf 42/21-1 well on the north flank of the St George's Channel Basin (Figure 6.7) Pliensbachian–Toarcian rocks are in the early part of the oil window (Corcoran and Clayton, 1999). Murphy *et al.* (1995) recorded a 73 m interval of oil-prone source rocks from the early Toarcian interval of well 42/21-1. The Lower Jurassic sequence is considered by Corcoran and Clayton (1999) to have attained peak palaeotemperatures after the Bathonian, probably in the Late Cretaceous, prior to Cenozoic inversion. Minor hydrocarbon shows have been recorded within the Jurassic section in a number of wells (Tappin *et al.*, 1994). However, the Liassic shales in the Mochras borehole are immature and have low organic content (Barr *et al.*, 1981). In the HGB 107/21-1 well on the southern basin flank, the entire Jurassic section is immature for hydrocarbon generation. Whilst the Upper Jurassic has poor source potential, the Lower and Middle Jurassic sections contain moderate quality gas-prone source sections. These would require 1000 m of further burial to achieve maturity, but could provide hydrocarbon charge in the deeper parts of the basin. On the same southeastern flank of the basin, in the Arco 106/24-1 well (Figure 6.7), the Jurassic section is mature and marginally prospective for heavy oil from the upper Kimmeridgian and for gas condensate from the Bathonian to lower Kimmeridgian section (Tappin *et al.*, 1994).

In a reconstructed thermal profile for the Gulf 42/21-1 well (drilled to the Pliensbachian) Green *et al.* (2001), using combined apatite fission track and vitrinite reflectance data, also showed a palaeotemperature maximum at the end of the Cretaceous time (58–70 Ma), with the subsequent cooling episode terminating the main phase of hydrocarbon generation. The same authors pointed out that this event may mask earlier (e.g. Variscan or Early Cretaceous) palaeo-thermal events, provided that the maximum attained temperatures were lower than those at end-Cretaceous. This result is considered by Green *et al.* (2001) to require the depostion of 2000 m of post-Oxfordian sediments by the end of the Cretaceous, with subsequent removal of 1500 m of section in the Paleocene. A further 500 m was then deposited in the period to the early Miocene, before the remaining total of 1000 m of section was removed by uplift in the Neogene. Corcoran and Clayton (1999), using vitrinite reflectance, estimated *c.*1600 m of net exhumation for the Liassic section in the 42/21-1 well. The Gulf 42/21-1 thermal profile contrasts with results from wells (e.g. HIL 42/12-1, Conoco 42/16-1 and Marathon 42/17-1) in the Central Irish Sea Basin (Figure 6.8) or on the St Tudwal's Arch extension, which began to cool from maximum post-depositional palaeo-temperatures in Early Cretaceous time (120–115 Ma). This contrast was thought by Green *et al.* (2001) to reflect the different structural regime operative in the Celtic Sea basins south of the St Tudwal's Arch extension.

Reservoir rocks and structures

Reservoir rocks are present at several levels in the basin. The Sherwood Sandstone Group, unconformable on potential Carboniferous source sequences and sealed by the Mercia Mudstone Group evaporites, provides one target. Up to 100 m of Triassic sandstones, with porosities of up to l0%, are recorded in the UK sector of the St George's Channel Basin (Barr *et al.*, 1981), suggesting the possibility of similar reservoir targets in the Irish sector of the basin. The Sherwood interval had excellent reservoir characteristics in the HIL 42/12-1&2 wells, but was thinner and mud-prone in the Conoco 42/16-1 well (Figure 6.8). There is a possibility of westward improvement in reservoir quality towards the Irish Massif. However, as noted above, some workers believe that hydrocarbon expulsion occurred at the end of the Carboniferous, prior to the deposition of Mesozoic reservoirs and the formation of

syn-rift traps. Sourcing from Lower Jurassic levels would require juxtaposition across fault zones in the deeper parts of the basin. Sinemurian sandstones occur within the Lower Jurassic sequence (*see* Chapter 7) which also includes source rocks likely to be mature for oil in parts of the basin. Marginal marine Middle Jurassic sandstones and Lower Cretaceous Greensand horizons also provide secondary reservoir targets.

As mentioned above, Upper Carboniferous strata may be extensive in this area. Judging from drilling in the Central Irish Sea Basin, sandstone units in the sequence may range up to 14 m in thickness, with porosities in the range 4–16%, and could constitute a valid target sealed and sourced by intra-Carboniferous shales and coals. Clayton *et al.* (1986) also described sandstone units up to 8 m in thickness in boreholes immediately onshore, in the Wexford Syncline. However, robust Carboniferous structural targets with adequate seal will be difficult to locate and constitute a high risk play. Also, if hydrocarbon generation occurred in Late Carboniferous time, a long sequestration period is required in a tectonically active region.

Structures in the basin are fault-related, sometimes with related salt movement. Barr *et al.* (1981) reported that faults near the HGB/Deminex 106/28-1 and Arco 106/24-1 wells (Figure 6.7) have associated salt intrusion. Tilted fault-block and roll-over anticlinal traps can be anticipated. However, several phases of basin inversion, together with synsedimentary accommodation on major faults, represent a considerable addition to the exploration risk. There have been relatively few wells drilled in these basins, and the deeper parts of the St George's Channel Basin, where Jurassic source rocks are mature, are regarded as having remaining potential. However, in late 1994, the Marathon 103/1-1 well in the UK sector of the St George's Channel Basin, adjacent to the Irish/UK boundary, reported the first interesting, and potentially significant, gas discovery in Jurassic sandstones in the basin, demonstrating the presence of a working petroleum system in the basin.

North Celtic Sea Basin
Basin framework

This is the largest of the Celtic Sea basins and has an ENE–WSW orientation. The North Celtic Sea Basin is structurally linked to the Fastnet Basin that lies to the west and links eastwards to the St George's Channel and Cardigan Bay basins. Deformed Variscan basement (Devono-Carboniferous) is overlain by Triassic and

Jurassic strata. Up to 600 m of Triassic has been drilled in wells on the southern margins of the basin (Esso 56/20-1 and Conoco 57/9-1) and has also been interpreted on seismic profiles (Shannon and MacTiernan, 1993). This succession is typically marly with minor evaporitic mudstones. The Lower and Middle Jurassic are broadly similar to the succession encountered in the adjacent Fastnet Basin, consisting of a basal limestone package overlain by predominantly marine organic-rich mudstones, with occasional sandy units. In contrast to the Fastnet Basin, a thick (*c.*1 km) syn-rift Upper Jurassic succession is present in the basin. Fluvial Callovian to Oxfordian sandstones more than 200 m thick (Caston, 1995) are overlain by muddy, silty and calcareous Oxfordian and younger Jurassic strata. The Cretaceous is similar in lithology and stratigraphy to, but generally thicker than, the succession in the Fastnet Basin. The Cenozoic is thin and only patchily preserved in the basin, and has been eroded following phases of Paleocene and Oligo-Miocene inversion (Murdoch *et al.*, 1995).

Deep structural control has been demonstrated on the trend and development of the Celtic Sea basins. As would be expected, the main basin margins in the Celtic Sea are seen on many seismic sections to be fault related. This is particularly the case for the lower stratigraphic horizons, as illustrated on published sections and maps (Figures 10.3B, 10.6 and 10.7). Nevertheless, onlap and pinch-out of Triassic and Jurassic units is also seen on both the northern and southern margins of the North Celtic Sea Basin. There are also basin margin locations where Triassic–Jurassic sequences are overlain with angular unconformity by Cretaceous strata, or where Triassic, Jurassic and Cretaceous sections subcrop thin Cenozoic sediments, clearly indicating that these older units originally extended beyond their current fault-preserved margins.

The geoseismic profile shown in Figure 10.3B is based on seismic line BP87-50-32 (which is of only moderate quality). It shows the Mesozoic sequence faulted against older sequences, with only thin Cenozoic cover extending northwards onto older rocks. The line shows a major southeast-dipping low-angle fault in the southern portion of Block 41/27 that steepens upwards into a complex fault zone near the seabed. Older rocks towards the northwest are juxtaposed across the fault against Mesozoic and older successions. The fault thus forms a distinct northern margin to the North Celtic Sea Basin. Line BP87-50-32 is close to the northern portion of SWAT- 4 deep seismic profile, mentioned previously, and probably intersects the same basin margin fault. The fault can be traced on the SWAT profile down into the lower crust at 20 km and the upper ramp appears to have a local strike of about 100°. Correlation with the Marathon 50/3-3 well suggests that much of the Mesozoic section south of the boundary fault is of Lower Jurassic age, resting on a relatively thin Triassic succession. There is no marked angular discordance at the Variscan surface and the underlying sequence probably comprises Upper and Lower Carboniferous units. North of the boundary fault the data show little character and may represent basement rocks. The rocks immediately north and onshore of the northern end of Line BP87-50-40 are Lower Palaeozoic and Precambrian in age. However, it is possible that Carboniferous rocks are present in part of this area, masked by poor data quality. The Triassic units on both geoseismic sections in Figure 10.4 show no appreciable shoreward thinning, supporting the view of Keeley (1995) that a substantial thickness of Permo-Triassic strata was deposited onshore in southernmost Ireland.

Exploration potential

Source rocks

Coal-prone Westphalian C–D (Moscovian stage: Bosovian–Asturian sub-stages) successions have been recorded in the St George's Channel Basin (Quadrant 42: Maddox *et al.* 1995) and may extend westwards along the northern margin of the North Celtic Sea Basin to link with the Westphalian (Bashkirian–Moscovian) sequence encountered in the Marathon 50/3-3 well (Figure 10.1). This well penetrated 212 m of Westphalian strata comprising interbedded sandstone, siltstone, claystone and minor coals. Onshore boreholes immediately east of Block 41/22 encountered previously unknown Westphalian and probable Namurian (Serpukhovian–Bashkirian) rocks (Clayton *et al.*, 1986). However, Carboniferous rocks encountered in drilling further west in the basin have generally been indurated and of poor source quality.

A number of source rock intervals have been identified within the Mesozoic sequences of the North Celtic Sea and Fastnet basins. The thick Toarcian sequence, which is in excess of 600 m in the main depocentres, is an important basin-wide source rock (Figure 10.4). Murphy *et al.* (1995) calculated that up to 137 m of strata with lithology capable of generating liquid hydrocarbons are present in the basin centre, including Toarcian source rocks rated as very good to rich (pyrolysis S2 yields >5.0 mg/g). The burial history curve presented by the same authors indicates maximum generation and expulsion of oil and gas during the period

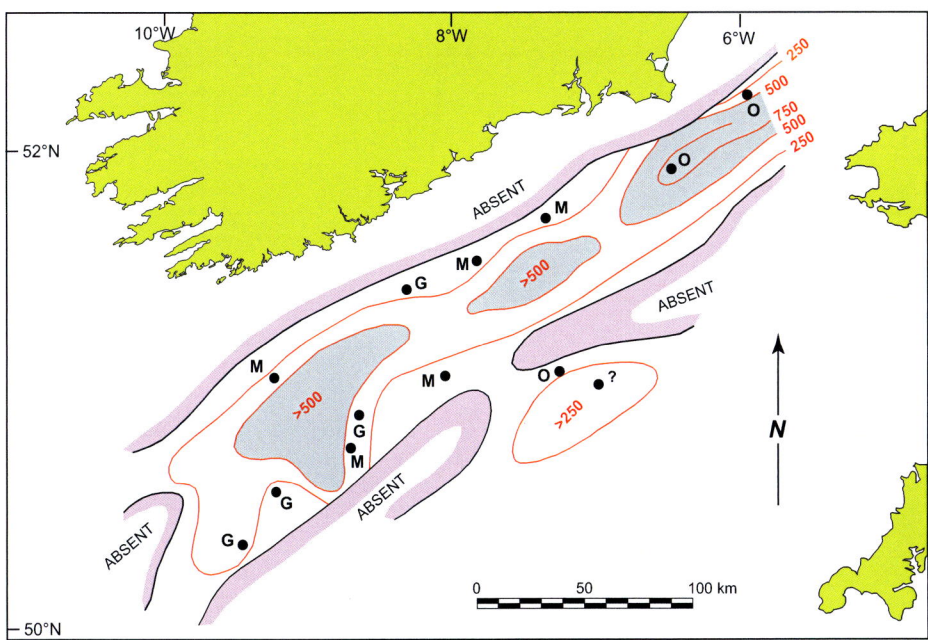

Figure 10.4. Thickness distribution of Toarcian interval rocks in the Celtic Sea region based on the control wells shown, allied with regional seismic mapping. The hydrocarbon potential of the source sequences within the Toarcian interval at the wells is indicated: G, gas-prone; O, oil-prone; M, mixed potential. Modified after Murphy *et al.* (1995).

of greatest burial in Late Cretaceous time. Caston (1995) reported that the best Lower Jurassic source interval in the Helvick oil discovery area was Upper Sinemurian–Lower Pliensbachian in age (TOC 1.52-2.11%), although the Toarcian also had potential. However, the source rocks are in an early stage of maturity, whilst the oil in the accumulation was derived from medium mature source rocks, suggesting a complex migration and maturation process in the basin. The same author also identified the 'Portlandian–Purbeckian' (late Tithonian–Berriasian) between 1058–1067 m in the Gulf 49/9-2 well as a lacustrine source rock of high potential (TOC 1.36-3.93%). Howell and Griffiths (1995) produced a maturity map at top Liassic level for the central basin area immediately west of the Kinsale Head Gasfield, together with a burial history model for the Gulf 48/19-1 well. The same authors indicated that the main pulse of hydrocarbon generation probably occurred during Late Cretaceous to early Cenozoic times. However, oil to source studies in this part of Quadrant 48 indicate a locally mature 'Portlandian–Purbeckian' (late Tithonian–Berriasian) source for oil encountered in the Esso wells (48/24-1, 48/24-2 and 48/28-1) drilled on the Seven Heads structure.

The origin of the gas in the Kinsale Head Gasfield (Figure 10.1) is a matter of debate. Initial work (Colley *et al.*, 1981) explained the 'dry' nature and high methane content (commonly 99%) of the gas as resulting from the freshwater flushing of paraffinic oil that had migrated into the Wealden, producing methane that subsequently

was mixed with thermally-generated Jurassic methane. This composite gas then charged the overlying Aptian–Albian sand reservoirs by migration along faults. Howell and Griffiths (1995) reported that 'carbon isotope data indicated that the gas was evolved from a mature source and/or from the cracking of heavier liquid hydrocarbons'. However, Taber *et al.* (1995) modelled the Lower Jurassic interval of the Marathon 48/25-1 well near the south flank of the field as being in the dry gas window (>2.0 % *Ro*), with expulsion from the Lower Jurassic during Early Cretaceous time. These authors suggested that the bulk of the gas was produced from thermally mature Lower Jurassic source rocks, although the possibility of blending with biogenic components was recognised. Migration into the reservoir was envisaged to have taken place during Palaeogene uplift and trap formation.

Murdoch *et al.* (1995) presented burial curve models for wells along the northern flank of the basin (Figure 10.5, D–F), together with estimated maturity maps for the basin at Top Jurassic and Top Lower Jurassic levels. Top Jurassic level strata are shown as being in the oil window only in a few deep depocentres. Their map at Top Lower Jurassic level shows the source rocks to be in the oil window in a zone along the axis of the basin, and in the gas window in the deepest parts. There are differences in detail between this map and that of Howell and Griffiths (1995: Figure 9 *Top Liassic Maturity Map*). Part of the problem here may be the lack of regional control on Jurassic interval thicknesses for modeling purposes due

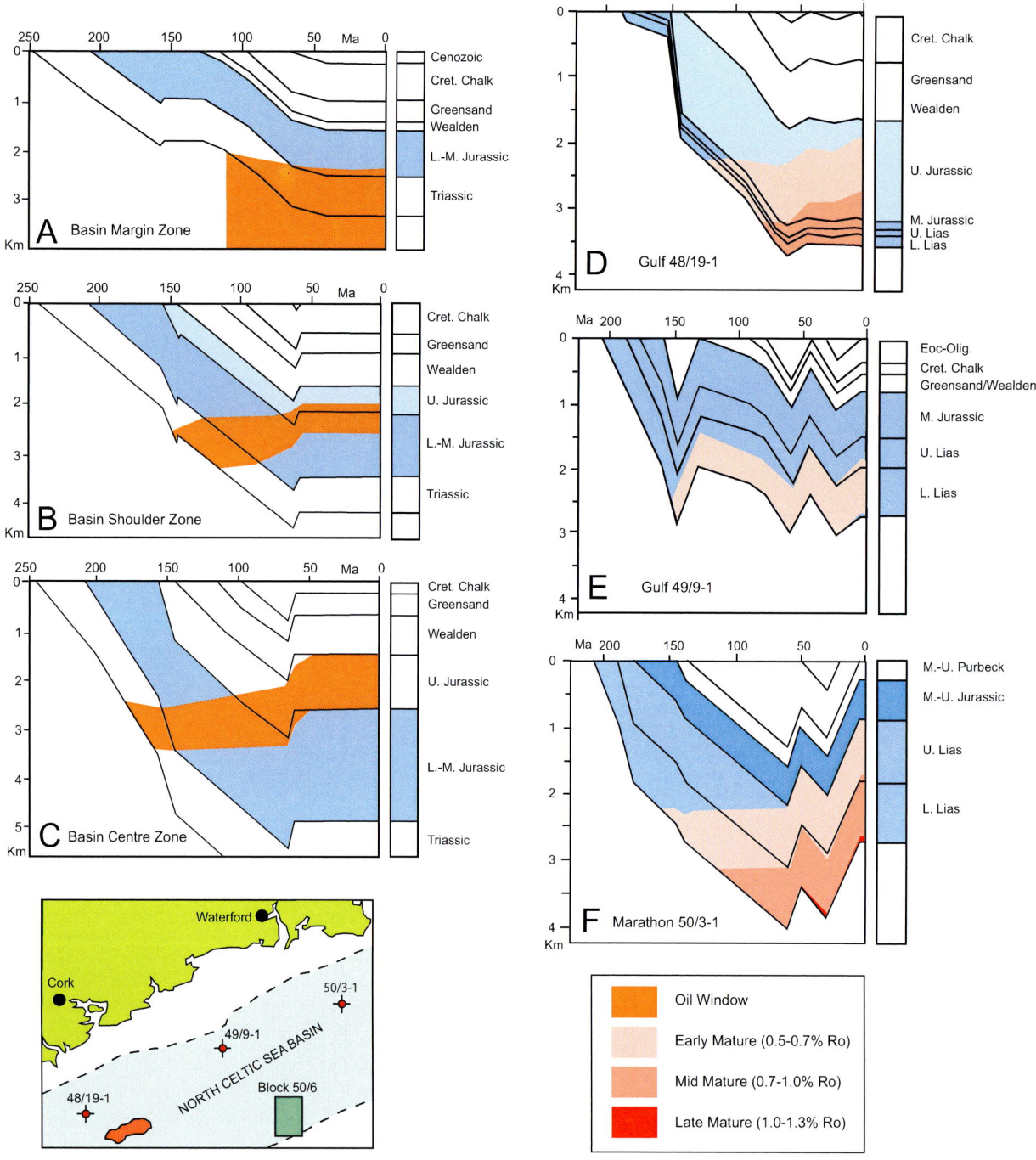

Figure 10.5. Burial history models for parts of the North Celtic Sea Basin. A-C: Block 50/16 profiles on the southern flank of the basin (after Shannon and MacTiernan, 1993). D-F Wells 48/19-1, 49/9-1 and 50/3-1 along the northern flank of the basin (after Murdoch *et al.*, 1995).

to poor seismic resolution and lack of well control points.

On the south flank of the North Celtic Sea Basin Shannon and MacTiernan (1993) modelled burial curves in Quadrant 50 (Block 50/17). These indicate that Lower and Middle Jurassic source rocks are currently mature

in the basin centre, immediately northwest of the Basin Shoulder Zone (Figure 10.6). Based on these reconstructions the same authors concluded that the onset of oil generation was in Late Jurassic time in the basin centre and in Early Cretaceous time in the shoulder

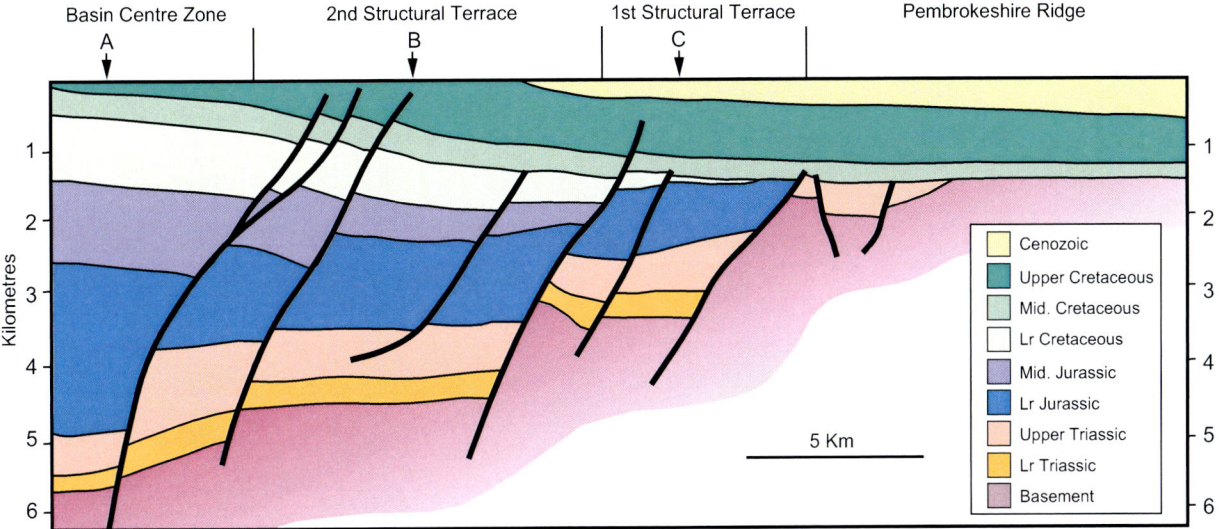

Figure 10.6. Geoseismic section showing the main structural zones on the north flank of the Pembrokeshire Ridge. Profile based on a NNW-SSE seismic line crossing block 50/16. After Shannon and MacTiernan (1993).

zone. Although there are differences in detail between the various published burial models (compare the profiles for the Gulf 48/19-1 in Howell and Griffiths, 1995 and Murdoch *et al.*, 1995), a number of conclusions can be drawn (Howell and Griffiths, 1995):

◆ The Top Lower Jurassic horizon is mature for oil generation in the central portion of the basin along its length. Lower Jurassic source rocks could have produced liquid hydrocarbons prior to maximum burial, but had entered the gas window in the central basin areas by maximum burial (Late Cretaceous).

◆ The Top Jurassic horizon had entered the oil window by the time of maximum burial and reservoirs could have been charged from Upper Jurasssic source rocks.

◆ Gas generation occurred in the Cenozoic burial episodes in the central parts of the basin.

Craven (1995) reported on source potential within the northwest lobe and depocentre of the North Celtic Sea Basin, referred to as the Mizen Basin. Seismic interpretation allied to studies of source rock potential and hydrocarbon shows in the Esso 56/12-1 well (2674 m TD in the Valanginian) within the sub-basin, suggest that Liassic source rocks attained optimum maturity in mid-Late Cretaceous times, and Upper Jurassic source rocks in the early Cenozoic.

Reservoir rocks and structures

Potential reservoir sequences are found at a number of stratigraphic levels throughout the Mesozoic sequence of the North Celtic Sea Basin. The potential of the

poorly known, and largely untested, Triassic section was identified by Shannon and MacTiernan (1993). Only a few wells have drilled through the Jurassic and none of these is located on the northern margin of the basin. Seismic interpretation suggests that a Triassic succession is present along this margin (Figure 10.7) and both the Mercia Mudstone and Sherwood Sandstone groups are widespread within the basin. Wells that have penetrated through the Triassic section to basement have found Upper Triassic mudrocks resting on Carboniferous, whilst in other instances a lower Sherwood Sandstone sequence was also present. In a seismic interpretation study of Triassic prospectivity along the southern basin margin in Quadrant 50, Shannon and MacTiernan (1993) recognised two structural zones – the Basin Shoulder and Basin Margin Zones – between the basin centre and the Pembrokeshire Ridge (Figure 10.6). The Wealden section pinches out in the Basin Margin Zone, with the Jurassic thinning by onlap and erosion across the zones. The underlying Triassic, divisible into two seismic units, also thins southwards. The lower seismic package, interpreted as a sand-prone Sherwood Sandstone sequence, pinches out at the basin edge and is overstepped by the upper unit of shale-prone Mercia Mudstone strata that extends onto the basement of the Pembrokeshire Ridge.

Reservoir sections have been identified and tested in the Jurassic of the basin. Upper Sinemurian sandstones occur in the east of the basin and were derived from the Leinster massif to the north (Kessler and Sachs, 1995: also *see* Chapter 7). The reservoir potential of the interval

NW   **SE**

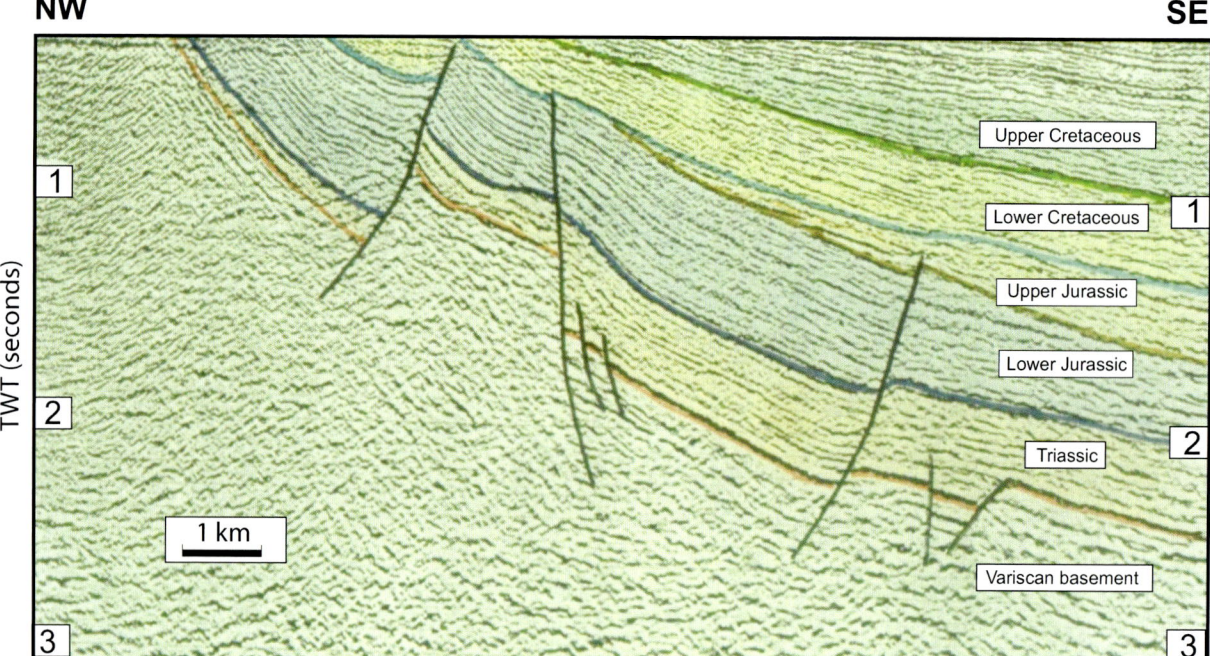

Figures 10.7. Seismic profile showing the rapid thickening of the Triassic across a pre-Jurassic fault on the northern margin of the North Celtic Sea Basin. The Cretaceous succession onlaps northwards onto the underlying Jurassic succession. Interpretation after Shannon (1991b).

was demonstrated by the Marathon Oil drilling in Block 50/3. Well 50/3-1 encountered a 63.1 m thick marine shelf sandstone succession, while 67.4 m was proved in well 53/3-3. Kessler and Sachs (1995) interpreted the sand sequence as a complex shelf sand ridge system in which sediment transfer was to the southwest. They considered, but rejected, the concept that the deposit is the product of a transgressive shoreface.

The discovery of the Helvick oil accumulation in Block 49/9 (Caston, 1995) focused attention on reservoir potential within the Middle and Upper Jurassic (*see* Figure 7.11). The discovery well, Gulf 49/9-2, drilled a small hanging wall high on an extensional fault, as did the immediately adjoining Gulf 49/9-4 well on a separate structure. Four intervals tested oil: three sandstones spanning the Callovian/Oxfordian boundary, and a Bathonian limestone. The latter is up to 48 m thick, with porosities in the range 11–16%, and comprises a pelletal/oolitic limestone with thin calcareous sandstone interbeds, immediately overlain by a shallow-marine sandstone. Effective porosity is limited to specific horizons, notably the sandstones, and is variable. The most important reservoir in the Helvick discovery – the 'Main Sand' – is near the base of the Wexford Formation (Figure 7.10) and comprises a fine- to very coarse-grained sandstone

with conglomerates and thin mudstones. The stacked fining-upwards sequences are interpreted as the products of a high-energy, braided river system. Petrographic studies (Caston, 1995) indicate derivation in the main from the Palaeozoic rocks of the Irish mainland, rather than the Leinster or Carnsore granites or the Rosslare metamorphic complex (*see* Chapter 1). Sandstones in the section overlying the 'Main Sand' (up to the lowest Oxfordian) were tested in well 49/9-2, but are generally thin and have poor reservoir characteristics. Although the 'Main Sand' has been traced between wells in Block 49/2 for a distance of 6.5 km, it is likely that the other sands have very limited lateral extent. Caston (1995) listed the gross sandstone thickness in the 'Main Sand' reservoirs as 15.85 m, with a net/gross ratio of 75% and an average effective porosity of 19.5%.

The main reservoir intervals within the Lower Cretaceous have been discussed in Chapter 8. They comprise the upper sandstones within the Wealden group and the Albian 'A Sand' productive in the Kinsale Head gas field. In discussing the Mizen (sub-)Basin, Craven (1995) cited sandstone reservoir levels at Triassic, Lower Jurassic, Upper Jurassic and Lower Cretaceous levels. Seals for the reservoirs were seen as being provided by Upper Triassic, Lower and Upper Jurassic and Lower

Cretaceous claystones. In the same area, McCann and Shannon (1993) mapped a number of WSW-trending faults with recurrent movement during Triassic to Early Cretaceous rift episodes. A series of half-grabens resulted, with seismic units thickening southeastwards into the controlling faults. Fault movement mostly pre-dated the Base Wealden reflector, leaving fault-generated depocentres that were passively infilled by Lower Cretaceous sediments. The footwall highs provided sediment into the hanging wall depocentres, producing complex potential reservoir targets (McCann and Shannon, 1993).

Fastnet Basin

Basin framework

The Fastnet Basin (Figure 10.1) is the westernmost of the Celtic Sea basins and is a contiguous narrow extension of the North Celtic Sea Basin, with which it shares similarities of stratigraphy and development. Lying to the north of the Labadie Bank–Pembrokeshire Ridge basement high, the basin has a NE–SW orientation (Figure 7.2A). Structure within the basin, and at its margins, is dominated by NE–SW oriented faults. Many of the faults mapped at Liassic level are also present on the Base Upper Cretaceous Chalk horizon, but the effect at the higher level is usually slight. Where significant throw is recorded in the Upper Cretaceous on a few faults, the sense of displacement is usually the reverse of that seen in the Liassic, probably due to an early Cenozoic inversion along faults that had an earlier extensional history. A later set of NW–SE faults transect the basin and are interpreted as frequently having sinistral strike-slip movement (Robinson et al., 1981). Two fault zones with this orientation have an important effect on the basin. The first is a centrally located composite zone that crosses blocks 55/30 and 64/1 (Figure 7.2A). There is a swing in basin trend across this zone from NNE–SSW in the north to NE–SW in the south. The central fault zone also has a close spatial association with dolerite sills of late Bajocian age (Robinson et al., 1981). The structural nature of the basin margins at Liassic level shows reversal across the zone, such that the Liassic limestone marker terminates against the basin margin fault to the northwest in the northern zone and to the southeast in the south. At the opposing basin margins, in each case, the Liassic limestone subcrops the Upper Cretaceous Chalk (Figure 7.2A). The second of the important fault zones crossing the basin is a major WNW–ESE structure near its southwestern limit. To the south of this zone of dislocation the basin at Liassic limestone level is reduced to a

shallow northeast-plunging syncline that is overstepped by the Upper Cretaceous Chalk. Smith (1995) views this zone as a northwesterly extension of the Cockburn Fault from the Cockburn Basin (see below and Figure 10.1). In contrast to the Fastnet Basin, where the fault had clear influence on the Mesozoic development of the basin, movement on the fault zone in the Cockburn Basin is entirely Cenozoic in age. This is ascribed to the possible presence of a reactivated deep NE–SW Variscan thrust beneath the Fastnet Basin, but not beneath the Cockburn Basin, which resulted in the segmentation of the Cockburn–Goban Fault, with different responses to stress between the segments (Smith, 1995).

Up to 4.5 km of post-Variscan strata occur within the Fastnet Basin. Devonian red-beds and Mississippian shelf limestones have been encountered by exploration wells in the basin (Figure 7.2B: Robinson et al., 1981). Permian strata appear to be absent and a widespread Triassic succession, approximately 600 m thick, consists of a lower sandy and an upper marl and mudstone evaporitic succession. There is no clear evidence for active syn-rift Triassic faults within the basin, although it is likely that the boundary faults were active at that time (Robinson et al., 1981; Figure 7.2B). The generally uniform thickness of the Triassic succession in the basin (Shannon, 1995) is consistent with development of an early post-orogenic wide-rift basin as suggested by Štolfová and Shannon (2009). A base Jurassic unconformity is seen on the west side of the basin, indicating that this area remained positive whilst the normally conformable Rhaetian marine transgression progressed elsewhere in the basin and throughout much of the region, with the development of a transgressive marine limestone and mudstone Liassic succession, including a locally thick Sinemurian regressive deltaic sandy succession. The basin underwent widespread uplift and erosion during the end Jurassic Cimmerian episode. Close to 2 km of Liassic strata have been drilled in the basin (Robinson et al., 1981). While some Middle Jurassic mudstones and Middle Jurassic igneous rocks have been encountered in drilling (Caston et al., 1981), Upper Jurassic strata are largely absent, with inverted Liassic strata typically overlain unconformably by Lower Cretaceous sediments (Figure 10.8). Erosion cut down to the lower Bajocian in the basin centre and to the Toarcian on the basin shoulders. The Cretaceous, up to 1 km thick in wells, comprises a sandy and muddy fluvial Wealden succession, a sandy and glauconitic marine Aptian–Albian Greensand and an upper Chalk Group succession. Movement on faults continued in the Early Cretaceous, resulting in thinner

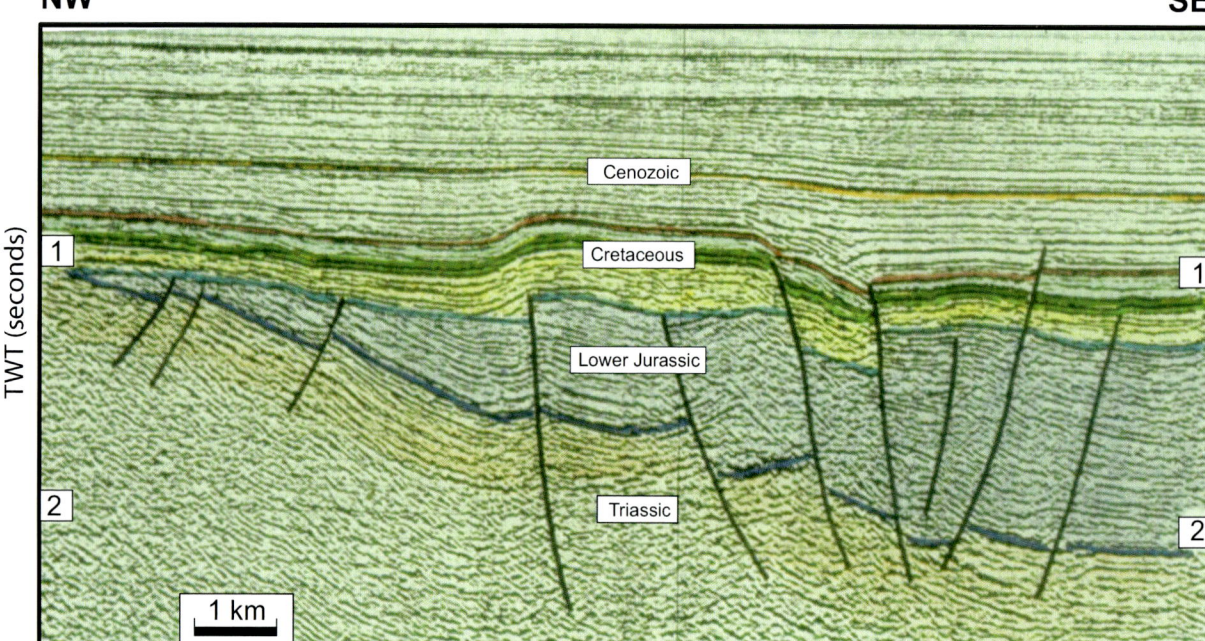

Figure 10.8. Seismic profile from the Fastnet Basin showing pre-Cretaceous fault reactivation superimposed by younger (probably earliest Cenozoic) inversion. Interpretation after Shannon (1991b,c).

sequences on uplifted blocks, and some movement continued into the Late Cretaceous. General inversion of the Fastnet Basin occurred in the post-Maastrichtian to pre-mid-Eocene interval, with non-deposition or erosion of the lowermost Cenozoic and the upper part of the Chalk. In the region of block 63/8 extreme uplift resulted in the removal of the entire Chalk section. Cenozoic sediments are typically up to 1 km thick and comprise a mid-Eocene to Oligocene limestone/chalk succession overlain by Miocene and Pliocene marine claystones, with a veneer of Pleistocene sediments.

Exploration potential

The Liassic section is regarded as containing intervals with good to very good source potential. This has previously been discussed in the context of a wider regional source potential of the interval in the discussion of the North Celtic Sea Basin (above). However, as pointed out by Robinson *et al.* (1981), the source rocks are thermally mature in only limited areas of the Fastnet Basin. Reservoir rocks, on the other hand, are present at least at four stratigraphic levels:

◆ Greensand/Wealden, with net sandstone thicknesses in the range 50–140 m penetrated in wells. Porosities are generally in the range 25–30% and occasionally as high as 40%.

◆ In excess of 20 m of Middle Jurassic sand penetrated in one well in the southwest of the basin, with porosities in the range 10–15%.

◆ Good quality thick (11–70 m) Sinemurian sandstones occur over the central and southern parts of the basin, with porosities in the range 19–38%. The most significant hydrocarbon indications were encountered at this level in the Cities Service 63/10-1 well which flowed small quantities of oil on test, and in the Elf 64/2-1 well which encountered oil shows.

◆ Sandstones (sometimes in excess of 50 m) in the Lower Triassic, with porosities generally up to 13%. In places, some faults juxtapose Liassic source rocks against potential Triassic reservoir sandstones and afford migration pathways.

The lack of an extensive source kitchen downgrades the exploration potential of the Fastnet Basin. Nevertheless, Robinson *et al.* (1981) argued that the deeper southeastern part of the basin, where the Liassic source rocks are likely to be mature, and fault-related structures may be available, still has unexplored potential. Wells in the basin were drilled in the 1970s, and based on variable quality seismic data. In the light of improved seismic technology the area is probably due further examination.

South Celtic Sea Basin

Basin framework

The NE–SW trending South Celtic Sea Basin lies to the south of the Labadie Bank–Pembrokeshire Ridge positive element. Deep seismic profiling (Coward and Trudgill, 1989) shows SE-dipping deep reflectors comparable to those beneath the northern basin, and thought to represent the main detachment structure beneath the basin. North-dipping faults are antithetic to the controlling structure. The basin is smaller and narrower than the North Celtic Sea Basin. The axial trend changes gradually eastwards and the basin merges with the east–west trending Bristol Channel Basin (Figure 10.9). The boundary between the two basins is taken at the NW-trending Cambeak or West Lundy Fault, which is parallel to the Sticklepath fault system. Both basins are bounded to the south by the Cornubian Platform. Only the western part of the South Celtic Sea Basin is in Irish waters. The basin fill consists of up to 3000 m of (Permo)-Triassic beds resting with marked unconformity on basement, conformably overlain by a Jurassic–Cretaceous sequence that locally exceeds 1500 m in thickness. In the Bristol Channel Basin Upper Triassic rocks rest unconformably on deformed Devonian and Carboniferous strata, Permian and Lower Triassic rocks being absent (Kamerling, 1979). However, the possible existence of small pockets of Permian strata in the South Celtic Sea Basin was suggested by Petrie et al. (1989) and UK well 93/6-1 on the south basin flank in UK waters penetrated

70 m of undated beds that may be Permian in age, resting unconformably on the Palaeozoic (Tappin et al., 1994). The Sherwood Sandstone Group is relatively thin and sand poor. Only two wells within Irish waters, Marathon 49/29-1 and Marathon 58/3-1, have penetrated the Triassic (Shannon, 1995). The Sherwood Sandstone group was not reached in 49/29-1 and was only poorly developed in well 58/3-1 (Figure 10.9). The latter well, however, penetrated the pre-Mesozoic, encountering Carboniferous fine-grained clastics similar to those in the North Celtic Sea Basin (Higgs, 1983).

Thick salts, interbedded with red mudstones, are developed in the Upper Triassic of the South Celtic Sea Basin, in contrast to its northern counterpart where salt is developed only in a narrow band along the southern margin. Periods of distal clastic input into a hypersaline shallow perennial lake, alternating with periods of evaporation, may explain the observed facies patterns (Shannon, 1995). The Penarth Group is thin and characterised by grey marls with minor interbedded limestones. Salt pillows are developed (Figure 10.10), and some salt movement may have taken place in Early Jurassic time, resulting in thicker sequences on the flanks of the features (Shannon, 1991b). A later stage of salt movement occurred in the Cenozoic, resulting in broad domal structures. Lower Jurassic sequences similar in thickness to those of the St George's Channel Basin are interpreted on the seismic sections, whereas the Middle Jurassic is generally thinner. The sequence is dominated by mudrocks

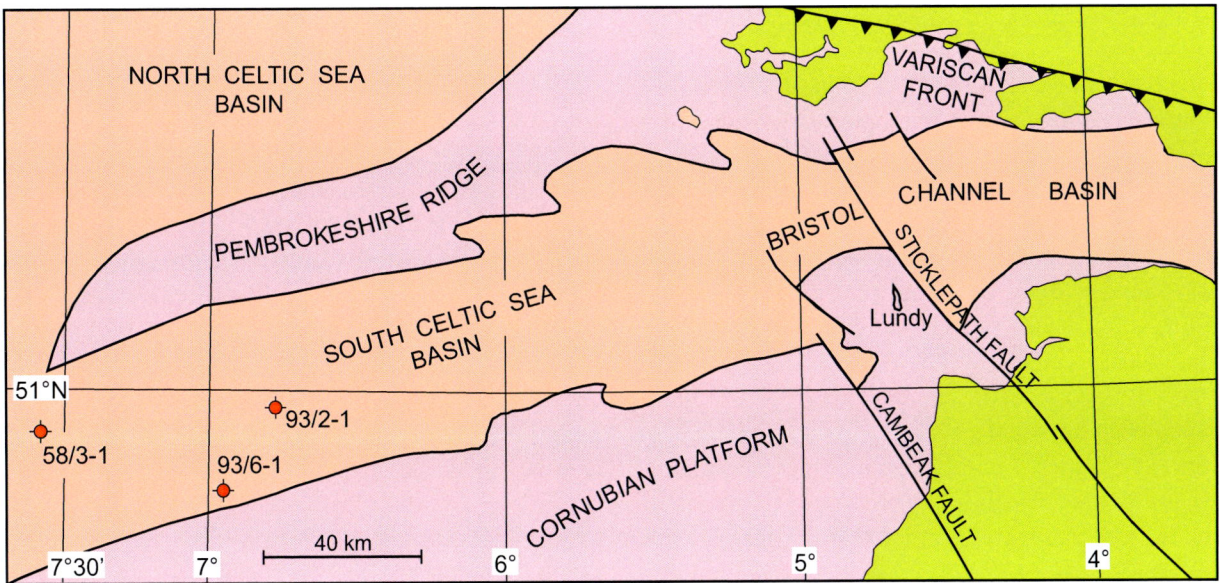

Figure 10.9. Generalised map of the South Celtic Sea and Bristol Channel area, showing the outline of Triassic-Jurassic basins with intervening pre-Mesozoic highs, and wells mentioned in the text (modified after Van Hoorn, 1989).

NW   **SE**

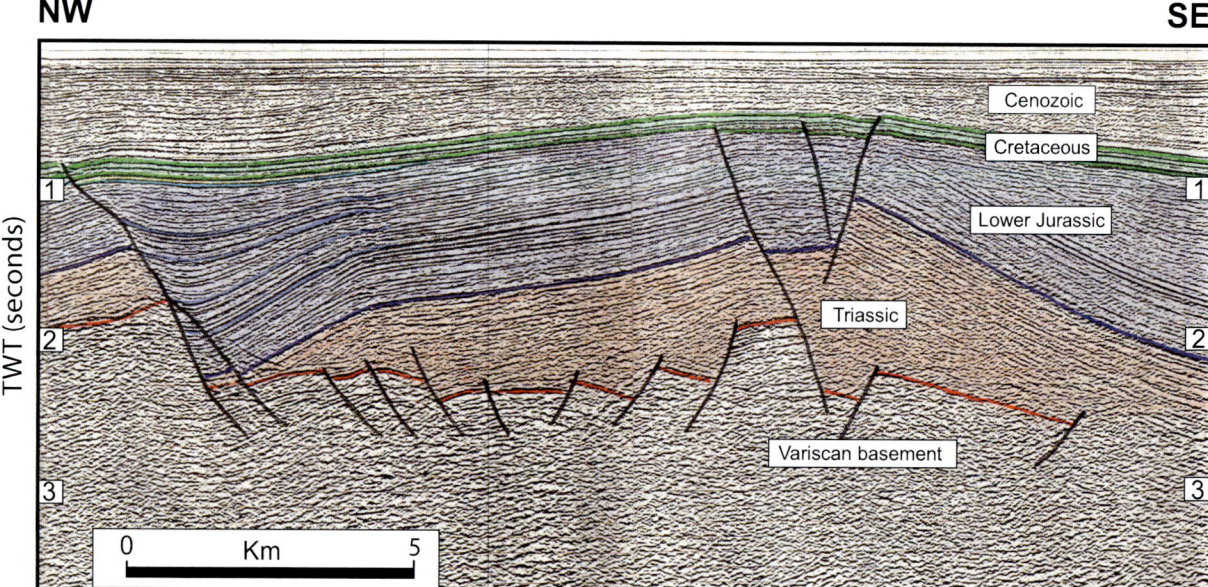

Figure 10.10. Seismic profile in the South Celtic Sea Basin showing a Triassic salt pillow structure. The salt movement took place in mid-Lower Jurassic time coincident with growth fault movement. Interpretation after Shannon (1991b,c).

with interbedded limestone units. Mid-Cimmerian uplift affected the region and Upper Jurassic strata are preserved only within the Bristol Channel Basin in the east. The Late Cimmerian phase of uplift was also important, and resulted in erosion that was particularly severe at the basin margins. As much as 3000 m of strata may have been removed from the southern basin margin (Van Hoorn, 1987). Salt movement may also have taken place during this phase. The Lower Cretaceous non-marine Wealden deposits rest unconformably on pre-Cretaceous strata of varying age – the youngest being the Bathonian encountered in UK well 93/2-1 (Figure 10.9; Tappin et al., 1994). The Gault–Greensand section comprises shallow-marine glauconitic mudstones, siltstones and sandstones and is conformably overlain by the Chalk (Cenomanian to Maastrichtian). Cenozoic inversion of the basin, as in the North Celtic Sea Basin, produced a gentle upwarp of the depocentre (*see* Shannon, 1991b: Figure 8), with associated reversal of the faults on the southern basin margin.

Exploration potential

The South Celtic Sea Basin is generally regarded as being less prospective than its larger counterpart to the north, and the exploration wells drilled to date have been dry. The Lower Cretaceous sequence, productive in the northern basin gasfields, is here thinner and less prospective. The Aptian–Albian sands are only thinly developed (a

maximum penetrated thickness of 18 m in UK well 93/2-1). Lower Triassic sandstones probably afford the best reservoir targets, sealed by Mercia Mudstone Group halites.

Source rocks are a general problem in the basin. The Lower Jurassic source horizons developed in the North Celtic Sea Basin are here generally of poor quality (Tappin *et al.*, 1994). The Upper Jurassic is missing and possible younger source rocks are immature. The nature of the Carboniferous sequence is largely unknown, but intersections in wells within the North Celtic Sea Basin, and in the Marathon 58/3-1 well in the northern part of the South Celtic Sea Basin, have shown poor source potential and evidence of low grade metamorphism. Even if source rocks are present it is likely that they attained maturity at the end of the Carboniferous.

Potential structures are likely to be fault- and salt-related. Salt domes and diapiric structures may provide structuring in the Jurassic and lowermost Cretaceous succession. Salt movement has occurred at different times – probably during the Early Jurassic, and accompanying the Late Cimmerian and Cenozoic inversion phases. This, and associated fault movement, will inevitably have modified existing structures.

Cockburn and Little Sole basins
Basin framework

The small Cockburn Basin lies between the Fastnet Basin to the northwest and the Western Approaches Basin in

UK waters to the southeast (Figure 10.1). The basin has a NE to NNE alignment, in contrast to the main Celtic Sea basins, with the ENE–WSW Caledonian structural influence less evident. The basin margins are not obviously fault-controlled, and typically show older Mesozoic strata truncated beneath the Aptian unconformity. Seismic mapping (Smith, 1995) shows the strong structural influence of a series of NW–SE faults. In particular, the Cockburn Fault that crosses the south central part of the basin has a mapped dextral displacement of four to five kilometres, with the movement dated as late Palaeogene. Although the basin is undrilled, comparison of seismic profiles with those of the adjacent basins suggests a broadly similar but thinner succession to those encountered in the main Celtic Sea basins (Smith, 1995). Results from the nearest wells, in the Goban Spur, Fastnet and Haig Fras basins, indicate that basement beneath the Cockburn Basin probably comprises Upper Palaeozoic sediments, although the Amerada Hess 73/2-1 well near the northern margin of the Western Approaches Basin to the southeast encountered amphibolite gneiss of possible Cadomian affinity (Evans, 1990). A widespread interpreted Triassic succession overlies Variscan basement and, in a small half-graben close to the northwest basin margin, a wedge of Permian sediments is postulated with angular contact beneath the base Triassic unconformity. This can be interpreted as the remains of possible Permian intermontane basin, resulting from orogenic collapse following Variscan mountain-building, although as noted elsewhere, no proven Permian strata have been encountered in wells within the Celtic Sea region. The Triassic sequence is interpreted on seismic character to consist of a lower sand packet overlain by claystones and siltstones (Smith, 1995). The Jurassic succession in the Cockburn Basin is relatively thin and, by comparison with the Fastnet Basin, probably consists of Liassic limestones, marls and organic-rich mudstones. The Cretaceous, resting unconformably on the Liassic, is also relatively thin, with presumed Lower Cretaceous strata restricted to the axis of the basin. An angular unconformity is also seen beneath the Aptian, and the overlying Chalk sequence probably ranges up to Campanian in age. The Cenozoic is comparable in thickness, and probably in lithology, to that of the Fastnet Basin.

The relatively unknown Little Sole Basin, situated at the shelf edge southwest of the Cockburn Basin, has a NNW orientation, comparable to structural elements within the Goban Spur province immediately to the west. The basin is essentially a down-to-the WSW half graben developed against a NNW-trending eastern bounding fault. A thickness of 1800 ms TWT of strata overlying the Base Penarth Group reflector is developed in limited pockets against the boundary fault (Smith, 1995). A Lower Cretaceous syn-rift succession, comparable to that in the Cockburn Basin, rests unconformably on a thin Lower Jurassic and basal Triassic sequence. Basin development was probably synchronous with that of the Goban margin basins where a major North Atlantic rifting episode began in late Hauterivian or early Barremian time and was terminated by seafloor spreading initiation in the early Albian (Chapter 8).

Exploration potential

The post-Variscan basin fill of the Cockburn Basin exceeds 2 sec. TWT only in the axial portion of the basin. The shallowness of the basin means that the Liassic source rocks (Base Penarth Group seismic horizon at a maximum of 1750 ms TWT in the basin centre: Fig. 8 of Smith, 1995) are probably immature, except at the basin centre (Shannon, 1991b). The restricted area of the potential source kitchen adversely affects the prospectivity of the basin, and this also applies to the adjacent Little Sole Basin. This perception explains the lack of exploration wells, despite the possibility of fault-controlled traps and the probable presence of reservoir rocks at Lower Triassic and Lower Cretaceous levels.

Summary of regional traps and play types

The North Celtic Sea Basin, the largest of the basins in the Celtic Sea region, is regarded as the most prospective. It has the thickest and most complete geological succession, and is also the only one of the basins in the region to contain producing fields. While the Fastnet and Cockburn basins are proven to contain reservoirs at Triassic, Lower Jurassic and Lower Cretaceous levels, together with Triassic and Jurassic tilted fault-block structures, a major problem appears to be the lack of a regionally-mature source rock (Robinson et al., 1981) The Upper Jurassic succession is largely absent from the basins, and the Lower Jurassic shales are only locally within the oil window. The South Celtic Sea Basin is likewise viewed as being less prospective than the North Celtic Sea and St George's Channel Basins. Once again, the Upper Jurassic is not widespread in the basin, while the Lower Cretaceous succession is thinner than in the North Celtic Sea Basin. All the reservoirs in the Celtic Sea region are generally overlain by adequate thicknesses of shales to provide efficient seals. Source rocks in the region occur in the Lower Jurassic (oil and gas), Middle and Upper Jurassic (oil), and locally in lowermost Lower Cretaceous shales (oil). The Lower Jurassic, with mixed oil and gas potential, is likely to be within the gas window in the central parts of the North Celtic Sea and St George's Channel basins, but is probably only marginally mature in the Cockburn Basin and much of the Fastnet and South Celtic Sea basins. The Lower Jurassic entered the oil window during the Late Jurassic in the central part of the North Celtic Sea Basin, and in the Early Cretaceous on the basin margins. The Upper Jurassic entered the oil window during Late Cretaceous time in the basin centre and Early Cenozoic time along the margins. The oils encountered in various accumulations in the North Celtic Sea indicate derivation primarily from Lower Jurassic and/or latest Jurassic–earliest Cretaceous sources. From the descriptions given above, it is clear that a wide variety of regional traps and play types exist in the Celtic Sea region (Figure 10.11). A brief summary of these is as follows:

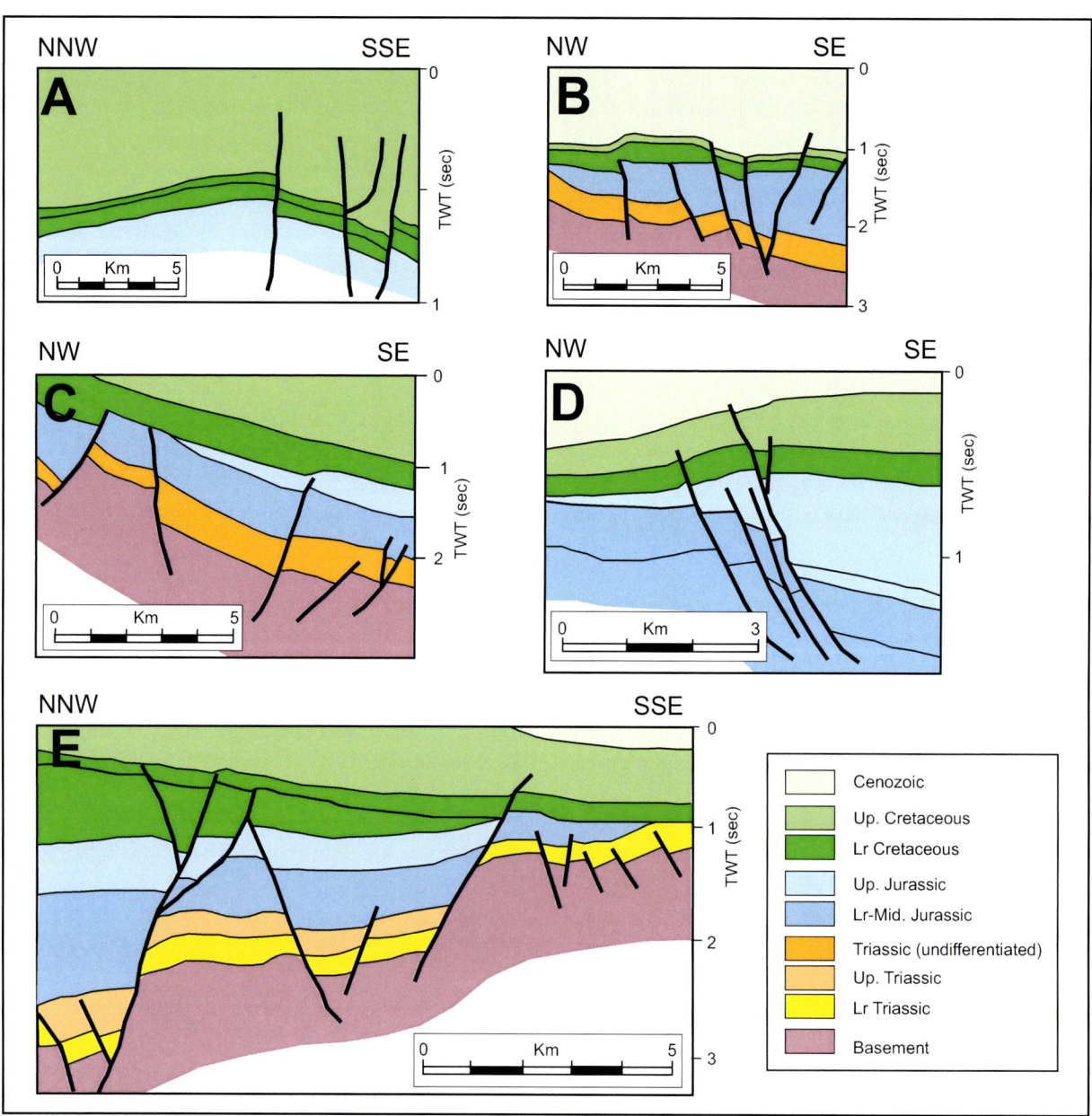

Figure 10.11. Schematic illustration of play and trap types in the Celtic Sea basins (after Shannon and Naylor, 1998).

Early Cenozoic inversion traps

These structures (e.g. Figures 10.11A, B and D) are mainly located towards the centre of the North Celtic Sea and Fastnet basins. Most of the obvious large structures have already been drilled, with notable success at Kinsale Head, Ballycotton and Seven Heads. Apparently similar structures drilled during the early phases of exploration were found to be either water-wet or contained heavy and biodegraded oil. This is probably due to the fact that the timing of structuring and petroleum migration was closely related. The major phase of oil and gas generation and migration in the region took place in Late Cretaceous to Early Cenozoic times, while the main phase of inversion structuring is of slightly later Palaeogene age. Inversion led to a major phase of freshwater flushing which resulted in the biodegradation of trapped oil (Howell and Griffiths, 1995). The successful petroliferous structures may be the fortuitous consequence of the breaching of older structures and the remigration of hydrocarbons into late inversion structures. Furthermore, many of the structures are shallow, and crestal faults frequently extend close to the sea floor. The main reservoirs for the play are Lower Cretaceous (Greensand and Wealden) sandstones. Oil source rocks are likely to be in the Lower and Upper Jurassic. The gas source is more uncertain, but is likely to be primarily provided by the Lower Jurassic shales in the deep, central part of the basin. A few such structures remain to be drilled. In addition, with the dramatic improvement in seismic data quality within the past decade (Figure 10.12), more subtle inversion structures can now be identified more accurately, while important elements of stratigraphic trapping within these structures, especially at Lower Cretaceous level, which were hitherto unseen, can now be mapped.

Figure 10.12. Improvement of seismic data quality in the North Celtic Sea Basin over time. Three vintages of seismic data (A:1977; B:1986 and C:2008), along an approximately coincident line through the Esso 44/22-1A well indicate how little structure was imaged on the old data and how much detail is now seen on the latest seismic data. Profiles A and B are courtesy of the Petroleum Affairs Division while C is published with the permission of Landsdowne Oil and Gas.

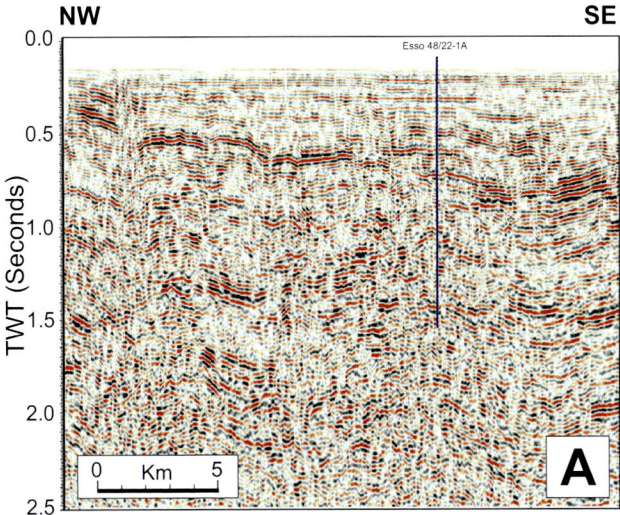

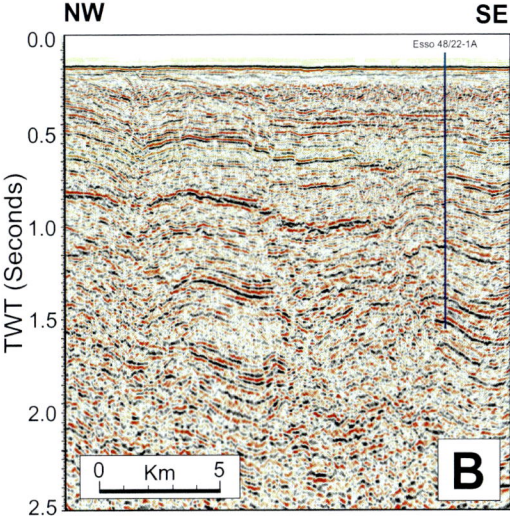

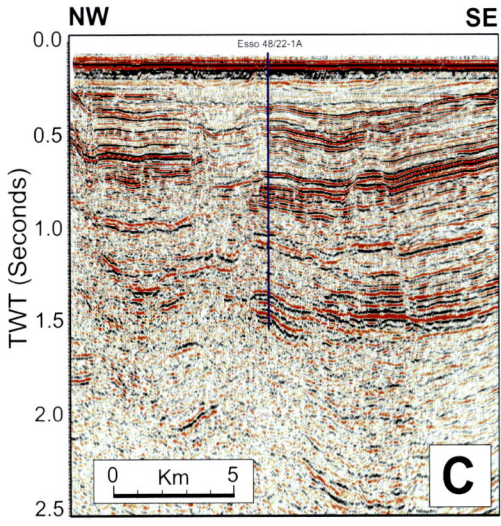

Tilted Jurassic fault-block traps

These (e.g. Figure 10.11B, C and D) are best seen in the North Celtic Sea and Fastnet basins. Drilling of such structures in the Fastnet Basin was largely unsuccessful due to a combination of insufficiently mature source rock and inadequate migration pathways from source kitchen to trapping structures. The Helvick oil accumulation on the northern margin of the North Celtic Sea Basin is trapped in a Jurassic syn-rift tilted fault-block structure (Caston, 1995). To date, all of the accumulations of this trap type are small. This is probably due to a combination of the structural complexity of the basin margin and reactivation and breaching of such syn-rift structures during Early Cenozoic basin inversion. The main Jurassic reservoir fairway for the play lies along the northern margin of the North Celtic Sea Basin, with the principal sediment source area probably lying to the north, onshore in Ireland and on the shelf immediately south of the Irish coast. The Upper Jurassic provides the major source rock for the play. The syn-rift Jurassic sandstones appear to decrease in thickness towards the basin centre and this, together with the burial and resulting porosity occlusion, makes the centre of the basin less prospective for the play. The largest and most obvious of these structures have been drilled, but some remain undrilled adjacent to the basin margins.

Tilted Triassic fault-block traps

Most of the wells drilled to date in the Celtic Sea region terminated within the Jurassic, and the regional reservoir architecture of the Triassic succession is still fairly poorly understood. However, tilted fault-block structures have been identified at Triassic levels (e.g. Figure 10.11E), and some recent exploration has begun to focus upon such targets in areas away from the inversion axis of the North Celtic Sea Basin. Those in the basin centre are likely to be too deep to image clearly, and also likely to have been affected by later Jurassic and Cenozoic structuring. Structures were developed during Triassic syn-rift extension, with some later reactivation during Late Jurassic extension. The most likely reservoir fairways appear to lie along the southern shoulder zone of the North Celtic Sea Basin, with sediment shed northwards from the Pembrokeshire Ridge; and in the eastern and NE parts of the basin, where the basin is relatively narrow and sediment may have been derived from the Caledonian granites and metasediments in SE Ireland and from the Pembrokeshire Ridge. Source rocks which occur at these levels are likely to be the Upper Jurassic (oil) and Lower Jurassic (gas) shales in the main depocentre. A number of such structures remain to be drilled in several of the basins.

Halokinetic traps

Thick Triassic salt deposits occur in the St George's Channel and South Celtic Sea Basin, and are locally developed at the southern margin of the North Celtic Sea Basin (e.g. Figure 10.10). Salt dome and diapiric structures in these areas may provide structuring in the Jurassic and lowermost Cretaceous succession in which reservoir potential exists. Source and cap rock mudstones for the plays also occur at Jurassic and Lower Cretaceous levels.

Stratigraphic traps

The Lower Cretaceous succession in the region contains a large range of depositional facies from fluvial to shallow-marine, and offers potential stratigraphic traps at various levels. These include sandstone drapes across Jurassic structures, lateral facies variations and pinch-outs within fluvial and shallow-marine successions. Source rocks for the plays may lie either within the Lower Cretaceous or in the underlying Upper Jurassic. In common with all the Irish basins, stratigraphic traps have rarely been tested in the Celtic Sea, largely due to the historically poor seismic data quality. With improvement in acquisition and processing technology, such traps are likely to be increasingly important in the region.

References

Barr, K.W., Colter, V.S., and Young, R. (1981) The geology of the Cardigan Bay–St. George's Channel Basin. *In*: Illing, L.V. and Hobson, G.D. (eds), *Petroleum Geology of the Continental Shelf of North-West Europe*. Heyden and Son Ltd., London, 432–443.

BIRPS and ECORS (1986) Deep seismic reflection profiling between England, France and Ireland. *Journal of the Geological Society, London*, **143**, 45–52.

Brück, P.M., Colthurst, J.R.J., Feely, M., Gardiner, P.R.R., Penney, S.R., Reeves, T.J., Shannon, P.M., Smith, D.G., and Vanguestaine, M. (1979) South-east Ireland: Lower Palaeozoic stratigraphy and depositional history. *In*: Harris, A.L., Holland, C.H., and Leake, B.E. (eds), *The Caledonides of the British Isles: reviewed*. Geological Society, London, Special Publications, **8**, 533–544.

Caston, V.N.D. (1995) The Helvick oil accumulation, Block 49/9, North Celtic Sea Basin. *In*: Croker, P.F. and Shannon, P.M. (eds), *The Petroleum Geology of Ireland's Offshore Basins*. Geological Society, London, Special Publications, **93**, 209–225.

Caston, V.N.D., Dearnley, R., Harrison, R.K., Rundle, C.C., and Styles, M.T. (1981) Olivine-dolerite intrusions in the Fastnet Basin. *Journal of the Geological Society, London*, **138**, 31–46.

Clayton, G., Sevastopulo, G.D., and Sleeman, A.G. (1986) Carboniferous (Dinantian and Silesian) and Permo-Triassic rocks in South County Wexford, Ireland. *Geological Journal*, **21**, 366–374.

Colley, M.G., McWilliams, A.S.F., and Myers, R.C. (1981) Geology of the Kinsale Head gas field, Celtic Sea, Ireland. *In*: Illing, L.V. and Hobson, G.D., (eds), *Petroleum Geology of the Continental Shelf of North-West Europe*. Heyden and Son Ltd., London, 504–510.

Corcoran, D.V. and Clayton, G. (1999) Interpretation of vitrinite reflectance profiles in the Central Irish Sea area: implications for the timing of organic maturation. *Journal of Petroleum Geology*, **22**, 261–286.

Corcoran, D.V. and Clayton, G. (2001) Interpretation of vitrinite reflectance profiles in sedimentary basins, onshore and offshore Ireland. *In*: Shannon, P.M., Haughton, P.D.W., and Corcoran, D.V. (eds), *The Petroleum Exploration of Ireland's Offshore Basins*. Geological Society, London, Special Publications, **188**, 61–90.

Coward, M.P. and Trudghill, B. (1989) Basin development and basin structure of the Celtic Sea basins (SW Britain). *Bulletin de la Societe Géologique de France*, **3**, 423–436.

Craven, J.E. (1995) The tectonic evolution, stratigraphy and petroleum potential of the Mizen Basin, southwest Celtic Sea. *In*: Croker, P.F. and Shannon, P.M. (eds), *The Petroleum Geology of Ireland's Offshore Basins*. Geological Society, London, Special Publications, **93**, 277.

Croker, P.F. and Shannon, P.M. (1995) The petroleum geology of Ireland's offshore basins: introduction. *In*: Croker, P.F. and Shannon, P.M. (eds), *The Petroleum Geology of Ireland's Offshore Basins*. Geological Society, London, Special Publications, **93**, 1–8.

Dobson, M.R. and Whittington, R.J. (1987) The geology of Cardigan Bay. *Proceedings of the Geologists' Association*, **98**, 331–353.

Dunford, G.M., Dancer, P.N., and Long, K.D. (2001) Hydrocarbon potential of the Kish Bank Basin: integration within a regional model for the Greater Irish Sea Basin. *In*: Shannon, P.M., Haughton, P.D.W., and Corcoran, D.V. (eds), *The Petroleum Exploration of Ireland's Offshore Basins*. Geological Society, London, Special Publications, **188**, 135–154.

Evans, C.D.R. (1990) *United Kingdom offshore regional report: The geology of the Western English Channel and its Western Approaches*. London HMSO for the British Geological Survey.

Ford, M., Klemperer, S.L., and Ryan, P.D. (1992) Deep structure of southern Ireland: a new geological synthesis using BIRPS deep reflection profiling. *Journal of the Geological Society of London*, **149**, 915–922.

Green, P.F., Duddy, I.R., Bray, R.J., Duncan, W.I., and Corcoran, D.V. (2001) The influence of thermal history on hydrocarbon prospectivity in the Central Irish Sea Basin. *In*: Shannon, P.M., Haughton, P.D.W., and Corcoran, D.V. (eds), *The Petroleum Exploration of Ireland's Offshore Basins*. Geological Society, London, Special Publications, **188**, 171–188.

Higgs, K. (1983) Palynological evidence for the Carboniferous strata in two wells drilled in the Celtic Sea area. *Bulletin of the Geological Survey of Ireland*, **3**, 107–112.

Holford, S.P., Turner, J.P. and Green, P.F. (2005) Reconstructing the Mesozoic–Cenozoic exhumation history of the Irish Sea basin system using apatite fission track analysis and vitrinite reflectance data. *In*: Doré, A.G. and Vining, B.A. (eds), *Petroleum Geology: North-West Europe and Global Perspectives. Proceedings of the 6th Petroleum Geology Conference*. Geological Society, London, 1095–1107.

Howell, T.J. and Griffiths, P. (1995) A study of the hydrocarbon distribution and Lower Cretaceous Greensand prospectivity in Blocks 48/15, 48/17, 48/18 and 48/19, North Celtic Sea Basin. *In*: Croker, P.F. and Shannon, P.M. (eds), *The Petroleum Geology of Ireland's Offshore Basins*. Geological Society, London, Special Publications, **93**, 261–275.

Izzat, C., Maingarm, S. and Racey, A. (2001) Fault distribution and timing in the Central Irish Sea Basin. *In*: Shannon, P.M., Haughton, P.D.W., and Corcoran, D.V. (eds), *The Petroleum Exploration of Ireland's Offshore Basins*. Geological Society, London, Special Publications, **188**, 155–169.

Kamerling, P. (1979) The geology and hydrocarbon habitat of the Bristol Channel Basin. *Journal of Petroleum Geology*, **2**, 75–93.

Keeley, M.L. (1995) New evidence of Permo-Triassic rifting, onshore southern Ireland, and its implications for Variscan structural inheritance. *In*: Boldy, S.A.R. (ed.), *Permian and Triassic Rifting in Northwest Europe*. Geological Society, London, Special Publications, **91**, 239–253.

Kessler, L.G. and Sachs, S.D. (1995) Depositional setting and sequence stratigraphic implications of the Upper Sinemurian (Lower Jurassic) sandstone interval, North Celtic Sea/St George's Channel Basins, offshore Ireland. *In*: Croker, P.F. and Shannon, P.M. (eds), *The Petroleum Geology of Ireland's Offshore Basins*. Geological Society, London, Special Publications, **93**, 171–192.

Maddox, S.J., Blow, R., and Hardman, M. (1995) Hydrocarbon prospectivity of the Central Irish Sea Basin with reference to Block 42/12, offshore Ireland. *In*: Croker, P.F. and Shannon, P.M., (eds), *The Petroleum Geology of Ireland's Offshore Basins*. Geological Society, London, Special Publications, **93**, 59–77.

McCann, T. and Shannon, P.M. (1993) Lower Cretaceous seismic stratigraphy and fault movement in the Celtic Sea Basin, Ireland. *First Break*, **11**, 335–344.

McGeary, S., Cheadle, M.J., Warner, M.R., and Blundell, D.J. (1987) Crustal structure of the continental shelf around Britain derived from BIRPS deep seismic profiling. *In*: Brooks, J. and Glennie, K.W. (eds), *Petroleum Geology of North-West Europe*. Graham and Trotman, London, 33–41.

Murdoch, L.M., Musgrove, F.W., and Perry, J.S. (1995) Tertiary uplift and inversion history in the North Celtic Sea Basin and its influence on source rock maturity. *In*: Croker, P.F. and Shannon, P.M. (eds), *The Petroleum Geology of Ireland's Offshore Basins*. Geological Society, London, Special Publications **93**, 297–319.

Murphy, N.J., Sauer, M.J., and Armstrong, J.P. (1995) Toarcian source rock potential in the North Celtic Sea Basin, offshore

Ireland. *In*: Croker, P.F. and Shannon, P.M. (eds), *The Petroleum Geology of Ireland's Offshore Basins*. Geological Society, London, Special Publications, **93**, 193–207.

Murphy, F.X. (1990) The Irish Variscides: a fold belt developed within a major surge zone. *Journal of the Geological Society of London*, **147**, 451–460.

Naylor, D. and Shannon, P.M. (1982) *The Geology of Offshore Ireland and West Britain*, Graham and Trotman, London, 161pp.

Naylor, D. and Shannon, P.M. (1999) The Irish Sea region: why the general lack of exploration success? *Journal of Petroleum Geology*, **22**, 363–370.

PESGB (2007) Structural framework of the North Sea and Atlantic margin: 1:750,000 scale map, 2005 edition. Petroleum Exploration Society of Great Britain, London.

Petrie, S.H., Brown, J.R., Granger, P.J., and Lovell, J.P.B. (1989) Mesozoic history of the Celtic Sea Basins. *In*: Tankard, A.L. and Balkwill, H.R. (eds), *Extensional tectonics and stratigraphy of the North Atlantic Margins*. American Association of Petroleum Geologists Memoir, **46**, 433–444.

Readman, P.W., O'Reilly, B.M., Edwards, J.W.F. and Sankey, M.J. (1995) A gravity map of Ireland and surrounding waters. *In*: Croker, P.F. and Shannon, P.M. (eds), *The Petroleum Geology of Ireland's Offshore Basins*. Geological Society, London, Special Publications, **93**, 9–16.

Readman, P.W., O'Reilly, B.M., and Murphy, T. (1997) Gravity gradients and upper-crustal tectonic fabrics, Ireland. *Journal of the Geological Society of London*, **154**, 817–828.

Robinson, K.W., Shannon, P.M., and Young, D.G.G. (1981) The Fastnet Basin: an integrated analysis. *In*: Illing, L.G. and Hobson, G.D. (eds), *Petroleum Geology of the Continental Shelf of North-West Europe*. Heyden and Son Ltd, London, 444–454.

Shannon, P.M. (1991a) The development of Irish offshore sedimentary basins. *Journal of the Geological Society, London*, **148**, 181–189.

Shannon, P.M. (1991b) Tectonic framework and petroleum potential of the Celtic Sea, Ireland. *First Break*, **9**, 107–122.

Shannon, P.M. (1991c) Irish offshore basins: geological development and petroleum plays. *In*: Spencer, A.M. (ed.), *Generation, accumulation, and production of Europe's Hydrocarbons*. Special Publication of the European Association of Petroleum Geoscientists, **1**, 99–109.

Shannon, P.M. (1995) Permo-Triassic development of the Celtic Sea region, offshore Ireland. *In*: Boldy, S.A.R. (ed.), *Permian and Triassic Rifting in Northwest Europe*. Geological Society, London, Special Publications, **91**, 215–237.

Shannon, P.M. and MacTiernan, B. (1993) Triassic prospectivity in the Celtic Sea, Ireland: a case history. *First Break*, **11**, 47–57.

Shannon, P.M. and Naylor, D. (1998) An assessment of Irish offshore basins and petroleum plays. *Journal of Petroleum Geology*, **21**, 125–152.

Smith, C. (1995) Evolution of the Cockburn Basin: implications for the development of the Celtic Seea basins. *In*: Croker, P.F. and Shannon, P.M. (eds), *The Petroleum Geology of Ireland's Offshore Basins*. Geological Society, London, Special Publications, **93**, 279–295.

Štolfová, K. and Shannon, P.M. (2009) Permo-Triassic development from Ireland to Norway: basin architecture and regional controls. *Geological Journal*, **44**, 652–676.

Taber, D.R., Vickers, M.K., and Winn, R.D. Jr. (1995) The definition of the Albian 'A' Sand reservoir fairway and aspects of associated gas accumulations in the North Celtic Sea Basin. *In*: Croker, P.F. and Shannon, P.M. (eds), *The Petroleum Geology of Ireland's Offshore Basins*. Geological Society, London, Special Publications **93**, 227–244.

Tappin, D.R., Chadwick, R.A., Jackson, A.A., Wingfield, R.T.R., and Smith, N.J.P. (1994) *United Kingdom offshore regional report: The geology of Cardigan Bay and the Bristol Channel*. London HMSO for the British Geological Survey, 107pp.

Turner, J.P. (1997) Strike-slip fault reactivation in the Cardigan Bay basin. *Journal of the Geological Society, London*, **154**, 5–8.

Van Hoorn, B. (1987) The South Celtic Sea/Bristol Channel Basin: origin, deformation and inversion history. *Tectonophysics*, **137**, 309–334.

Welch, M.J. and Turner, J.P. (2000) Triassic-Jurassic development of the St. George's Channel basin, offshore Wales, UK. *Marine and Petroleum Geology*, **17**, 723–750.

Williams, G.A., Turner, J.P., and Holford, S.P. (2005) Inversion and exhumation of the St. George's Channel basin, offshore Wales, UK. *Journal of the Geological Society, London*, **162**, 97–110.

Woodland, A.W (ed.), (1971) *The Llanbedr (Mochras Farm) borehole*. Institute of Geological Sciences, Report **71/18**, 115pp.

Chapter 11

Atlantic margin basins

Regional Setting

A set of basins of various shapes, sizes and ages wrap around the western Irish Atlantic shelf (Figure 11.1). They lie in deep water, typically 300 m to more than 2 km. The larger basins are overlain by deep-water embayments, indicative of their sediment under-supplied nature in Neogene times. The smaller, narrower basins, typically located closer to the Irish mainland, have a less pronounced bathymetric expression reflecting a combination of less dramatic basin subsidence and a greater amount of Late Mesozoic and Cenozoic uplift, inversion and erosion.

The basins are typically elongate, and the orientations are somewhat variable along strike, reflecting the long and complex history of the region, both of the basement and of the sedimentary basin evolution. On a broad scale the basins can be considered to consist of a set of relatively small, elongate 'inboard' basins lying landward of the large, deep-water 'outboard' basins. Most of the basins are broadly NE–SW in orientation, following inherited and reactivated Caledonian structures and fabrics. These basins include the narrow Slyne and Erris basins, the large Rockall and Hatton basins, and the set of small basins aligned along both flanks of the Rockall Basin. A north–south orientation typifies the Porcupine Basin, parallel to that of the coeval large basins in the mid-Norway region (Møre and Vøring basins) as well as the Viking Graben in the northern North Sea. This orientation is orthogonal to the major extensional direction of Jurassic time when the basin inherited its N–S shape. An E–W basin orientation is seen in the Goban Spur region, the southernmost set of basins in the Irish Atlantic margin domain. This is parallel or sub-parallel to the Variscan structures that define the main structural orientation

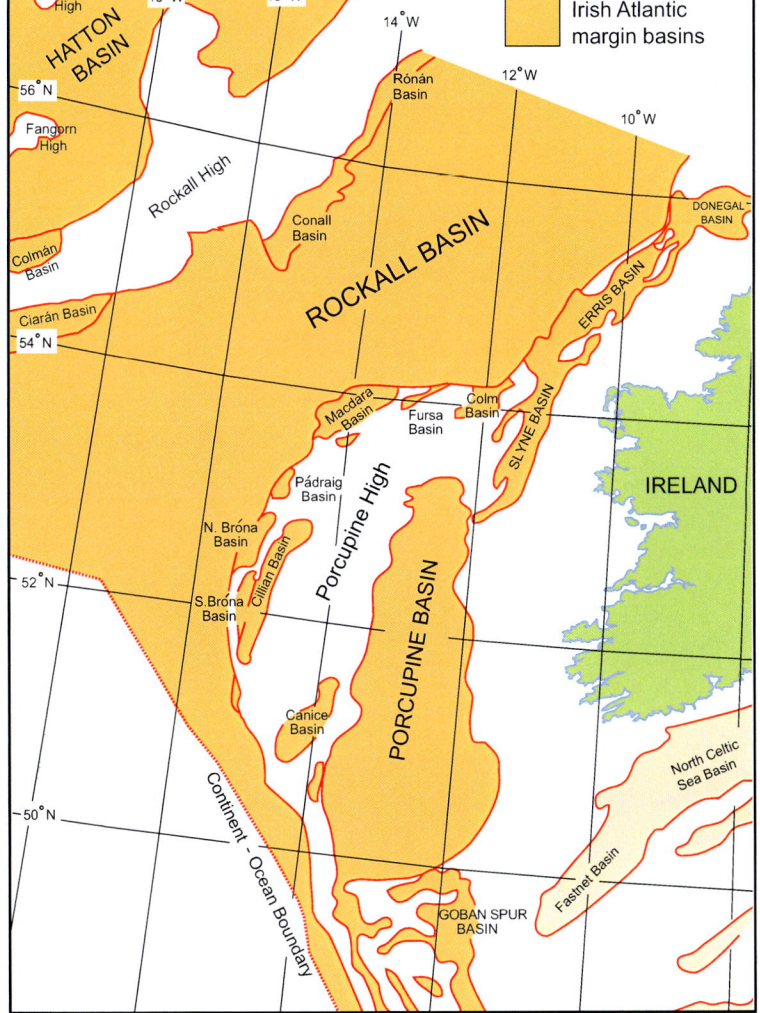

Figure 11.1. Location map showing the Atlantic margin basins. These comprise an inner set of narrow, elongate, sediment-filled basins (Slyne, Erris and Donegal) and an outer set of large sediment-undersupplied basins (Porcupine, Rockall and Hatton). Modified after Naylor *et al.*, 1999, 2002)

in the Celtic Sea region along strike to the east, as described in Chapter 10. An east–west orientation is also seen within the Rockall Basin where it impinges against the northern margin of the Porcupine High. The overall result is a major strike swing in the basin and a cross-cutting of the N–S and NE–SW orientations. The basin then reverts to a NE–SW orientation further north to parallel the narrow Slyne and Erris basins. This brings the eastern margin of the Rockall Basin closer to the Irish shore and results in very steep slopes immediately west of the Slyne and Erris basins.

The crustal structure appears to have played a significant, albeit complex and poorly understood, role in the orientation and development of the basins. The pronounced NE–SW orientation of the Slyne and Erris basins, of the Rockall Basin north of 54°N, and of the set of small 'perched' presumed early Mesozoic basins along both margins of the main Rockall Basin (Figure 11.2) is suggestive of an underlying inherited Caledonian structural grain. The east–west alignment of the eastern margin of the Rockall Basin at 54°N, the long-standing east–west Finian's Spur separating the North Porcupine Basin from the main Porcupine Basin to the south at 53°N, the east–west shape of the Carboniferous Clare Basin (Croker 1995; Naylor *et al.*, 1999) and the pronounced east–west basin and high structures in the Goban Spur region (Naylor *et al.*, 2002) are likely to reflect a different structural control. It is tempting to suggest a Variscan influence on this orientation. However, north of the Goban Spur, the Carboniferous strata encountered in drilling show little if any thermal or structural effects of Variscan deformation. In addition, the putative trend of the Iapetus Suture at the mouth of the Shannon estuary, and the western parts of some major Caledonian structures such as the Clew Bay Fault (e.g. Naylor *et al.*, 1999), show an almost east–west orientation. These large-scale east–west features may therefore be of pre-Variscan age, with several phases of reactivation to account for the east–west alignment of the Carboniferous Clare Basin, the several phases of Mesozoic reactivation along the Finian's Spur (*see* Naylor *et al.*, 2002) and the cross-cutting relationship of the Cenozoic Rockall Basin with the early Mesozoic 'perched' basins.

The Atlantic margin basins have extensive and frequently thick Mesozoic and Cenozoic successions. Up to 12 km of Late Palaeozoic and younger sediments are interpreted on seismic profiles in some of the largest basins and thick successions have been drilled by exploration wells. In general, the sedimentary succession is thicker and younger than that in the Celtic Sea and Irish Sea basins. A general younging and thickening of post-Palaeozoic sediments occurs from east to west across the Irish region. This reflects

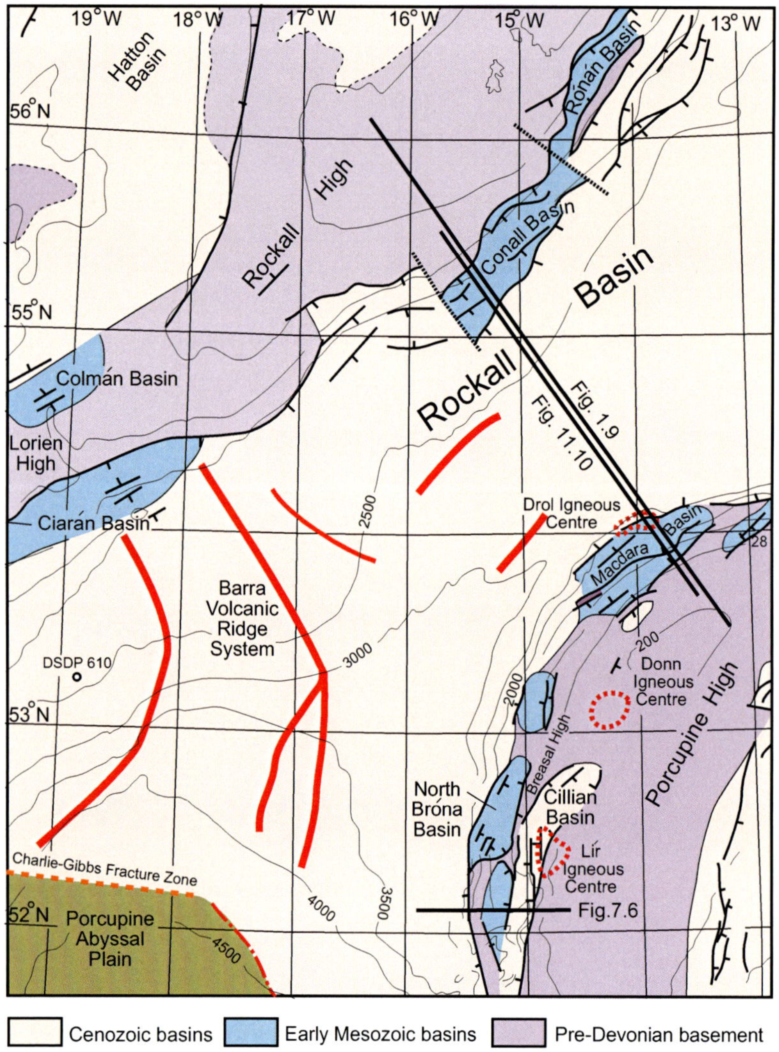

Figure 11.2. Basins and structural features in the Rockall region (after Naylor *et al.*, 1999 and Naylor and Shannon, 2009). A series of elongate 'perched' basins, of presumed Permo-Triassic to Early Jurassic age, lie along both the western and eastern margins of the large Rockall Basin.

the migration, through Mesozoic time, of crustal extension and rifting westward toward the site of crustal rupture and the onset of seafloor spreading in the North Atlantic to the west of the Hatton continental margin.

The Atlantic margin basins show a range of structural and stratigraphic architectures that suggests a composite development through a series of Mesozoic predominantly fault-related rift episodes, interspersed with phases of differential non-faulted thermal subsidence and sedimentation. The interplay between these events gave rise to temporal and spatial variations through the region. Relatively little is known about the Late Palaeozoic basin development style, but interpretation of the preserved strata suggests that sedimentation in the Porcupine region took place in an east–west elongate subsiding depocentre (the Clare Basin), as outlined by Croker and Shannon (1987). A major regional unconformity is recorded between the Pennsylvanian strata and the overlying Mesozoic, and the regional distribution of Mesozoic strata suggests an initial patchy infill of an irregular topography following uplift and erosion of variable amounts of Pennsylvanian strata (Robeson *et al.*, 1988). Permo-Triassic rift strata are patchily preserved in the region, typically in NE–SW oriented, fault-controlled basins. The thickest and best constrained successions are in the Slyne, Erris and North Porcupine basins (Chapman *et al.*, 1999; Dancer *et al.*, 2005), with seismic evidence indicating the preservation of Permo-Triassic strata in the 'perched' basins along the flanks of the Rockall Basin. Their geometry suggests that the original deposition of Permo-Triassic strata was more widespread and that these perched basins simply reflect preserved remains of an early Mesozoic succession deposited in a series of linked depocentres as part of a wide-rift basin system (Štolfová and Shannon, 2009). Syn-rift Permo-Triassic strata are also inferred in the Goban Spur Basin (Naylor *et al.*, 2002).

The clearest evidence of fault-related rifting and crustal extension is seen in the Jurassic succession in the region. The main rifting appears to be of Middle and especially Late Jurassic age. Tilted fault blocks, with variations in coeval facies from fluvial to basin floor fans, are spectacularly developed on the flanks of the Porcupine Basin (Croker and Shannon, 1987; MacDonald *et al.*, 1987; Naylor and Anstey, 1987; Sinclair *et al.*, 1994; Naylor *et al.*, 2002) where a Late Jurassic age is constrained by drilling. Rifting may have commenced slightly earlier (Middle Jurassic) in the Slyne and Erris basins (Cunningham and Shannon, 1997; Chapman *et*

al., 1999). Late Jurassic syn-rift faulting waned during earliest Cretaceous time and was followed by rapid marine incursion across the partly peneplaned Late Cimmerian surface (Naylor and Shannon, 2005). Jones *et al.* (2001) suggested that transient uplift of up to 700 m occurred at this time, followed by subsidence of up to 500 m, coeval with the onset of seafloor spreading along the Goban margin. A regressive episode of deltaic progradation in the northeastern part of the Porcupine Basin has been interpreted as a minor rift episode that interrupted the regional thermal subsidence following the main Jurassic rifting (Croker and Shannon, 1987; Sinclair *et al.*, 1994). The Cretaceous succession is several kilometres thick in the Porcupine Basin (*see* Naylor *et al.*, 1999, 2002) – much thicker than would be expected from a thermal subsidence following the magnitude of Jurassic rifting interpreted from seismic profiles. Naylor and Shannon (2005) interpreted it as the product of differential subsidence associated with mantle circulation, resulting in dynamic support of the basin margin highs and the thermal collapse of the basinal areas underlain by significantly thinned crust (*see* below). Such thick, deepwater marine strata and catenary-shaped regional stratal geometries are typical of basins developed above hyperextended crust in passive margin settings. Regional eustatic sea level rise, combined with thermally-driven subsidence of the basin depocentres, resulted in the widespread development of Upper Cretaceous chalk and marls throughout the Atlantic margin basins.

The Early Cenozoic in the region is marked by a sea level fall and marine regression, a change from carbonate to coarse-grained clastic deposition, widespread igneous activity and another interruption in thermal subsidence. A similar pattern is observed in the North Sea, suggesting a regional control. Stoker *et al.* (2005a) and Praeg *et al.* (2005) proposed a deep-seated mantle thermal convection cell control, with differential uplift of the thick crustal regions coeval with subsidence in areas of thinner crust beneath the major sedimentary basins. Similar controls are invoked to explain flank uplift on the margins of the Rockall Basin, thereby eroding late Mesozoic strata from the Slyne and Erris basins.

The interpreted absence of a thick late Mesozoic succession from the Hatton Basin region has been explained by the permanent uplift and buoyancy provided by an elongate underplated igneous body at the base of the crust adjacent to the continental–oceanic crustal boundary (Vogt *et al.*, 1998; Shannon *et al.*, 1995, 1999). This represents the terminal phase of rifting and continental crustal extension that migrated westwards through a series of failed rifts until final crustal separation occurred west of the Hatton region. Crustal thinning beneath the nearshore basins (e.g. Slyne and Erris) appears to be slight, while the crust is severely attenuated beneath the large Porcupine and Rockall basins. Wide-angle seismic data from the latter basins (Hauser *et al.*, 1995; Morewood *et al.*, 2005; O'Reilly *et al.*, 2006) show that the continental crust is less than 5 km, and in places less than 2 km thick beneath the centres of the basin. It has been suggested that this was achieved, without the creation of oceanic crust and the onset of sea-floor separation, by differential stretching facilitated by a lower crustal detachment (Hauser *et al.*, 1995). Upper and middle crustal stretching was more severe than lower crustal and mantle lithospheric stretching. The less attenuated lower crust coupled to less stretched mantle lithosphere resulted in an increase in the overall lithospheric strength, thereby inhibiting crustal breakup. The overall crustal stretching led to serpentinisation of the upper mantle in the Rockall Basin (O'Reilly *et al.*, 1996) and probably in the Porcupine Basin, and also to the major vulcanism in Cretaceous and Cenozoic times.

The areas of maximum crustal thinning (Rockall and Porcupine) are coincidentally overlain by bathymetric embayments. These reflect the outstripping of sedimentation by subsidence during the Neogene. A regional unconformity in latest Eocene to earliest Oligocene times resulted in a major change from downslope sediment transport to alongslope reworking in deep-water basins. The onset of contourite currents in turn contributed to the construction, and later localised mass failures, of steep slopes along the basin margins, especially the eastern margin of the Rockall Basin. These were later modified, with subsequent local failures, in Pleistocene time, when a combination of localised sediment input through melting glaciers and ice sheets, together with ocean current activity and scouring during glacial times, finally shaped the present dramatic physiography in the deep-water Atlantic margin.

Key regional observations regarding the Atlantic margin basins are as follows:

a) the continental crust beneath the largest basins (Porcupine and Rockall) is extremely thin (2–5 km) in places, while thinning is less pronounced beneath the smaller basins;

b) the depositional basins appear to have had different orientations through time;

c) the basins developed through phases of rifting in the Permo-Triassic and Jurassic with interspersed periods of thermal subsidence and occasional episodes of local and regional inversion;

d) the Cretaceous and Cenozoic successions in the large basins are thick but generally devoid of evidence of major fault-related rifting; and

e) major igneous activity occurred periodically during the history of the basins, linked to both passive margin development and to more regional lithospheric processes.

Regional Exploration History

The first licences were granted in the Irish Atlantic basins in 1976 as part of the first formal Licensing Round in the Irish offshore (*see* Chapter 3). Blocks were awarded to a small number of large multinational companies and consortia in the Porcupine, Slyne, Erris and Donegal basins, as well as the Kish Bank Basin east of Ireland (*see* Chapter 12) and the Fastnet Basin in the western part of the Celtic Sea (*see* Chapter 10). The first well in the region, Shell 35/13-1, was drilled in 1977. This was in the northern part of the Porcupine Basin and encountered a thick Cenozoic and Cretaceous succession with some oil shows. Drilling in the Atlantic Margin basins reached a peak in 1978 when eight wells were drilled. However, seismic data quality in the basins at this time was generally poor, leading to lack of definition on the structures within the basins, and consequently to generally disappointing results from the drilling. While the broad outlines of large Jurassic tilted fault blocks could be imaged on the flanks of the Porcupine Basin, little detail could be resolved of the pre-Late Cretaceous succession in the deep parts of the basin, away from the margins in areas where the succession was at its thickest. Seismic data quality was also poor in the Slyne, Erris and Donegal basins. Here the difficulties were the result of a combination of relatively extensive Early Cenozoic sills and lavas, Permo-Triassic salt movement, and structural complexities resulting from a combination of basin edge inversion on the margins of the Rockall Basin, and to strike-slip and transfer fault systems within the basins. Only a small amount of regional seismic data was available from the Rockall Basin, as this region was too distant from shore to warrant exploration interest. In addition, the seismic data quality was of even poorer quality than the other nearby basins, due largely to the impedance difficulties resulting from Early Cenozoic sills and lavas. Little, if any, structure or stratigraphy could be imaged with any confidence beneath the Upper Palaeogene strata (Naylor,

1972). This was to remain a problem in the Rockall Basin in particular until the late 1990s.

As a consequence of the relatively poor quality of seismic data in the region, the early exploration wells in frontier regions tended to focus on the obvious and large structural traps. The majority of the early wells (to the mid-1980s) were drilled on tilted fault-block structures on the northern margins of the Porcupine Basin and to a lesser extent on similar structures in the shallower waters of the Slyne, Erris and Donegal basins. Analogues were sought for the tilted fault-block structures that were successful in the North Sea.

The few early wells in the Slyne, Erris and Donegal basins (four wells between 1978 and 1995), drilled on large structural highs, were generally unsuccessful. While reservoir intervals were encountered in the Middle and Upper Jurassic succession, and source rocks at Lower Jurassic and Upper Carboniferous levels, none of the wells flowed oil or gas on test. However, the early drilling led to some technical successes in the Porcupine, although none of these has yet led to commercial production. Most of the wells drilled in the basin recorded shows (Croker and Shannon, 1995). These were found in different stratigraphic horizons, including Middle and Upper Jurassic, Lower and Upper Cretaceous and Lower Cenozoic levels. Four of the early wells in the Porcupine Basin flowed significant quantities of good quality (32–41° API) petroleum from Jurassic and Lower Cretaceous reservoirs. The Phillips 35/8-1 well flowed at a rate of 730 barrels of oil per day (bopd) in 1978, from poor quality thin Lower Cretaceous turbiditic sandstone reservoirs. BP drilled oil discovery wells 28/28-1 and 26/28-2 in 1979 and 1980. These flowed 5589 bopd and 1550 bopd respectively (MacDonald *et al.*, 1987) from Upper Jurassic fluvial sandstones within a structurally complex tilted fault-block structure. They were the discovery wells for the Connemara oil accumulation. MacDonald *et al.* (1987) estimated that the western segment of the structure contained approximately 120 MMbo in place, while an additional 75 MMbo is possibly in place within the eastern segment. However, despite significant appraisal drilling and the acquisition of 3D seismic data during the past thirty years, the accumulation still remains undeveloped although it is currently under licence to Island Oil and Gas Ltd. In 1981, the Phillips 35/8-2 exploration well encountered a gas condensate accumulation that flowed oil and gas at rates of 925 bopd and 4.853 MM scfd respectively from Upper Jurassic turbiditic sandstones (Shannon, 1993; Robinson

and Canham, 2001) in a Jurassic tilted fault-block structure draped by Lower Cretaceous marine mudstones. Robinson and Canham (2001) estimated mean recoverable reserves of up to 50 million barrels of condensate and 764 bcf of gas. However, the seismic data quality from this part of the basin is generally very poor, due in significant part to the presence of igneous intrusions. Although many of the wells drilled in the early phase of exploration in the Atlantic Margin basins encountered hydrocarbons, the lack of extensive, thick Jurassic reservoirs gave rise to generally disappointing results. The discovery block 35/8 is currently licensed to Providence Resources, and they have named the Phillips 35/8-1 well the Burren Discovery, and the Phillips 35/8-2 well is now called the Spanish Point Discovery. New seismic data have been acquired across both prospects and are being assessed prior to a decision on further drilling of the discoveries. Industry reports suggest that the median resource estimates have been upgraded to 160 MMbo and 1.4 Tcf of gas.

A significant amount of good quality seismic data was shot in the Porcupine Basin in 1981, improving the imaging of deeper structures in the basin. This was in advance of the Second Licensing Round in 1982, which had a major focus on the Porcupine Basin. A number of wells resulted from drilling on the blocks awarded, and these were largely focused on large tilted fault blocks in the northern part of the basin. The award of blocks in the Goban Spur led to the drilling by Esso of the first (and only) exploration well to be drilled in the basin. This deep-water well penetrated a Cenozoic, Cretaceous and Jurassic succession but encountered disappointingly poor quality reservoirs at the Jurassic target level.

A lull in exploration interest in the region during the mid-1980s and early 1990s was reversed with the First and Second Frontier Licensing Rounds in 1994 and 1995, which focused on acreage in the Slyne, Erris and Porcupine basins (*see* Chapter 3). This led to the drilling of two wells in the Erris Basin in 1996, one of which (18/20-1) was the discovery well for the Corrib Gasfield, and two wells in the Porcupine Basin in 1997. These wells began to test more complex targets that could not be imaged with any confidence during the earlier phases of exploration. Wells tested salt-related structures in the Erris Basin and more subtle Early Cenozoic stratigraphic submarine fan targets in the Porcupine Basin.

In 1998 to 2001 Enterprise Oil drilled a number of successful appraisal wells on the Corrib Gasfield, with wells flowing at rates of 33–66 MMscfd (Dancer *et al.*,

2005). The same authors suggested that the field contains approximately 1.2 Tcf of gas in place, with approximately 870 bcf recoverable. In 2001 the first exploration well (Enterprise 5/22-1) was drilled in the Irish sector of the Rockall Basin but was plugged and abandoned as a dry hole. However, the following year Enterprise Energy Ireland Ltd drilled well 12/2-1, a tilted fault-block structure (the Dooish prospect) on the eastern margins of the Rockall Basin, and announced encouraging results at Permo-Triassic level, confirming the presence of a working petroleum system. The Dooish discovery well was re-entered and deepened in 2003 and confirmed the existence of a substantial gas condensate column.

In 2007 two appraisal/development wells were drilled by Shell (who had by now taken over Enterprise Oil) on the Corrib Gasfield in the Slyne Basin. In 2008 four licences, involving blocks in the northern and central parts of the Porcupine Basin and in the Goban Spur, were awarded to three groups under the Seventh Frontier Licensing Round. Statoil drilled well 19/8-1 in the Erris Basin and Shell drilled a further well (12/2-2) on the Dooish structure in the Rockall Basin.

2009 was a relatively quiet year for exploration in the Atlantic margin basins. Providence Resources shot a 3D seismic survey in the Block 35/8 region, encompassing the Spanish Point and Burren accumulations. However, well 27/4-1, drilled by Serica Energy on the Bandon prospect in the Erris Basin, yielded encouraging and potentially significant results. The well was drilled on a salt-related structural high adjacent to the western margin of the basin. It resulted in the discovery of oil in the basin, with a reservoir in the Jurassic, sourced from Lower Jurassic mudstones, and proved up a new play in the region. However, no drill stem tests were carried out, so the extent or likely productivity of the discovery is still uncertain. Nonetheless, combined with the Corrib Gasfield in the northern Slyne Basin, it indicates the presence of working oil and gas systems in the narrow inboard Slyne–Erris basin system.

A summary of the structural and stratigraphic framework of each of the major basins in the area, and a discussion of their exploration potential follows.

Goban Spur
Basin framework
The Goban Spur, lying to the south of the Porcupine Seabight and west of the Celtic Platform, is a remote plateau area on the continental margin. Bathymetrically it is a westward-projecting plateau area comprising a smooth platform that slopes gently westwards to depths of 2000 m. It is a structurally complex region (Figure 11.3) containing a number of fault-bounded basement ridges with intervening small sedimentary basins (Naylor *et al.*, 2002). The largest of these are the Goban Spur Basin, drilled by the Esso 62/7-1 well, and the Goban Graben, with smaller elongate depocentres in the Pendragon, Shackleton, Merlin and King Arthur basins. The Goban Spur province is separated from the Celtic Sea–Fastnet basin system to the east by the Fastnet High, and from the Porcupine Basin to the north by the ENE-striking Porcupine Fault (Dingle and Scrutton, 1979). The western boundary of the Goban Spur is marked by a faulted structural high, west of which lies oceanic crust of latest Cretaceous age. In contrast to the broadly north–south orientation of the bounding faults and structures in the Porcupine Basin to the north, the structure of the Goban Spur region is controlled by an interplay between NW to NNW-oriented faults, parallel to the oceanic–continental boundary immediately to the west and to the well-defined ENE–WSW to east–west fault orientations that control the Celtic Sea structural architecture.

Understanding of the basin framework of the Goban Spur region is based largely on gravity, magnetic and reflection seismic data (Dingle and Scrutton, 1977, 1979; Scrutton, 1979; Roberts *et al.*, 1981), and on results from the Deep Sea Drilling Programme (Montadert *et al.*, 1979; Masson *et al.*, 1985; Graciansky *et al.*, 1985). Additional ground-truth geological data have also come from dredging and gravity coring (Auffret *et al.*, 1979) and from submersible dives and dredging (Masson *et al.* 1989). Importantly, the results from the one deep oil exploration well (Esso 62/7-1, drilled in 1027 m of water to a depth of 4671 m, together with the synthesis of additional geophysical and palaeontological data (Cook, 1987; Colin *et al.* 1992) have helped constrain the geological development of the region (Figures 7.4 and 11.4).

Basement in the region is likely to comprise Carboniferous and older granites and metasediments. Late Variscan granodiorites (250–291 Ma), metamorphic and sedimentary rocks (black shales, sandstones and shallow-water limestones: probably of Carboniferous age) were retrieved by dredging prior to the DSDP programme (Auffret *et al.*, 1979). Subsequent submersible dives and dredging on the Pendragon Escarpment at 49°30'N (Masson *et al.*, 1989) recovered a sample of sandstone of possible Devonian age near the base of the escarpment, overlain by Barremian limestones.

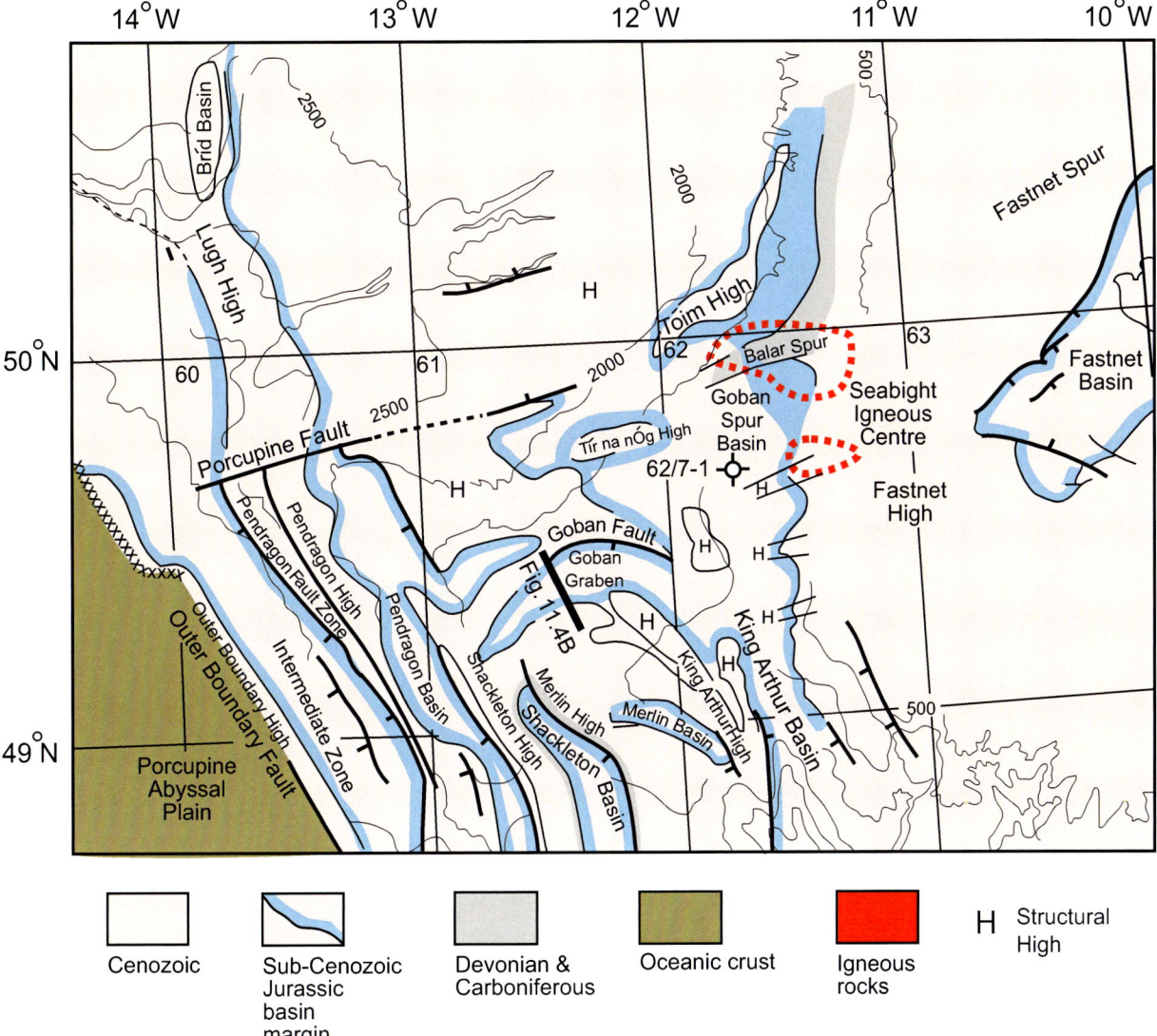

Figure 11.3. Structural feature of the Goban Spur region (after Naylor *et al.*, 2002) showing the system of elongate basins and structural highs in the region. Two major structural trends reflect the Caledonian/Variscan (ENE-WSW) and Atlantic opening (NW-SE).

Up to 6 km of post-Variscan strata are interpreted from seismic and well data in the Goban Spur region. Cook (1987), using seismic data, suggested that the oldest post-Variscan strata in the region are of mid-Triassic age, with a possible evaporitic succession overlying a clastic succession. Triassic strata were not penetrated in the Esso 62/7-1 well, which bottomed in probable lower Sinemurian strata (Colin *et al.*, 1992). Permo-Triassic strata, deposited during fault-controlled rifting, are also interpreted from the Goban Graben (Figure 11.4). The Jurassic shale-prone succession, approximately 1500 m thick, ranged up to Callovian–Bathonian in age, with Upper Jurassic strata absent. The Jurassic section predominantly comprised siltstones and claystones with thin limestone interbeds deposited in inner shelf to outer neritic environments. At the top of the sequence Callovian–Bathonian claystones and shoreface sandstones were overlain by 214 m of Mid-Jurassic porphyritic basaltic to andesitic and basalt flows (Figure 7.4). Jurassic strata are also interpreted from seismic data in the Goban Graben, developed within a half-graben basin during an extended period of fault movement on the east–west Goban Fault (Figure 11.4). The Late Cimmerian unconformity is interpreted in the Esso 62/7-1 well and from seismic data in the region by erosion and onlap of the overlying Lower Cretaceous strata.

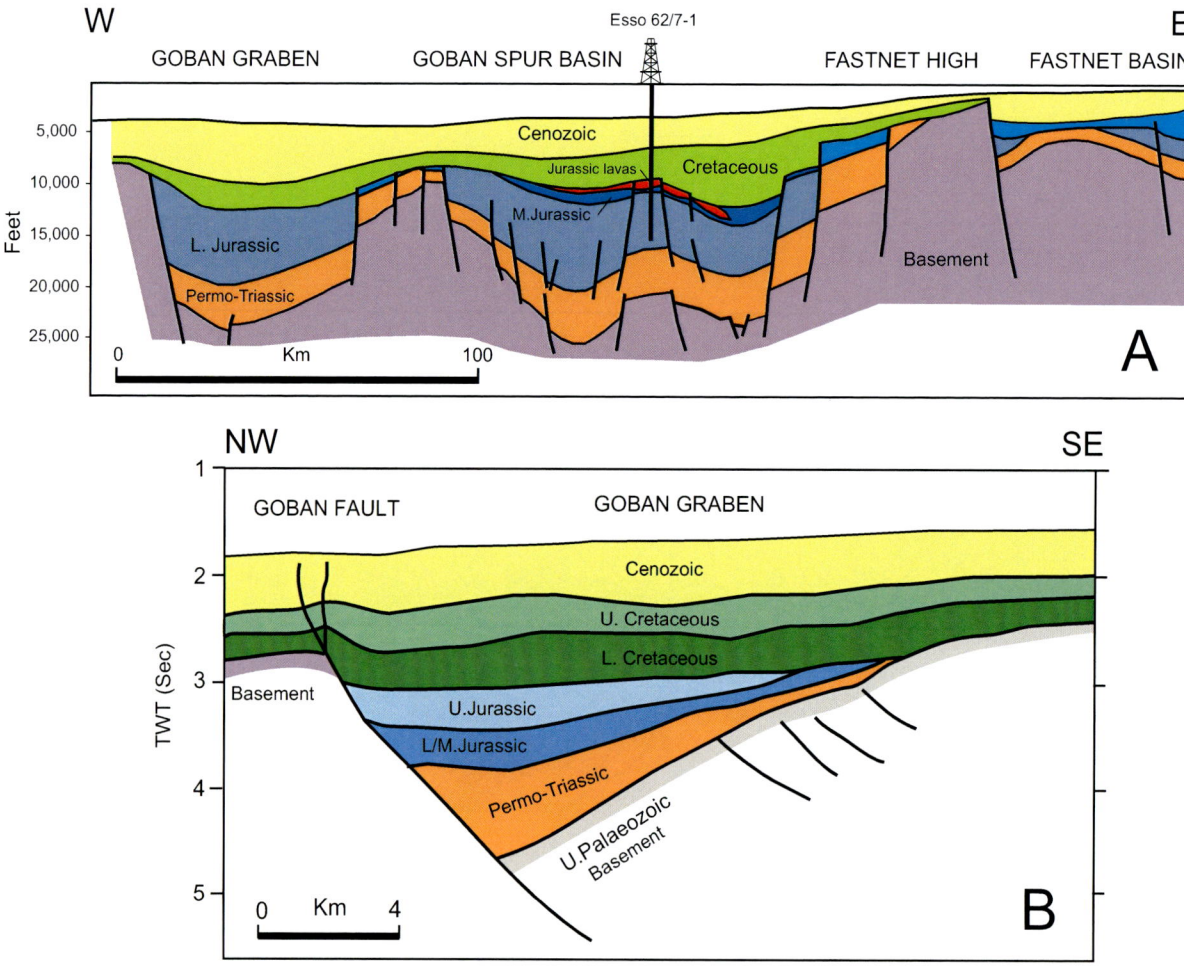

Figure 11.4. A. Geoseismic section showing the post-Carboniferous succession in the Goban Spur Basin (from Cook, 1987). **B.** Geoseismic section showing the interpreted succession in the Goban Graben (from Naylor *et al.,* 2002).

Littoral–sublittoral Barremian–?Hauterivian mudstones, limestones and silty clays are interpreted as synrift deposits. At the Esso 62/7-1 well location possible Neocomian marine sandstones and cherty limestones containing terrestrial palynomorphs are overlain by shallow-marine Barremian to lower Aptian limestones and claystones. On Goban Spur and in the Western Approaches Basin a widespread unconformity separates the syn-rift from the drift sequence. On Goban Spur this surface has been dated as Aptian to pre-Aptian in age. Cook (1987) interpreted extensive Aptian–Albian volcanic rocks at the eastern margin of the Goban Spur Basin, based on gravity, magnetic and seismic data. These are likely to be coincident with the initiation of seafloor spreading. Late Albian basaltic pillow lavas were penetrated in the DSDP drilling at Site 550 southwest of the Goban Spur (Figure 3.4). In the Esso 62/7-1 well there

is a marked facies change at the top of the lower Aptian sequence. The overlying Cenomanian to Maastrichtian limestone-dominant sequence was deposited in outer shelf and upper bathyal environments. Turonian sediments are absent in the well, interpreted by Colin *et al.* (1992) as due to a major mid-Cretaceous eustatic event. On the seaward edge of Goban Spur, at DSDP Site 549, the Upper Cretaceous sequence of Cenomanian to Maastrichtian nano-chalk, a section disrupted by hiatuses, rests unconformably on middle Albian siltstones. Seafloor spreading in the Atlantic domain continued through the Cretaceous, with increasing water depths and relative sediment starvation in the Goban Spur area. The uppermost Maastrichtian is also missing in the 62/7-1 well. This is probably due to the effects of latest Cretaceous–Palaeogene uplift, similar to that interpreted further east in the Celtic Sea region (*see* Chapter 10).

The end-Cretaceous tectonism, uplift and basin inversion had a widespread influence throughout the basins of the Northwest European shelf. On Goban Spur a succession of Eocene to Oligocene nano-chalks rest unconformably on Upper Cretaceous sediments. Graciansky *et al.* (1985) identified four major hiatuses in the Cenozoic section at DSDP Site 548 (early Paleocene, early Eocene/late Eocene, early Oligocene/late Oligocene, and middle Miocene/late Miocene). At the Esso 62/7-1 well location, where almost 1 km of Cenozoic strata occur, the base of the sequence consists of upper Paleocene–lower Eocene claystones; the younger Cenozoic succession was not sampled. Seismic data reveal a major mid-Oligocene erosive unconformity with onlap above, resulting from regional uplift (Cook, 1987). Subsidence throughout the Cenozoic, combined with limited sediment supply, has resulted in the present-day water depths.

Subduction of Biscay oceanic crust beneath the northern margin of the Iberian peninsula, which terminated in the late Eocene, affected the area from Galicia Bank to Goban Spur (Boillot *et al.*, 1979; Montadert *et al.*, 1979). Late Eocene compression formed the narrow east–west folds that run along the Meriadzek segment of the southern margin of the Spur (Masson and Parson, 1983). Early Cenozoic inversion features are also seen in the southern part of the Porcupine Basin (Masson and Parson, 1983) and at the southeastern margin of the Porcupine High.

Overall, the geohistory of the Esso 62/7-1 well (Colin *et al.*, 1992) indicates three major episodes of basin evolution (Figure 11.5), broadly similar to the evolution of the Fastnet Basin to the east. A syn-rift phase of basin initiation occurred during the Triassic and Early Jurassic, characterised by rapid subsidence. This was followed by a quiescent period throughout the remainder of the Jurassic and Early Cretaceous, following the Late Cimmerian unconformity, which was characterised by low basin subsidence and sedimentation rates. A mid-Cretaceous (Albian) phase of post-rift subsidence and basin deepening followed, with accelerated subsidence during the Early Cenozoic.

Exploration potential
Source Rocks

The pre-Jurassic succession, comprising Variscan-deformed basement and interpreted Triassic red-bed facies, is unlikely to contain any source potential. However, the Sinemurian to Lower Pliensbachian unit in the Esso 62/7-1 well comprises predominantly marine claystones with thin interbedded limestones, deposited in an outer neritic environment, suggestive of rapid basin subsidence that exceeded clastic input. Vitrinite reflectance ($R_o=0.7$) and spore coloration data suggest that the claystones are within the oil generation window (Cook, 1987). However, they have an average Total Organic

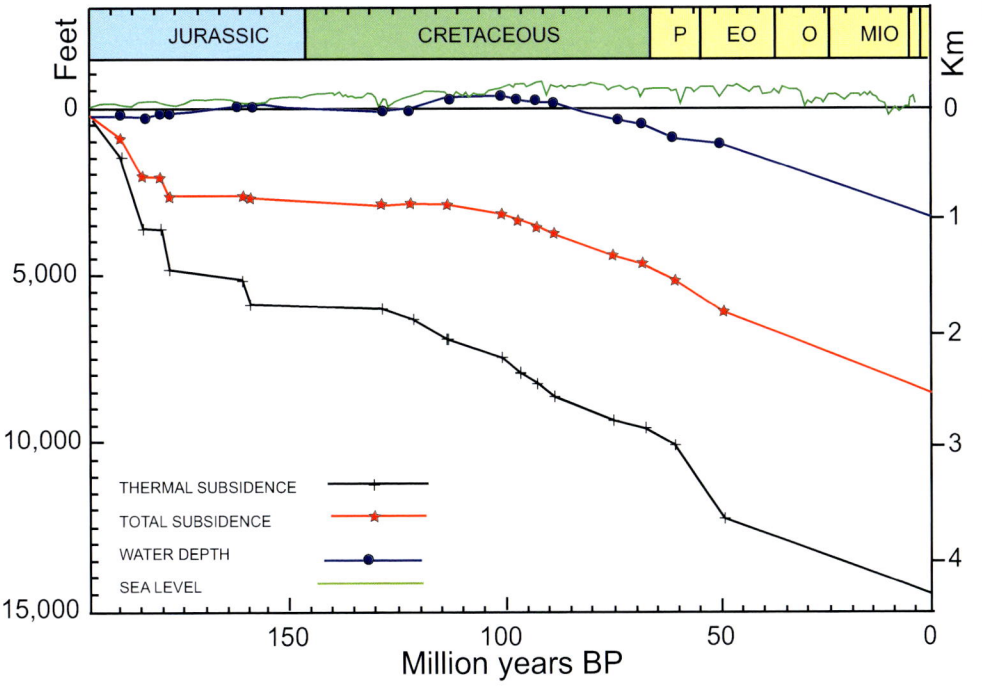

Figure 11.5. Geohistory plot for the Esso 62/7-1 well in the Goban Spur Basin showing three phases of major basin development (from Colin *et al.*, 1992). A Permo-Triassic to Early Jurassic syn-rift phase of rapid subsidence was followed by a quiescent period of low basin subsidence until mid-Cretaceous times. This was followed by rapid post-rift subsidence, accelerated during Early Cenozoic times.

Carbon (TOC) value of 1.5% and comprise mainly inertinitic and gas-prone woody organic matter, with little potential for the generation of significant quantities of liquid hydrocarbons. The Toarcian succession contains a similar lithology, with an average TOC of 1% and a woodier kerogen. This succession is marginally mature for oil at the well location, but has low liquid hydrocarbon potential. The Late Toarcian–Bajocian unit has an average TOC of 1.2% but has similar kerogen types to the underlying units. Overall therefore, while the well proved the presence of thermally mature Lower Jurassic source rocks, they contain only moderate amounts of predominantly terrestrial organic matter. However, away from the basin margin, especially in the centre of the basin depocentres, it is possible that the Lower Jurassic claystones and mudstones may be richer in oil-prone kerogens, similar to those proven in the Fastnet and North Celtic Sea basins to the east.

Reservoir rocks and structures

Triassic strata were not encountered in the well, but have been interpreted from seismic data. The presence of such strata in the Fastnet Basin to the east lends confidence to this interpretation and suggests that thick sandstones, of good reservoir quality, are likely to be present in the Goban Spur region. Although the thick Sinemurian potential reservoir sandstones found in the Fastnet Basin (Robinson *et al.*, 1981) were not encountered in the well, they may be developed in other parts of the Goban Spur basins. The Esso 62/7-1 well encountered a Callovian–Bathonian succession that was the primary reservoir objective in the well (Figure 7.4). This contains 30 m of massive sandstones (Cook, 1987). These are fine- to medium-grained, well sorted, laminated subarkoses, interpreted as stacked sequences within the surf zone of a high energy beach. Dipmeter data analysis of this unit indicates a NE–SW palaeo-shoreline running through the well location and separating an open shelf marine environment to the northwest from a possible deltaic to lagoonal environment to the southeast. A primary local sedimentary provenance is interpreted, and Cook (1987) suggested that this source may be Devonian arkosic sandstones to the southwest. Overall these strata have excellent reservoir characteristics, with average log porosities of 25% and core permeabilities of 830 mD (Cook, 1987). While this succession is thinner than was prognosed, it may be thicker in other parts of the region, as erosion of the uppermost part of the succession may have taken place at the well location prior to the extrusion of the thick lava unit. Structures in the basin are typically fault-related, especially at Jurassic and deeper levels, with tilted fault blocks likely to represent the main targets. However, stratigraphic components of trapping, as either drapes above fault blocks, or as carbonate buildups within the Cretaceous succession (Cook, 1987) have also been indicated from drilling or seismic data.

Porcupine Basin
Basin framework

The Porcupine Seabight is a large (320 x 240 km) north–south oriented deep-water area lying north of the Goban Spur (Figure 2.14). At its southern end it merges southwestwards into the Porcupine Abyssal Plain. It overlies a large and complex basin that contains up to approximately 10 km of Upper Palaeozoic to Cenozoic sediments. The north–south shape of the Porcupine Basin is in sharp contrast with other basin orientations along the Atlantic margin. Croker and Shannon (1987), Shannon (1991a, b) and others have suggested that the early Mesozoic basin development in the region was broadly northeast–southwest in orientation, possibly guided along reactivated Caledonian fabrics. This resulted in a set of small, elongate basins within the Porcupine region, lying beneath the younger north–south large basin. The main north–south orientation of the basin appears to have developed during Middle and Upper Jurassic times, broadly coincident with the main phase of syn-rift basin development. The Jurassic facies distribution (Croker and Shannon, 1987), with a major basin-edge alluvial clastic fan complex seen in the Shell 34/19-1 well, supports this model. However, the general north–south shape of the basin is interrupted by several spaced NW–SE gravity lineaments. These correspond to offsets in the basin margins, and also to changes in the sedimentation style and geometry and to variations in the crustal thickness (Readman *et al.*, 2005). Their regional extent, with some of them extending into the Celtic Sea domain, suggests that they are deep-seated and long-standing inherited crustal structures.

Approximately 1500 m of Pennsylvanian fluvial to deltaic and brackish sandstones, siltstones, mudstones and coals have been drilled on the eastern margin of the basin (Croker and Shannon, 1987; Robeson *et al.*, 1988). These Upper Palaeozoic sediments are largely undeformed where drilled, showing none of the effects of Variscan deformation seen in the coeval rocks of the Celtic Sea and Goban Spur regions to the south.

Upper Palaeozoic strata are overlain locally by poorly dated Permo-Triassic shallow-marine sandstones and evaporitic mudstones, and more regionally by extensive Middle and Upper Jurassic fluvial to transgressive shallow-marine sandstones, mudstones and thin limestones, with more than 1000 m drilled in places (Croker and Shannon, 1987). Figure 7.5 shows the general southward thickening of strata from south to north in the basin. Permo-Triassic strata have only been encountered by drilling in the North Porcupine Basin and in a couple of isolated wells in the main Porcupine Basin. Lower Jurassic limestones and marine mudstones, similar to those in the Celtic Sea region to the east, and in the Slyne and Erris basins to the northeast, were encountered and appear to be conformable with the underlying Upper Triassic succession (Croker and Shannon, 1987).

Middle Jurassic strata are widely developed in the Porcupine Basin and typically rest upon the Upper Carboniferous succession, with no significant angular unconformity. The succession comprises sandy braided fluvial deposits, interpreted by Sinclair et al. (1994) as the product of an onset warp phase of tectonism immediately prior to the onset of the Late Cimmerian rifting. Upper Jurassic strata within the basin reflect deposition in a syn-rift setting, with the development of a range of lithologies and facies (Chapter 7). These range from basin-edge alluvial fans and braided to meandering fluvial strata to deep marine submarine fans (Croker and Shannon, 1987; MacDonald et al., 1987).

The syn-rift Jurassic succession is unconformably overlain by a thick (more than 1 km drilled in wells) Cretaceous succession of mudstones and local marine and deltaic sandstones, overlain in turn by a thick Chalk succession that onlaps the rifted Jurassic margins of the basin (Figure 11.6). Lower Cretaceous strata in the Porcupine Basin represent the product of two rift episodes – the residual phase of Late Jurassic rifting and locally an Aptian–Albian rift phase. The Late Jurassic rifting waned during the early part of the Cretaceous and a major unconformity or set of close-spaced unconformities marks the approximate Jurassic/Cretaceous boundary (Naylor and Anstey, 1987). The Cretaceous throughout most of the Atlantic margin region consists of shale-prone marine strata, but in the Porcupine Basin shale deposition was interrupted by deltaic sandstones that reflect Aptian–Albian rifting.

Up to 2 km of Cenozoic mudstones, sandstones and thin limestones have been drilled in the Porcupine Basin, with late Paleocene to Eocene deltaic sandstones in the north of the basin giving way southwards to deep-water equivalents. Johannessen and Steel (2005), using 3D seismic data from the basin, illustrate the complex sedimentological relationships in the basinward-prograding, clinoform depositional system, including the interpreted presence of localised, channelised slope sandstones and basin floor sandstones. The sandy Eocene succession in deltaic to submarine fan facies (Moore and Shannon, 1992; Shannon, 1992) was initially interpreted (Shannon et al., 1993) as resulting from ridge-push effects caused by seafloor spreading and oceanic crustal development in the region to the west of Rockall. However, more recently it has been suggested to represent the effects of lithospheric thermal convective cells (Praeg et al., 2005). Flank uplift on the margins of the Porcupine Basin associated with these processes produced sediment source areas, while coeval subsidence in the basin provided the accommodation space for the deposition of the thick clastic deposits.

A basinwide unconformity of latest Eocene to early Oligocene age marks a change in sedimentation style and rate, with the basin becoming dominated by deep-water contouritic siltstones and mudstones, and the products of downslope debris flows. Volcanics (lavas and intrusions) of Cretaceous and Cenozoic age occur in parts of the basin (Naylor et al., 2002).

Exploration potential
Source rocks
The flows and shows of oil, gas and condensate from wells in the Porcupine Basin indicate the presence of a number of working petroleum systems. Source rock potential exists in the Late Carboniferous, Middle and Late Jurassic successions, and these are generally mature throughout most of the basin. Some oil and gas potential also exists in the Cretaceous and Cenozoic successions but these are immature where drilled in the basin. However, they may reach marginal maturity in the deeper undrilled parts of the basin.

The Westphalian (Pennsylvanian: upper Baashkirian–Moscovian) succession, encountered in many of the wells drilled in the basin, contains thin coal beds with good potential for gas and condensate (Croker and Shannon, 1987). In addition, occasional sapropel-rich shale successions have moderate potential for oil towards the basin margins, and for gas in the deeper central parts of the basin. The same authors suggested that the Late Carboniferous source rocks reached thermal maturity in pre-Cretaceous (possibly

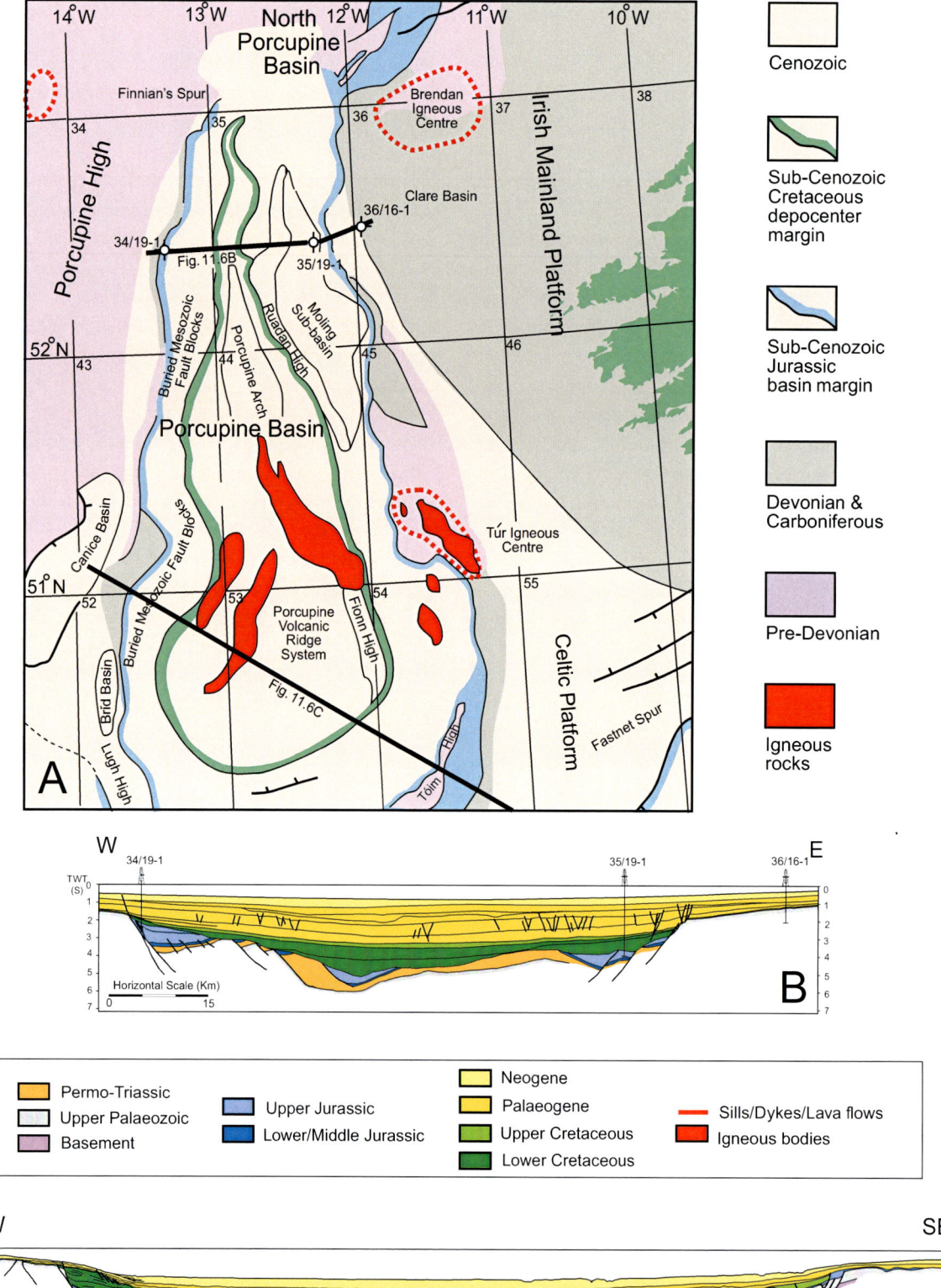

Figure 11.6. Map of the Porcupine Basin (A) and geoseismic sections (B and C) illustrating Jurassic syn-rift tilted faults beneath a regionally-extensive, largely thermal subsidence Cretaceous and Cenozoic succession (after Naylor *et al.,* 1999 and Naylor and Shannon, 2009).

pre-Mesozoic) time but were uplifted during the Late Mesozoic, before later reburial in Palaeogene times to resume hydrocarbon generation. The present oil generating threshold for the post-Carboniferous succession in the basin is at an approximate depth of 2500 m, while with peak oil generation is occurring at a depth of approximately 3000 m (Croker and Shannon, 1987).

The Middle Jurassic succession in the basin has variable source potential. Shales drilled on the northwestern flank (Deminex 34/15-1 and Deminex 35/6-1) have locally good oil and gas potential and have Total Organic Carbon (TOC) values in the range 1-1.85%. These show a non-marine (lacustrine) geochemical signature (Butterworth et al., 1999), and correlate with some of the oils encountered in the Middle Jurassic reservoirs in the Phillips 35/8-1 oil discovery (Croker and Shannon, 1987). These Middle Jurassic source rocks are thought to be mature throughout most of the basin.

The Late Jurassic, and especially the Kimmeridgian, succession has variable source potential. It is a good proven source in the northern (Chevron 26/27-1B, BP 26/28-1, Phillips 35/2-1), western (Deminex 34/15-1) and central (Phillips 35/8-2) parts of the basin, and is regarded as the single most important source rock interval in the basin (Croker and Shannon, 1987). Marine shale units in the Kimmeridgian and occasionally the Tithonian (e.g. Elf 35/2-1) contain good to excellent oil and gas source potential. The TOC values in the richest horizons are in the range 3–4% and pyrolysis yields sometimes exceed 7 kg/tonne. However, Butterworth et al. (1999) showed that the geochemical signature of the Upper Jurassic marine source rocks in the basin is similar to those of the Kimmeridgian Egret Member in the Jeanne d'Arc Basin offshore Newfoundland, but differs from the Upper Jurassic Kimmeridge Clay Formation source rocks of the North Sea. The section, although immature in the northern part of the basin, is mature throughout most of the basin. Modelling results by Naeth et al. (2005) showed that Jurassic and older source rocks in the Porcupine Basin are mature to overmature; Vitrinite reflectance (Ro) values range from 0.7 to >4.7%, and present-day temperatures are from 95 to 260°C. Maturity values are low on the flanks of the basin.

MacDonald et al. (1987) suggested that the geochemical characteristics of the oil in the Jurassic reservoirs of the Connemara oil accumulation in Block 26/28 cannot be unambiguously correlated with any single known potential source horizon found in these wells. The oils have gravities in the range 32.4–38.2° API and are light, paraffinic, waxy and low in sulphur. The carbon isotope ratio lies between - 28.5 and - 29.2 ppm. MacDonald et al. (1987) described the geochemical characteristics as suggesting an origin from a source rock containing dominantly non-marine kerogen, with a minor contribution from marine kerogen. This mixed character is consistent with the source coming partly from the Middle Jurassic non-marine and partly from the Upper Jurassic marine mudstones in the northern part of the Porcupine Basin.

The Early Cretaceous succession in the basin has source potential in two main intervals, as documented by Croker and Shannon (1987). Ryazanian to Aptian marine shales have poor source potential in the northern part of the Porcupine Basin, but increase in source rock quality southwards in the basin. In the Phillips 35/8-2 well, the interval contains amorphous oil-prone kerogen and has TOC values up to 1.8%. It is likely to have been the predominant source for the oil in the Phillips 35/8-2 (Spanish Point) gas condensate discovery. Aptian–Albian shales in the 35/8-2 well also have fair oil-generating potential and contain TOC values up to 2.7%. A burial history plot for the central part of the Porcupine Basin (Figure 11.7) indicates rapid periods of subsidence during the Early Cretaceous and during the mid to Late Cenozoic. The Early Cretaceous section is immature on the basin margins and throughout the North Porcupine Basin, but is at an early mature stage in the 35/8-2 well. It is likely to be fully mature in the deeper parts of the basin. Modelling by Naeth et al. (2005) showed that Cretaceous strata are immature to mature in the central part of the Porcupine Basin, and immature on the basin flanks. Vitrinite reflectance ranges between 0.3 and 4.6 % Ro, and present-day temperatures are between 40° and 240°C.

Eocene coaly, lignitic and shaly deltaic sequences in the basin have moderate to very good source potential for gas, and occasionally for oil and condensate. TOC values of up to 47.3% were recorded in the Deminex 35/6-1 well and up to 50.5% in lignites in the Chevron 26/27-1B well, with pyrolysis yields up to 163.1 kg/tonne. However, the succession is immature around the basin margins, and is likely to have reached maturity only where deeply buried in the central part of the basin. Modelling results from Naeth et al. (2005) showed that the Cenozoic sequence remains immature over the entire basin. Vitrinite reflectance Ro reaches a maximum of 0.4%, and present-day temperature ranges are between 4° and 70°C.

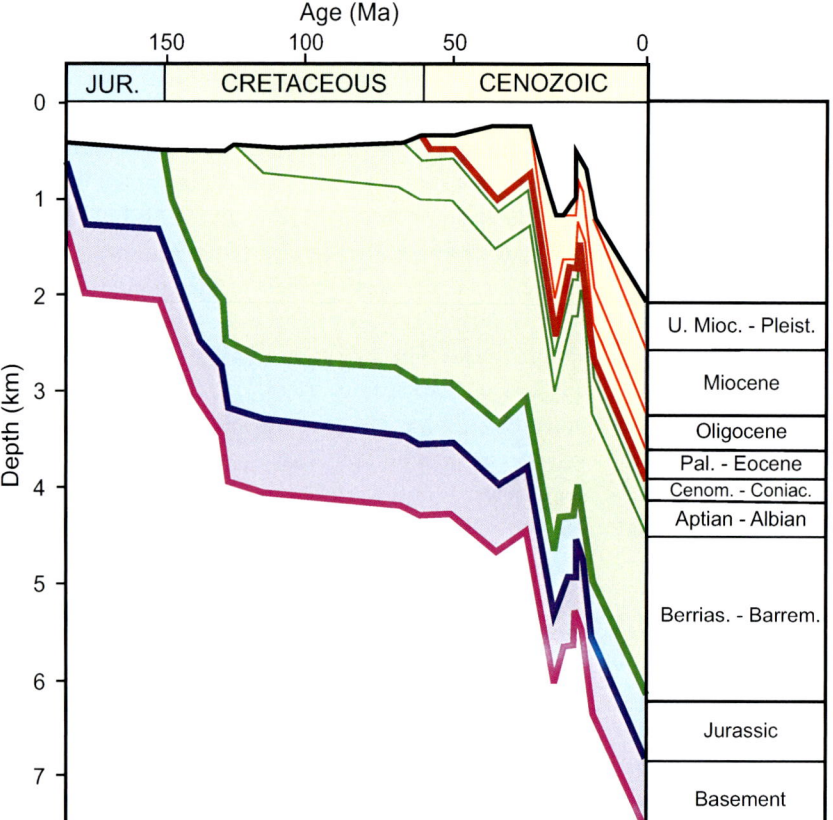

Figure 11.7. Burial history diagram for the central part of the Porcupine Basin (modified from Naeth *et al.,* 2005).

Reservoir rocks and structures

The Porcupine Basin contains a broad spectrum of proven and potential reservoir successions. Thick units of reservoir quality Aptian–Albian and Eocene deltaic sandstones have been drilled in the northern part of the basin. Middle and Upper Jurassic sandstones are also present and are oil-bearing in places. However, a major uncertainty within the basin is the extent of these reservoirs in the deeper and southern parts of the basin, where large Jurassic tilted fault-block structures occur. The basement and pre-Upper Carboniferous rocks are unlikely to contain any significant reservoir potential except perhaps locally on the basin margin, if fractured and where capped by onlapping Cretaceous and Early Cenozoic thermal subsidence mudstones. The Lower Jurassic succession is very limited in extent in the basin, and is virtually devoid of sandstones and possesses little if any reservoir potential. The principal reservoir targets in the basin are outlined below. Where tested by drilling the pre-Cretaceous reservoirs have generally been located in fault-block structures, while the younger strata may have elements of potential stratigraphic trapping.

Upper Carboniferous deltaic sequences, drilled in many of the wells in the basin, contain substantial thicknesses of sandstones with moderate reservoir potential. Up to 200 m of sandstones have been drilled on the eastern side of the basin, and these have net/gross ratios typically within the range of 0.1–0.5. Porosities are in the range 5–15% but occasionally range up to 20% (Croker and Shannon, 1987).

Triassic sandstones have been drilled in the North Porcupine Basin and are thought to have only limited extent within the main Porcupine Basin. However, they offer locally good reservoir potential. In the Gulf 26/21-1 well 150 m of net sandstones were recorded in a 171 m gross interval. They have an average porosity of 22% and have permeabilities of up to 65 mD (Croker and Shannon, 1987).

Middle Jurassic fluvial sandstones are regionally developed within the northern part of the basin. Wells contain gross thicknesses of 250–400 m in the northern and marginal areas of the basin. Individual fluvial packages are sandstone-prone with typically high net/gross ratios, and overall reservoir quality is good. Porosities average 17–22% and are occasionally as high as 30%

(Croker and Shannon, 1987). Permeabilities are typically of the order of tens to hundreds of millidarcies. While the succession thickens southwards, the overall net/gross ratios of 0.16–0.25 in the northern part of the basin decrease in a southward direction.

The Upper Jurassic succession contains a large range of proven and potential reservoirs, in tilted fault-block structures. These reflect the active syn-rift tectonic setting of the time, with the development of local provenances and a wide range of clastic depositional facies. Oxfordian to lower Kimmeridgian continental to marginal marine packages have been drilled along the basin margins. These are around 200 m thick with a net/gross ratio of approximately 0.25. Porosities average 19% and permeabilities are typically tens to hundreds of millidarcies, with occasional streaks of several darcies (Croker and Shannon, 1987). MacDonald *et al.* (1987) described the characteristics of the individual main Upper Jurassic sandstone reservoir packages in the Connemara oil accumulation (Block 26/28) as having typical thicknesses of less than five metres, with none of the stacked packages being thicker than 10 m. Individual sandstones are thought likely to be laterally discontinuous, particularly along depositional strike. The overall low net/gross ratio suggests that vertical communication between reservoir sand bodies is likely to be poor, while the extensive later faulting within the trap structure is likely to have reduced further the inter-sandstone communication. Thick Tithonian clastic fans have been drilled in the northern part of the basin (e.g. BP 26/28-4A) and are very sandy, with net/gross ratios sometimes in excess of 0.5 within a 350 m interval. Porosities (12–15%) and permeabilities (2–3 mD) tend to be rather low in the proximal parts of the fans. In the more central parts of the basin, submarine fan sandstones occur, and were encountered in the Phillips 35/8-2 well, where they occur in a tilted fault-block structure draped by Early Cretaceous mudstones. In this well three main sandy intervals were cored, logged and tested. These provided a cumulative gross sandstone thickness of 150 m with a net sandstone thickness (7% porosity cutoff) of 109.3 m (Robinson and Canham, 2001). While average porosities were relatively low (*c.*12%) these may improve in other parts of the fan complex.

The Lower Cretaceous succession also contains a wide range of proven and potential reservoir horizons developed in different facies. Late Ryazanian to Early Aptian shoreface to shallow-marine sandstones with 22.5 m of the net sandstones and porosities averaging 15% were recorded from the Gulf 26/21-1 well in the North Porcupine Basin (Croker and Shannon, 1987), and similar sequences may occur elsewhere close to the basin margins. Deep-water sandstones of Barremian to Aptian age, in a possible stratigraphic trap, were encountered in the more central parts of the basin (Phillips 35/8-1) where 10 m of net sandstones in a 46 m gross interval have porosities averaging 11% but with occasional streaks of up to 20%. However, the thickest and most important sequence of reservoir rocks in the Cretaceous succession comprises Upper Aptian to Albian deltaic and overlying shallow-marine sandstone-prone strata that occur in areas adjacent to the basin margins. Drilled thicknesses range from 94 m to 366 m, while net/gross ratios typically exceed 0.5 and porosities average 25–30%.

Thick packages of sandstones occur in the Eocene deltaic and associated beach ridges, barrier bars and shallow-marine bar facies. The deltaic units are sometimes in excess of 200 m of net sandstones and have porosities up to 39%. The shallow-marine glauconitic units contain up to 120 m of net sandstones and have average porosities of 30%. The Paleocene–Eocene deeper water systems are also likely to contain reservoir systems ranging from thin and relatively complex channelised slope sandstones to thick and extensive basin floor fan sandstones. The deep-water, shale-prone strata that characterise the Oligocene and younger succession in the basin contains a number of deep marine channel and contourite deposits. Some of these may offer possible reservoir potential.

Slyne-Erris-Donegal basins
Basin framework

The Slyne, Erris and Donegal basins (Figure 11.8) are narrow, interconnected half-graben basins lying on the continental shelf east of the Rockall Basin and to the northeast of the Porcupine Basin. Water depths in the Slyne Basin, the southernmost of the three linked basins, are approximately 200 m, while water depths in the Erris Basin increase oceanwards from 200 m up to 1500 m and locally to more than 2000 m. Water depths in the Donegal Basin range from less than 200 m through most of the basin to approximately 500 m at its western limit.

The Slyne Basin has a NNE–SSW orientation and is an elongate (140 km × 25 km) half-graben system that changes fault polarity along strike, resulting in the formation of three discrete half-graben sub-basins: the southern, central and northern Slyne Basin (Trueblood and

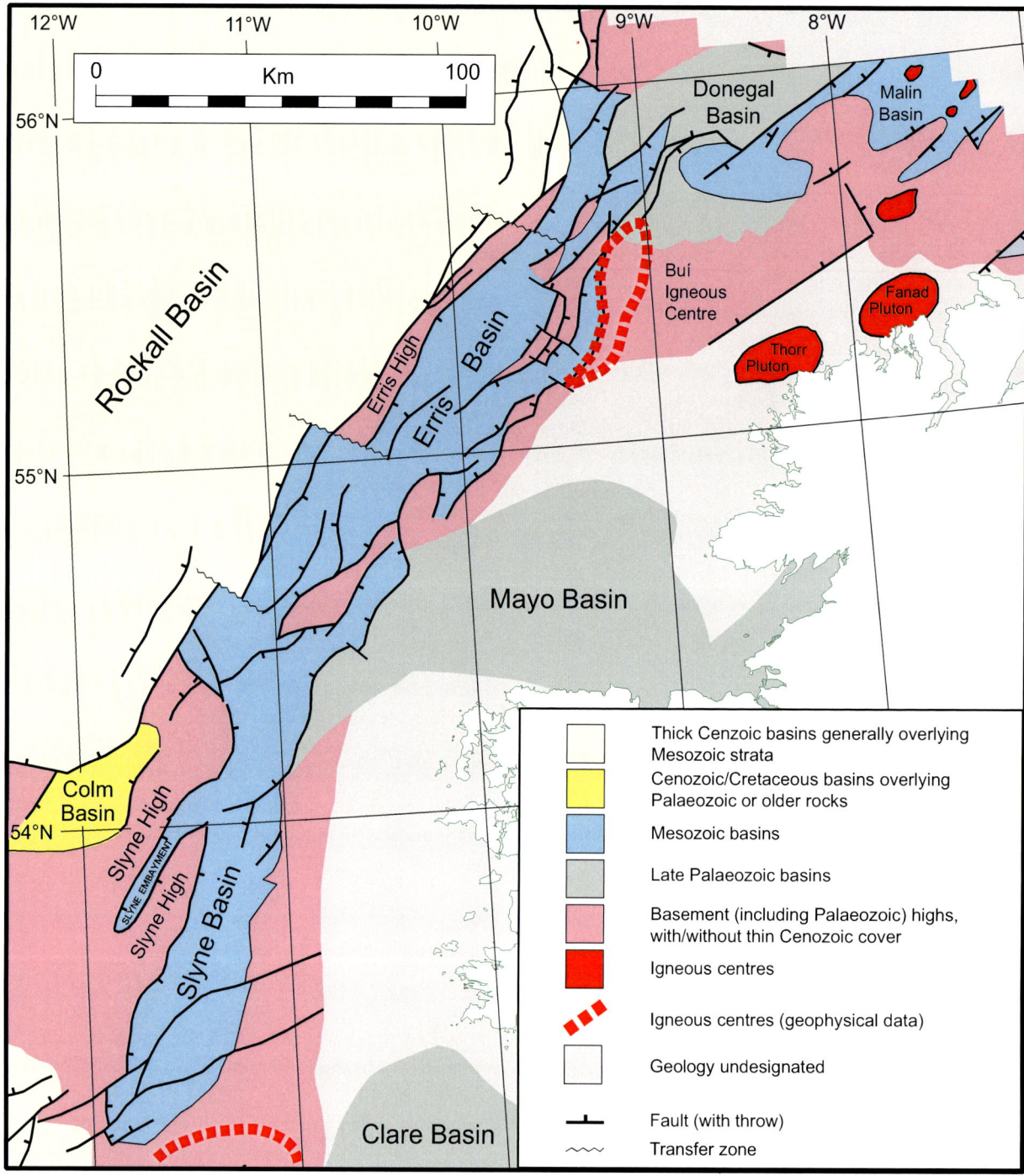

Figure 11.8. Structural elements in the Slyne, Erris and Donegal basins (after Naylor *et al.*, 1999).

Morton, 1991; Scotchman and Thomas, 1995; Dancer *et al.*, 1999). Each shows evidence of long-standing growth faulting from Permo-Triassic through to Late Jurassic times (Figure 11.9). The Erris Basin lies to the northeast of the northern Slyne Basin and the boundary between the basins is marked by a strike change to a more

NE–SW orientation in the Erris Basin. A diffuse faulted basement high at approximately 54°30′N separates the Erris Basin from the northern Slyne Basin (Chapman *et al.*, 1999). Like the Slyne Basin it is a Mesozoic half-graben. A complex buried structural high, the Erris High (Cunningham and Shannon, 1997), defines the western

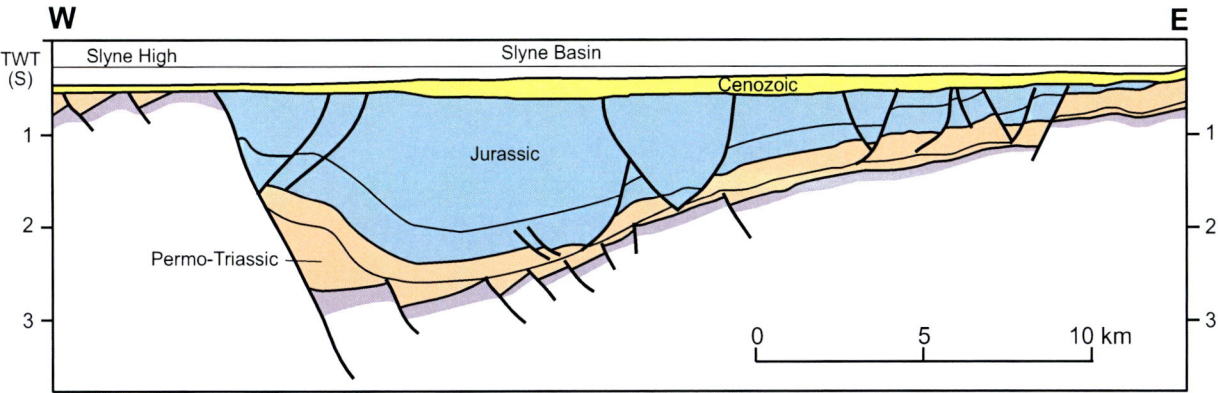

Figure 11.9. Geoseismic cross-section through the Slyne Basin showing asymmetric strata geometry indicative of long-standing Permo-Triassic through Jurassic growth faulting (after Naylor *et al.,* 1999).

boundary of the basin. The Donegal Basin (80x60 km) has an ENE trend, oblique to the NE–SW alignment of the Slyne–Erris trend, and runs approximately parallel to the north coast of Ireland in the area of the Malin Sea.

Pennsylvanian (Westphalian A, B and C: Langsettian to Bolsovian) Coal Measures comprising siltstones, claystones, thin coals and sandstones have been encountered in drilling in the Slyne, Erris and Donegal basins (Tate and Dobson, 1989; Dancer *et al.*, 2005). More than 1500 m of Upper Mississippian and Pennsylvanian fluvial and shallow-marine sandstones were encountered in the Amoco 19/5-1 well in the Erris Basin and these strata probably have a widespread, regional extent. They are broadly similar in age and facies to the succession recorded in the Porcupine Basin to the south, and suggest the presence of a large fluvial to deltaic depocentre prior to the creation and incision of the fault-controlled younger, narrow rift basins. As in the Porcupine Basin, the Carboniferous succession shows little evidence of any Variscan deformation. Halite-rich units, sometimes thick, have been encountered in both the Slyne and Erris basins (e.g. Enterprise 27/5-1 and Amoco 19/5-1 wells), resting unconformably on Upper Carboniferous strata. In both basins they are overlain by clastic-dominant red-bed facies presumed to be Triassic in age (Figure 7.7). A sandy Sherwood Sandstone Group succession and a muddy to silty Mercia Mudstone Group succession is recorded in both the Slyne and the Erris basins. Up to 700 m of Permo-Triassic strata (including Zechstein evaporites), consisting largely of fluvial and locally aeolian sandstones overlain by playa mudstones were penetrated by the Amoco 12/13-1A well in the Erris Basin (Murphy and Croker, 1992; Chapman *et al.*, 1999). The Permo-Triassic succession is likely to represent the deposits of early rifting and the initiation of the narrow rift basins. Lower Jurassic strata have been drilled in both the Slyne (e.g. Elf 27/13-1 and the Corrib gasfield wells) and Erris (e.g. 12/13-1A and 19/5-1 wells) basins. They are predominantly fine-grained clastics, carbonates and anhydrites, similar to the succession in the Hebrides Basin (Trueblood and Morton, 1991). On seismic sections the Lower Jurassic, especially in the Slyne Basin, appears to thicken towards the main bounding faults, suggesting continued rifting and half-graben basin formation. Middle Jurassic strata rest unconformably on the Lower Jurassic strata in the Slyne Basin (Figure 7.7), with the seismic boundary corresponding to a Bajocian limestone. The succession is thick, with more than 2.5 km of fluvial and estuarine sandstones, siltstones and mudstones estimated in the basin centre (Dancer *et al.*, 1999). These have been correlated to the Great Estuarine Group that crops out on the Isle of Skye (Trueblood and Morton, 1991). On seismic profiles the Middle Jurassic succession shows sedimentary thickening towards the basin-bounding master faults, suggestive of continued rifting. The Middle Jurassic section represents a major shift in sedimentation conditions from marine conditions of the Early Jurassic to a coastal plain-dominated depositional environment in the Middle Jurassic. Permo-Triassic to Jurassic strata may also be inferred from seismic character in the structurally deeper parts of the Donegal Basin, analogous to the proven Mesozoic occurrences in the adjacent Malin and Hebridean basins to the northeast.

The Upper Jurassic and Cretaceous succession is generally thin to absent throughout much of the Slyne–Erris–Donegal basin region, largely due to uplift and erosion along the flanks of the Rockall Basin. Up to 500 m of Upper Jurassic varicoloured, silty claystones, interbedded with limestone stringers, sandstone beds and thin coal beds are locally preserved in the Corrib gasfield region of the Slyne Basin (Dancer *et al.*, 2005). Further north, in well Amoco 12/13-1A, approximately 600 m of Lower Cretaceous (Ryazanian–Albian) sub-marine fan sand-prone strata were drilled in the Amoco 12/13-1A well in the Erris Basin. These progressively onlap and thin northwestwards onto the Erris Ridge (Cunningham and Shannon, 1997).

A thin layer of hard white Upper Cretaceous chalk is preserved in the northern part of the Slyne Basin. This was encountered in the Enterprise 18/20 well (Dancer *et al.*, 1999) where it rests unconformably on the Lower Cretaceous and Middle Jurassic strata. The Cenozoic succession in the region rests with a marked unconformity on the Mesozoic strata. It typically consists of marine sandstones, mudstones and volcanics (Trueblood, 1992). The volcanic rocks, dated as Eocene in age (40–43 Ma) in the 18/20-1 well, consist mainly of basaltic lavas intercalated with tuffs and ignimbrites, while the overlying Oligocene to Recent strata are predominantly unconsolidated clays and sands (Dancer *et al.*, 1999).

Exploration potential

Source rocks

Potential oil and gas source rocks are present through the Carboniferous and Jurassic successions of the Slyne, Erris and Donegal basin system. However, the major source rocks are Upper Carboniferous coals and deltaic mudstones, and Lower Jurassic marine mudstones. The presence of the Corrib Gasfield and the recent oil shows reported from the Middle Jurassic sandstones in the Serica 27/4-1 well attest to the presence of working petroleum systems and at least two petroleum source rock successions. Some source potential may lie in the locally preserved Cretaceous to Cenozoic succession, but the generative potential of these is negligible because of the low level of organic maturity resulting from the limited burial history of these sediments.

The source rock in the Corrib Gasfield is assumed to be the Pennsylvanian (Westphalian) Coal Measures, similar to those drilled by the Enterprise 27/5-1 well. The gas is consistent with a Type III humic source rock (Dancer *et al.*, 2005). Serpukhovian to Moscovian (Namurian and Westphalian A–D) shales and coals are potential oil- and gas-prone source rocks in the Erris Basin. In the Amoco 19/5-1 well thinly interbedded grey shales and coals of Langsettian–Duckmantian age (Westphalian A–B: Robeson *et al.*, 1988) have TOC

values ranging from 1 to 52 wt%, and have pyrolysis S2 yields of 0.2 to 262.8 mgg^{-1} and hydrogen indices (HI) of 14 to 679 (Chapman *et al.*, 1999). While the predominant Type III kerogen composition suggests primarily a gas source potential, the presence of Type I and Type II kerogens in thinly bedded exinite-rich coals and oil shales indicates that the Upper Carboniferous section may also have some oil-source potential.

A number of potential oil-prone source rocks occur in the Lower Jurassic succession of the Slyne–Erris region. The Elf 27/13-1 in the Slyne Basin encountered minor biodegraded and water-washed shows in Bathonian–Bajocan sandstones and in Lower Jurassic shales, which were correlated by biomarkers to mature Lower Jurassic shales in deeper parts of the Slyne Basin (Scotchman and Thomas, 1995; Chapman *et al.*, 1999). This well penetrated two intervals of rich oil-prone source rocks in the Lower Jurassic succession. Here the Toarcian Portree Shale Formation interval is approximately 84 m thick (Butterworth *et al.*, 1999) and contains predominantly sapropelic organic matter Type II kerogens that have significant oil generative capacity (Chapman *et al.*, 1999). Measured TOC contents for the interval range from 3.4 to 7.2 wt%, with pyrolysis S2 yields in the range 11.2–39.6 mgg^{-1} and HI values ranging from 329 to 555. The deeper Sinemurian Pabba Shale interval predominantly contains mixed Type II/Type III kerogens and is more gas prone (Scotchman and Thomas, 1995). It is approximately 250 m thick, with TOC values for this interval in the range 2.6–6.5 wt%, while S2 pyrolysis yields are 5.6–24.5 mgg^{-1} and the HI range is 205–377. Basin modelling of the Slyne Basin suggests two phases of hydrocarbon generation (Scotchman and Thomas, 1995). The first occurred in the latest Jurassic and was associated with rifting and rapid burial. This was interrupted by regional uplift and erosion in the Early Cretaceous and resumed following renewed burial in the Late Cretaceous to Early Cenozoic. Basin modelling for the Erris Basin (Chapman *et al.*, 1999) suggests that this latter phase of burial was when the main Lower Jurassic source rocks entered the oil window and generated oil during the Early Eocene to Early Miocene, due partly to burial depths and partly to the elevated geothermal gradient during the Palaeogene. The Upper Carboniferous source rocks may have generated hydrocarbons at various stages during Mesozoic to Cenozoic times, influenced by the multiphase inversion history of the region.

Reservoir rocks and structures

Significant reservoir potential occurs in the Slyne, Erris and Donegal basins at Upper Carboniferous, Triassic, Middle Jurassic and Early Cretaceous levels, while some local reservoir possibilities exist in the Palaeogene succession. Most wells drilled to date have been targeted on four-way dip-closed tilted fault-block footwall structures. Some reservoirs have been proven by drilling in the basins, while further reservoir potential is indicated by regional geological comparisons.

While Devonian strata have not been encountered by drilling to date in the basins, fluvial sandstones similar to those occurring onshore in Ireland, and also in the Orcadian Basin of northern Scotland and in the West Shetland Basin (Clair Oilfield) could be present in the deep parts of the Slyne, Erris and Donegal basins. However, Serpukhovian–Bashkirian (Namurian) fluvial and shallow-marine sandstones, with net thicknesses of almost 70 m and with an average porosity of 22% (Murphy and Croker, 1992) were penetrated in the Amoco 19/5-1 well (Figure 3.7) in the Erris Basin. These sandstones can be expected to have a regional extent through the basins.

The most important reservoirs proven to date in the region are the Triassic Sherwood Sandstone equivalents, encountered in several parts of the Slyne and Erris basins. These fluvial sandstones are thick, extensive and have excellent reservoir characteristics. In the Erris Basin, well Amoco 19/5-1 encountered a net thickness of 167 m with porosities of 20–30%, while Amoco 12/13-1A (Figure 3.7) drilled 100 m of net sandstones with porosities in the range 15–30% (Murphy and Croker, 1992). In the Slyne Basin, this interval represents the main reservoir in the Corrib Gasfield. It is approximately 400 m thick, consisting of high net/gross low-sinuosity braided fluvial channel sandstones with minor sand-flat and playa mudstones. On a large scale it is a uniform sandstone with only subtle facies variations, although on a small scale variations occur in grain size and cementation characteristics that affect fluid flow and will influence productivity (Dancer *et al.*, 2005). Average porosities in the Enterprise 18/20-1 discovery well were low (7%), but improved in some of the later appraisal wells, with porosities of up to 16% and permeabilities in excess of 100 mD. Overall, the field-wide net/gross ratio is 72.1%, the average porosity is moderate to low at 8.5%, and permeabilities average 15.2 mD (Dancer *et al.*, 2005). The highest permeabilities, streaks of up to 736 mD, are recorded in high-stage bar channel facies. The Corrib Gasfield structure was

originally mapped as a tilted fault block, with an associated rollover structure, located in the eastern footwall of the main Rockall Basin. However, following newer and better quality seismic data, the structure was shown to be a relatively simple faulted anticline with a more complex Jurassic and younger succession that is detached from the Triassic reservoir following movement of Triassic salt (Dancer *et al.*, 2005)

The Jurassic (Lower and Middle) also contains proven reservoir potential in the Slyne Basin. The best reservoir section encountered in the Elf 27/13-1 well was a 60 m net sandstone within the Bathonian Great Estuarine Group with average porosities of 20%. Trueblood (1992) correlated this package of sandstones with the Elgol Sandstone Formation in the Hebrides, suggesting regional extent for the succession throughout the Slyne and Erris basins. The Hettangian to Sinemurian Broadford Beds Formation in the Slyne Basin also has reservoir potential, containing sandstones in the range 7–12%. Trueblood (1992) also indicated that the Bajocian Bearreraig Sandstone Formation was not of reservoir quality in the 27/13-1 well, largely due to the high carbonate content of the sandstones. However, comparisons with the Hebridean region suggest that it may be developed in a more sandy facies with reservoir potential in other parts of the Slyne basin or elsewhere in the region. A similar argument was used to indicate possible reservoir potential for the Pliensbachian Scapa Sandstone Formation, which is developed in a sandy siltstone facies in the 27/13-1 well.

The thick Lower Cretaceous (Hauterivian) submarine fan facies in the Amoco 12/13-1A well in the Erris Basin contained in excess of 120 m of net sandstones with porosities in the range 14–35% (Murphy and Croker, 1992). These offer significant reservoir potential, especially towards the margins of the Erris Basin. Within the Porcupine Basin, reservoir quality Aptian–Albian deltaic sandstones, and Eocene deltaic and submarine fan sandstones have been recorded (the latter also recorded in the Faroe Basin) and similar facies may be developed in places in the Slyne, Erris and Donegal basins.

Rockall and Hatton basins
Basin framework

The Rockall and Hatton basins are two of the largest deep-water basins in the Irish offshore (Figure 11.1). Rockall Basin lies oceanwards of the Porcupine Basin and of the Slyne–Erris–Donegal basin system, while the

Hatton Basin lies further west, adjacent to the continental/ocean boundary of the Hatton Continental Margin. Both basins underlie bathymetric troughs, reflecting relative sediment starvation especially during Neogene time. Because they lie in deep water (up to 3 km for much of the Rockall Basin) and due to their great distance from shore, the stratigraphy and detailed geological development of the basins are poorly constrained by drilling. However, recent results from the Dooish gas condensate discovery well (Shell 12/2-1z) on the eastern margin of the Rockall Basin (Tyrrell *et al.*, 2010), together with results from well BP 132/15-1 and others in UK waters (e.g. Musgrove and Mitchener, 1996) help to constrain the deeper, Mesozoic geology of the basin, while shallow boreholes along the eastern margin (Haughton *et al.*, 2005) provide constraints on the shallower, Cenozoic succession in the Irish sector of the basin. Wide-angle seismic data, integrated with seismic reflection data (e.g. Morewood *et al.*, 2004, 2005) have provided an insight into the large-scale sedimentary architecture and the deeper, crustal control on the basin. DSDP and BGS boreholes (Hitchen, 2004) have yielded information on the succession in the Hatton Basin.

While seismic reflection data images only the uppermost (largely Cenozoic and Upper Cretaceous) strata through the central part of the Rockall Basin, dipping syn-rift strata on the basin margins are interpreted as Permo-Trassic to Jurassic sediments (Figure 11.10). However, integration of wide-angle seismic data and occasional good quality normal incidence seismic data suggests that the Rockall Basin contains up to 7 km of Upper Palaeozoic to Holocene rocks (Morewood *et al.*, 2004, 2005), with up to 4.5 km of such strata in the Hatton Basin (Shannon *et al.*, 1995). Although Carboniferous to Jurassic strata have been inferred and modelled in these basins, until recently their presence and distribution were largely conjectural. Recent results from the Shell 12/2-1z Dooish condensate exploration/

appraisal well (Figure 3.7) provide important new information on the pre-Cenozoic succession (Figure 11.11). Tyrrell *et al.* (2010) described a 100 m section of Pennsylvanian (Westphalian) sandstones and shales at the base of the well, at a depth of *c*.4.5 km. This is succeeded by a succession, approximately 180 m thick, of fluvial channel and aeolian sandstones, with lacustrine mudstones, interpreted as being of Lower Permian (Asselian) age. This, in turn, is overlain by an almost 50 m thick predominantly sandy section interpreted as probably Permo-Triassic in age. A thick (128 m) feldspathic sandstone unit of probable Middle Jurassic age overlies the Permo-Triassic succession and consists of fluvial feldspathic sandstones and conglomerates, with calcretes, capped by a thin marine sandstone. Similar successions in neighbouring basins developed in a rift basin setting, and regional comparisons would suggest a similar origin for the Permo-Triassic and Middle Jurassic successions in the Rockall Basin. The seismic architecture for the Middle Jurassic (Tyrrell *et al.*, 2010), showing significant thickness variations, is also consistent with development in a rift setting.

No evidence of Upper Jurassic strata was reported from the Shell 12/2-1z well, where a volcanic unit, of probable Early Cretaceous age, succeeds the Middle Jurassic sandy unit and is, in turn, capped by a thin (*c*.30 m) Lower Cretaceous deep-water mudstone succession, overlain by a thick Upper Cretaceous succession. Musgrove and Mitchener (1996) reported that the BP 132/15-1 well in the UK sector of the eastern Rockall margin encountered a syn-rift Cretaceous (Hauterivian to Cenomanian) succession of mudstones with interbedded limestones lying on crystalline basement, and overlain by a predominantly post-rift mud-dominated Upper Cretaceous succession. This is consistent with the draping and sheet-like geometry of the Upper Cretaceous in the Dooish region, which is also seen in the thick overlying Cenozoic succession (Tyrrell *et al.*, 2010).

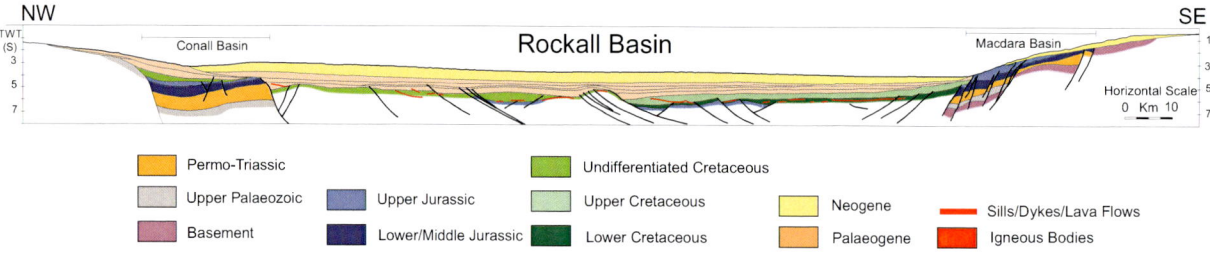

Figure 11.10. Geoseismic cross-section across the Rockall Basin showing evidence of interpreted Jurassic syn-rift blocks beneath an extensive Cretaceous and Cenozoic succession (after Naylor *et al.*, 1999). Location is shown on Figure 11.2.

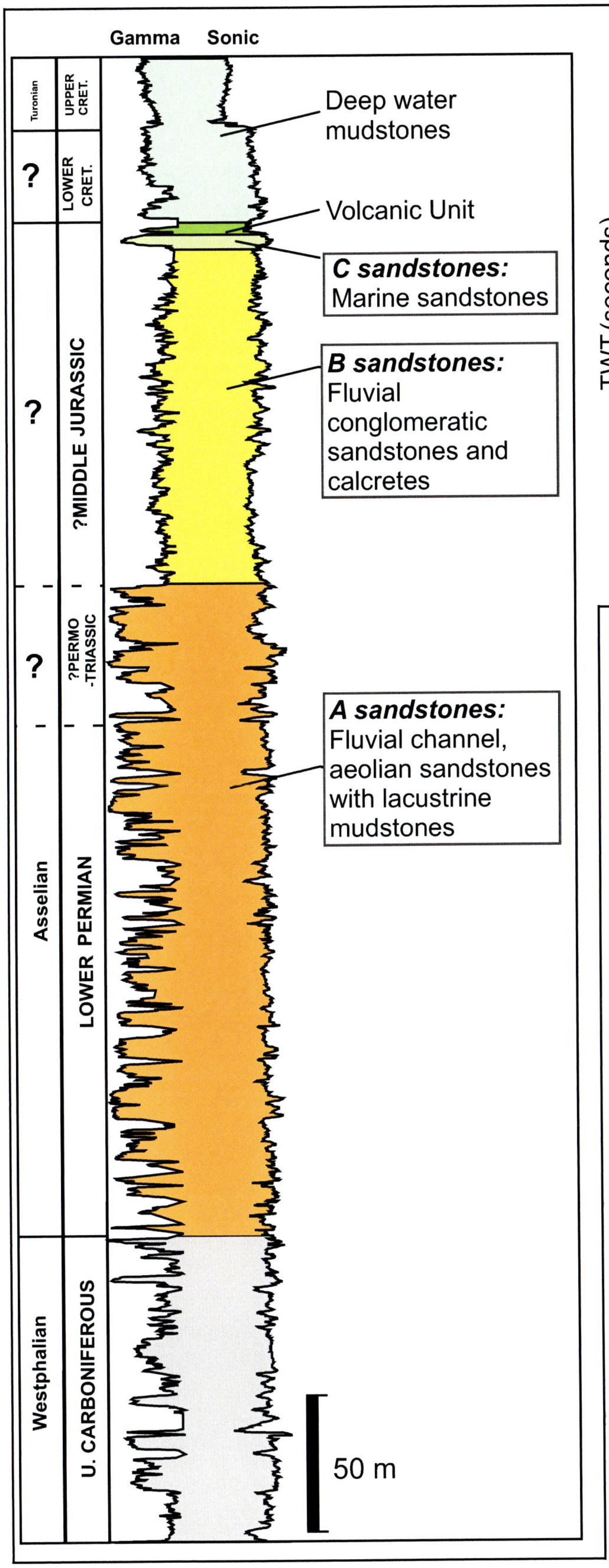

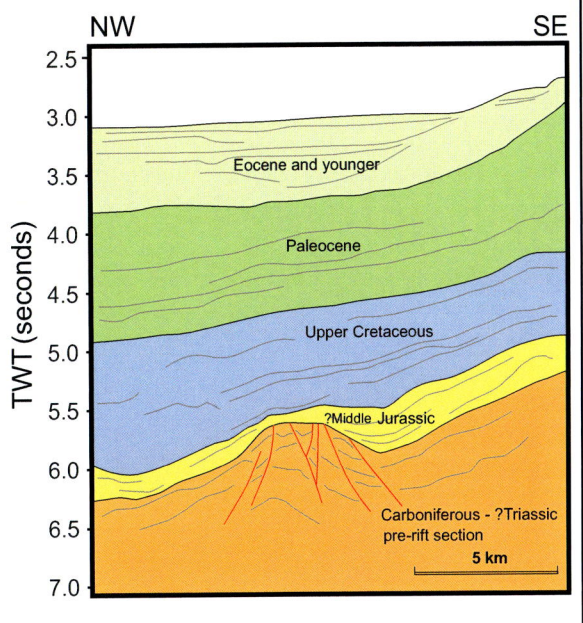

The well log column (left side) shows the following stratigraphic column from top to bottom:

Age	Formation	Description
Turonian / UPPER CRET.		Deep water mudstones
? / LOWER CRET.		Volcanic Unit
? / ?MIDDLE JURASSIC (Asselian / LOWER PERMIAN)		**C sandstones:** Marine sandstones
		B sandstones: Fluvial conglomeratic sandstones and calcretes
? / ?PERMO-TRIASSIC		
? / Asselian / LOWER PERMIAN		**A sandstones:** Fluvial channel, aeolian sandstones with lacustrine mudstones
Westphalian / U. CARBONIFEROUS		

Gamma   Sonic

50 m

Figure 11.11. Dooish well 12/2-1z and geoseismic section (from Tyrrell *et al.,* 2010).

A series of fault-bounded, 'perched' basins, lying on the slopes of the later Cretaceous to Cenozoic Rockall Basin, have been interpreted as containing thick Permo-Triassic to Jurassic strata (Naylor *et al.*, 1999; Walsh *et al.*, 1999). Upper Jurassic strata were proven at the base of the 83/20-sb01 borehole, unconformably overlain by Cretaceous sediments, and may also occur in the more equivocally dated 83/24-sb02 borehole (Figure 9.2) further southwestwards along the margin (Haughton *et al.*, 2005).

A Cenozoic succession, up to 2 km thick, has been inferred in the Rockall Basin on the basis of seismic interpretation and comparison with the succession in the nearby Porcupine Basin (McDonnell and Shannon, 2001). The interpretation is supported by evidence from the BP 132/15-1 well, in which a mud-dominated Paleocene sequence is followed by an Early Eocene sandy succession (McDonnell and Shannon, 2001), overlain in turn by deep marine mudstones above a major early Oligocene unconformity. The latter corresponds to a major deepening of the basin and the creation of steep slopes on the basin (*see* Chapters 2 and 9). Eocene to Pleistocene strata have been drilled on the eastern margin of the Rockall Basin (Haughton *et al.*, 2005), with Neogene deep-marine mud-prone strata drilled at various locations within the Rockall Basin (Stoker *et al.*, 2005a, b).

The geology of the Hatton Basin is poorly constrained by deep wells and much of the understanding of the basin development has come from interpretation of wide-angle seismic data (Shannon *et al.*, 1995). As in the Rockall Basin, a Late Palaeozoic to Jurassic rift succession is suggested, although the Hatton Basin suffered less extension, with rifting in the Jurassic and possibly locally in the Permo-Triassic. The strata occur in two probably fault-controlled sub-basins, with a thickness of up to 3 km. Boreholes on Hatton Bank also drilled Cretaceous (Aptian) shales and sandstones (Hitchen, 2004), suggesting the likelihood of such strata also in the nearby Hatton Basin.

An upper seismic layer, slightly thicker than 1 km, shows a broadly uniform geometry across the basin with a marked velocity contrast and unconformity with the underlying interpreted syn-rift largely Mesozoic succession. This was interpreted by Shannon *et al.* (1995) as due to the inversion and erosion related to the thermal effects of an underplated igneous body at the nearby continental/oceanic boundary to the west. An Eocene to Recent age is suggested for the younger layer on the basis of DSDP boreholes in the basin. Site 116 on the southeast flank of the Hatton Basin penetrated 808 m of Oligocene and Mio-Pliocene ooze and chalk with chert, and 46 m of Upper Eocene carbonate ooze (Naylor and Shannon, 1982). Site 117 on the southeastern margin of the Hatton Basin penetrated 152 m of Oligocene cherty limestones and ooze overlying 152 m of Lower Eocene ooze, sandstones and claystones.

Exploration potential
Source rocks
The lack of deep exploration wells in the Rockall and Hatton basins means that few firm data on source rock potential are available. However, the Dooish gas condensate discovery on the eastern margin of the Rockall Basin demonstrates the presence of a working petroleum system. Similar source rocks to those encountered in the nearby Porcupine, Slyne and Erris basins, and along the Hebrides Shelf in UK waters can be anticipated with reasonable confidence.

Some rich potential source rocks have been sampled on the Hebrides Shelf adjacent to the eastern margin of the Rockall Trough (Hitchen and Stoker, 1993). Middle

Jurassic and Ryazanian (Lower Cretaceous) shales have total organic carbon (TOC) values in the range 3–14% and have pyrolysis yields of up to 80,000 ppm. While these are immature on the Hebrides Shelf, they could be buried to sufficient depth in the Rockall Basin to be mature. Middle Jurassic lacustrine mudstones comprise a major source rock in the Porcupine Basin (*see* above) and could also be anticipated in the Rockall Basin, possibly associated with the fluvial succession similar to that interpreted from the Shell 12/2-1z well. While Upper Jurassic strata have only been encountered rarely along the margins of the basin (Haughton *et al.*, 2005), comparison with nearby basins also lends support to a more widespread presence within the Rockall Basin, and also probably the Hatton Basin. Marine source rocks at this level could therefore be anticipated, and are likely to be thermally mature.

Hitchen (2004) discussed the results from synthetic aperture radar survey of oil slicks produced from natural oil seeps in the Hatton Basin. Several slicks were associated with a gravity low and were interpreted as strong evidence that the low anomaly is due to a sub-basalt Mesozoic (or possible older) basin containing oil-prone source rocks. Hitchen (2004) also documented the presence in a shallow borehole of Albian freshwater or lacustrine mudstones with TOC values of up to 2.04%, S2 yields of up to 4.4 kg/tonne and a HI of 216. While these were immature at the borehole location, they indicate the presence of potential source rocks in the basin.

Reservoir rocks and structures
Little information is available on reservoir characteristics from the Rockall or Hatton basins. However, the Dooish gas condensate discovery indicates the presence of Permo-Triassic and Middle Jurassic reservoirs in a fault-block structure. In addition, reservoirs at other levels (e.g. Early Cretaceous and Early Cenozoic) can be predicted with some confidence on the basis of their occurrence in other nearby Atlantic margin basins (*see* above). The wireline logs of the Permo-Triassic sandstones in the Shell 12/2-1z well (Figure 11.11) suggest a high net/gross ratio. Where cored, they are reported (Tyrrell *et al.*, 2010) as being well sorted medium- to coarse-grained sandstones and could be anticipated to have good porosity and permeability. The Middle Jurassic sandstone section also has a high net/gross ratio and comprises coarse pebbly to fine-grained sandstones, with a higher feldspar content than the underlying Permo-Triassic sandstones. Lower Cretaceous marine fan sandstones,

similar to those drilled in the Amoco 12/13-1A well in the adjacent Erris Basin, may have developed along the margins of the Rockall Basin. Shannon *et al.* (1999) interpreted a series of potential Cretaceous reservoir sequences on the eastern margin of the Rockall Basin. These included a prograding marginal clastic wedge and a potentially sand-prone succession (high seismic amplitudes) that infilled an irregular topography above the Base Cretaceous unconformity. The same authors also interpreted Lower Cenozoic (probably Eocene) fan deposits that may also have reservoir potential towards the base of the slope on the eastern margin of the Rockall Basin (Shannon *et al.*, 1999). These represent potential stratigraphic traps.

Summary of regional traps and play types

Overall, the results from drilling to date have shown that the Irish Atlantic Margin basins contain the geological requirements for the generation and entrapment of oil and gas. Working petroleum systems have also been demonstrated in all of them, with shows or flows of oil, gas and condensate recorded in several instances. A wide range of petroleum plays has been identified. Mature source and reservoir rocks occur at various levels in each of the basins, as described above. Source rocks entered the oil- and gas-generating windows at various times from Jurassic through to Cenozoic times in the different basins, so that hydrocarbons have a long history of generation and migration. Consequently, structural and stratigraphic traps formed from Mesozoic syn-rift through to mid-Cenozoic thermal subsidence and inversion episodes have the potential to be hydrocarbon-charged. Cap rocks (typically mudstones and occasionally salt) are widespread and have generally not been identified as being problematical in any of the basins. However, in a number of places along the margins of the Porcupine Basin, potentially sandy reservoir facies within the Cenozoic run close to the surface and there is a risk of trap failure due to an insufficient thickness of overlying compacted mudstones. In addition, the unknown and untested nature of reservoirs in the southern part of the Porcupine Basin and the Goban Spur remains a major risk factor. The nature of source and reservoir rocks is an unknown exploration factor in the Rockall Basin, whilst the structural complexity of the Slyne and Erris basins is probably the major risk in these basins.

Most of the exploration to date in the basins has been on large structural traps (typically Jurassic tilted fault

blocks), with only a few wells drilled on the many stratigraphic traps which have been identified in the basins. Exploration during the past decade in the Slyne, Erris and Donegal basins has concentrated almost exclusively on structural traps, due at least in part to the difficulty in seismic resolution caused by the extensive sills and lavas in the basins. Likewise exploration has been hindered by seismic data quality through much of the Rockall Basin, although here additional barriers to exploration are posed by the deep water and distance from shore of drilling targets. Exploration is still in its infancy in the Rockall region, where it is likely that both structural and stratigraphic traps occur. Exploration in the Porcupine Basin within the past few years has begun to target some of the large stratigraphic targets at Cretaceous and Early Cenozoic levels. Some of the major traps and play types in the Atlantic margin basins, described by Shannon and Naylor (1998), are shown in Figure 11.12. A brief summary of these follows.

Carboniferous, Triassic and Jurassic tilted fault-block traps

These are most clearly seen along the basin margins of the Porcupine Basin and also in the Slyne Erris and Goban Spur basins. Large fault-block structures with potential Carboniferous reservoirs are best developed along the eastern margin of the Porcupine Basin. Upper Carboniferous deltaic sandstones are likely to be sealed by interbedded shales or by overlying Cretaceous mudstones in the Porcupine Basin, while Carboniferous and Permo-Triassic reservoirs are likely in such structures in the Slyne, Erris and Donegal basins. Source rocks for the Carboniferous tilted fault-block play may be either Carboniferous gas-prone coals or Upper Jurassic oil-prone shales. Jurassic syn-rift tilted fault-block structures occur along the western basin margin and towards the northern part of the Porcupine Basin. While most of the obvious large structures have been drilled in the northern part of the basin, others remain undrilled in the deeper southern region. The Connemara oil accumulation (MacDonald *et al.*, 1987) in Block 26/28 represents a good example of such a structure, with steeply-dipping Jurassic strata capped by marine Cretaceous mudstones (Figure 11.13).

A more subtle Jurassic tilted fault-block structure was the main target in the Phillips 35/8-2 (Spanish Point) gas condensate discovery (Figure 11.14). The BP 43/13-1 well targeted a large tilted fault block with growth faulting in the Upper Jurassic succession (Figure 11.15).

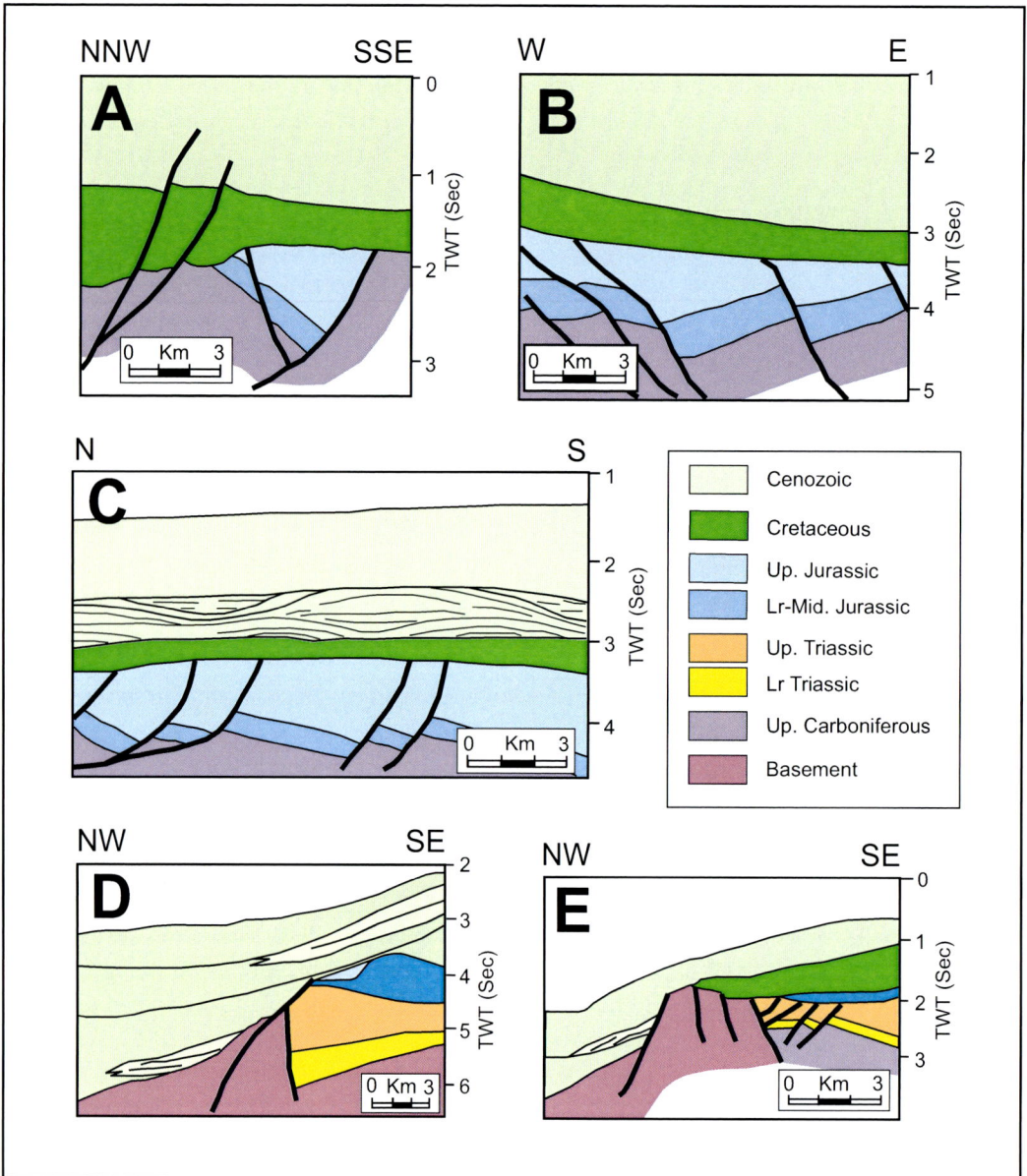

Figure 11.12. Plays and prospects in the Atlantic margin basins (from Shannon and Naylor, 1998). A. Upper Jurassic fault blocks with Middle and Upper Jurassic reservoirs sealed by Upper Jurassic and Cretaceous mudstones. B. Basin edge Jurassic tilted fault blocks with Middle and Upper Jurassic reservoirs sealed by Upper Jurassic and Cretaceous mudstones. C. Palaeogene strati-graphic traps with submarine fan sandstone reservoirs sealed by mudstones. D. Cenozoic stratigraphic traps and Triassic/Jurassic tilted fault blocks. E. Fractured basement horst, Permo-Triassic fault block and Cenozoic stratigraphic traps. Seals are Triassic salts and marls and Cenozoic mudstones.

Unfortunately, while oil shows were recorded, the quality of the Jurassic sandstones was relatively poor. The Corrib Gasfield in the Slyne Basin is also trapped in a large tilted fault-block structure, with a complex history (Corcoran and Mecklenburgh, 2005), probably created in part by footwall uplift associated with the eastern flank of the Rockall Basin. The Dooish gas condensate discovery on the eastern margin of the Rockall Basin is a large tilted fault-block structure sealed by post-rift Upper Cretaceous mudstones (Tyrrell *et al.*, 2010). Similar fault-block structures trapping Jurassic reservoirs may also be present along both margins of the Rockall Basin and also in the Hatton Basin (Shannon *et al.*, 1995). Reservoir units are predicted to consist mainly of Middle and Upper Jurassic fluvial sandstones and submarine fans in the Porcupine Basin, and of Triassic, Middle, and locally Lower, Jurassic

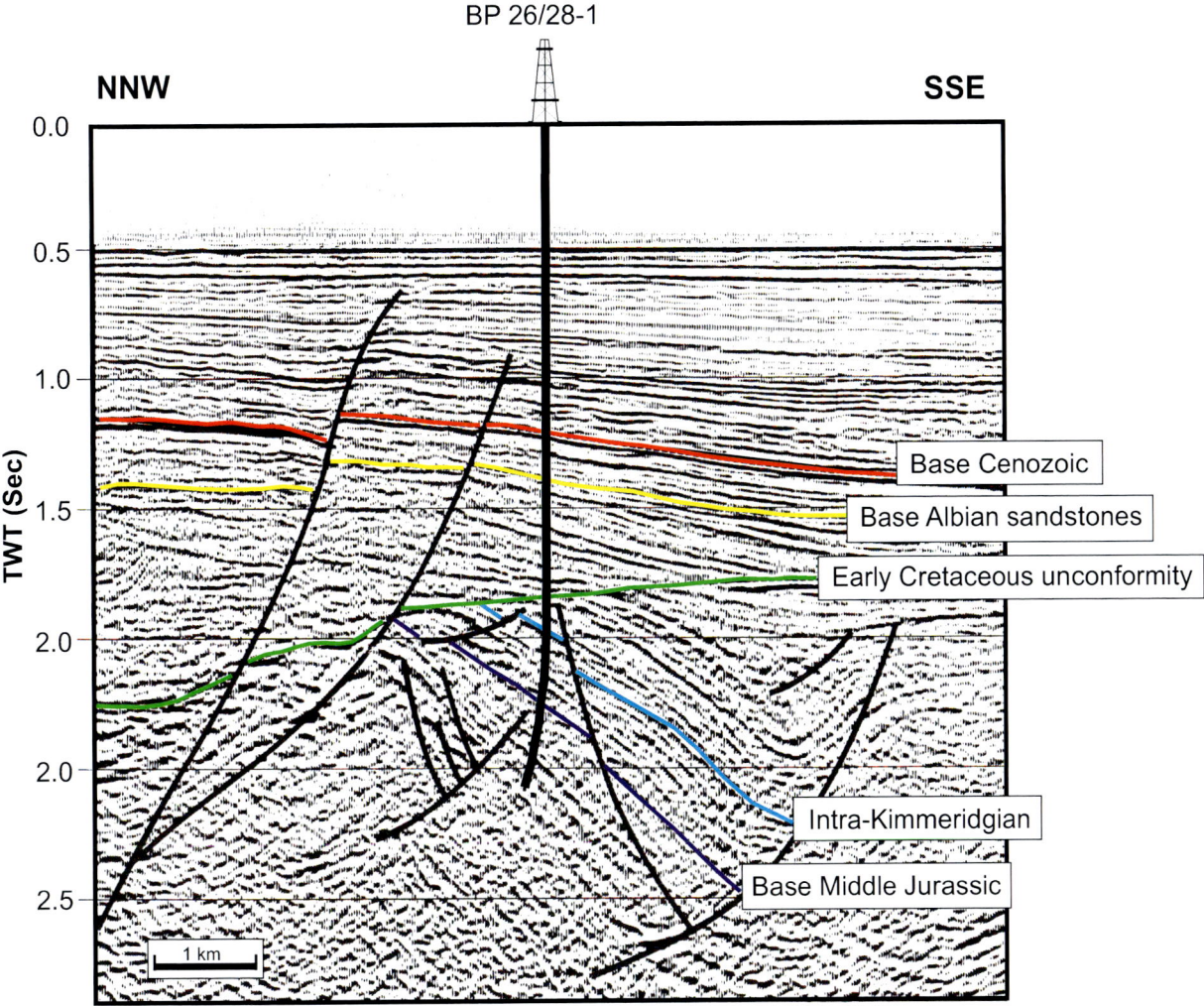

Figure 11.13. Seismic profile through the Connemara oil discovery on Block 26/28 in the Porcupine Basin (from MacDonald *et al.,* 1987).

sandstones in the Slyne and Erris basins. Cap rocks are likely to be the interbedded and also the overlying Lower Cretaceous marine mudstones.

Jurassic, Cretaceous and Cenozoic submarine fans
These are interpreted at various levels within the Porcupine Basin, Slyne, Erris and Donegal basins and along the eastern margin of the Rockall Basin. While those at Jurassic, and sometimes Lower Cretaceous, levels are difficult to image clearly on seismic sections, mounded and probably sand-prone stratigraphic features are identified within the Cenozoic of many of the basins. They are likely to be analogous in age and form to recent stratigraphic discoveries in the North Sea. Upper Jurassic fault-controlled fan deltas are developed along the basin margins and intrabasinal highs in the Porcupine Basin. Closure is either stratigraphic or is fault-related.

Sourcing is likely to be from Upper Jurassic mudstones with topseal provided by Lower Cretaceous mudstones. The location of submarine fans at Cretaceous level in the Porcupine Basin is often topographically controlled, with sand-prone sediment gravity flows preferentially focused into the topographic residual footwall lows developed during early Cretaceous thermal subsidence in the basin. Late Paleocene to Eocene submarine fans in the Porcupine Basin occur as a series of stacked mounds (Figure 11.12C) that prograde from the basin margins. Smaller submarine fans, possibly somewhat more mud-prone, also occur at Miocene level within the basin. Submarine fans are likely to be charged from the basin centre by mature Jurassic and Cretaceous mudstones or Carboniferous coals, with top seal provided by distal fan or overbank mudstones. Trapping is likely to be stratigraphic. Hydrocarbon migration may be facilitated along

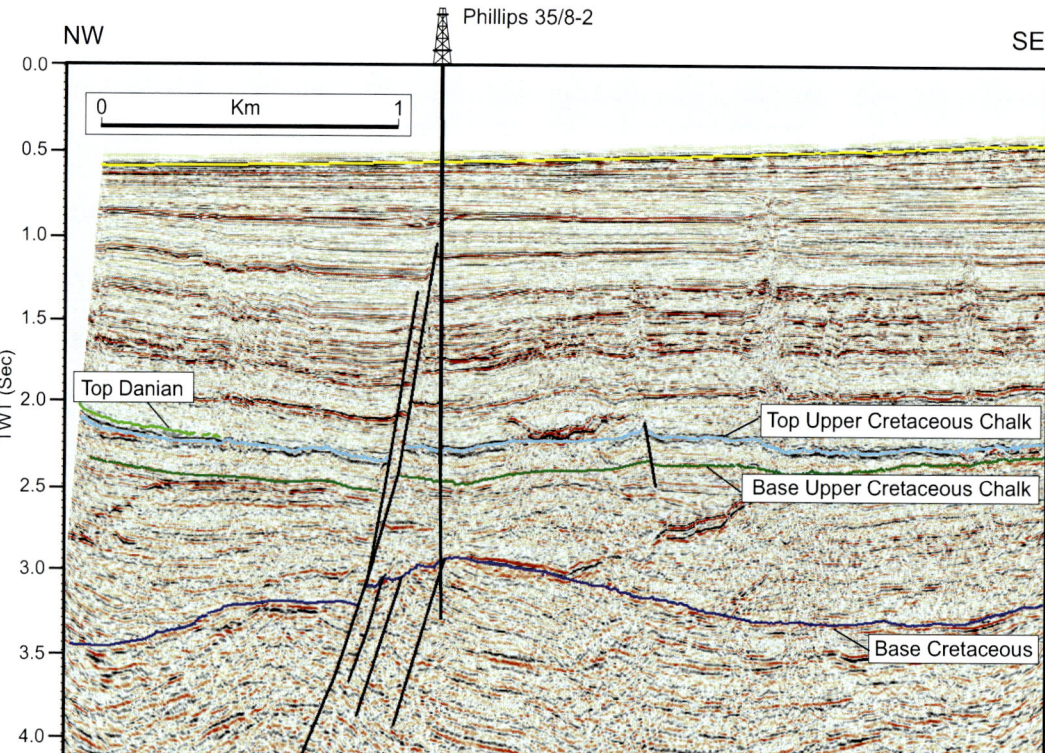

Figure 11.14. Seismic profile through the Phillips 35/8-2 well (Spanish Point gas condensate discovery) showing tilted fault blocks structures beneath onlapping Lower Cretaceous marine mudstones. Seismic profile is published with the permission of Providence Resources plc.

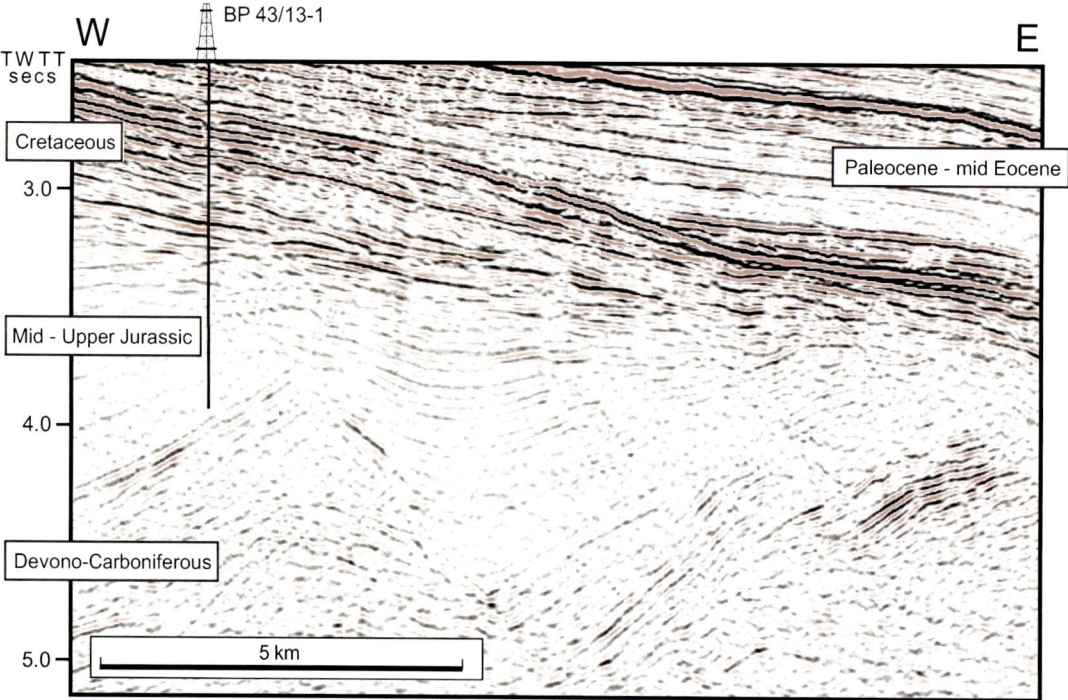

Figure 11.15. Seismic section on the western margin of the Porcupine Basin showing Jurassic tilted fault block structures overlain by wedge-shaped Cretaceous strata and draped by Upper Cretaceous and Cenozoic deep marine successions. The structure was drilled by the BP 43/13-1 well.

small faults or by seismic pumping along microfractures. A number of submarine-fan stratigraphic traps are likely in the Slyne, Erris and Rockall basin, with westward-dipping prograding submarine fans of Cretaceous and Palaeogene ages shed westwards from the Erris Ridge or the Irish mainland during thermal subsidence of the Rockall Basin (Figure 11.12 D, E).

Cretaceous and Cenozoic deltas and clinoforms

Thick Lower Cretaceous (Aptian–Albian) deltaic structures are especially well developed in the northern and eastern parts of the Porcupine Basin. Extensive and thick Lower Cenozoic (Paleocene–Eocene) deltaic sequences are also developed in the northern part of the basin, coeval with the submarine fans further south (Figures 9.3 and 9.4). Both sets of deltaic deposits may have updip fault seal with top-seal provided by overlying delta-top and muddy marine strata. Mature Jurassic or Cretaceous mudstones in the central part of the basin could provide the source. In addition, Johannessen and Steel (2005), using amplitude mapping on 3D seismic data, interpreted elongate, channelised sandy depositional systems on the large Early Cenozoic clinoforms in the Porcupine Basin.

These are likely to be encapsulated in muddy sealing slope facies.

Upper Cretaceous stratigraphic traps

A series of mounded, circular to slightly elongate structures occur at the top of the Upper Cretaceous chalk succession within the Porcupine Basin. They show evidence of bidirectional downlap on seismic sections, are onlapped and draped by uppermost Cretaceous and lowermost Cenozoic strata, and are interpreted as possible biohermal reefal buildups. Such structures were also interpreted in the Goban Spur by Cook (1987). The Upper Cretaceous chalk succession in the Porcupine Basin also contains structures, interpreted as calciturbidites (Moore and Shannon, 1995), which are capped by Lower Cenozoic mudstones. These structures could be sourced by migration along small faults or microfractures by seismic pumping from deeper, mature Jurassic or Lower Cretaceous mudstones.

Basement margin and intrabasinal highs

An intermittent basement ridge separates the Rockall Basin from the Erris Basin and a play, somewhat

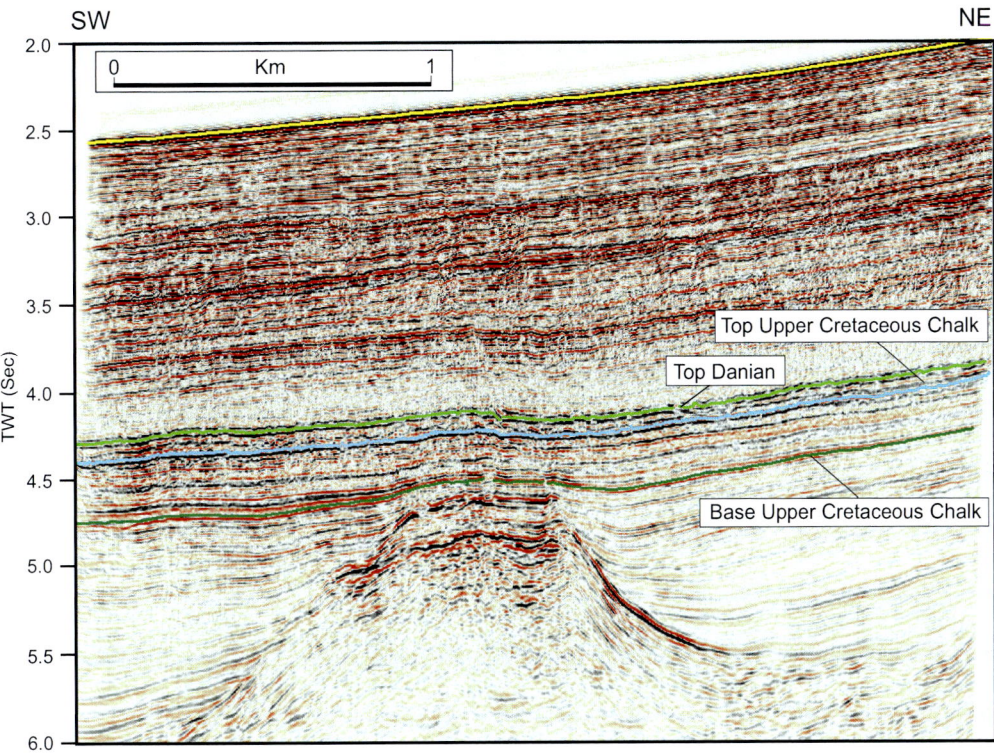

Figure 11.16. Large structural high (Dunquin Prospect) in the central part of the Porcupine Basin (Quadrant 44), variously interpreted as a tilted fault block crest, a volcanic ridge or a serpentinite ridge. It is progressively onlapped and capped by Cretaceous strata, presumed to be predominantly muddy in nature. The top of the structure appears to show some amplitude variations that could be interpreted as fringing reef deposits. Seismic profile is published with the permission of Providence Resources plc.

analogous to that of the Clair Oilfield play on the Rona Ridge in the West Shetland Basin, can be anticipated in the Erris High region (Figures 11.8 and 11.12E). Potential reservoirs include fractured basement, or Carboniferous to Mesozoic sandy drape structures sealed by overlying Cretaceous and Cenozoic shales. The play could be sourced from oil or gas-prone mudstones in either the Erris or the Rockall basins. Occasional large intrabasinal structural highs are seen within the Porcupine Basin. The most obvious of these is in the deep water of the central part of the basin (Quadrant 44). The main structural high as been variously interpreted as a tilted fault-block crest, a volcanic ridge or a serpentinite ridge (*see* Chapter 8). It appears to be of Early Cretaceous (or older) age and is progressively onlapped and capped by Cretaceous strata, presumed to be predominantly muddy in nature. The top of the structure appears to show some amplitude variations (Figure 11.16) that could be interpreted as fringing reef deposits. If so, and if these contain porosity and permeability, then they may represent a reservoir for a very extensive accumulation. However, if the main structural high has an igneous origin with a high heat flow, then the capping and draping potential reservoir target may have had any porosity occluded by hydrothermal cementation. Potential reservoirs could be sourced from Jurassic mudstones downdip or beneath the structure. An intriguing and tantalisingly large prospect (the Dunquin Prospect), with two deep-water targets, has been identified and is receiving close scrutiny since being licensed to Providence Resources, with ExxonMobil later farming in as operator. It has the potential to be a giant field if filled with hydrocarbons. Very little well control is available to constrain the stratigraphy of the structure of this region of the basin. The prospect lies in very deep water (*c*.1600 m) but is sufficiently large that sooner or later it is likely to be drilled.

References

Auffret, A., Pastouret, L., Cassat, G., de Charpal, O., Cravatte, J. *et al.* (1979) Dredged rocks from the Armorican and Celtic margins. *Initial Reports of the Deep Sea Drilling Project*, **48**. U.S. Government Printing Office, Washington, 995–1014.

Boillot, G., Dupeuble, P.A., and Malod, J. (1979) Subduction and tectonics on the continental margin off northern Spain. *Marine Geology*, **32**, 53–70.

Butterworth, P., Holba, A., Hertig, S., Hughes, W., and Atkinson, C. (1999) Jurassic non-marine source rocks and oils of the Porcupine Basin and other North Atlantic margin basins. *In*: Fleet, A.J. and Boldy, S.A.R. (eds), *Petroleum Geology of Northwest Europe: Proceedings of the 5th Conference.* Geological Society, London, 471–486.

Chapman, T.J., Broks, T.M., Corcoran, D.V., Duncan, L.A., and Dancer, P.N. (1999) The structural evolution of the Erris Trough, offshore northwest Ireland, and implications for hydrocarbon generation. *In*: Fleet, A.J. and Boldy, S.A.R. (eds), *Petroleum Geology of Northwest Europe: Proceedings of the 5th Conference.* Geological Society, London, 455–469.

Colin, J.P., Ioannides, N.S., and Vining, B. (1992) Mesozoic stratigraphy of the Goban Spur, offshore south-west Ireland. *Marine and Petroleum Geology*, **9**, 527–241.

Cook, D.R. (1987) The Goban Spur: exploration in a deep-water frontier basin. *In*: Brooks, J. and Glennie, K.W. (eds), *Petroleum Geology of North-West Europe.* Graham and Trotman Ltd, London, 623–632.

Corcoran, D.V. and Mecklenburgh, R. (2005) Exhumation of the Corrib Gas Field, Slyne Basin, offshore Ireland. *Petroleum Geoscience*, **11**, 239–256.

Croker, P.F. (1995) The Clare Basin: a geological and geophysical outline. *In*: Croker, P.F. and Shannon, P.M. (eds), *The Petroleum Geology of Ireland's Offshore Basins.* Geological Society, London, Special Publications, **93**, 327–339.

Croker, P.F. and Shannon, P.M. (1987) The evolution and hydrocarbon prospectivity of the Porcupine Basin, offshore Ireland. *In*: Brooks, J. and Glennie, K.W. (eds), *Petroleum Geology of North-West Europe*, Graham and Trotman, London, 633–642.

Croker, P.F. and Shannon, P.M. (1995) The petroleum geology of Ireland's offshore basins: introduction. *In*: Croker, P.F. and Shannon, P.M. (eds), *The Petroleum Geology of Ireland's Offshore Basins.* Geological Society, London, Special Publication, **93**, 1–8.

Cunningham, G.A. and Shannon, P.M. (1997) The Erris Ridge: a major geological feature in the NW Irish Offshore Basins. *Journal of the Geological Society, London*, **154**, 503–508.

Dancer, P.N., Algar, S.T., and Wilson, I.R. (1999) Structural Evolution of the Slyne Trough. *In*: Fleet, A.J. and Boldy, S.A.R. (eds), *Petroleum Geology of Northwest Europe: Proceedings of the 5th Conference.* Geological Society, London, 445–453.

Dancer, P.N., Kenyon-Roberts, S.M., Downey, J.W., Baillie, J.M., Meadows, N.S., and Maguire, K. (2005) The Corrib gas field, offshore west of Ireland. *In*: Doré, A.G. and Vining, B.A. (eds), *Petroleum Geology: North-West Europe and Global Perspectives: Proceedings of the 6th Petroleum Geology Conference*, Geological Society, London, 1035–1046.

Dingle, R.V. and Scrutton, R.A. (1977) Continental margin fault patterns mapped south-west of Ireland. *Nature*, **268**, 720–722.

Dingle, R.V. and Scrutton, R.A. (1979) Sedimentary succession and tectonic history of a marginal plateau (Goban Spur, south-west of Ireland). *Marine Geology*, **33**, 45–69.

Graciansky, P.C. de, Poag, W.C., Cunningham, R. *et al.* (1985) The Goban Spur transect: geologic evolution of a sediment-starved passive continental margin. *Bulletin of the Geological Society of America*, **96**, 58–76.

Haughton, P., Praeg, D., Shannon, P.M., Harrington, G., Higgs, K., Amy, L., Tyrrell, S., and Morrissey, T. (2005) First results from shallow stratigraphic boreholes on the eastern flank of the Rockall Basin, offshore western Ireland. *In*: Doré, A.G. and Vining, B.A. (eds), *Petroleum Geology: North-West Europe and Global Perspectives – Proceedings of the 6th Petroleum Geology Conference*, Geological Society, London, 1077–1094.

Hauser, F., O'Reilly, B.M., Jacob, A.W.B., Shannon, P.M., Makris, J., and Vogt, U. (1995) The crustal structure of the Rockall Trough: differential stretching without underplating. *Journal of Geophysical Research*, **100**, 4097–4116.

Hitchen, K. and Stoker, M. S. (1993) Mesozoic rocks from the Hebrides Shelf and implications for hydrocarbon prospectivity in the northern Rockall Trough. *Marine and Petroleum Geology*, **10**, 246–254.

Hitchen, K. (2004) The geology of the UK Hatton–Rockall margin. *Marine and Petroleum Geology*, **21**, 993–1012.

Johannessen, E.P. and Steel, R.J. (2005) Shelf-margin clinoforms and prediction of deepwater sands. *Basin Research*, **17**, 521–550.

Jones, S.M., White, N., and Lovell, B. (2001) Cenozoic and Cretaceous transient uplift in the Porcupine Basin and its relationship to a mantle plume. *In*: Shannon, P.M., Haughton, P.D.W., and Corcoran, D.V. (eds), *The Petroleum Exploration of Ireland's Offshore Basins*. Geological Society, London, Special Publications, **188**, 345–360.

MacDonald, H., Allan, P.M., and Lovell, J.P.B. (1987) Geology of oil accumulation in Block 26/28, Porcupine Basin, offshore Ireland. *In*: Brooks, J. and Glennie, K.W. (eds), *Petroleum Geology of North-West Europe*. Graham and Trotman Ltd, London, 643–651.

Masson, D.G. and Parson, L.M. (1983) Eocene deformation on the continental margin SW of the British Isles. *Journal of the Geological Society, London*, **140**, 913–920.

Masson, D.G., Montadert, L., and Scrutton, R.A. (1985) Regional geology of the Goban Spur Continental Margin. *In*: De Graciansky, P.C., Poag, C.W. *et al.* (eds), *Initial Reports of the Deep Sea Drilling Project*, **80**. U.S. Government Printing Office, Washington D.C., 1115–1139.

Masson, D.G., Dobson, M.R., Auzende, J.-M., Cousin, M., Coutelle, A., Rolet, J., and Vaillant, P. (1989) Geology of Porcupine Bank and Goban Spur, Northeastern Atlantic – preliminary results of the CYAPORC submersible cruise. *Marine Geology*, **87**, 105–119.

McDonnell, A. and Shannon, P.M. (2001) Comparative Tertiary basin development in the Porcupine and Rockall basins. *In*: Shannon, P.M., Haughton, P.D.W., and Corcoran, D.V. (eds), *The Petroleum Exploration of Ireland's Offshore Basins*. Geological Society, London, Special Publications, **188**, 323–344.

Montadert, L., Roberts, D.G., Auffret, G.A., Bock, W.D., du Peuble, P.A., Hailwood, E.A., Harrison, W.E., Kagami, H., Lumsden, D.N., Muller, C.M., Thompson, R.W., Thompson, T.L., and Timofeev, P.P. (1979) *Initial Reports of the Deep Sea Drilling Project*, Washington, U.S. Government Printing Office, **48**, 1183pp.

Moore, J. and Shannon, P.M. (1992) Palaeocene–Eocene deltaic sedimentation, Porcupine Basin, offshore Ireland: a sequence stratigraphic approach. *First Break*, **10**, 461–469.

Moore, J.G. and Shannon, P.M. (1995) The Cretaceous succession in the Porcupine Basin, Offshore Ireland: facies distribution and hydrocarbon potential. *In*: Croker, P.F. and Shannon, P.M., (eds), *The Petroleum Geology of Ireland's Offshore Basins*. Geological Society, London, Special Publications, **93**, 345–370.

Morewood, N.C., Mackenzie, G.D., Shannon, P.M., O'Reilly, B.M., Readman, P.W., and Makris, J. (2005) The crustal structure and regional development of the Irish Atlantic margin region. *In*: Doré, A.G. and Vining, B. (eds), *Petroleum Geology: North-West Europe and Global Perspectives – Proceedings of the 6th Petroleum Geology Conference*. Geological Society, London, 1023–1034.

Morewood, N.C., Shannon, P.M., Mackenzie, G.D. (2004) Seismic stratigraphy of the southern Rockall Basin: a comparison between wide-angle seismic and normal incidence reflection data. *Marine and Petroleum Geology*, **21**, 1149–1163.

Murphy, N.J. and Croker, P.F. (1992) Many play concepts seen over wide area in Erris, Slyne troughs off Ireland. *Oil and Gas Journal*, Sept. 14, 92–97.

Musgrove, F.W. and Mitchener, B. (1996) Analysis of the pre-Tertiary rifting history of the Rockall Trough. *Petroleum Geoscience*, **2**, 353–360.

Naeth, J., di Primio, R., Horsfield, B., Schaefer, R.G., Shannon, P.M., Bailey, W.R., and Henriet, J.P. (2005) Hydrocarbon seepage and carbonate mound formation: a basin modelling study from the Porcupine Basin (offshore Ireland). *Journal of Petroleum Geology*, **28**, 147–166.

Naylor, D. (1972) Hydrocarbon potential of offshore Western Britain and Ireland. *Financial Times Second North Sea Conference*. I.P.C. Industrial Press Limited, 130–137.

Naylor, D. and Anstey, N.A. (1987) A reflection seismic study of the Porcupine Basin, offshore West Ireland. *Irish Journal of Earth Sciences*, **8**, 187–210.

Naylor, D. and Shannon, P. M. (1982) *The Geology of Offshore Ireland and West Britain*. Graham and Trotman, London, 161pp.

Naylor, D. and Shannon, P. M. (2005) The structural framework of the Irish Atlantic Margin. *In*: Doré, A.G. and Vining, B.A. (eds), *Petroleum Geology: North-West Europe and Global Perspectives – Proceedings of the 6th Petroleum Geology Conference*. Geological Society, London, 1009–1021.

Naylor, D. and Shannon, P. M. (2009) Geology of offshore Ireland. *In*: Holland, C.H. and Sanders, I.S. (eds), *The Geology of Ireland*. Second Edition. Dunedin Academic Press, Edinburgh, 405–460.

Naylor, D., Shannon, P.M., and Murphy, N. (1999) *Irish Rockall Basin region – a standard structural nomenclature system*. Petroleum Affairs Division, Dublin, Special Publication **1/99**, 42pp.

Naylor, D., Shannon, P.M., and Murphy, N. (2002) *Porcupine-Goban region – a standard structural nomenclature system*. Petroleum Affairs Division, Dublin, Special Publication **1/02**, 65pp.

O'Reilly, B.M., Hauser, F., Jacob, A.W.B., and Shannon, P.M. (1996) The lithosphere below the Rockall Trough: wide-angle seismic evidence for extensive serpentinisation. *Tectonophysics*, **255**, 1–23.

O'Reilly, B.M., Hauser, F., Ravaut, C., Shannon, P.M., and Readman, P.W. (2006) Crustal thinning, mantle exhumation and serpentinisation in the Porcupine Basin, offshore Ireland: evidence from wide-angle seismics. *Journal of the Geological Society, London*, **163**, 775–787.

O'Reilly, B.M., Readman, P.W., and Murphy, T. (1996) The gravity signature of Caledonian and Variscan tectonics in Ireland. *Physics and Chemistry of the Earth*, **21**, 299–304.

Praeg, D., Stoker, M.S., Shannon, P.M., Ceramicola, S., Hjelstuen, B.O., and Mathiesen, A. (2005) Episodic Cenozoic tectonism and the development of the NW European 'passive' continental margin. *Marine and Petroleum Geology*, **22**, 1007–1030.

Readman, P.W., O'Reilly, B.M., Shannon, P.M., and Naylor, D. (2005) The deep structure of the Porcupine Basin, offshore Ireland, from gravity and magnetic studies. *In*: Doré, A.G. and Vining, B. (eds), *Petroleum Geology: North-West Europe and Global Perspectives – Proceedings of the 6th Petroleum Geology Conference*. Geological Society, London, 1047–1056.

Roberts, D.G., Masson, D.G., Montadert, L., and De Charpal, O. (1981) Continental margin from the Porcupine Seabight to the Armorican marginal basin. *In*: Illing, L.V. and Hobson, G.D. (eds.), *Petroleum Geology of the Continental Shelf of North-West Europe: Proceedings of the 2ⁿᵈ Conference*. Heyden and Son, London. 455–473.

Robeson, D., Burnett, R.D., and Clayton, G. (1988) The Upper Palaeozoic geology of the Porcupine, Erris and Donegal Basins, offshore Ireland. *Irish Journal of Earth Sciences*, **9**, 153–175.

Robinson, A.J. and Canham, A.C. (2001) Reservoir characteristics of the Upper Jurassic sequence in the 35/8-2 discovery, Porcupine Basin. *In*: Shannon, P.M., Haughton, P.D.W., and Corcoran, D.V. (eds), *The Petroleum Exploration of Ireland's Offshore Basins*. Geological Society, London, Special Publications, **188**, 301–321.

Robinson, K.W., Shannon, P.M., and Young, D.G.G. (1981) The Fastnet Basin: an integrated analysis. *In*: Illing, L.G. and Hobson, G.D. (eds), *Petroleum Geology of the Continental Shelf of North-West Europe*. Heyden and Son Ltd, London, 444–454.

Scotchman, I.C. and Thomas, J.R.W. (1995) Maturity and hydrocarbon generation in the Slyne Trough, northwest Ireland. *In*: Croker, P.F. and Shannon, P.M. (eds), *The Petroleum Geology of Ireland's Offshore Basins*. Geological Society, London, Special Publications, **93**, 385–411.

Scrutton, R.A. (1979) Structure of the crust and upper mantle at Goban Spur southwest of the British Isles – some implications for margin studies. *Tectonophysics*, **59**, 201–215.

Shannon, P.M. (1991a) The development of Irish offshore sedimentary basins. *Journal of the Geological Society, London*, **148**, 181–189.

Shannon, P.M. (1991b) Tectonic framework and petroleum potential of the Celtic Sea, Ireland. *First Break*, **9**, 107–122.

Shannon, P.M. (1992) Early Tertiary submarine fan deposits in the Porcupine Basin, offshore Ireland. *In*: Parnell, J. (ed.), *Basins on the Atlantic Seaboard: Petroleum Geology, Sedimentology and Basin Evolution*. Geological Society, London, Special Publications, **62**, 351–373.

Shannon, P.M. (1993) Submarine Fan Types in the Porcupine

Basin, Ireland. *In*: Spencer, A.M. (ed.), *Generation, Accumulation and Production of Europe's Hydrocarbons. III*. Special Publication of the European Association of Petroleum Geoscientists 3, Springer-Verlag, Berlin, 111–120.

Shannon, P.M. and Naylor, D. (1998) An assessment of Irish offshore basins and petroleum plays. *Journal of Petroleum Geology*, **21**, 125–152.

Shannon, P.M., Moore, J.G., Jacob, A.W.B., and Makris, J. (1993) Cretaceous and Tertiary basin development west of Ireland. *In*: Parker, J.R. (ed.), *Petroleum Geology of North-west Europe: Proceedings of the 4th Conference*. Geological Society, London, 1057–1066.

Shannon, P.M., Jacob, A.W.B., Makris, J, O'Reilly, B., Hauser, F., Readman, P.W. and Makris, J. (1999) Structural setting, geological development and basin modelling in the Rockall Trough. *In*: Fleet, A.G. and Boldy, S.A.R (eds), *Petroleum Geology of Northwest Europe: Proceedings of the 5th Conference*. Geological Society, London, 421–431.

Shannon, P.M., Jacob, A.W.B., Makris, J., O'Reilly, B., Hauser, F., and Vogt, U. (1995) Basin development and petroleum prospectivity of the Rockall and Hatton region. *In*: Croker, P.F. and Shannon, P.M. (eds), *The Petroleum Geology of Ireland's Offshore Basins*. Geological Society, London, Special Publications, **93**, 435–457.

Sinclair, I.K., Shannon, P.M., Williams, B.P.J., Harker, S.D., and Moore, J.G. (1994) Tectonic control on sedimentary evolution of three North Atlantic borderland Mesozoic basins. *Basin Research*, **6**, 193–218.

Stoker, M.S., Praeg, D., Hjelstuen, B.O., Laberg, J.S., Nielsen, T., and Shannon, P.M. (2005b) Neogene stratigraphy and the sedimentary and oceanographic development of the NW European Atlantic margin. *Marine and Petroleum Geology*, **22**, 977–1005.

Stoker, M.S., Praeg, D., Shannon, P.M., Hjelstuen, B.O., Laberg, J.S., van Weering, T.C.E., Sejrup, H.P., and Evans, D. (2005a) Neogene evolution of the Atlantic continental margin of NW Europe (Lofoten Islands to SW Ireland): anything but passive. *In*: Doré, A.G. and Vining, B. (eds), *Petroleum Geology: North-West Europe and Global Perspectives – Proceedings of the 6th Petroleum Geology Conference*. Geological Society, London, 1057–1076.

Štolfová, K. and Shannon, P.M. (2009) Permo-Triassic development from Ireland to Norway: basin architecture and regional controls. *Geological Journal*, **44**, 652–676.

Tate, M.P. and Dobson, M.R. (1989) Pre-Mesozoic geology of the western and north-western Irish continental shelf. *Journal of the Geological Society of London*, **146**, 229–240.

Trueblood, S.P. and Morton, N. (1991) Comparative sequence stratigraphy and structural styles of the Slyne Trough and Hebrides Basin. *Journal of the Geological Society, London*, **148**, 197–201.

Trueblood, S. (1992) Petroleum geology of the Slyne Trough and adjacent basins. *In*: Parnell, J. (ed.), *Basins of the Atlantic Seaboard: Petroleum Geology, Sedimentology and Basin Evolution*. Geological Society, London, Special Publications, **62**, 315–326.

Tyrrell, S., Souders, A.K., Haughton, P.D.W., Daly, J.S., and Shannon, P.M. (2010) Sedimentology, sandstone provenance and palaeodrainage on the eastern Rockall Basin margin: evidence from the Pb isotopic composition of detrital K-feldspar. *In*: Vining, B.A. and Pickering, S.C. (eds), *Petroleum Geology: From Mature Basins to New Frontiers – Proceedings of the 7th Petroleum Geology Conference.* Geological Society, London, 937-952.

Vogt, U., Makris, J., O'Reilly, B.M., Hauser, F., Readman, P.W., Jacob, A.W.B., and Shannon, P.M. (1998) The Hatton Basin and continental margin: crustal structure from wide-angle seismic and gravity data. *Journal of Geophysical Research*, **103**, 12545–12566.

Walsh, A., Knag, G., Morris, H., Quinquis, H., Tricker, P., Bird, C., and Bower, S. (1999) Petroleum geology of the Irish Rockall Trough. *In*: Fleet, A.J. and Boldy, S.A.R. (eds), *Petroleum Geology of Northwest Europe: Proceedings of the 5th Conference.* Geological Society, London, 433–444.

Chapter 12

Northern Ireland and Irish Sea basins

Regional setting

The basins along the northern and eastern margins of Ireland differ from those of the other offshore regions in that the basin fill largely comprises Permian and Triassic strata. This is in direct contrast to the thick Cretaceous and Cenozoic sequences of the Atlantic margin basins, or the basins in the Celtic Sea region that contain little Permian, but have thick Triassic and Jurassic successions. In consequence, the main petroleum play in the region is the classic search for hydrocarbons in Permo-Triassic reservoirs sourced from underlying Carboniferous sequences. Permian sandstones sealed by the Upper Permian Belfast Harbour Group, and Sherwood Sandstone Group reservoirs sealed by the overlying Mercia Mudstone Group are the main targets. A secondary play exists in Carboniferous sandstone reservoirs, but modest porosities and difficulty in defining targets beneath the younger basins are negative factors.

The marked Caledonian (NE–SW) orientation of some Carboniferous basins, notably the Larne–North Channel and Solway–Peel basins, points to reactivated Caledonian control on Carboniferous deposition. Late Carboniferous inversion and erosion, partly controlled by the Caledonian faults, probably resulted in the removal of large thicknesses of Upper Carboniferous strata from the Irish Sea basins. Corcoran and Clayton (1999) and Quirk et al. (1999) emphasised the importance of Early Permian thermal uplift and erosion in the central Irish Sea area. Erosion of the Carboniferous sequences at the basin margins was also widespread during the Triassic, as evidenced by re-worked Carboniferous spores (Warrington, 1994; Newman, 1999; Naylor and Clayton, 2000).

A number of authors have offered observations regarding basin development and the pattern of preserved stratigraphy of the Irish Sea region in terms of phases of exhumation, erosion or non-deposition (e.g. Cope, 1984, 1997; Green et al., 1997, 2001). There is general agreement that multiphase basin inversion has played a significant role in the region. Apatite fission-track analysis (AFTA), vitrinite reflectance (VR) studies and sonic velocity profile data, together with tectonic modelling, have been used to estimate the timing and the degree of exhumation of the inversion events. The stratigraphic control on exhumation timing, as well as the amount of available and published data, differs from basin to basin. The deep erosion of Mesozoic stratigraphy in the Irish Sea has been interpreted as due to crustal uplift and denudation resulting from plume-related transient dynamic uplift under the Irish Sea during the Palaeocene (Cope, 1994, 1997), or caused by igneous underplating (Brodie and White, 1995; White and Lovell, 1997). Green et al. (1997) identified at least four palaeo-thermal events that can be recognised throughout the region, based on AFTA and VR studies. These are:

- pre-Permian (>290 Ma);
- Late Permian to mid- Triassic (260–220 Ma);
- Early Cretaceous (140–110 Ma); and
- Early Palaeogene (prior to ~60 Ma).

Ware and Turner (2002) speculate that 'the present basin-and-high configuration of the Irish Sea area is a product of Neogene compressional reactivation of Palaeozoic basement faults beneath a once-continuous basin system'. However, modelling of AFTA and VR data by Allen at al. (2002) for onshore Ireland points to a more complex history. For example, the models indicate significant denudation during the Cretaceous in east-central Ireland, posing the possibility that the absence of Cretaceous rocks in the adjacent central Irish Sea may be at least partly due to Cretaceous non-deposition, rather than solely to erosion during a Paleocene crustal uplift event. The interaction between the various factors operating in the region differed from basin to basin, so that each has a different development history. In turn, this has impacted on the exploration potential, which also varies across the region.

Key observations regarding the stratigraphy of the region are:

a) Permian and Triassic strata were deposited on the eroded Variscan surface;

b) no Jurassic strata younger than Lower Jurassic have been proven north of the Cardigan Bay–St George's Channel basins; and

c) no Cretaceous strata are preserved in the region between the onshore Larne Basin and Cardigan Bay Basin.

Loch Indaal, Foyle and Rathlin basins
Basin framework

This group of basins contain up to 2 to 3 km of Permo-Triassic sediments. The thickest sections occur against the controlling faults of the half-grabens – against the Leannan Fault in the Loch Indaal Basin, as shown in offshore boreholes, against the Lough Foyle Fault in the case of the small Foyle Basin, and in the hanging wall of the Tow Valley Fault of the Rathlin Basin (Figure 6.1A).

The Loch Indaal half-graben is entirely offshore, abutting the Leannan Fault to the northwest and onlapping the Middle Bank positive element to the southeast. More than 2 km of dominantly red-bed Permo-Triassic sediments are postulated against the Leannan Fault (Evans *et al.*, 1980; Figure 6.1). The Foyle Basin, located at the mouth of Lough Foyle, has the same sense of polarity but extends onshore where it was drilled by the Magilligan borehole (Geological Survey of Great Britain, 1965; Wilson, 1983). The borehole proved a thick Mesozoic sequence, but found the Sherwood Sandstone resting directly on Carboniferous, without intervening Permian. In the Rathlin Basin, with which the Foyle Basin is certainly contiguous at upper levels of Mesozoic stratigraphy, the deep Port More borehole (Figure 6.1) terminated in conglomeratic sandstones believed to be Permian, and assigned to the Enler Group. The cross-section shown in Figure 12.1B (after McCann, 1988) is in several respects at variance with the gravity model in Figure 12.2 (after Reay, 2004a). The gravity model suggests a shallowing to almost one kilometre of the base Permo-Triassic in the coastal zone between the Corbally and Magilligan boreholes, and a deepening of the same surface westwards from Magilligan to the Foyle Fault. Also, the gravity model shown indicates that the base Permo-Triassic in the vicinity is at a depth approaching 3.5 km, which suggests a thick Permian sequence. However, McCann (1988), who used the gravity profile for guidance, suggested a thick Carboniferous section beneath the base of the Port More borehole (Figure 12.1B).

In this group of basins, as in the onshore Larne and Lough Neagh basins, the remnant stratigraphy places only wide constraints on events. Permian or Triassic rocks rest unconformably on Carboniferous or older strata, indicating late Carboniferous to early Permian inversion and erosion. There is no preserved (or at least proven) stratigraphic record in these basins for the interval between Liassic (Pliensbachian) and the Hibernian Greensand–Ulster White Limestone formations (Cenomanian to early Maastrichtian). The Paleocene Antrim Lava Group (Preston, 2001) rests unconformably on the karstified limestone surface.

Exploration potential

The problems of imaging Mesozoic and older sediments on seismic data beneath a thick basalt layer are further exacerbated in the Rathlin–Lough Foyle basins by reduced impedance contrast at Top Sherwood Sandstone level and lack of carbonate development ('Magnesian Limestone') within the Belfast Group. Both these horizons provide reasonable seismic reflectors elsewhere in the region. As a result, little is known regarding the subsurface structure within the two basins.

Source rocks

The potential source rocks of the Brigantian to Pendleian Ballycastle Group (Figure 4.11) were deposited in a separate fault-controlled basin north of the Highland Boundary Ridge. The group is exposed along the coastline around Ballycastle, as well as being known from mining activities, and a similar sequence was also penetrated in the Magilligan borehole in the Lough Foyle Basin (Figure 6.1). Conditions of deposition ranged from shallow-marine to deltaic and coal swamp environments, with brief marine transgressions giving rise to thin carbonates. The original depositional basin probably extended from Fair Head in the east to Lough Foyle in the west (Mitchell, 2004), although the present disposition of Carboniferous rocks beneath the Rathlin Basin is unknown. The coals in the sequence at Ballycastle contain mainly gas-prone organic material, although cannel coals and oil shales (TOC >12%) may have oil potential. If these source rocks are present beneath the Rathlin Basin, then modelling suggests (Figure 12.3C, after Reay, 2004b) that in the deeper parts of the basin they entered the oil window in Jurassic time and may have

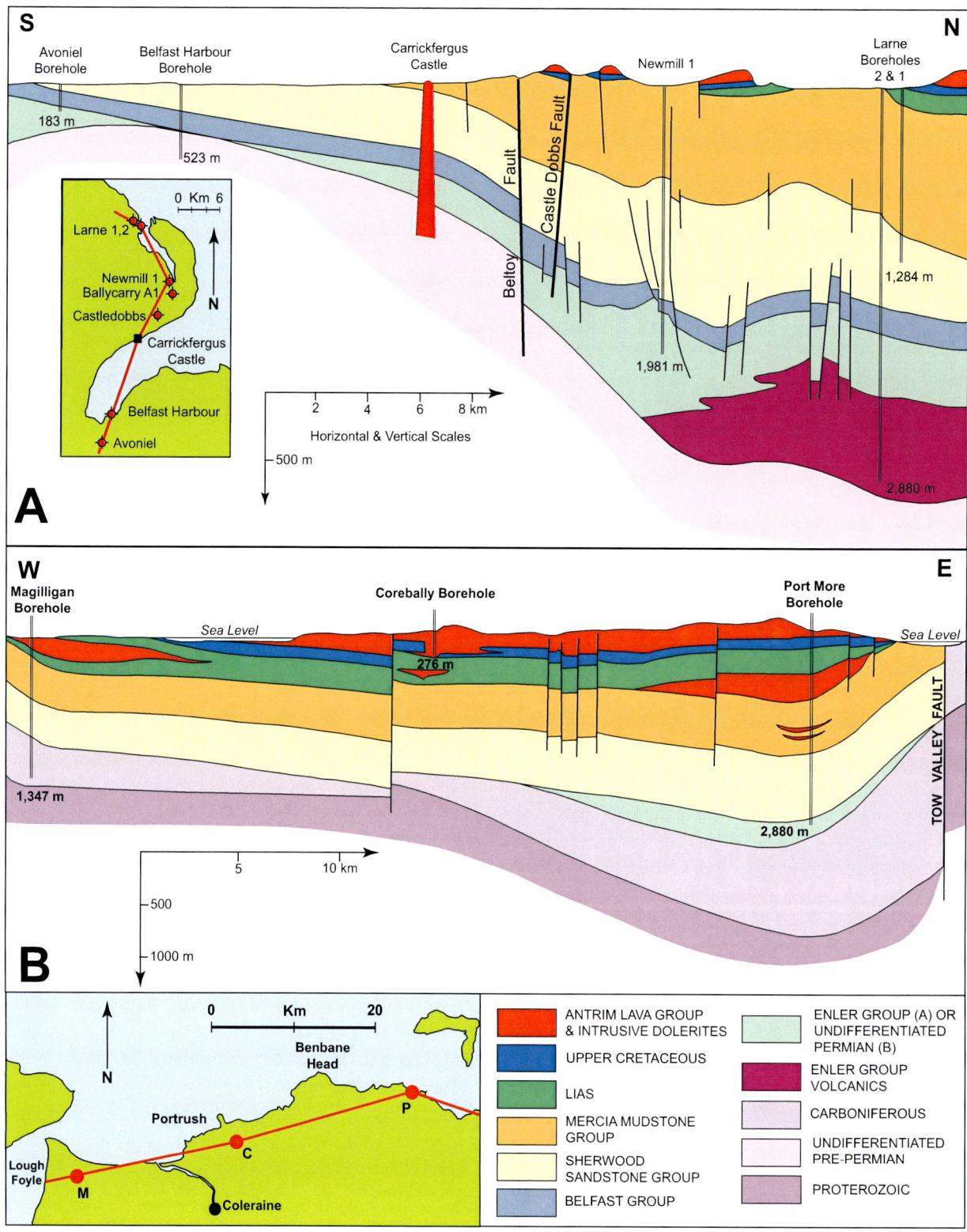

Figure 12.1. A. Diagrammatic geological cross-section of the Larne Basin. Simplified after McCaffrey and McCann (1992). **B.** Diagrammatic geological cross-section of the Rathlin Basin. Well depths in metres. Simplified after McCann (1988).

generated hydrocarbons in a number of pulses since that time. The same is possibly true for the thicker sequences in the hanging wall of the Foyle Fault. Middleton *et al.* (2001) reported that palaeotemperatures, derived from fluid inclusion studies, were persistently higher than those anticipated from burial curves or recorded from vitrinite reflectance. This was attributed to short-lived episodes of high temperature fluid movement through the basin. However, these hot fluid flow episodes may have been too short to affect hydrocarbon generation from source rocks in the basin. The same authors (Middleton *et al.*, 2001) reported on apatite fission track analysis (AFTA) results from the Magilligan and Port More boreholes (Figure 12.1B) – the former showing a cooling from 100–110°C at some time between 230 and 135 Ma, and the latter from >105°C between 200 and 100 Ma. The cooling is thought to result from uplift and exhumation associated with structuring events such as basin inversion and, if of significant magnitude, may cause the cessation of oil or gas generation until further, later burial. AFTA results from an offshore borehole (BGS 73/28) in the northern part of the Rathlin Basin close to Kintyre, indicate two cooling episodes – from 75–90°C some time between 180 and 35 Ma, and from temperatures of 30–70°C some time between 45 and 0 Ma. Although these are wide time bands, they do not preclude a similar sequence of events to those suggested for the Larne and Lough Neagh basins below.

Reservoir rocks and structures

The sandstone sequences of the Permian Enler Group and the Sherwood Sandstone Group provide adequate reservoir potential. Early grain coatings and cements probably reduced permeability, but dissolution of these has produced secondary porosity even at depth in the boreholes (Fitzsimons and Parnell, 1995). However, it is doubtful whether a seal exists for Permian reservoirs. In the Triassic, the transition between the Sherwood Sandstone and Mercia Mudstone groups is gradational. Nevertheless, the overlying thick Mercia Mudstone sequences should provide an adequate seal, in the correct structural setting.

As indicated above, the detailed structural configuration of the basins is not known. No petroleum exploration wells have been drilled, and in general the Rathlin and Foyle basins are less well known than the Larne and Lough Neagh basins. It may be anticipated, however, that there will be fault-related structures similar to those in Larne and Lough Neagh basins (*see* below). The

discussion of hydrocarbon plays and exploration risks with regard to the Larne and Lough Neagh basins generally apply also to the Rathlin and Lough Foyle basins.

Larne and Lough Neagh Basins
Basin framework

The onshore Larne Basin extends southwards from the Highland Border Ridge to the Longford-Down Massif and westward from the North Channel to the Lough Neagh Basin (Figure 6.2). It is known from deep boreholes and seismic data to contain in excess of 3 km of post-Carboniferous strata over much of the onshore area (Manning and Wilson, 1975; Penn, 1981; McCann, 1990). As in the Rathlin Trough, the Permo-Triassic red-beds pass conformably upwards to a thin Liassic sequence that is unconformably overlain by the Hibernian Greensand and Ulster White Limestone formations. Thick Paleocene basalts were extruded onto the karstified surface of the limestones. The Upper Cretaceous chalk of the Ulster White Limestone is considerably more indurated than its English counterpart, and Simms (2000) considered this to have been an early diagenetic cementation feature that allowed karstification and cave formation prior to extrusion of the basalts (but *see* below for further comment).

The dominant caledonoid NE–SW structures exercised control on Permo-Triassic depositional patterns. Within this framework, however, the basins were further partitioned by re-activation of other Caledonian NE–SW or NNW–SSE structural lineaments, or prominent north–south structures. The Sixmilewater Fault is an important NE–SW intrabasinal fault, NNW–SSE structures have controlled the North Channel and Portpatrick offshore basins, whilst the Ballytober Horst and Toome Graben provide examples of north–south structural features.

A number of post-Carboniferous structural phases can be identified within the region (Johnston, 2004; Reay, 2004b), the principal of which are:

◆ *End Carboniferous*: Variscan deformation phase with transpression, uplift, faulting and erosion;
◆ *Permian to Early Jurassic*: subsidence in a regime of ENE–WSW extension;
◆ *Late Jurassic to Early Cretaceous*: Cimmerian uplift and erosion, with ESE–WNW compression;
◆ *Late Cretaceous to Palaeocene*: Hebridean phase with extension, thermal doming and basalt extrusion;
◆ *Miocene*: Alpine phase with NNW–SSE compression. Fault re-activation, uplift and erosion.

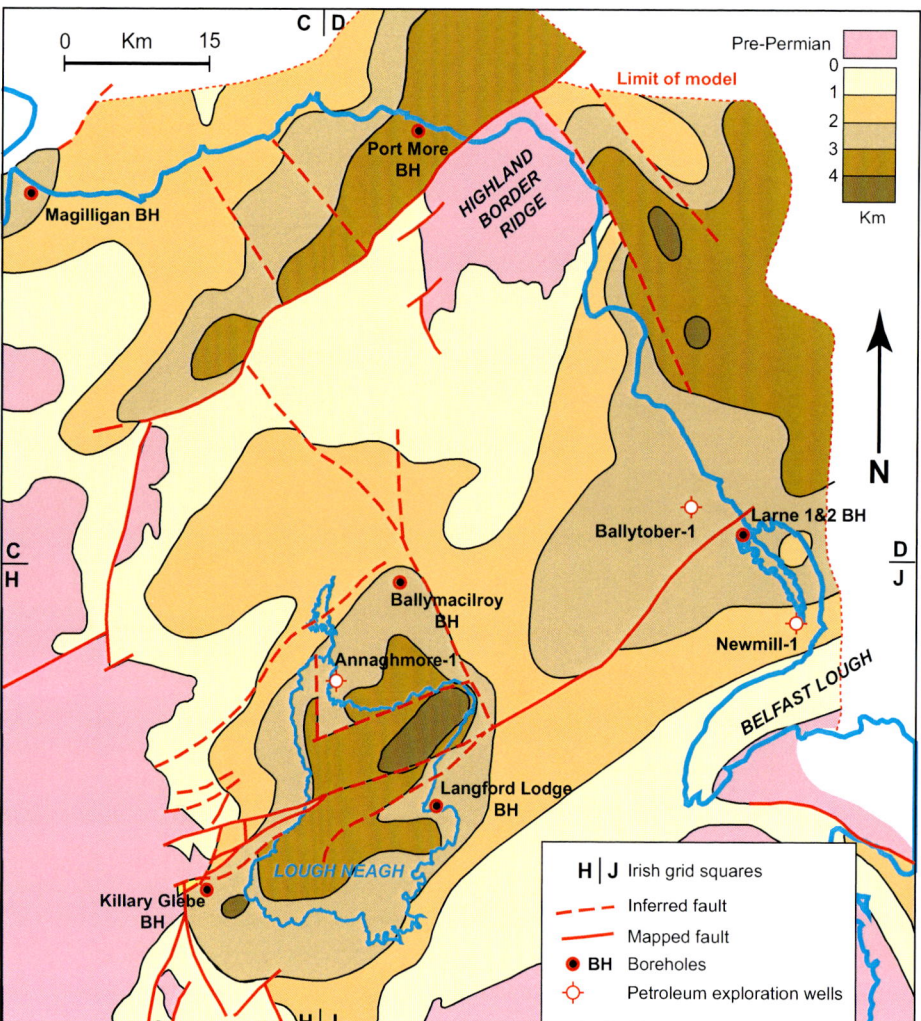

Figure 12.2. Ulster Basin 3-D gravity model. Depth below sea-level to base Permo-Triassic. Simplified after Reay (2004 a).

The only substantial deposits younger than the Antrim Lavas are the onshore Lough Neagh Clays (>350 m thick) – a dominantly lacustrine sequence of Oligocene age. Shelton (1997) suggested that accommodation space for this sequence was probably created by localised extension around the Lough Neagh depocentre, in contrast to the adjoining Larne Basin, which was characterised by uplift during that period.

Exploration potential

The difficulty of obtaining good quality seismic data through a thick basalt pile with internal weathered layers has severely hampered exploration in the onshore Permo-Triassic basins. Data quality to date has been fair to moderate at best – *see* Papworth (1985) for discussion. There is no doubt that more exploration wells would have been drilled in these basins if it were not for this problem. The outcome of the Department of Commerce and petroleum

industry drilling (Table 3.1) has been that, except near the basin margins, only one hole – Langford Lodge (Figure 6.4) – penetrated beneath the Permian into the Variscan floor. Combined with the difficulty in obtaining high quality deep seismic data, this means that detail of the subcrop pattern and distribution of potential source rocks on the pre-Permian basin floor is largely unknown and thus presents a significant exploration risk. A major segmentation of the basin by phases of reactivation on major faults can be anticipated. When details of the seismic structure are available – for example, in the area west of Lough Neagh (Illing and Griffith, 1986) where the Belfast Group map shows faulted compartments and the seismic sections reveal angular unconformity at the Variscan surface – it is clear that the sub-Permian surface beneath the basins is probably quite complex. This problem was realised at an early stage by Wright (1919) who proposed that the possible location of concealed coal strata beneath the basins could be controlled by NNW–SSE structural elements.

Overall, a picture emerges in the two basins of continued re-activation of older faults, with the sense of fault movement sometimes reversed. Faults with caledonoid trend exerted a strong influence on Upper Palaeozoic deposition (as on the Fintona Block: Mitchell and Owens, 1990) and during the Variscan episode. Important cross faults have further segmented the basin. As a result, a complex subcrop pattern may be expected on the basin floor, perhaps limiting the extent of Carboniferous source rock preservation. Fault reactivation continued at intervals through the Permo-Triassic, and it is likely that considerable variation in stratigraphy occurs across many of the faults within the basin.

Source rocks and burial history
The prospectivity and source potential of the Carboniferous strata of northeast Ireland have been discussed earlier, in Chapter 4. Source rocks occur in the Viséan, Serpukhovian and Bashkirian sequences, including coals in the upper part of the section. In the Lough Allen area the gas-prone source sequences within the Carboniferous appear to have generated most of their potential hydrocarbons. However, the maturity of Carboniferous source rocks decreases eastwards between the Lough Allen Basin and the Coalisland area (Johnston, 1979) on the west side of Lough Neagh (Reay, 2004b). Modelled burial histories for the Carboniferous source sequences beneath the Permo-Triassic basin have been presented by Illing and Griffith (1986) and Reay (2004b). These models, of course, incorporate critical assessment of various parameters, in particular the amount of uplift and erosion that has occurred at several points in the geological history. In turn, this controls the point at which hydrocarbon formation was initiated and the time that it was terminated, and whether this happened on more than one occasion. The models presented by Illing and Griffith (1986) and by Reay (2004b; Figure 12.3) suggest that the Carboniferous source sequences beneath the Rathlin, Larne and Lough Neagh basins were mature for oil generation, certainly by the Jurassic and possibly earlier. The burial graphs also indicate that maximum burial depths were attained in the Palaeogene, with phases of oil and gas generation in Late Cretaceous–early Palaeogene time and in the Oligocene (Reay, 2004b). This model thus allows for the late charging or re-charging of traps after the main phases of exhumation.

Naylor *et al.* (2003) identified fault reactivation in the Lough Neagh area in Late Permian and possibly Late Triassic times. They considered it likely that the main phase of uplift and erosion of the Lower Jurassic and older section occurred during the Late Cimmerian phase at the end of the Jurassic. The end Cretaceous uplift, although producing a well-defined unconformity surface exhibiting features of exposure and erosion, did not remove significant section, as evidenced by the widespread preservation of the thin Ulster White Limestone Formation. Four samples at intervals down the Annaghmore-1 well were submitted for apatite fission track analysis (AFTA: performed by Geotrack International). The AFTA results were not capable of unique solution but, allied with regional information (Green *et al.*, 1997), require at least three palaeothermal events; in the Jurassic (200–140 Ma), Early Cenozoic (65–60 Ma) and Late Cenozoic (30–20 Ma). The Palaeozoic thermal history was unconstrained by the data. In detail, there may be more than one Mesozoic episode (as suggested by Green *et al.*, 2001, their Figure 7), but these could not be resolved from the available data (Naylor *et al.*, 2003). Naylor and Shannon (1999) suggested that the main phases of hydrocarbon generation probably occurred in Late Carboniferous and ?Early Jurassic times. However, the decreased maturation levels of Carboniferous source rocks on the western side of Lough Neagh, discussed above, suggest that although these may have produced an increment of their hydrocarbon potential in Late Carboniferous time, they were probably capable of generating hydrocarbons during later burial episodes. Green *et al.* (1997, their Figures 11 and 13) showed the Larne Basin as having peak hydrocarbon generation in pre-Cretaceous episodes. Shelton (1997), however, suggested that the AFTA data from the Larne-2 borehole are dominated by the Cenozoic heating event. The data were interpreted as indicating uplift and erosion of the order of 450 m (with a large error bar) during Cenozoic time (probably in the Miocene) at Larne-2. Allen *et al.* (2002) modelled values of 320 m to >1000 m of denudation between 66–10 Ma across the Larne and Lough Neagh basins, with values generally increasing from the coast inland.

In a recent paper Holford *et al.* (2009) presented a detailed exhumation analysis of the Larne-2 borehole (Figure 12.3d), based on AFTA and rock compaction studies, but their results are at variance with the results presented by Reay (2004b; Figure 12.3a–c). The authors concede that the reconstruction is non-unique, but consider that it is a 'best-fit' solution to the measured data,

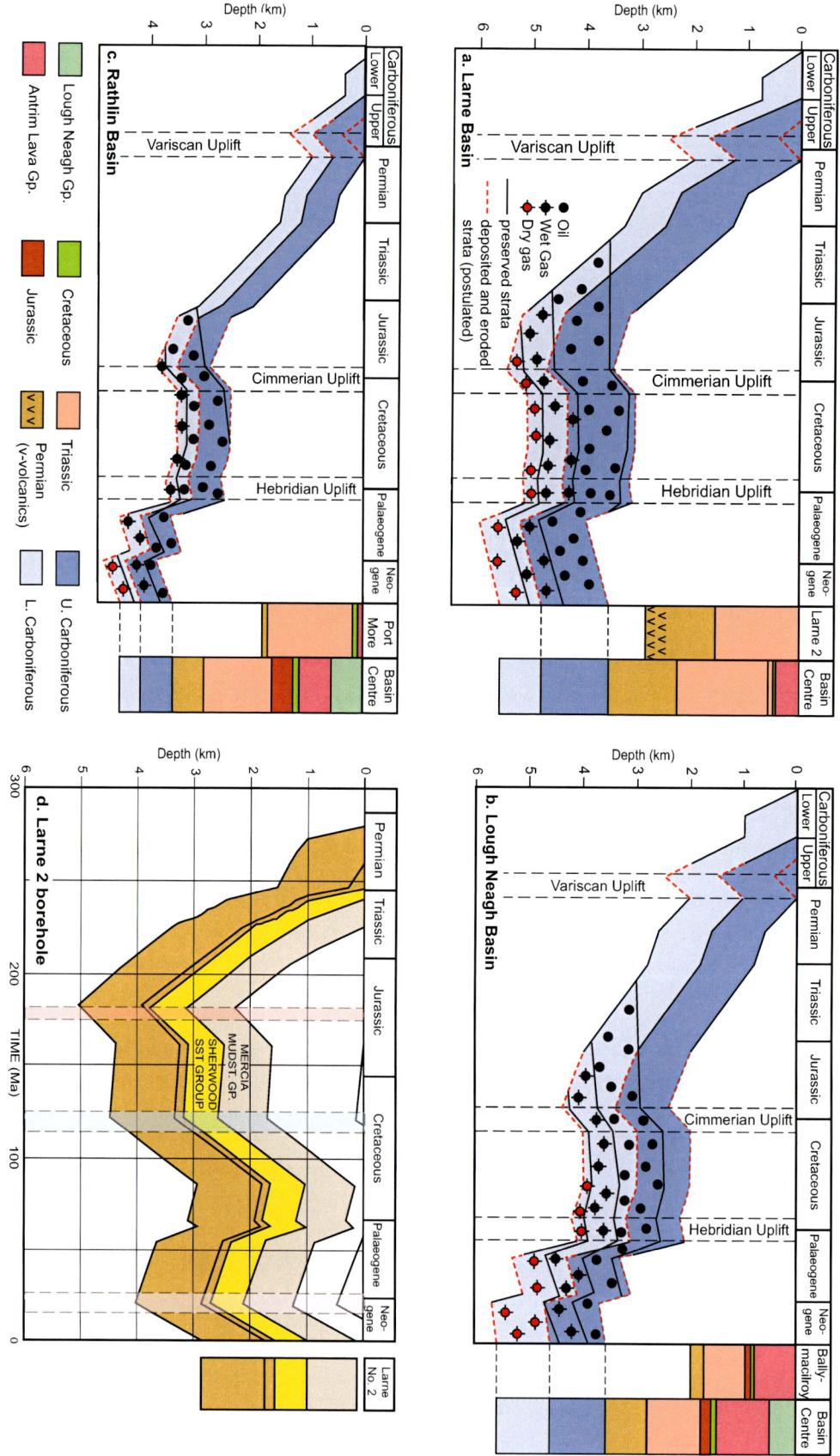

Figure 12.3. Burial history plots in northeast Ireland showing modelled generation of oil and gas from Carboniferous source rocks (after Reay, 2004b). a. Larne Basin b. Lough Neagh Basin c. Rathlin Basin. Published with the permission of the Director of the Geological Survey of Northern Ireland. d. Reconstructed burial and uplift history for the Larne No. 2 borehole (simplified after Holford *et al*, 2009). The model is constrained by AFTA and compaction data and assumes a temporally invariant geothermal gradient of 24°C/km. Vertical colour bars represent estimates of the onset of exhumation-related cooling episodes (i.e. 180-170 Ma, 125-110 Ma and 25-15 Ma).

the borehole geology and regional stratigraphy. The analysis suggests that a further 2.3 km of Upper Triassic and Lower Jurassic strata were deposited above the existing Permo-Triassic sequence. Maximum burial was attained prior to the mid-Jurassic exhumation phase beginning at 180 Ma, which removed 0.7 km of the additional section. Accumulation of a thin (100 m) Upper Jurassic–Lower Cretaceous sequence (no longer represented onshore) was followed by an exhumation episode (120 Ma onset) that removed 1.6 km of section. After deposition of the Upper Cretaceous Hibernian Greensand and Ulster White Limestone formations there was a brief period of early Palaeogene exhumation. This clearly did not lead to significant erosion, given the preservation of the thin Upper Cretaceous chalk sequence. Data from the borehole suggest a further 1.3 km of burial prior to the onset of the Neogene exhumation (20 Ma). Onset of Paleocene volcanism resulted in the accumulation of a 0.7 km sequence of flood basalts (Antrim Lava Group) on the Antrim plateau. Holford *et al.* (2009) suggested that the basalt sequence was formerly of wider extent and could comprise part (*c.*700 m) of the additional Cenozoic sequence modelled at Larne-2. The remainder of the section may equate with the Oligocene Lough Neagh Group to the west. At the Larne-2 location Neogene erosion is considered to have removed the entire Cenozoic section, together with 100 m of the underlying Mesozoic. The additional Cenozoic section, including the basalts, is regarded as contributing to the hardness of the Ulster White Limestone Formation across the region, in addition to the early diagenetic effects mentioned above. The NNW–SSE trending anticline at Larne is thought by Holford *et al.* (2009) to be a product of Neogene compression that resulted in the localised removal of up to 1.3 km of section. It is notable that the Penarth Group and Liassic strata encountered in the nearby Larne-1 borehole have been eroded at the Larne-2 location (Figure 6.3). The same authors cite regional variation of structure contours on the base of the basalts and warping of the Lough Neagh Group sequence as evidence of widespread Neogene compression. One product of this could be to produce heterogeneous exhumation across the region, with more section being removed in some areas, such as the Larne anticline.

The model for Larne-2 proposed by Holford *et al.* (2009) has implications for hydrocarbon prospectivity in the region. The main phase of hydrocarbon generation from potential source rocks (coals and organic-rich mudstones) in the Sepukhovian and Pennsylvanian section probably occurred prior to mid-Jurassic onset of exhumation. As Holford *et al.* (2009) pointed out, this is earlier than the inferred hydrocarbon generation phase in the Slyne Basin (Corrib Gasfield; Corcoran and Mecklenburgh, 2005) or the East Irish Sea Basin (Morecambe Gasfield: Jackson *et al.*, 1995; Green *et al.*, 1997). Minor gas shows were encountered in the Ballytober Sandstone Formation (Enler group: Permian) in the Newmill-1 and Larne-2 wells, and oil-staining in Annaghmore-1 (Reay, 2004b). These indicate that a petroleum system operated, but that any charge in the structures was subsequently lost during later inversion or compressional phases that involved fault reactivation or trap modification.

Overall, therefore, the balance of maturation, apatite fission track and modelling data suggests that the main phase of hydrocarbon generation within the basins occurred during the Jurassic. This phase was probably terminated by the onset of mid-Jurassic exhumation (Holford *et al.*, 2009), rather than by end Jurassic Late Cimmerian uplift and erosion, as envisaged by Naylor *et al.* (2003). The possibility of later phases of hydrocarbon generation is open, but on balance appears unlikely – even assuming that the source rocks still retained some potential after the earlier episodes. If this is the case there is obviously a greater risk of trap rupture. Even if hydrocarbon generation continued into the Cenozoic, the basins appear to lack widespread thick halites that could provide seal following trap rupture (as is the case in the Morecambe Bay Gasfield) and allow re-charge.

Reservoir rocks

In a detailed study of the reservoir potential in the Permo-Triassic sequences of the onshore basins, Parnell (1992) concluded that the potential of the sandstones was 'analogous to that in the Morecambe gasfield'. Although porosity is occluded by a variety of cements, there are significant regional variations and zones of high secondary porosity. Log analysis calculations on the 286 m thick Ormskirk Sandstone Formation (Sherwood Sandstone Group) in the Larne-2 well (Figure 6.3) suggest a net:gross ratio of 65%, with porosities >20% (Maddox *et al.*, 1997). Core data near the base of the interval show porosities of 23% and permeabilities of 13–63 mD. The same authors quote a core close to the top of the formation in the Newmill-1 well as yielding permeabilities averaging 306 mD, porosities of 21% and a net:gross ratio of 90%. Studies of the Lower Permian sandstones (Enler Group) penetrated by the two wells (Maddox *et al.*, 1997) gave porosities of 7%, with permeabilities ranging 0.2 mD to 1.0 mD from core

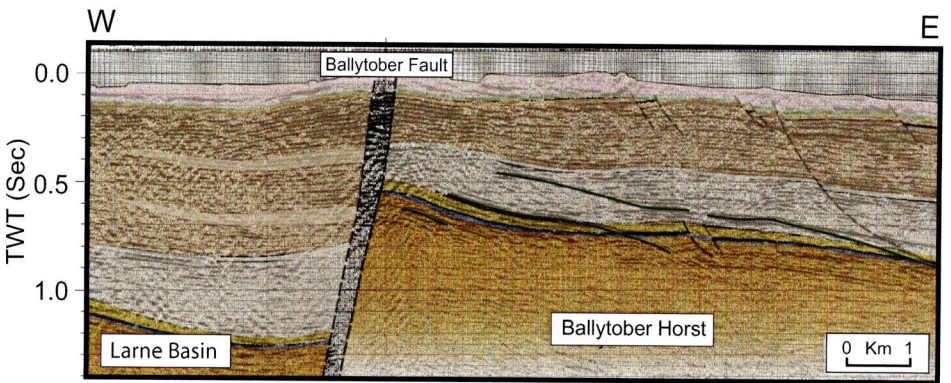

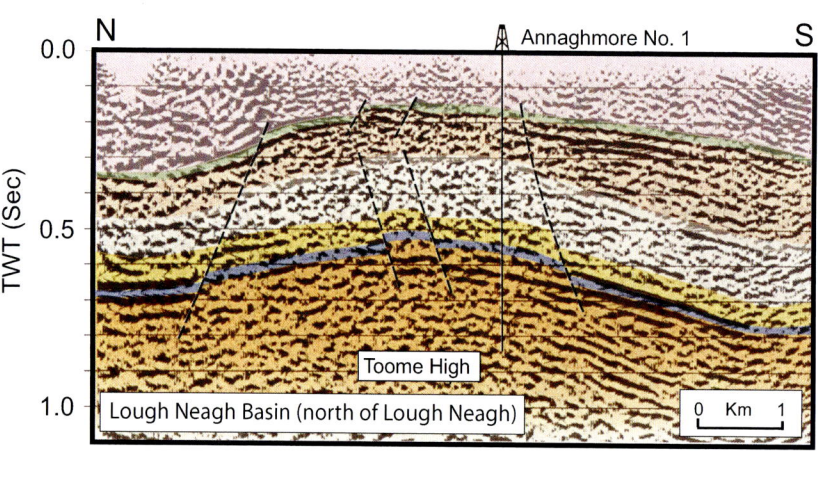

Figure 12.4. Interpreted seismic sections across the Larne and Lough Neagh basins (after Reay, 2004b). Published with the permission of the Director of the Geological Survey of Northern Ireland.

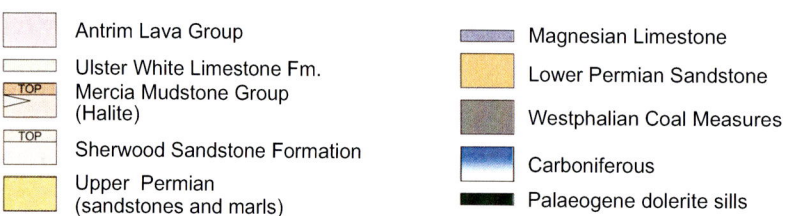

Antrim Lava Group	Magnesian Limestone
Ulster White Limestone Fm.	Lower Permian Sandstone
Mercia Mudstone Group (Halite)	Westphalian Coal Measures
Sherwood Sandstone Formation	Carboniferous
Upper Permian (sandstones and marls)	Palaeogene dolerite sills

in Larne-2. These contrast with average porosities of 16% and permeabilities of 212 mD (ranging up to 1 Darcy) in Newmill-1. Parnell (1992) pointed out that Permian sandstones in the basins are cemented by clay minerals, possibly in part derived from the underlying volcanic pile, which contains minerals readily altered to clays and iron oxide. This may explain the somewhat lower potential of the Permian sandstones in some areas.

Exploration drilling results
As described in Chapter 3, a number of deep boreholes (Table 3.1) have been drilled in the basins in the search for concealed coal deposits and to study geothermal potential. The boreholes were not located on closed structures or by using seismic surveys. Only four exploration wells have been drilled in the two basins in search of hydrocarbons. It is worth examining the location and results of these wells, to better understand the lack of exploration success to date.

The Shell Newmill-1 well was drilled in 1971 under a farm-out from Marathon (Figure 6.2). Although some seismic lines had been shot in the immediate offshore area, the onshore well was sited by surface mapping. Geoseismic interpretations of regional seismic lines acquired later (Illing and Griffith, 1986) indicate that neither Newmill-1 nor Larne-2 drilled into closure at potential reservoir level. The Fynegold Petroleum Ballytober-1 well, drilled in 1991, was located on a seismically-defined structure (Figures 12.2 and 12.4). The well was drilled on a north-trending horst, the bounding faults of which show evidence of syndepositional movement during Mercia Mudstone Group time (Ruffell and Shelton, 1999; Shelton, 1997), and probably also in Sherwood Sandstone Group time (Figure 12.4). Several authors (Musgrove *et al.*, 1995; Ruffell and Shelton, 1999) have also pointed to the importance in a wider context of synsedimentary faulting during deposition of the Mercia Mudstone Group.

Naylor *et al.* (2003) reported on two closely spaced exploration wells drilled on the northwest shore of Lough Neagh, Northern Ireland – Annaghmore-1 and Ballynamullan-1 (Figure 6.4). Vibroseis data delineated a fault-bounded structure whose outline and main faults are shown in Figure 12.5. In this interpretation, a north–south oriented graben occurs in Toome Bay, and closure of the Annaghmore-1 structure is afforded by the eastern bounding fault of the graben. Closure to the north, east and south is provided by dip. The inadequacy of the seal to the west against the graben boundary fault (A, Figure

12.5), combined with a subsidiary parallel fault (B), was recognised as a significant risk to the prospect, particularly if late rejuvenation of the faults had occurred. Post-Paleocene rejuvenation of N–S and NNW–SSE-trending faults is a common feature of the Antrim Plateau. There had been significant erosion of the structure before deposition of the Chalk, resulting in a relatively thin Mercia Mudstone sequence. The primary targets were sandstones in the Permian, with shallow reservoirs in the Sherwood Sandstone as possible secondary targets. The structure was tested (1993) in a near-crestal position by the Annaghmore-1 well, drilled vertically to a depth of 5100 ft (1554.5 m) and plugged and abandoned as a dry hole. The follow-up deviated Ballynamullan-1 well (P&A, 1994) was drilled to a vertical depth of 4500 ft (1371.6 m) on a seismic prospect, based on the possible identification of gas (a zone of bright reflections) within Permian sandstones trapped up-dip against faults B and C (Figure 12.5). However, the seismic anomaly was probably a product of the unexpected development of anhydrite in the Permian east of the faults.

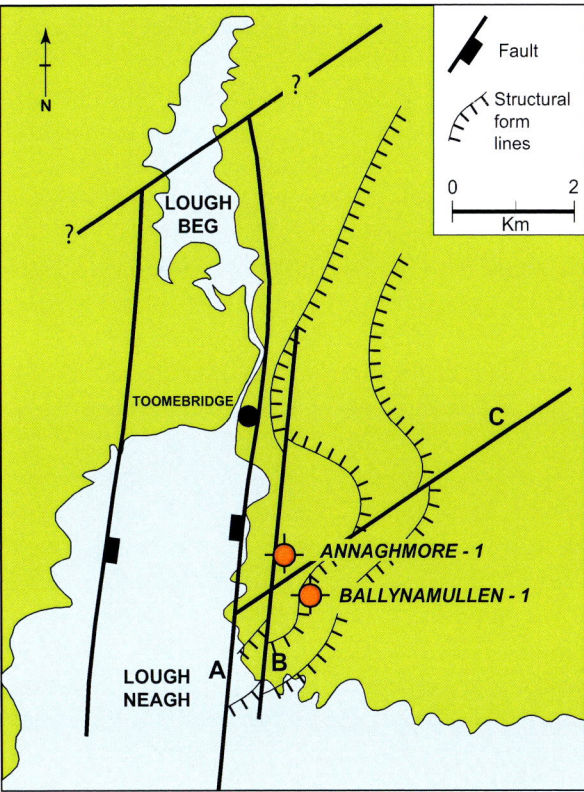

Figure 12.5. Sketch map of the geological structure immediately north of Lough Neagh, with the locations of the Annaghmore-1 and Ballynamullen-1 wells (after Naylor *et al.*, 2003). Faults A, B and C are mentioned in the text. See Figure 12.2 for wider location.

The structural concept at the Annaghmore-1 location appears to have been validated by the drilling, and it was located on a large regional structure where a substantial element of closure was reliant on a major bounding fault showing multiple phases of syndepositional movement. The seismic data are of reasonable quality and adequate for the recognition and mapping of the main reflectors. The stratigraphic identification of the reflectors was substantiated by the wells, within the range of variation normal in depth conversions. Nor was lack of reservoir a problem, since adequate porous intervals were encountered at the target levels. Questions must be asked, therefore, of the seal and source elements of the exploration play. The seal on the structure to the west was dependent on Fault A, the bounding fault of the Toome Bay Graben. The graben was possibly initiated during the Permian, but the faults had clearly been reactivated during the Jurassic, and possibly during the Cenozoic. Vertical seal on the reservoir horizons was anticipated from units within the Belfast Harbour Evaporite Formation or the Mercia Mudstone Group.

Maturation levels in Annaghmore-1 (Naylor *et al.*, 2003), based on spore and pollen colour and fluorescence characteristics, indicate that the section is probably sub-mature to mature with respect to oil generation to a depth of *c.*4000 ft (1219 m). The interval from 4299 ft (1310.3 m) to 4687 ft (2438.6 m) appears to be below the 'oil window' but is mature for dry gas. Any Pennsylvanian source rocks underlying the Permo-Triassic section penetrated by the well should be at least mature in terms of dry gas generation. It is unlikely, given the poor reservoir properties of the sandstones and the poor seismic definition at depth, that the Carboniferous section will provide adequate targets to justify drilling beneath the deeper parts of the Permo-Triassic basin.

There were surprising differences in stratigraphic unit thicknesses, lithologies and diagenesis between the closely spaced Annaghmore-1 and Ballynamullan-1 wells. The most likely explanation for this is to be found in the intervening faults, particularly the NE–SW-trending Fault C (Figure 10.5). The Mercia Mudstone Group and several units in the Permian section are thicker in the Ballynamullan-1 well, and sandstone units are better cemented. The evaporitic upper member of the Belfast Harbour Evaporite Formation is not present in Annaghmore-1. The data suggest syndepositional fault activity during Late Permian time, and probably also in Mercia Mudstone time.

As indicated on Figure 6.4, by comparison with the Ballymacilroy borehole (Thompson, 1979), the area at the northern margin of Lough Neagh suffered post-Mercia Mudstone erosion. The upper part of the Triassic section, together with the Lower Jurassic, were probably deposited in this area, but subsequently eroded. Up to 1450 ft (442 m) of strata are absent compared with Ballymacilroy-1. Reactivation of faults with caledonoid trend probably took place during the Late Jurassic Cimmerian tectonic episode. To the northeast of the drill sites, the Carnlough Fault (Geological Survey of Northern Ireland, 1997) is on trend with the north-west corner of Lough Neagh. A fault-controlled structural high, shaped by north–south cross-faults, may have existed at the head of Lough Neagh in Early Cretaceous time, which then suffered erosion and peneplanation before deposition of the Ulster White Limestone. The Ulster White Limestone is also thinner in the two wells than in Ballymacilroy-1, which may mean that the area continued to be somewhat positive during the Cretaceous. All sign of this positive element was lost following extrusion of the Antrim Lava Group basalts and the subsequent Cenozoic subsidence of the Lough Neagh Basin.

Positive exploration features of the basins are:
◆ Early fault-dependent structures were generated during the Permo-Triassic;
◆ adequate reservoir and seal sequences were available;
◆ maturation of Carboniferous source rocks giving rise to hydrocarbon generation during Jurassic time into the early structures.
The major exploration risks are:
◆ Possible lack or patchy distribution of the source sequences beneath the basin. Development of thick Permian volcanic sequences (Penn *et al.*,1983) may also prove a barrier to hydrocarbon migration;
◆ multiple re-activation phases on faults, some in the Cenozoic, with late breach of reservoirs being a substantial risk.

North Channel and Portpatrick Basins
Basin framework

The results from deep drilling on the Larne coast, allied with modelling of the gravity data (Figure 12.2), makes it clear that the Larne Basin extends into the contiguous offshore region. There has been some variation of use in the literature of the names for the offshore basins. Here, the term 'North Channel Basin' is used for the depocentre immediately offshore from Larne, and 'Portpatrick Basin' for the offset depocentre lying south of the Southern Uplands Fault Zone (SUFZ) extension (Figure 6.1). The strong element of NW–SE and NNW–SSE structural control in the offshore region is a further reason for distinguishing the basins from the onshore Larne Basin.

The North Channel Basin and Portpatrick basins lie between the Antrim coast and the Rhinns of Galloway in Scotland. The gravity data (Figure 12.2) suggest basin depths of 3 km not far offshore, increasing to 4 km in the North Channel Basin centre, which is elongated in a NNW–SSE direction. The Permo-Triassic succession in the North Channel Basin thickens westwards and the half-graben appears to be controlled offshore by a NNW trending fault running parallel to the coast (Figure 12.6A). The basin sequence thins against the Highland Border Ridge to the north and also by overlap across a series of fault blocks towards the eastern basin margin (Shelton, 1995). A similar succession is anticipated to that encountered in the Larne boreholes (Figure 6.3). Gravity, aeromagnetic and seismic data provide clear

imaging of the SUFZ across the North Channel seaway – *see* for example Figure 3 in Maddox *et al.*, 1997. Tertiary intrusive bodies with NW–SE trend are known from aeromagnetic data (Maddox *et al.*, 1997) and from sidescan sonar surveys (Caston, 1975) and are common in both basins. They may pose a significant exploration risk.

The Portpatrick Basin is a NNW-trending half-graben of reversed polarity lying south of the line of the SUFZ, with a thick Permian and Triassic section developed in the east against the boundary fault along the Lower Palaeozoic positive element of The Rhinns of Galloway (Figure 12.6B). Possible halokinetic structures are evident on seismic records, suggesting that thick halite units may be developed in the offshore depocentres. The basin axis is offset to the northeast with respect to that of the North Channel Basin. The SUFZ appears to have acted as a transfer zone that accommodated the change in polarity between the basins (Maddox *et al.*, 1997) during ENE–WSW extension in Permian and Triassic times, and is responsible for the lateral offset of

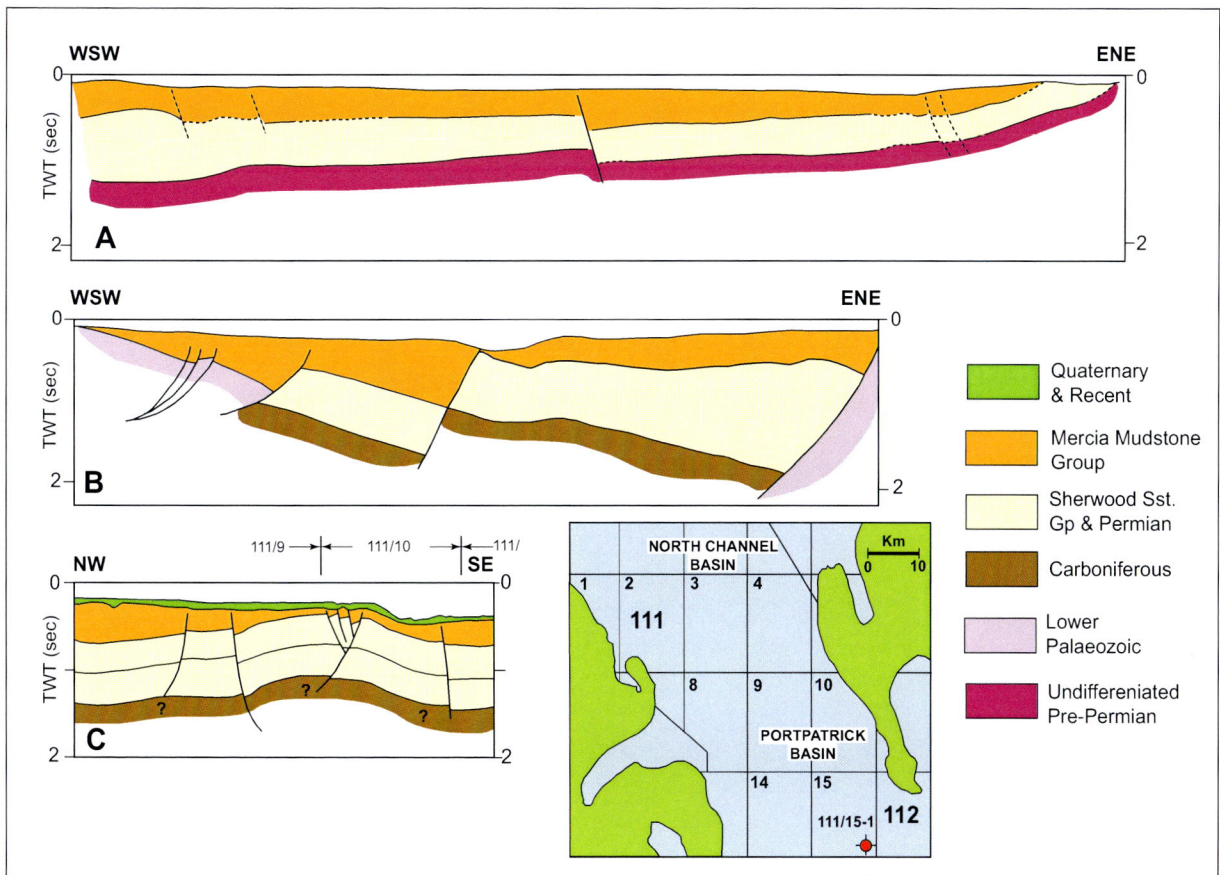

Figure 12.6. Simplified geoseismic sections in the North Channel region (not precisely located). **A.** WSW-ENE cross-section from the depocentre of the North Channel Basin towards the Rhinns of Galloway (after Shelton, 1995). **B.** ENE-WSW cross-section through the Portpatrick Basin (after Shelton, 1995). **C.** NW-SE strike section along the Portpatrick Basin (based on Maddox *et al.*, 1997).

the basin axes. Southwards, the basin extends beyond the offshore projection of the Orlock Bridge Fault at the northern margin of the onshore Longford-Down Lower Palaeozoic massif. The British Gas 111/15-1 well (Figure 12.6), near the southern basin limit, targeted a rollover anticline in the hanging wall of a listric fault known as the Laggantulloch prospect (Reay, 2004b), but was plugged and abandoned with the Ormskirk Sandstone equivalent (Sherwood Sandstone Group) reported to be water-bearing (Quirk *et al.*, 1999).

The seismic data (Maddox *et al.*, 1997; Shelton, 1997) provide no evidence for the preservation of post-Triassic rocks in either of the offshore basins. The widespread Palaeogene lavas of the onshore region are absent offshore. This is the case also in the offshore Rathlin Basin, and Fyfe *et al.* (1993) suggested that they were probably not originally deposited. However, Shelton (1995) attributed the lack of younger Mesozoic strata and Palaeogene lavas in the North Channel to Cenozoic (Tertiary) uplift, and Maddox *et al.*, (1997) thought that 'several uplift and erosional events in the late Mesozoic and Tertiary' were responsible. Since Lower Jurassic and Upper Cretaceous sediments are deposited in the immediate onshore area, it would be surprising if this deposition did not originally extend over the offshore basins. The lack of Antrim Lava Group basalts, however, may indicate different conditions in Paleocene time.

Exploration potential

The primary play in these basins is for hydrocarbons from Carboniferous source sequences trapped in Sherwood Sandstone Group sandstone reservoirs and sealed by evaporitic shales and halite of the Mercia Mudstone Group. Exploration potential also exists in Permian sandstones fed from the same source, although seal could be a risk factor in some areas. At this time the lack of data regarding the distribution and structure of the Carboniferous means that the play for hydrocarbons trapped within the Carboniferous can be discounted as a realistic target. The discussion of the Permo-Triassic play given above with respect of the Larne and Lough Neagh basins applies equally for the North Channel and Portpatrick basins, and will not be repeated here. However, the following points can be made regarding the offshore basins:

◆ The lack of significant drill information, allied with the lack of post-Triassic section, make it difficult to model the basin development with any confidence.

◆ Although a Carboniferous section can be anticipated

beneath parts of the offshore basin, details of the subcrop pattern are unknown. In particular, the southern part of the Portpatrick Basin may lack Carboniferous source rocks. Onshore the Carboniferous section beneath the Permian on the south side of Belfast Lough at Cultra is only 280 m thick, and rests on the Lower Palaeozoic rocks at the northern margin of the Longford-Down massif. As noted above, the Portpatrick Basin extends south of the offshore projection of this line and a Carboniferous source rock section may not be present in this area.

◆ Significant tilted fault-block and rollover anticlines have been identified on seismic sections at the target reservoir levels in the Permo-Triassic (Maddox *et al.*, 1997).

◆ As in the onshore basins there are risks related to the timing of hydrocarbon generation and trap formation, and significant risks related to multiple, including late, re-activation of faults and the possible negative influence of Cenozoic intrusions.

◆ The optimum exploration targets are probably structures having four-way closure that is not fault-dependent, with halite units in the overlying sequence to act as cap rocks.

Solway and Peel Basins
Basin framework

The Solway and Peel basins have a strong inherited Caledonian NE–SW trend, with NW–SE cross-faults. They lie between the Lower Palaeozoic Southern Uplands to Longford-Down massif to the north, and the Ramsey–Whitehaven ridge to the south (Figures 5.2 and 6.5). In these basins Permo-Triassic strata rest unconformably on eroded Carboniferous sequences (Holkerian to Pendleian). The Lower Jurassic is locally preserved (Newman, 1999), but younger strata are unknown. Four wells have been drilled in the basins – Elf 112/19-1 (Solway Basin), Elf 111/25-1A and Elf 111/29-1 (Peel Basin), the results of which have been reported by Newman (1999), and Esso 112/15-1 (Solway Basin), the results of which have not been released.

North–south directed extension during the Carboniferous re-activated the existing NE–SW lineaments (originally Caledonian thrusts) and NW–SE faults. Some faults that displace the Carboniferous are seen on seismic sections (Figure 12.7) to terminate upwards at the Variscan surface. Sedimentation occurred within a NE–SW basin, perhaps contiguous with the Northumberland Trough, onshore UK. The pattern onshore is of thick basinal Carboniferous sequences with different onlapping and thinner facies on the intervening highs. A similar picture is imaged on seismic profiles offshore between the Ramsey–Whitehaven Ridge and the Solway Basin (Newman, 1999). The seismic data suggest that up to 3000 m of Carboniferous strata are developed beneath the Solway Basin, but lack of basal Carboniferous reflector makes the picture less certain in the case of the Peel Basin. However, drilling, allied with seismic interpretation, has shown that Pennsylvanian strata have been eroded over much of the basin areas. Newman (1999) suggested that this resulted from basin inversion under the influence of a Late Variscan north–south compressive phase, possibly with reverse movement on the southern basin margin fault, followed by erosion.

The preserved drilled Carboniferous sequence in the wells comprises Late Brigantian to Pendleian deltaic clastics and shallow-marine carbonates in the Solway Basin and Holkerian to Asbian shelf carbonates in the Peel Basin. Stress patterns in the Permian gave rise to east–west extension, the reactivation of earlier NE–SW and NW–SE fault lines, and perhaps the initiation of north–south faults. Further fault reactivation occurred at intervals in Mesozoic and Cenozoic times. Many of the faults affecting the Triassic sequence are listric in form and sole out in either Mercia Mudstone or Permian evaporites. The developing basins received relatively thin Upper Permian sediments, with only the Elf 111/25-1A well encountering Lower Permian Collyhurst sandstones. The Upper Permian comprises, in ascending order, carbonates, evaporitic mudstones and mudstones, typical of the Cumbrian Coast Group in the adjacent UK onshore area.

There is a regional, probably depositional, thinning of the Triassic sequence southwestwards from the Solway Basin across the Peel Basin. The Lower Triassic Sherwood Sandstone Group is 855 m thick in the Elf 112/19-1 (Solway Basin) well, 746 m thick in Elf 111/25-1A, and 585 m thick in Elf 111/29-1 (both Peel Basin). The two constituent formations of the group in the East Irish Sea Basin, the St. Bees Sandstone Formation and the overlying Ormskirk Formation, are also identifiable here (Newman, 1999). It is interesting to note that regional seismic lines demonstrate that the Sherwood Sandstone Group maintains its thickness over regional highs, suggesting a later development of these features. The Mercia Mudstone Group was penetrated in the three wells and comprises shales and evaporites. The regional thinning in the Sherwood Sandstone Group, outlined above, continued into Mercia Mudstone Group succession, so that it thins from 724 m in 112/19-1 to only 174.5 m in well 111-29-1, with an increase in sand content. As noted further north in the region, synsedimentary fault movement has occurred during Mercia Mudstone Group time (Ruffell and Shelton, 1999). Some of the development of intra-basinal highs may have occurred during this interval, as may the features referred to informally by Newman (1999) as 'collapse trenches', in which narrow grabenal features, where the Sherwood Sandstone is structurally thinned, are infilled by Mercia Mudstone sediments (*see* Figure 12.7).

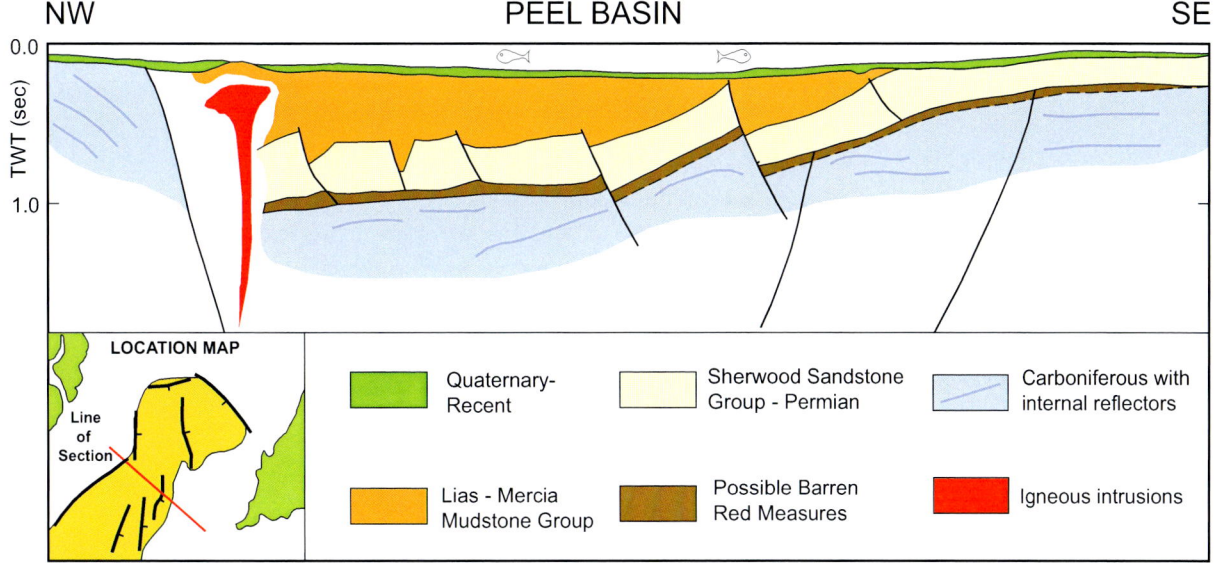

Figure 12.7. Geoseismic section across the Peel Basin (based on Newman, 1999).

The Elf 111/29-1 well in the Peel Basin encountered 52 m of claystones, dated as Rhaetian to Hettangian, above the Mercia Mudstone Group and overlain in turn by Quaternary deposits (Newman, 1999). Further north in the same basin the Elf 111/25-1A did not encounter the Lower Jurassic, suggesting that a thin post-Triassic layer may be preserved only in the central and south-western parts of the basin. The two wells drilled in the Solway Basin also failed to encounter the Lower Jurassic, but seismic evidence suggests that strata of this age are developed in the basin centre and may be up to 800 m thick (Newman, 1999).

Exploration potential

The play in these basins is for Ormskirk Sandstone Formation reservoirs sealed by Mercia Mudstone Group mudstones. Reservoir characteristics vary according to detailed facies type within the fluvial–aeolian sequence, but are generally good (Newman, 1999). Seismic interpretation has revealed structural traps associated with NE–SW and north–south faults, and the largest of these were targeted in the four wells. As noted above, Variscan erosion had removed the Serpukhovian–Pennsylvanian sequence with its potential source rocks. Analysis of the Mississippian sequence reported by Newman (1999) indicates poor source rock potential. The only hydrocarbon indications in the four wells were minor bitumen traces in the Ormskirk sandstones of the Elf 112/19-1 well in the Solway Basin, possibly derived from thin lagoonal or algal carbonates in the Carboniferous. The Elf 112/19-1, Elf 11/25-1A and Elf 111/29-1 wells are reported by Newman (1999) to have tested valid structural traps at top Ormskirk Sandstone level. A secondary play may exist in the Lower Permian Collyhurst Sandstone Formation, encountered with poor reservoir characteristics in Elf 111/25-1A. Sandstone porosities of <10% were recorded and this is probably as a result of carbonate and silica cementation. However, the sandstone interval was absent in the Elf 111/29-1 and Elf 112/19-1 wells and the distribution of the unit elsewhere in the basin is not well known. In addition, the sealing properties of the Upper Permian probably deteriorate southwestwards towards the Elf 111/29-1 well. The exploration potential of the Carboniferous is regarded as low, due to poor reservoir properties and poorly imaged traps.

Carboniferous maturity levels of around Ro=1.0% (vitrinite reflectance) in the basins are thought to have resulted from a period of Mesozoic burial, with modelling suggesting the onset of hydrocarbon generation from any potential source rocks in the Early Jurassic (Newman, 1999). Maturation levels in the East Irish Sea Basin, south of the Ramsey–Whitehaven Ridge, are considerably higher (Ro >1.3%) as a result of greater Mesozoic burial, with source rocks there being in the gas window. However, the lack of post-Liassic rocks means that geologically the Mesozoic and Cenozoic exhumation episodes are largely unconstrained.

Using apatite fission track and vitrinite reflectance data, Green et al. (1997) demonstrated the considerable heterogeneity of thermal histories in the basins of the Irish Sea, and this is discussed below. Specifically in relation to the basins under discussion they noted an eastward variation along the southern coast of Scotland from the Stranraer Basin in the west, where continuous cooling commenced from a thermal high in the period 260–220 Ma, through an intermediate area in which maximum post-Palaeozoic temperatures were attained in the Cretaceous, to the Solway Basin in the east where maximum temperatures were reached in the early Cenozoic. The apatite fission track data also support the view that the Southern Uplands massif and the Ramsey–Whitehaven Ridge were uplifted by Early Cretaceous (Late Cimmerian) movements. Newman (1999) reported on apatite fission track data derived from the 112/19-1, 111/25-1A and 111/29-1 wells. Two phases of cooling are indicated in the Cenozoic – a rapid phase during the Paleocene (~60 Ma) and a slower phase during the Miocene (20 Ma). The earlier of the two phases (Laramide) was related to the development of the Tertiary Igneous Province and gave rise to WNW–ESE fractures that are often intruded by dykes. There is an anomalous situation regarding the apatite fission-track results in these wells, in that the data indicate that elevated temperatures of >110° C were attained in the Late Cretaceous in the Sherwood Sandstone Group and Carboniferous sections, prior to cooling at 60 Ma. However, the Lower Jurassic interval in 111/29-1 reached a maximum temperature of 100° C prior to Late Cimmerian uplift and cooling. Newman (1999) explains this discrepancy by invoking the circulation of hot hydrothermal fluid through the Sherwood reservoirs associated with the igneous province that effectively obscures the normal burial curve, but the Lower Jurassic was separated from this event by the seal afforded by the Mercia Mudstone Group. However, Floodpage et al. (2001) presented new data suggesting that the Permo-Triassic of the Peel Basin well Elf 11/29-1 achieved maximum palaeotemperatures during Early Cretaceous time.

The lack of exploration success and low exploration potential in the Peel and Solway basins results from a number of factors:

◆ It is difficult to avoid the conclusion reached by Newman (1999) that the main reason is probably the lack of Carboniferous source rocks due to inversion and erosion during the Variscan episode.

◆ Timing of hydrocarbon generation may have predated the main phase of trap formation. Against this there is some evidence for fault movements in Mercia Mudstone Group time that could have provided subtle, low-amplitude structural traps.

◆ As with the basins in the northern part of the Irish Sea, the faults have suffered multiple phases of movement, including phases in the Cenozoic. As a result there was a high risk of trap rupture. The Mercia Mudstone Group seal on the main potential reservoir lacks thick annealing halites and, particularly in the Peel Basin, is relatively thin.

Kish Bank Basin
Basin framework

The Kish Bank Basin is a relatively small (32 x 48 km), isolated, deep sedimentary trough, elongated NE–SW, that is separated from the Central Irish Sea Basin to the southeast by an area of outcropping Carboniferous and older strata (Figure 6.5). The presence of the basin was initially indicated on gravity data (Bott, 1964) and later confirmed by speculative and proprietary seismic surveys. The basin is developed in the hanging walls of two orthogonal faults, the Bray and Dalkey–Lambay faults. A major dextral strike-slip fault, the NW–SE trending Codling Fault, bisects the basin. Four wells have targeted the basin – Amoco 33/22-1 (1977), Shell 33/21-1 (1979), Fina 33/17-1 (1986) and Enterprise 33/17-2A (1997). The Amoco 33/22-1 well was drilled on the southeastern rim of the basin, outside the areal limit of Permo-Triassic strata, and penetrated a Carboniferous section unconformably overlying presumed Lower Palaeozoic metasediments (Jenner, 1981). The other three wells drilled the Permo-Triassic basin-fill sequence, but without reaching the base of the Permian. The outline stratigraphy of the four wells is shown in Figure 6.6.

The Amoco 33/22-1 well penetrated a Pennsylvanian (Duckmantian to Asturian) section c.740 m thick. The sequence is unconformable on undated chloritic slates (110.6 m penetrated) of possible Lower–Middle Cambrian (Jenner, 1981) or Upper Cambrian–Arenig (Naylor et al., 1993) age. The Pennsylvanian is of 'coal measure' facies and includes numerous seams of medium to high volatile bituminous coal. Jenner (1981) gives a cumulative thickness of 11 m for the coal seams encountered by the Amoco 33/22-1 well. The uppermost 114 m of the Carboniferous comprises an undated sequence of red shales, brown sandstones and grey siltstones, considered by Jenner (1981) to be of possible Stephanian age. However, they may represent a red (or reddened) unit of Asturian age (previously Westphalian D) comparable to those developed in North Wales and the West Midlands of England.

The Permo-Triassic succession is similar to that in the East Irish Sea Basin and includes thick halite horizons within the Mercia Mudstone Group. The Fina 33/17-1 well penetrated a section of the Lower Permian Collyhurst Sandstone Formation. Good porosity in the dominantly aeolian sands is secondary following dissolution of earlier cements (Naylor et al., 1993). However, the overlying Upper Permian Cumbrian Coast Group marls are only 97 m thick and have thin sandstone and siltstone interbeds. The Sherwood Sandstone Group is divisible into the St Bees Sandstone Formation and overlying Ormskirk Sandstone Formation. The sandstones in the sequence are variably porous, with visible porosity ranging from poor to good. However, the Ormskirk Sandstone Formation would provide good potential as a gas reservoir throughout, particularly in the Enterprise 33/17-2A well where the Ormskirk Sandstone Formation contains a higher percentage of probable aeolian sandstones (Dunford et al., 2001). Porosity averages 17.7% in the Ormskirk Sandstone interval of the Enterprise 33/17-2A well. Reservoir parameters range from 66% net/gross and average log porosity 14% in Fina 33/17-1 to corresponding values of 80% and 12% in Shell 33/21-1 (Dunford et al., 2001). Primary depositional texture exercised a major control on diagenetic history. The lack of early pore-filling cements in muddier sandstones allowed compactional fabrics to dominate, and permeabilities are low as a result. In the originally mud-poor sandstones the main controls are interpreted to be early cement formation, reducing the main effects of compaction, formation of platy illite within the secondary pores, followed in some cases by secondary dissolution.

The Mercia Mudstone Group comprises predominantly marls, with thick intervals of halite. The Shell 33/21-1 and Enterprise 33/17-2A wells, west of the Codling Fault, contain more halite in the sequence (30%) than does the Fina 33/17-1 well (16%) lying to the east (Dunford et al., 2001). This, together with general thickening of the Mercia Mudstone sequence, suggests that the basin was deeper in the west, controlled by movement on the bounding faults. Synsedimentary fault movements in this period have been noted in the other basins of the Irish Sea region. The Mercia Mudstone Group, with its contained halite beds, is considered to afford adequate seal on potential traps across the basin.

Although the four wells in the basin did not penetrate Jurassic strata, pockets of apparently conformable Lower Jurassic strata have been interpreted on seismic data (Jenner, 1981; Naylor *et al.*, 1993; Dunford *et al.*, 2001). Lower Liassic rocks were sampled from the seabed at the northern margin of the basin (Etu-Efeotor, 1976; Dobson, 1977a, b). Broughan *et al.* (1989) interpreted up to 2700 m of Lower Jurassic rocks preserved against the Bray and Dalkey bounding faults. Ammonites in glacial deposits on the immediately adjacent coastline, possibly derived from the Kish Bank area, were dated as Lower, Middle and Upper Liassic in age (Broughan *et al.*, 1989).

Exploration potential

The poor exploration outcome in the Kish Bank Basin was considered by Naylor *et al.* (1993) to result from a lack of source rocks beneath the basin. Inversion and erosion during the Variscan episode was cited as the reason, and drilling subsequently showed this to be the case in the Solway and Peel basins (Newman, 1999). However, interpretation of better quality seismic data (Dunford *et al.*, 2001) suggests that Viséan, Serpukhovian and Pennsylvanian rocks are widespread beneath the Kish Bank Basin. Accepting this interpretation, then the lack of success must be sought in other factors.

The Kish Bank Basin appears to have been a depositional centre during Late Carboniferous time, with the Pennsylvanian strata overstepping the Mississippian limits onto the flank of the Mid Irish Sea Uplift (well Amoco 33/22-1). There is no evidence of major fault movement during the Permian, and onlap of some reflectors within the sequence against the Variscan surface can be seen on a number of seismic sections. However, north–south fault trends may have been established at this time, or formed by reactivation of basement lineaments. There is more direct evidence of fault activity during the Triassic, particularly during Mercia Mudstone time. The Liassic followed conformably on the Triassic, as elsewhere in the region, and there is no certain evidence of post-Liassic pre-Cenozoic rocks in the basin. Jurassic structural phases are unconstrained, but regional data suggest that Late Jurassic rifting and faulting was followed by uplift and erosion during the Early Cretaceous. NE–SW directed extension during the rift phase probably gave rise to normal fault movement on the western boundary faults and on the Codling Fault complex. The different phases of extension in the Permian to end Jurassic time interval probably produced low-amplitude tilted fault-block closures. The probable Late Cretaceous deposition of a glauconitic sandstone and Upper Cretaceous Chalk sequence, similar to that in the Larne Basin, was followed by later uplift and erosion that has removed all evidence of Cretaceous strata from the central portion of the Irish Sea.

A summary of Enterprise Oil proprietary apatite fission-track (AFTA) data provided by Dunford *et al.* (2001) suggests that there was a marked period of uplift and erosion in the Cenozoic, commencing at 60 Ma (Paleocene). The same authors quoted estimates of amounts of erosion during the Cenozoic at the four well locations based on shale velocity analysis – 1500 m at Shell 33/21-1, 960 m at Amoco 33/22-1, 875 m at Enterprise 33/17-2A and 350 m at Fina 33/17-1 – a decrease towards the northeast. These figures are less than those derived from the AFTA data, a discrepancy Dunford *et al.* (2001) attributed to a hydrothermal heating pulse of the Sherwood Sandstone Group during the Paleocene – as discussed earlier for the Peel and Solway basins. The Cenozoic uplift occurred under NNW–SSE compression and WSW–ENE extension and was probably multi-phase, in some cases continuing up to the present time. Jenner (1981) argued, using examples of similar structures in the Irish Sea region, that the main phase of dextral movement on the Codling Fault took place in Eocene–Oligocene times. The stress regime was oblique to the Codling Fault and this produced 6 km of lateral displacement along the fault itself, and 9 km on the shear zone complex overall (Dunford *et al.*, 2001). Most faults in the basin were probably reactivated during the Cenozoic, including earlier north–south fractures, with new fault splays formed, particularly in the wider Codling Fault zone. These movements would clearly have affected earlier fault blocks and closure, causing tilting and trap rupture, and may have formed new, late structures. The fold closure drilled by the Shell 33/21-1 well was probably produced as a late inversion structure during the Cenozoic (Naylor *et al.*, 1993).

Naylor *et al.* (1993) reported vitrinite reflectance (Rm) levels ranging from *c.*0.8% to 1.3% in the Carboniferous section of the Amoco 33/22-1 well (Figure 5.3). Maturation levels in the Permo-Triassic section range from *c.*1.3% to 1.6% Rm, suggesting that any underlying Pennsylvanian source rocks in the deeper parts of the basin would be mature to marginally post-mature with respect to dry gas generation. The same authors estimated that a further 1.3 km of cover would have been required to account for maturation levels in the preserved Carboniferous section. They considered that this section was eroded during Late Jurassic uplift, whereas Corcoran and Clayton (1999) saw this as an end-Carboniferous event. Although the section in the 33/22-1 well has only moderate gas generation capability, the Pennsylvanian section is interpreted as thickening beneath the basin. Also underlying Viséan–Serpukhovian rocks, absent in 33/22-1 at the basin margin, are exposed onshore in the Dublin Basin and are judged to have good oil-source potential. Modelling of the burial data (Dunford *et al.*, 2001) indicates generation from oil source rocks prior to the Variscan episode, and a further phase initiated in Late Triassic time. Oil potential is judged to have been exhausted before the end of the Jurassic and prior to Cimmerian uplift. Gas would have been generated during Late Cretaceous burial before being switched off by uplift early in the Cenozoic. It follows from this scenario that structures either formed or highly modified during the Cenozoic would only be charged by hydrocarbons sequestered in earlier structures, and would therefore be high-risk exploration targets. This is somewhat at variance with the conclusions of Naylor *et al.* (1993) regarding the Pennsylvanian sequence, assuming that the gas generation potential of this section had not been exhausted during earlier burial phases. Primary migration of hydrocarbons from the generation kitchens in the deeper sections in the hanging walls of the boundary faults would have been up-dip to the southeast, within the Carboniferous section.

Dunford *et al.* (2001) cited a number of strands of evidence in support of active hydrocarbon generation during development of the basin:

◆ Dead oil shows in the top 5 m of the Sherwood Sandstone in Fina 33/17-1.
◆ Oil staining and fluorescence in the Pennsylvanian section of Amoco 33/22-1.
◆ Evidence of seeped hydrocarbons in seabed cores on the Enterprise 33/17-2A structure.
◆ 'Seepfinder' survey anomalies along the Codling Fault zone and to a lesser extent over the Dalkey and Lambay faults. A strong anomaly was also reported southwest of the Amoco 33/22-1 well, probably where Carboniferous source rocks subcrop.

◆ Numerous shallow gas features, often associated with faults, and including flat spots, gas chimneys, phase reversals and seabed mounds.

The 33/21-1, 33/17-1 and 33/17-2A wells apparently drilled valid structural targets within mapped closures and came in close to prognosis. The target structures are of different types, as illustrated in Figure 12.8. The Shell 33/21-1 well was drilled on a fold structure, although closure at Ormskirk Sandstone level was probably fault-dependent. However, as noted above, this structure is probably a late Cenozoic inversion feature, post-dating generation and migration of hydrocarbon in the basin. This structure would also have a limited drainage area from which to accumulate hydrocarbons. Fina 33/17-1 tested a quite different structure, an uplifted fault horst east of the Codling Fault. A dead oil zone at the top of the water-wet Sherwood Sandstone in the well suggests that hydrocarbons were initially trapped in this structure, which was probably the product of Triassic and Jurassic fault movement. Cenozoic reactivation of the bounding faults and possible tilting of the structure could have ruptured the trap or have caused lateral re-migration. It is also noteworthy that detailed seismic shows evidence of shallow gas in the Cenozoic section over the structure. The third of the wells – Enterprise 33/17-2A – tested anticlinal closure within a tilted fault structure immediately west of the Ulysses Fault, a splay of the Codling Fault. Dunford *et al.* (2001) suggested that hydrocarbons produced in the kitchen proximal to the Dalkey Fault would have migrated up-dip to the first major fault structure that could then provide a migration pathway to charge the Ormskirk Sandstone reservoirs of the 33/17-2A feature. Unfortunately the well proved the prospect to be dry, possibly due to development of the anticlinal element of closure during Cenozoic fault movement and after migration of the hydrocarbons.

It is clear that the Kish Bank basin contains the elements of a potential hydrocarbon play. There is a mature source rock section, good reservoir sandstones (Ormskirk Sandstone Formation) and an adequate halite-bearing seal. However, three valid structures have been drilled without success. Secondary target horizons are present in the Collyhurst Sandstone Formation and in sandstones within the Carboniferous, but both are high risk. The Collyhurst sandstones, although comprising a good gas reservoir, lack good top seal, whilst seal potential within the Carboniferous section on any target structure would be difficult to assess. The main problem with respect to the Ormskirk Sandstone play is that the main hydrocarbon

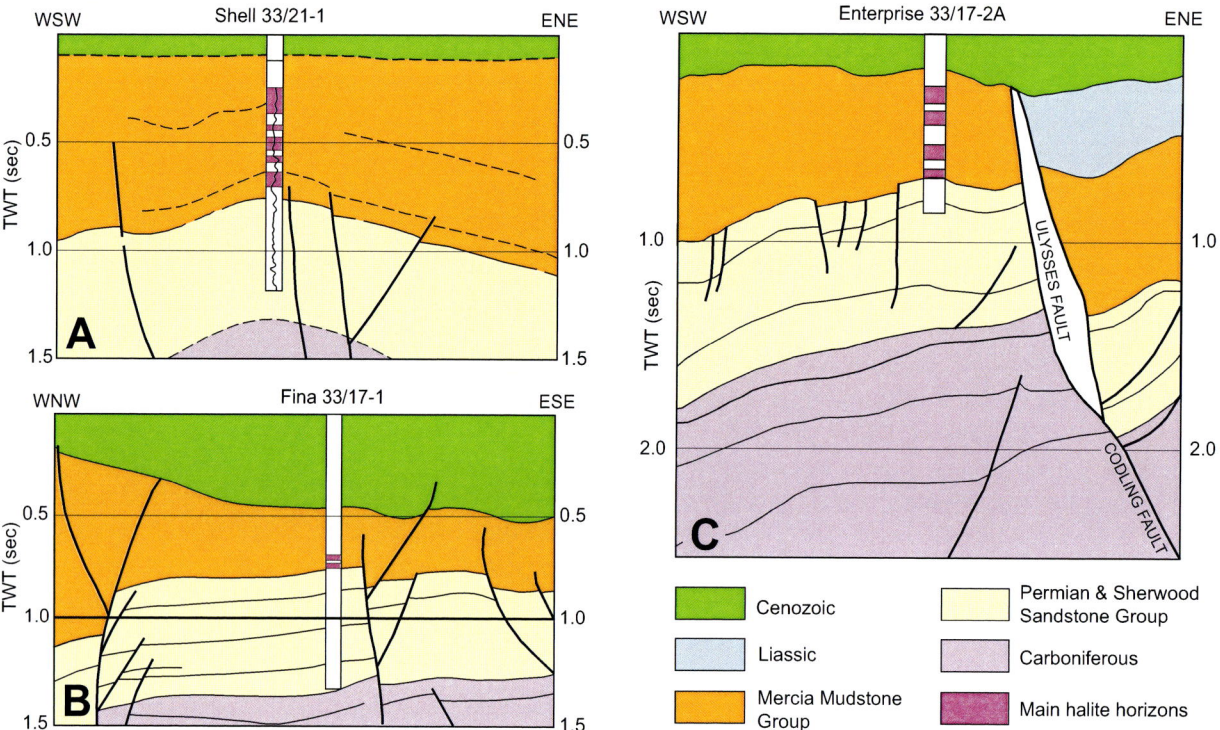

Figure 12.8. Structural styles in the Kish Bank Basin. Geoseismic profiles across three well locations. (A) and (B) after Naylor *et al.* (1993), (C) derived from Dunford *et al.* (2003).

charge periods predated the significant Cenozoic structural phase. Many of the potential structures are in close proximity to the major and active Codling Fault system. Earlier fault structures that may have trapped hydrocarbons had a high risk of rupture, tilting and modification in the Cenozoic. Hydrocarbon generation had been switched off prior to this phase, so that secondary charge could only occur through migration of previously sequestered hydrocarbons – obviously a high-risk situation. Late phases of reactivation on some faults adds further to this risk. Dunford *et al.* (2001) suggested that two of the drilled structures may still have some potential. Tilting of the Fina 33/17-1 structure during the Cenozoic could possibly have resulted in re-migration to a new culmination at the southern end of the horst. In the case of Enterprise 33/17-2A, the well tested the anticlinal closure that may have resulted from Cenozoic fault movement. The fault-block portion of the same structure, which is reported to have a seismic flat spot, remains untested. Despite this, both these prospects must be regarded as high risk, with possibly low reserve potential. In general terms the migration routes suggest that the east of the basin has higher potential, whilst any target structure would need to have avoided rupture and modification during the Cenozoic.

Central Irish Sea Basin
Basin framework

The Central Irish Sea Basin is a NE–SW oriented basin that is separated from the Cardigan Bay and St George's Channel basins to the south by the St Tudwal's Arch and its offshore extension (Figure 6.7). To the north the basin is separated from the Kish Bank Basin by the NE–SW trending Mid-Irish Sea High. The inherited Caledonian structural trend is clearly discernable and the older lineaments have undergone a number of phases of reactivation in the Carboniferous (Variscan), Mesozoic (Cimmerian) and Cenozoic (Laramide–Alpine). There is demonstrable influence by the major faults on facies and thickness development of sedimentary units at various levels. The basin is also segmented by important NW–SE and N–S faults at local and regional scales. Four wells have been drilled in the southeast part of the Central Irish Sea Basin and outline results published by Maddox *et al.* (1995). The two Hydrocarbons Ireland Ltd (HIL) wells, 42/12-1 and 42/12-2, were drilled on fault-related structures within the basin, while Marathon 42/17-1 and Conoco 42/16-1 were located on the St Tudwal's Arch extension (Figure 6.8). No significant shows were reported from any of the wells. Duncan *et al.* (1998) reported on two additional wells in the basin (BHP 42/8-1 and UK well 107/1-1) that tested structural closures at Sherwood Sandstone Group level.

Drilling has shown a Triassic succession resting unconformably on the Carboniferous, and thinning southwards towards the fault-bounded St Tudwal's Arch extension (Figure 6.8). No Permian strata have been encountered in drilling. The Sherwood Sandstone Group is thinner than further north in the region, ranging from a sand-prone 232 m in well HIL 42/12-1 to a mud-prone section of only 120 m in Conoco 42/16-1 (Maddox *et al.*, 1995). Dunford *et al.* (2001, their Figure 7) see this change as part of a regional picture in which the Central Irish Sea Basin is a re-entrant at the basin margin bypassed by the main fluvial system draining towards the northwest. In the Conoco 42/16-1 well a Stephanian miospore assemblage was recorded from sidewall core in the basal part of the Sherwood Sandstone Group, 21 m above a level taken on dipmeter evidence to represent the unconformity at the top of Stephanian red beds (Naylor and Clayton, 2000). This reworking probably represents local erosion of Stephanian outcrop on the Variscan landscape during the Early Triassic. The record is significant in proving the original deposition of a more or less complete Late Carboniferous sequence in this part of the basin. The reddened top of the Carboniferous is of varying age in the Irish Sea region; Bolsovian in Wales, Asturian in the Kish Bank Basin, and Stephanian in parts of the Central Irish Sea Basin. Part of the Carboniferous sequence (Serpukhovian and Bashkirian: previously Namurian) is absent in the Central Irish Sea Basin wells, and Duckmantian to Stephanian coal measures rest unconformably on the Mississippian. Whether this is a result of non-deposition, intra-Carboniferous inversion and erosion, or whether the missing sequence is present in the deeper undrilled parts of the basin, is not known (Maddox *et al.*, 1995).

The Mercia Mudstone Group succession comprises anhydritic mudstones and minor sandstones interbedded with thick halites. Only minor halokinetic structures are seen on the seismic data. The unit is 1425 m thick in HIL 42/12-1 and thins to 776 m in Conoco 42/16-1. The main faults were probably active at this time, as documented in other basins in the region.

Only Lower Jurassic strata have been proved by drilling in the Central Irish Sea Basin, although Maddox *et al.* (1995) suggested, on the basis of unpublished seismic interpretation, that younger Jurassic and Cretaceous sequences may be preserved in the deeper parts of the basin. As discussed in Chapter 10, exploration drilling in the Cardigan Bay and St George's Channel basins, south of St. Tudwal's Arch, has revealed an almost complete Lower Liassic to Lower Purbeckian succession (Barr *et al.*, 1981 and Figure 10.2A). The George's Channel–Cardigan Bay basins, with extensive Jurassic sequences, are within the Celtic Sea structural domain, rather than with the Permo-Triassic dominated basins of the Irish Sea.

Exploration potential

The main exploration target in the Central Irish Sea Basin, as elsewhere in the Irish Sea, is for Sherwood Sandstone reservoirs sealed by the Mercia Mudstone Group. Of the five wells drilled in the Irish section of the basin, at least three tested valid traps (Floodpage *et al.*, 2001). Reasons for the lack of exploration success need to be examined, and a number of factors have been cited by different workers.

Apatite fission track analysis has been used to study the multiphase inversion history of the basin. Duncan *et al.* (1998) and Green *et al.* (2001) recognised three palaeo-thermal episodes using AFTA and vitrinite studies of the 42/12-1, 42/16-1 and 42/17-1 wells in the Central Irish Sea Basin and on its southern margin. Cooling from temperature maxima occurred in Early Cretaceous (120–115 Ma), Late Cretaceous–Early Cenozoic (70–55 Ma), and Late Cenozoic 25–0 Ma). Maximum palaeo-temperatures were achieved in the Early Cretaceous and maturation levels in the Carboniferous source sequence are considered to relate to this. Green *et al.* (2001) accept that the Early Cretaceous maximum in the Central Irish Sea Basin wells may mask earlier lower palaeo-temperature events, and specifically an event related to the Variscan episode. Results from another well in the general region – Gulf 42/21-1 – showed a marked difference (*see* Chapter 10 for further discussion). Here there was no evidence of the Early Cretaceous event, but the thick Jurassic section in the well began cooling from maximum palaeo-temperatures in Late Cretaceous to Early Cenozoic time. Green *et al.* (2001) pointed out that published apatite fission-track results from the North Celtic Sea Basin (Murdoch *et al.*, 1995) show that the Early Cenozoic cooling episode is dominant in that basin. The Gulf 42/21-1 well lies in the western part of the St George's Channel Basin, and as noted above, this basin is best considered as part of the Celtic Sea region. In contrast, VR studies of the Central Irish Sea and Kish Bank basin wells led Corcoran and Clayton (1999) to conclude that Carboniferous sediments experienced peak maturation during the Variscan interval. This they ascribed to elevated heat flows in Stephanian to early Permian time, possibly related to the known volcanic activity during this period. They also postulated at least one further period of exhumation after the Variscan, some time between latest Jurassic and Early Cenozoic. In a later paper, the same authors (Corcoran and Clayton, 2001) favoured a Paleocene to Oligo-Miocene timing for this second exhumation episode.

A number of estimates have been produced for the amount of erosion that has occurred within the Central Irish Sea Basin at different times. Data modelled by Green *et al.* (2001) were interpreted as showing some *c.*3.0 km of deposition between latest Triassic to Early Cretaceous times, eroded later in the Early Cretaceous. Deposition of *c.*2 km of section in the Late Cretaceous was followed by erosion early in the Palaeogene, and a further *c.*1 km of Cenozoic sediments was eroded in the Neogene. Data presented by Floodpage *et al.* (2001) for the UK well 107/1-1 well within the basin also support the view that Triassic sediments achieved maximum palaeo-temperatures during Early Cretaceous time, but that Carboniferous source rocks reached the oil window in mid-Triassic time and the gas window in Early Jurassic time. A further 1 km of sediment was deposited in the Paleocene to Miocene interval, with the remaining 1.3 km of section removed since that time (Green *et al.*, 2001). Using a quite different approach with vitrinite reflectance data, Corcoran and Clayton (1999) proposed that at the HIL 42/12-1 well up to 2250 m of Stephanian–Early Permian sediments were eroded during mid-Permian uplift, prior to reburial beneath a minimum of 1768 m of Triassic and Mesozoic sediments. They suggested a period of Paleocene–Eocene exhumation, in line with that proposed for this location by Murdoch *et al.* (1995). The implications of the results from the AFTA and vitrinite reflectance data are discussed below.

With respect to source rocks in the Central Irish Sea Basin, Corcoran and Clayton (1999) conclude that the younger Pennsylvanian coal facies rocks have fair to good gas-prone potential and are in the dry gas generation window (HIL 42/12-1). However, Floodpage *et al.* (2001) questioned the efficacy of this source rock sequence in the basin, and consider the source rock potential to be limited to a few thin coal beds. Obviously, if the missing Sepukhovian-Bashkirian section is developed beneath any part of the basin, this would add considerably to the source rock possibilities. However, on the basis of seismic interpretation Floodpage *et al.* (2001) argued against the development of this unit in the basin. The Lower Jurassic sequence is organic-rich and oil-prone, but is immature in the Marathon 42/17-1 well on the St Tudwal's Arch extension.

As pointed out above, the target Sherwood Sandstone reservoir interval is thinner in the Central Irish Sea Basin than further north, and also is increasingly shale-prone southwards. Maddox *et al.* (1995) report 88–98% net/gross ratios over more than 200 m in the HIL 42/12-1 & -2 wells decreasing to only 54% net/gross in the Conoco 42/16-1 well, where the potential reservoir interval is only 120 m thick. Porosity values are 14%–18% in the northern wells where, as noted in the Kish Bank Basin, facies type has controlled diagenesis and reservoir quality. Permeability is good in some sections, but variable, possibly due to secondary illite development. Overall, however, the sequence is probably capable of serving as an adequate gas reservoir. The overlying Mercia Mudstone Group is anticipated to provide an adequate top seal and to provide side seal on tilted fault-block structures (Maddox *et al.*, 1995). However, Duncan *et al.* (1998) suggested that, on inversion, mudstones pass into the brittle tensional strength field and become more susceptible to failure. This poses a risk to seal efficiency, particularly in those parts of the basin where the Mercia Mudstone Group is less than 600 m thick (HIL 42/12-2 well). In addition, sandstones within the basal part of the Mercia Mudstone Group in the CISB may form a potential 'thief zone' between the reservoir and the lowest effective halite seal (Maddox *et al.*, 1995).

The model for hydrocarbon generation envisaged by Maddox *et al.* (1995) from Carboniferous source rocks is for a phase of oil generation in the Late Jurassic, and a later phase of gas expulsion in the Early Cenozoic. The source rocks are then believed to have been 'switched off' during the main Alpine inversion and exhumation. As Maddox *et al.* (1995) pointed out, this scenario means that at least one of the phases of hydrocarbon generation post-dates trap formation, which they see as occurring in Triassic–Jurassic times. These views contrast with those of Corcoran and Clayton (1999) in which the Carboniferous (Duckmantian–Stephanian) source sequence of the Central Irish Sea Basin wells was exposed to maximum palaeotemperatures prior to the Early Jurassic, and probably during Stephanian to Early Permian time. This effectively means that hydrocarbon generation from this source was 'switched off' during Variscan uplift. Lower Jurassic source rocks, on the other hand, probably reached peak palaeotemperatures in Late Cretaceous time and may have provided charge in the south of the area. A third view, from Green *et al.* (2001), is that maximum palaeotemperatures occurred in Early Cretaceous time (120–115 Ma) and the subsequent cooling episode represented the termination of the main phase of hydrocarbon generation.

Most of the potential trapping structures in the Central Irish Sea Basin are tilted fault-block closures associated with NE–SW faults, or rollover anticlines in the footwalls. These structures were probably formed during the Triassic–Jurassic tensional regime and then modified and tightened during later compression and inversion episodes. In addition to exhumation in the Cenozoic, there was significant movement also on boundary faults and major dextral strike-slip on NW trending structures such as the Codling and Sticklepath faults.

Although good gas shows were encountered in the Carboniferous section of the Conoco 42/16-1 well, targets within the Carboniferous are extremely high risk. Lack of detailed trap definition, low sandstone porosities and doubt concerning adequate seal are serious negative factors. Potential concerns regarding the Mercia Mudstone seal on Triassic reservoirs have been outlined above. There is doubt also as to whether some wells were drilled on-structure (Maddox *et al.*, 1995). However, the lack of shows in the Triassic points to more serious exploration problems within the basin. The timing of hydrocarbon generation suggested by Corcoran and Clayton (1999) would require sequestration within Carboniferous traps and later migration to Triassic reservoirs. This is clearly a high-risk assumption. The low source rock potential of the known Carboniferous sequence is a further negative factor. Maximum hydrocarbon generation in the mid-Triassic to Early Jurassic interval (Floodpage *et al.* (2001), Early Cretaceous (Green *et al.*, 2001), or Late Jurassic and Early Cenozoic (Maddox *et al.*, 1995) probably post-dates trap formation, but there was significant risk of trap rupture during the Cenozoic. The wells have been drilled in the southern part of the basin and Maddox *et al.* (1995) cited risks concerning the long distance to the main source kitchen in the basin centre, and doubts concerning migration pathways as further concerns. Nevertheless, the Central Irish Sea Basin remains under-explored, particularly in the basin centre and north, where additional source sequences may exist, and where there is the possibility of a Permian reservoir sequence.

Summary of regional traps and play types

The major factors controlling local prospectivity in each of the Irish Sea basins have been discussed. Attempts to understand these variations within a comprehensive view of the region have been made by a number of workers, most notably Green *et al.* (1997), Quirk *et al.* (1999), Corcoran and Clayton (1999), Naylor and Shannon (1999), Floodpage *et al.* (2001) and Holford *et al.*(2005). Gradual release of well, maturation, apatite fission track and seismic data in the last fifteen years has brought greater understanding of the main controlling factors, and some degree of agreement regarding reasons for the lack of exploration success in the region.

The only producing basin in the region, and one with significant hydrocarbon reserves, is the East Irish Sea Basin lying in UK waters. In summarising the potential plays in the region it is instructive to examine in outline the main phases of development of that basin, and attempt to understand the controlling factors that have favoured the generation and entrapment of hydrocarbons. In terms of source rocks, the productive Serpukhovian–Baskirian (Pendleian to Langsettian: formerly Namurian) source sequences extend from the Bowland Basin in northern England westwards to the Dublin Basin, and are present beneath the East Irish Sea Basin and probably the Kish Bank Basin. They are in less promising facies north and south of this belt (Jackson *et al.*, 1995). The distribution of Carboniferous units in the region was severely modified during the Variscan orogeny and the subsequent erosion. The Ormskirk Sandstone Formation provides an excellent reservoir in the East Irish Sea Basin, and the Mercia Mudstone Group, with thick halites, an effective seal. Modelling of apatite fission track data in the basin (Floodpage *et al.*, 2001) suggests only modest Late Cimmerian uplift in the Late Jurassic to halt oil generation that had begun earlier in the Jurassic. During subsequent burial in the Cretaceous oil, and then gas, generation was resumed, and may have been augmented by an Early Cenozoic heating event (Floodpage *et al.*, 2001). As a result of this late history of hydrocarbon generation it was possible for traps disrupted in early tectonic episodes to be recharged following annealing of the breaching faults by the Mercia Mudstone Group halites. Cenozoic recharge was also possible as a result of gas expansion during exhumation.

The other basins in the Irish Sea region lack one or more of the elements seen in the East Irish Sea Basin. In a regional sense the Sherwood Sandstone Group provides a good reservoir across the region, with adequate seals generally provided by the Mercia Mudstone Group. The original Triassic basin limits extended beyond the current subcrop limits, which are mainly the result of erosion during Cenozoic uplift. As mentioned previously, reservoir properties decline southwards in the Central Irish Sea Basin, accompanied by a probable reduction in the sealing capacity of the Mercia Mudstone sequence. A similar reduction in sealing capacity is probably also seen in the western part of the Peel Basin. Nevertheless the Sherwood/Mercia reservoir-seal couple is a low risk target in the region.

The Permian Collyhurst Sandstone Formation, whilst presenting good reservoir characteristics in some locations, is variable and difficult to predict, and often lacks an adequate seal. The Carboniferous is a similar high-risk target. Thus although some exploration targets exist in parts of individual basins, as outlined above, and the

region is by no means fully explored, no individual basin is likely to replicate the successes of the East Irish Sea Basin.

Overall, the main reasons for exploration failure include the following factors: (a) early periods of peak hydrocarbon generation, in some cases pre-dating trap formation, but also allowing for later trap rupture and modification, without the possibility of reservoir recharge, (b) demonstrable and frequent late movement on faults, and (c) lack of adequate source sequences due to non-deposition (?Central Irish Sea Basin) or erosion (Peel–Solway basins).

Nonetheless, a number of trap and play types offering exploration potential can be identified in the Irish Sea region (Shannon and Naylor, 1998). These are illustrated in Figure 12.9 and can be summarised as follows.

Permo-Triassic and Carboniferous fault-block traps: These are typically located adjacent to the basin-bounding

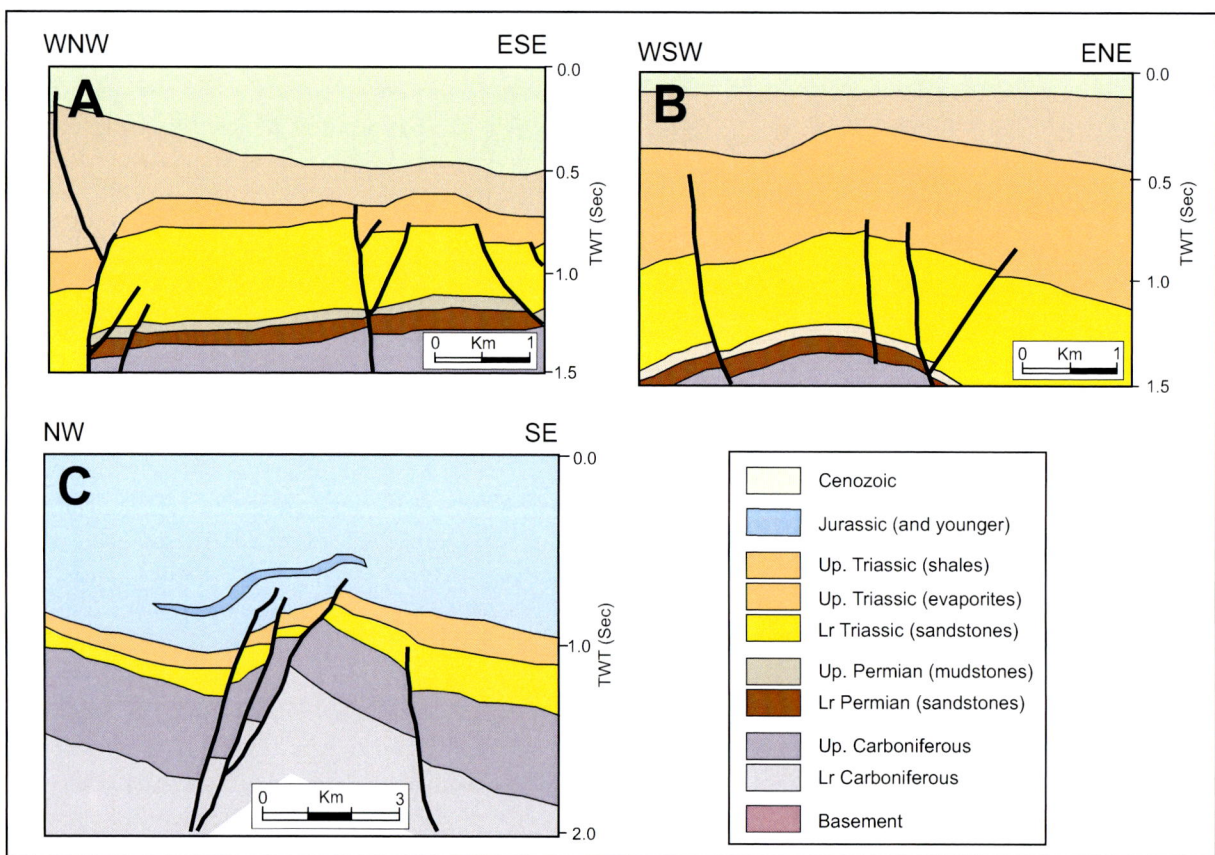

Figure 12.9. Schematic illustration of play and trap types in the Irish Sea basins (after Shannon and Naylor, 1998). **A.** Fault-block play with reservoirs in the Lower Permian (Collyhurst Sandstone) and Lower Triassic (Sherwood Sandstone Group, with seals in the Upper Permian (Manchester Marl), Upper Triassic (Mercia Mudstone Group) and source rocks in the Upper Carboniferous. Based on Naylor *et al.* (1993). **B.** Four-way dip-closed anticline with similar reservoirs, seals and source to (A) above. Based on Naylor *et al.* (1993). **C.** Fault-block, stratigraphic pinch-out and drape plays. Reservoirs are predicted in the Lower Triassic, which thins across the fault blocks and thins updip to the NW, and locally in the Lower Jurassic. Seals are provided by the Upper Triassic and Lower Jurassic. Source rocks lie in the Upper Carboniferous. Based on Shannon (1996).

faults or to intra-basinal faults (Figure 12.9A). High-sided fault closure is likely adjacent to the intra-basinal faults, with low-sided closure associated with the basin-bounding faults. Most of the structures probably developed in the Permo-Triassic, prior to hydrocarbon generation, with minor re-activation during the Cenozoic. The principal reservoirs for this play are in the Permo-Triassic, with secondary targets in the Carboniferous. The principal reservoirs are sealed by the mudrocks and evaporites of the Mercia Mudstone Group, and the secondary reservoirs by interbedded Pennsylvanian shales. The principal likely source rocks are Serpukhovian–Pennsylvanian coals or shales.

Four-way dip-closed traps: These may occur at several stratigraphic levels (Figure 12.9B). Reservoir intervals exist within the Serpukhovian–Pennsylvanian, Permo-Triassic (Collyhurst Sandstone Formation and Sherwood Sandstone Group) and locally in the Lower Jurassic. Caprocks are provided by Carboniferous shales, Permo-Triassic shales and evaporites, and by Lower Jurassic shales. Principal source rocks are Pennsylvanian coals, with secondary source potential in Serpukhovian–Pennsylvanian shales.

Drape structures and traps: Synsedimentary tectonism during the Permo-Triassic provides the potential for the development of drape or of combination traps involving a component of faulting and sedimentary draping. In addition, draping and differential compaction of Permo-Triassic strata across residual Variscan palaeotopography can be anticipated, particularly in the more southerly basins in the region. Sherwood Sandstone Group reservoirs, capped by Mercia Mudstone Group mudrocks and salt, are the probable targets, with Pennsylvanian coals as the principal source rock.

Stratigraphic pinch-out traps: Synsedimentary faulting during the Permian and Early Triassic probably produced pinch-out and other stratigraphic traps (Figure 12.9C). Permo-Triassic sediments were sourced primarily from the Irish and Welsh massifs, with local sediment input from the footwall crests of rotated fault blocks. Seismic evidence for marked Permo-Triassic erosion of the footwalls of intra-basinal faults is seen within the Central Irish Sea Basin (Shannon, 1996). Downlapping of reflectors from rotated footwall crests, together with the wedge-shaped geometry and seismic character, is interpreted (O'Reilly and Shannon, 1994) to indicate alluvial fan deposits that probably grade basinwards to well-sorted fluvial and aeolian sandstones with high porosity and permeability potential. These, in turn, are likely

to shale out into distal wadi and playa evaporitic shales. Pennsylvanian coals provide the most likely source rocks, with Serpukhovian–Pennsylvanian shales, if present, providing a potential local oil source. Stratigraphic pinch-outs of sandstones against basement horsts can also be prognosed at Carboniferous level, similar to the manner in which the Pennsylvanian onlaps Caledonian basement with angular unconformity at the Amoco 33/22-1 well on the south flank of the Kish Bank Basin. Seal would be provided by lateral pinch-out and by overlying shales. Interbedded coals and shales provide the probable source for this play.

References

Allen, P.A., Bennett, S.D., Cunningham, M.J.M., Carter, A., Gallagher, K., Lazzaretti, E., Galewsky, J., Densmore, A.L., Phillips, W.E.A., Naylor, D., and Solla Hach, C. (2002) The post-Variscan thermal and denudational history of Ireland. *In*: Doré, A.G., Cartwright, J.A., Stoker, M.S., Turner, J.P., and White, N. (eds), *Exhumation of the North Atlantic Margin: Timing, Mechanisms and Implications for Petroleum Exploration.* Geological Society, London, Special Publications, **196**, 371–399.

Barr, K.W., Colter, V.S., and Young, R. (1981) The geology of the Cardigan Bay–St. George's Channel Basin. *In*: Illing, L.V. and Hobson, G.D. (eds), *Petroleum Geology of the Continental Shelf of North-West Europe.* Heyden and Son Ltd., London, 432–443.

Bott, M.P.H. (1964) Gravity measurements in the north-eastern part of the Irish Sea. *Quarterly Journal of the Geological Society of London*, **120**, 369–396.

Brodie, J. and White, N. (1995) The link between basin inversion and igneous under-plating. *In*: Buchanan, J.G. and Buchanan, P.G. (eds), *Basin Inversion.* Geological Society, London, Special Publications, **88**, 21–38.

Broughan, F.M., Naylor, D., and Anstey, N.A. (1989) Jurassic rocks in the Kish Bank Basin. *Irish Journal of Earth Sciences*, **10**, 96–106.

Caston, G.F. (1975) Igneous dykes and associated scour hollows in the North Channel, Irish Sea. *Marine Geology*, **18**, M77–M85.

Cope, J.C.W. (1984) The Mesozoic history of Wales. *Proceedings of the Geologists' Association*, **95**, 373–385.

Cope, J.C.W. (1994) A latest Cretaceous hotspot and the south-easterly tilt of Britain. *Journal of the Geological Society of London*, **151**, 729–731.

Cope, J.C.W. (1997) The Mesozoic and Tertiary history of the Irish Sea. *In*: Meadows, N.S., Trueblood, S.P., Hardman, M., and Cowan, G. (eds), *Petroleum Geology of the Irish Sea and Adjacent Areas.* Geological Society, London, Special Publications, **124**, 47–59.

Corcoran, D.V. and Clayton, G. (1999) Interpretation of vitrinite reflectance profiles in the Central Irish Sea area: implications for the timing of organic maturation. *Journal of*

Petroleum Geology, **22**, 261–286.

Corcoran, D.V. and Clayton, G. (2001) Interpretation of vitrinite reflectance profiles in sedimentary basins, onshore and offshore Ireland. *In*: Shannon, P.M., Haughton, P.D.W., and Corcoran, D.V. (eds), *The Petroleum Exploration of Ireland's Offshore Basins*. Geological Society, London, Special Publications, **188**, 61–90.

Corcoran, D.V. and Mecklenburgh, R. (2005) Exhumation of the Corrib Gas Field, Slyne Basin, offshore Ireland. *Petroleum Geoscience*, **11**, 239–256.

Dobson, M.R. (1977a) The geological structure of the Irish Sea. *In*: Kidson, C. and Tooley, M.J. (eds), *The Quaternary History of the Irish Sea*. Seal House Press, Liverpool, 13–26.

Dobson, M.R. (1977b) The history of the Irish Sea basins. *In*: Kidson, C. and Tooley M.J. (eds), *The Quaternary History of the Irish Sea*. Seal House Press, Liverpool, 93–98.

Duncan, W.I., Green, P.F., and Duddy, I.R. (1998) Source rock burial history and seal effectiveness: key facets in understanding hydrocarbon exploration potential in the East and Central Irish Sea basins. *AAPG Bulletin*, **82**, 1401–1415.

Dunford, G.M., Dancer, P.N., and Long, K.D. (2001) Hydrocarbon potential of the Kish Bank Basin: integration within a regional model for the Greater Irish Sea Basin. *In*: Shannon, P.M., Haughton, P.D.W., and Corcoran, D.V. (eds), *The Petroleum Exploration of Ireland's Offshore Basins*. Geological Society, London, Special Publications, **188**, 135–154.

Etu-Efeotor, J.D. (1976) Geology of the Kish Bank Basin. *Journal of the Geological Society of London*, **132**, 708.

Evans, D., Kenolty, N., Dobson, M.R., and Whittington, R.J. (1980) The geology of the Malin Sea. Institute of Geological Sciences, Report **79/15**, 44pp.

Fitzsimons, S. and Parnell, J. (1995) Diagenetic history and reservoir potential of Permo-Triassic sandstones in the Rathlin Basin. *In*: Croker, P.F. and Shannon, P.M. (eds), *The Petroleum Geology of Ireland's Offshore Basins*. Geological Society, London, Special Publications, **93**, 91–105.

Floodpage, J., Newman, P., and White, J. (2001) Hydrocarbon prospectivity in the Irish Sea area: insights from recent exploration in the Central Irish Sea, Solway and Peel basins. *In*: Shannon, P.M., Haughton, P.D.W., and Corcoran, D.V. (eds), *The Petroleum Exploration of Ireland's Offshore Basins*. Geological Society, London, Special Publications, **188**, 107–134.

Fyfe, J.A., Long, D., and Evans, D. (1993) United Kingdom offshore regional report: *The geology of the Malin–Hebrides Sea area*. London HMSO for the British Geological Survey, 91pp.

Geological Survey of Great Britain (1965) Geological Survey of Norhern Ireland boreholes. *Summary of Progress 1964*. London HMSO for the British Geological Survey and Museum, 72–73.

Geological Survey of Northern Ireland (1997) *1:250,000 Northern Ireland: Solid Geology (2nd Edition)*. London HMSO.

Green, P.F., Duddy, I.R., and Bray, R.J. (1997) Variation in thermal styles around the Irish Sea and adjacent areas: implications for hydrocarbon occurrence and tectonic evolution. *In*: Meadows, N.S., Trueblood, S.P., Hardman, M., and Cowan, G. (eds), *Petroleum Geology of the Irish Sea and Adjacent Areas*. Geological Society, London, Special Publications, **124**, 73–93.

Green, P.F., Duddy, I.R., Bray, R.J., Duncan, W.I., and Corcoran, D.V. (2001) The influence of thermal history on hydrocarbon prospectivity in the Central Irish Sea Basin. *In*: Shannon, P.M., Haughton, P.D.W., and Corcoran, D.V. (eds), *The Petroleum Exploration of Ireland's Offshore Basins*. Geological Society, London, Special Publications, **188**, 171–188.

Holford, S.P., Green P.F., Hillis, R.R., Turner, J.P., and Stevenson, C.T.E. (2009) Mesozoic–Cenozoic exhumation and volcanism in Northern Ireland constrained by AFTA and compaction data from the Larne No. 2 borehole. *Petroleum Geoscience*, **15**, 239–257.

Holford, S.P., Turner, J.P., and Green P.F. (2005) Reconstructing the Mesozoic–Cenozoic exhumation history of the Irish Sea basin system using apatite fission track analysis and vitrinite reflectance data. *In*: Doré, A.G. and Vining, B.A. (eds), *Petroleum Geology: North-West Europe and Global Perspectives – Proceedings of the 6th Petroleum Geology Conference*. Geological Society, London, 1095–1107.

Illing, L.V. and Griffith, A.E. (1986) Gas prospects in the 'Midland Valley' of Northern Ireland. *In*: Brooks, J., Goff, J.C., and van Hoorn, B. (eds), *Habitat of Palaeozoic Gas in NW Europe*. Geological Society, London, Special Publications, **23**, 73–84.

Jackson, D.I., Jackson, A.A., Evans, D., Wingfield, R.T.R., Barnes, R.P., and Arthur, M.J. (1995) The geology of the Irish Sea. London HMSO for the British Geological Survey, 123pp.

Jenner, J. K. (1981) The structure and stratigraphy of the Kish Bank Basin. *In*: Illing, L.V. and Hobson, G.D. (eds), *Petroleum Geology of the Continental Shelf of North-West Europe*. Heyden and Son Ltd., London, 426–431.

Johnston, T.P. (1979) Preliminary report on the Killary Glebe No.1 borehole, Coalisland, Co. Tyrone. *Geological Survey of Northern Ireland Open File Report* **No. 63**, 23pp.

Johnston, T.P. (2004) Post-Variscan Deformation and Basin Formation. *In*: Mitchell, W.I. (ed.), *The Geology of Northern Ireland: Our Natural Foundation*. Geological Survey of Northern Ireland, Belfast, 205–210.

Maddox, S.J., Blow, R., and Hardman, M. (1995) Hydrocarbon prospectivity of the Central Irish Sea Basin with reference to Block 42/12, offshore Ireland. *In*: Croker, P.F. and Shannon, P.M. (eds), *The Petroleum Geology of Ireland's Offshore Basins*. Geological Society, London, Special Publications, **93**, 59–77.

Maddox, S.J., Blow, R.A., and O'Brien, S.R. (1997) The geology and hydrocarbon prospectivity of the North Channel Basin. *In*: Meadows, N.S., Trueblood, S.P., Hardman, M., and Cowan, G. (eds), *Petroleum Geology of the Irish Sea and Adjacent Areas*. Geological Society, London, Special Publications, **124**, 95–111.

Manning, P.I. and Wilson, H.E. (1975) The stratigraphy of the Larne Borehole, County Antrim. *Bulletin of the Geological Survey of Great Britain*, **50**, 1–27.

McCaffrey, R.J. and McCann, N. (1992) Post-Permian basin history of northeast Ireland. *In*: Parnell, J. (ed.), *Basins on the Atlantic Seaboard: Petroleum Geology, Sedimentology and Basin Evolution*. Geological Society, London, Special

Publications, **62**, 277–290.

McCann, N. (1988) An assessment of the subsurface geology between Magilligan Point and Fair Head, Northern Ireland. *Irish Journal of Earth Sciences*, **9**, 71–78.

McCann, N. (1990) The subsurface geology between Belfast and Larne, Northern Ireland. *Irish Journal of Earth Sciences*, **10**, 157–173.

Middleton, D.W.J., Parnell, J., Green, P.F., Xu, G., and McSherry M. (2001) Hot fluid flow events in Atlantic margin basins: an example from the Rathlin Basin. *In*: Shannon, P.M., Haughton, P.D.W., and Corcoran, D.V. (eds), *The Petroleum Exploration of Ireland's Offshore Basins*. Geological Society, London, Special Publications, **188**, 91–105.

Mitchell, W.I. (2004) Carboniferous. *In*: Mitchell, W.I. (ed.), *The Geology of Northern Ireland: Our Natural Foundation*. Geological Survey of Northern Ireland, Belfast, 79–116.

Mitchell, W.I. and Owens, B. (1990) The geology of the western part of the Fintona Block, Northern Ireland: evolution of Carboniferous basins. *Geological Magazine*, **127**, 407–426.

Murdoch, L.M., Musgrove, F.W., and Perry, J.S. (1995) Tertiary uplift and inversion history in the North Celtic Sea Basin and its influence on source rock maturity. *In*: Croker, P.F. and Shannon, P.M. (eds), *The Petroleum Geology of Ireland's Offshore Basins*. Geological Society, London, Special Publications, **93**, 297–319.

Musgrove, F.W., Murdoch, L.M., and Lenehan, T. (1995) The Variscan fold-thrust belt of southeast Ireland and its control on early Mesozoic extension and deposition: a method to predict the Sherwood Sandstone. *In*: Croker, P.F. and Shannon, P.M. (eds), *The Petroleum Geology of Ireland's Offshore Basins*. Geological Society, London, Special Publications, **93**, 81–100.

Naylor, D. and Clayton, G. (2000) Palynological and maturation data and their bearing on Irish post-Variscan palaeogeography. *Irish Journal of Earth Sciences*, **18**, 33–39.

Naylor, D. and Shannon, P. M. (1999) The Irish Sea region: why the general lack of exploration success? *Journal of Petroleum Geology*, **22**, 363–70.

Naylor, D., Philcox, M.E., and Clayton, G. (2003) Annaghmore-1 and Ballynamullan-1 wells, Larne–Lough Neagh Basin, Northern Ireland. *Irish Journal of Earth Sciences*, **21**, 47–69.

Naylor, D., Haughey, N., Clayton, G., and Graham, J.R. (1993) The Kish Bank Basin, offshore Ireland. *In*: Parker, J.R. (ed.), *Petroleum Geology of North-west Europe: Proceedings of the 4th Conference*. Geological Society, London, 845–855.

Newman, P. (1999) The geology and hydrocarbon potential of the Peel and Solway Basins, East Irish Sea. *Journal of Petroleum Geology*, **22**, 305–324.

O'Reilly, B.M. and Shannon, P.M. (1994) Fault analysis and modelling – an example from the Central Irish Sea – St George's Channel region. *6th Conference, European Association of Geoscientists and Engineers, Vienna. Extended Abstracts.* Paper P505.

Papworth, T.J. (1985) Seismic exploration over basalt covered areas in the U.K. *First Break*, **3**, 20–32.

Parnell, J. (1992) Hydrocarbon potential of Northern Ireland: 3. Reservoir potential of the Permo-Triassic. *Journal of Petroleum Geology*, **15**, 51–70.

Penn, I.E. (1981) Larne No.2 Geological well completion report. *Report of the Institute of Geological Sciences, London*, **81/6**, 58pp.

Penn, I.E., Holliday, D.W., Kirby, G.A., Soper R.A. *et al.* (1983) The Larne borehole No. 2: discovery of a new Permian volcanic centre. *Scottish Journal of Geology*, **19**, 333–46.

Preston, J. (2001) Tertiary igneous activity. *In*: C.H. Holland (ed.), *The Geology of Ireland*. Dunedin Academic Press, Edinburgh, 353–373.

Quirk, S.G, Roy, S., Knott, I, Redfern, J., and Hill, L. (1999) Petroleum geology and future hydrocarbon potential of the Irish Sea. *Journal of Petroleum Geology*, **22**, 243–260.

Reay, D.M. (2004a) Geophysics and Concealed Geology. *In*: Mitchell, W.I. (ed.), *The Geology of Northern Ireland: Our Natural Foundation*. Geological Survey of Northern Ireland, Belfast, 227–248.

Reay, D.M. (2004b). Oil and Gas. *In*: Mitchell, W.I. (ed.), *The Geology of Northern Ireland: Our Natural Foundation*. Geological Survey of Northern Ireland, Belfast, 273–290.

Ruffell, A. and Shelton, R. (1999) The control of sedimentary facies by climate during phases of crustal extension: examples from the Triassic of onshore and offshore England and Northern Ireland. *Journal of the Geological Society, London*, **156**, 779–89.

Shannon, P.M. (1996) Current and future potential of oil and gas exploration in Ireland. *In*: Glennie, K. and Hurst, A. (eds), *AD 1995 – NW Europe's Hydrocarbon Industry*. Geological Society, London, 51–62.

Shannon, P.M. and Naylor, D. (1998) An assessment of Irish Offshore Basins and petroleum plays. *Journal of Petroleum Geology*, **21**, 125–152.

Shelton, R. (1995) Mesozoic basin evolution of the North Channel: preliminary results. *In*: Croker, P.F. and Shannon, P.M. (eds), *The Petroleum Geology of Ireland's Offshore Basins*. Geological Society, London, Special Publications, **93**, 7–20.

Shelton, R. (1997) Tectonic evolution of the Larne Basin. *In*: Meadows, N.S., Trueblood, S.P., Hardman, M., and Cowan, G. (eds), *Petroleum Geology of the Irish Sea and Adjacent Areas*. Geological Society, London, Special Publications, **124**, 113–133.

Simms, M.J. (2000) The sub-basaltic surface in northeast Ireland and its significance for interpreting the Tertiary history of the region. *Proceedings of the Geologists' Association*, **111**, 321–336.

Thompson, S.J. (1979) Preliminary report on the Ballymacilroy No.1 borehole, Ahoghill, Co. Antrim. *Geological Survey of Northern Ireland Open File Report* **No. 63**, 17pp.

Ware, P.D. and Turner, J.P. (2002) Sonic velocity analysis of the Tertiary denudation of the Irish Sea basin. *In*: Doré, A.G., Cartwright, J.A., Stoker, M.S., Turner, J.P., and White, N. (eds), *Exhumation of the North Atlantic Margin: Timing, Mechanisms and Implications for Petroleum Exploration*. Geological Society, London, Special Publications, **196**, 355–370.

Warrington, G. (1994) *Palynology report on the Killary Glebe borehole near Coalisland County Tyrone*. Geological Survey of Northern Ireland Technical Report **GSNI/94-1**.

White, N. and Lovell, B. (1997) Measuring the pulse of a plume with the sedimentary record. *Nature*, **387**, 888–891.

Wilson, H.E. (1983) Deep drilling in Northern Ireland since 1947. *Irish Naturalists' Journal*, **21**, 160–163.

Wright, W.B. (1919) An analysis of the Palaeozoic floor of north-east Ireland, with predictions as to concealed coalfields. *Scientific Proceedings of the Royal Dublin Society*, **15**, 629–50.

Chapter 13

Future outlook

Introduction

Ireland produces only small volumes of natural gas and currently has no oil production. The country lies at the end of a very long supply chain. As mentioned in the Preface to the book, Ireland's position at the western end of the energy pipeline makes it particularly vulnerable to interruptions in oil or gas supply. In the 1970s, during the two major oil supply crises, the country witnessed the impact of oil shortages resulting in long queues for petrol, near panic as people drove from one petrol station to the next following rumours of new deliveries of petrol, and plans for the introduction of petrol rationing. In more recent times we have seen glimpses of the potential impact of gas shortages resulting from pipeline and payment disputes in Eastern Europe. Added to this is the inevitability of the rise in energy prices as 'easy oil' becomes a thing of the past. Worldwide it is becoming increasingly difficult and more expensive to even partially replace the dwindling oil and gas reserves. Compounding the problem is that as the world's population is predicted to grow from its present level by more than 40% to more than 9 billion people by 2050, the world's energy demand is also expected to almost double from its present level by 2050. It is difficult to see how Ireland is likely to be immune from the impact of such growth and demands. A further linked global challenge that is impacting on Ireland is the demand to lower greenhouse emissions (largely from fossil fuels) in order to attempt to halt or slow down major global temperature rises.

Ireland, like many other EU countries, has seen a continuous rise in its annual gas demand resulting from a switch away from oil and other fossil fuels with higher greenhouse gas emission levels (Figure 13.1). This rise, coupled with a drop in indigenous gas supplies as the Kinsale Head, Ballycotton and Seven Heads gas fields deplete, has resulted in an increase in imported gas (Figure 13.2). Ireland has a very high energy import dependency (90%), much larger than the EU average (c.52%), and this has risen more sharply than the EU average over the past 15 years (Sustainable Energy Ireland, 2009). The trends are shown in Figure 13.3. In recent years Ireland has taken a leading role in seeking to replace fossil fuels as an energy supply with renewable sources of energy, and has set an ambitious target of providing 40% of electricity demand from renewable sources by 2020. Most of this is from wind turbines, and already Ireland is one of the leading countries in terms of per capita production of wind energy. Even if this target is met, the majority of electricity, and other energy sources, will continue to rely on fossil fuels, and particularly from oil and gas. The search for indigenous sources of oil and gas will therefore be increasingly important in the medium future as the country strives for increased energy security and independence. To that end, the Irish Government's Energy White Paper of 2007 (Government White Paper, 2007) has a key strategic goal of creating a 'stable attractive environment for hydrocarbon exploration and production'. It also states that 'It is a key Government policy objective to encourage investment in oil and gas exploration off the Irish coast and to optimise the value of any oil and gas finds for Ireland'. While this will obviously centre on stimulating exploration for 'traditional' oil and gas, it is likely to encompass exploration for all sources of hydrocarbons, including the unconventional types including heavy oil and oil shales, shale gas and coalbed methane, and methane hydrates.

Exploration in the future is also likely to include an increased focus on identifying and de-risking reservoirs and traps to be used initially for temporary summer storage of gas that is used during the winter period of higher demand, and then used in the longer term for carbon dioxide storage or sequestration. The challenge of storing excess energy produced from renewable sources such as wind turbines so that it can be utilised during peak demand is increasingly important in the transition to a more carbon-free economy. The identification

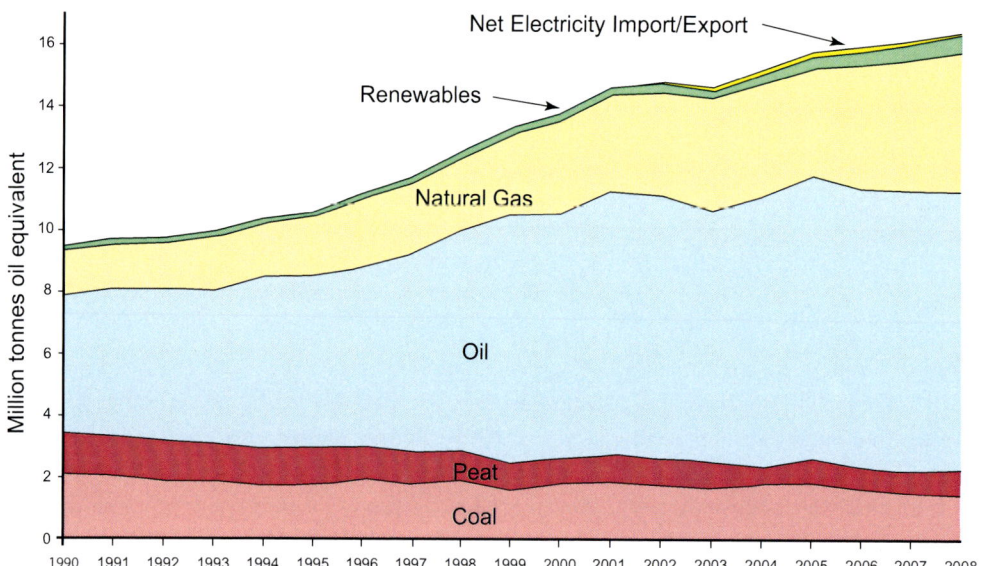

Figure 13.1. Ireland's primary total energy requirement. The majority of Ireland's energy needs are met from oil and natural gas. Gas has increased significantly, especially over the past 15 years. (Source: Sustainable Energy Ireland, 2009.)

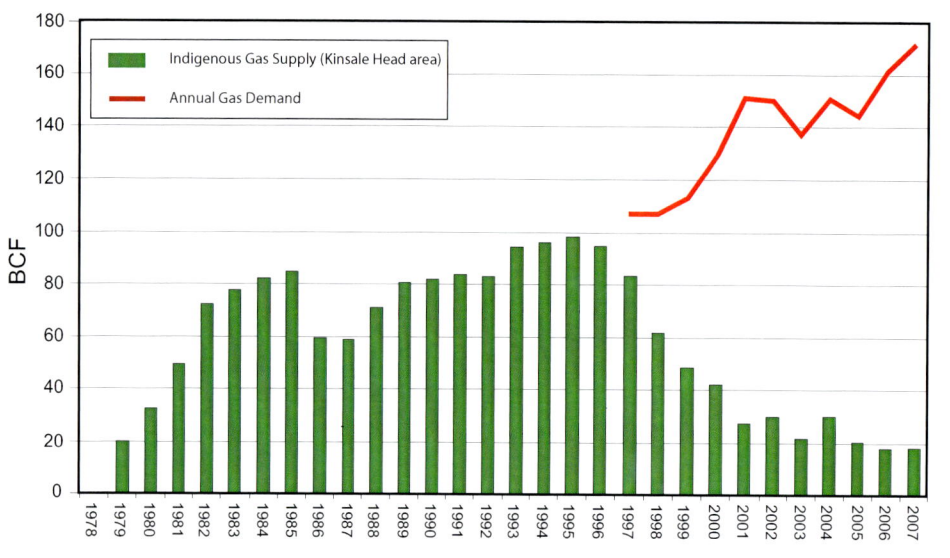

Figure 13.2. Ireland's annual gas demand and indigenous supply. While the gas demand has continued to increase, the indigenous supply has decreased sharply since the late 1990s, reflecting the rapid depletion in the Kinsale Head, Ballycotton and Seven Heads fields. (Source: Petroleum Affairs Division, DCENR website, 2010).

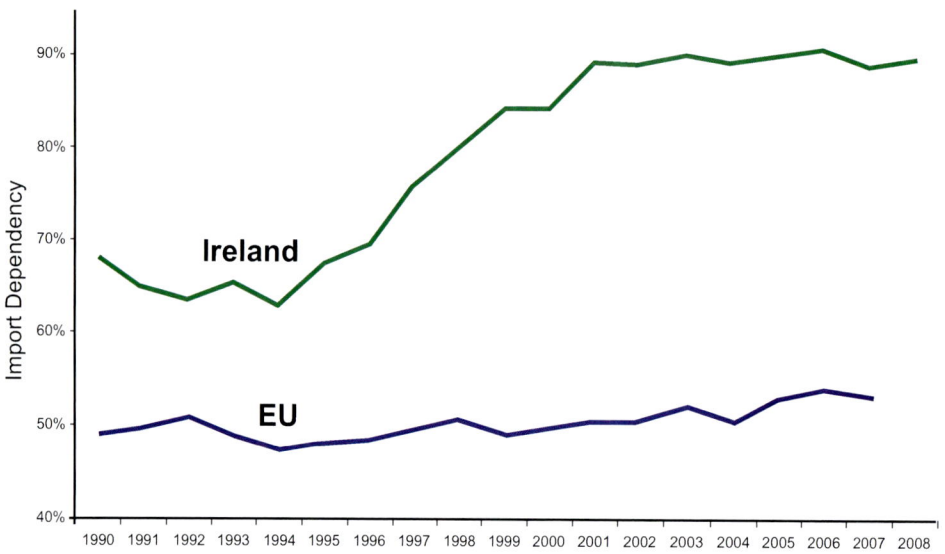

Figure 13.3. Energy import dependency trends. Ireland has become increasingly dependent (90%) while the EU average has shown a very slight gradual rise to approximately 52%. (Source: Sustainable Energy Ireland, 2009)

of subsurface energy storage reservoirs can be linked to aspects of petroleum exploration, particularly in onshore salt-prone regions. This aspect will also be touched upon in this wide-ranging chapter on the future outlook in the context of the petroleum geology of Ireland.

Present and future exploration situation

To date, 156 exploration and appraisal wells have been drilled in the Irish offshore, with mixed results. The total cost of these, in 2010 prices, exceeds €3000 million. Close to 375000 km of 2D, and in excess of 12000 km^2 of 3D data have been shot since 1965, at a cost running into many millions of Euro. Despite the relative lack of discoveries (although the number of wells drilled is small in comparison with producing basins elsewhere) there are encouraging prospects in most of the Irish offshore basins.

The distribution of wells drilled to date in the Irish offshore (Figure 3.5) is the cumulative product of the different waves of industry interest in the region since the 1960s. As discussed in Chapter 3, interest has focused at different times on such large-scale traps as anticlinal inversion structures similar to the Kinsale Head Gasfield (Celtic Sea), on tilted fault-block Brent Oilfield analogues (Porcupine Basin), and on the search for Morecambe Bay type gas fields (Irish Sea). These endeavours have met with only limited success, although the presence of active petroleum systems has been demonstrated in the North Celtic Sea, Porcupine, Slyne, and most recently the Rockall basins. The map of current licensed blocks shown in Figure 13.4 is a good indicator of the current focus of oil industry interest in the Irish offshore. It shows the focus on a NE–SW swath of licensed acreage through the eastern Rockall, the Slyne–Erris and Porcupine basins west of Ireland; a set of blocks centred on the main discoveries in the central part of the North Celtic Sea Basin and a small number of blocks in the eastern Celtic Sea and Irish Sea basins. A series of Frontier Licensing Rounds since 1993 has encouraged industry interest in the Atlantic margin basins. Additional impetus has been provided by improvements in deep-water technology in both the drilling and production sectors, and also by deep-water discoveries in other parts of the world, particularly in the South Atlantic. Negative factors operating against oil industry interest in the Irish Atlantic margin are the extreme physical environments encountered in the deep-water North Atlantic, together with the limited exploration successes to date. However, it is expected that a number of key commitment wells will be drilled

on deep-water licences in the near future – for example the Exxon-Mobil, ENI, Providence consortium holding the licence over the large deep-water Dunquin prospect in Quadrant 44 of the Porcupine Basin (*see* Chapter 11). Clearly, a major discovery would ignite exploration interest in the whole Atlantic margin.

Following the early discovery of the Kinsale Head Gasfield in 1971, subsequent exploration in the Celtic Sea basins has been largely disappointing (*see* Chapters 3 and 10). The nearby small Ballycotton and Seven Heads gas accumulations have been tied back to the Kinsale Head production facilities. Early enthusiasm following the Helvick oil discovery in the Jurassic (1983) was later tempered by the realisation that the trap was relatively small and complex, as also were similar structures seen on the improving seismic data. The Cenozoic inversion structures in the Celtic Sea region are typically shallow, have suffered rupture and/or biodegradation of contained hydrocarbons, whilst the region as a whole is structurally complex. For these reasons, together with the more detailed factors discussed in Chapter 10, new exploration drilling in the Celtic Sea basins has waned. Interest is largely limited at the present time (Figure 13.4) to re-examination by smaller independent oil companies of discoveries made in previous decades, but deemed non-commercial at the time. The most notable of these are the Ardmore (Block 49/14), Old Head of Kinsale (Block 49/23), Dunmore (Block 50/6) and Hook Head (Block 50/11) discoveries, outlined in Chapter 10. New completion technology, much improved seismic definition (Figure 10.12) and higher product prices make possible a re-appraisal of some of these subtle traps, both stratigraphic and low-amplitude anticlinal structures with Lower Cretaceous reservoirs. Some of the recent results lead to hopes for commercial development of some of these prospects.

The Porcupine Basin remains an area of exploration promise in a challenging physical environment, with up to 9 km of Cretaceous and Cenozoic strata preserved (Naylor *et al.*, 2002). The centre and southern parts lie in deep water, making exploration difficult. A major uncertainty within the basin is in the extent and nature of Jurassic reservoirs in the deep-water parts of the basin. In addition, the general southward dip of the basin means that a northward migration of hydrocarbons is likely to have taken place. The sandy nature of the sediments in the northern parts of the basin at Jurassic to Cenozoic levels increases the risk of lack of adequate cap rocks in the updip parts of the basin. A variety of untried plays

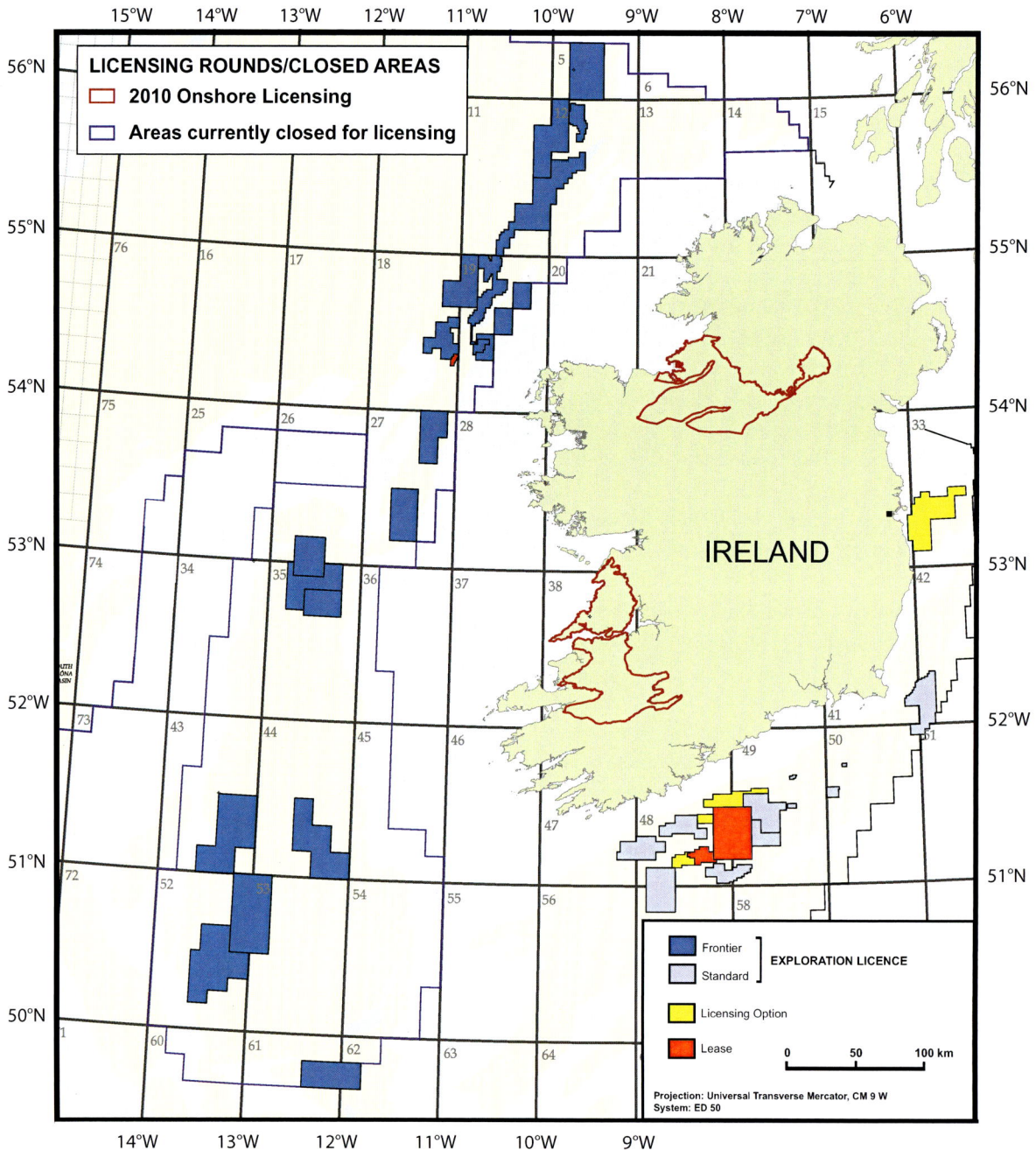

Figure 13.4. Republic of Ireland Petroleum exploration and Development Concession Map (March 2010). Published by permission of the Petroleum Affairs Division, Dublin.

and prospects remain to be drilled. The Connemara and Spanish Point accumulations in the Porcupine Basin are currently being re-evaluated, and the basin is likely to contain significant potential for stratigraphic and subtle traps, especially at Cretaceous and Cenozoic levels. The structural complexity of the Slyne, Erris and Donegal basins is probably the major risk in these narrow basins.

However, the discovery of the Corrib Gasfield in the Slyne Basin, which is being brought to production, gives encouragement to exploration in the basins. A number of structures, broadly similar to the Corrib structure, have been identified and still remain to be drilled. The promise of the Rockall–Hatton regions becomes increasingly obvious as more data are collected and published.

Rockall and Hatton basins have been shown to be fault-bounded basins with up to 5 km of sediments. Little exploration drilling has taken place in the region. The major exploration risks in these basins, which are a considerable distance from shore in harsh, deep Atlantic waters, are the poorly known nature of source and reservoir rocks. However, the Rockall Basin in particular appears to have the ingredients for a working petroleum system, as demonstrated by the recent Dooish gas condensate discovery on the eastern basin margin. A recent licensing initiative has involved the opening of all non-licensed acreage in the Atlantic margin (more than 1000 blocks and part-blocks) for a new frontier licensing round, with a closing date of 31st May 2011. A feature of this, the ninth Frontier Licensing Round, is that companies and consortia may be awarded a two-year Licensing Option on blocks in order to carry out early stage work prior to a longer, 15-year Frontier Exploration Licence with an agreed appropriate work programme.

In the Irish Sea basins the Permo-Triassic section has been the target for wells in the central and southern sections of the Irish Sea. Interest could spread northwards into the largely undrilled offshore area between Northern Ireland and Scotland, where speculative seismic surveys in the past have demonstrated the existence of basins with faulted structures. There are some remaining (mostly gas) prospects in the Central Irish Sea Basin and in the Kish Bank Basin, but in view of the lack of success to date in these basins they are regarded as relatively high risk and, in contrast to the other sets of basins, there is relatively little recent interest in the acreage. The search for oil or gas fields within the Permo-Triassic section of the under-explored onshore Larne, Lough Neagh and Rathlin basins, sourced from the underlying Carboniferous, still has potential. The most promising areas are probably those with thick halite development. Throughout the Irish Sea region the potentially negative geological factors appear to be generally early peak periods of hydrocarbon generation, the erratic distribution of Carboniferous source sequences, and demonstrable late fault movement and trap modification. Interestingly, a cumulative thickness of 11 m of coal in Pennsylvanian strata were proved on the southern margin of the basin by the Amoco 33/22-1 well (Jenner, 1981). While these offer gas source potential for the region, they have also attracted attention for gas generation of another type. An underground coal gasification project has been proposed for the Kish Bank Basin under a licence held by VP Power.

A recent initiative has been taken to encourage another look at the onshore Carboniferous basins. With the increased international interest in natural gas resources, the Minister for Communications, Energy and Natural Resources sought applications for Onshore Licensing Options over the Northwest Carboniferous Basin and the Clare (Shannon) Basin. The application deadline for licence applications was set for 11th June, 2010. An announcement on the results of the licensing competition is awaited at the time of going to press.

Unconventional oil and gas

Heavy oil: While the main focus of exploration in Ireland and its offshore is likely in the immediate future to concentrate on conventional oil and gas, it is worth examining briefly some possibilities for other forms of oil and gas. These tend to be difficult and expensive to produce, although with rising oil and gas prices there is a greater incentive to examine and review such possibilities. Heavy oil accumulations occur in a number of areas in the UK offshore, most notably the giant Clair Oilfield in the West Shetland Basin (Coney *et al.*, 1993). However, most of the oil shows and flows in the Irish offshore have been light, good quality oil. Nonetheless, some waxy and heavy oil has been encountered in the Celtic Sea, in a number of instances lying below a gas accumulation. Most notably, heavy oil (16° API gravity) has been reported beneath the gas cap in the Ardmore accumulation. As mentioned in Chapter 10, this has been the subject of some recent appraisal and re-examination by Providence Resources, who have suggested that the accumulation could be significant, with an in-place resource potential of up to *c.*230 million barrels of heavy oil. Bula Oil also encountered an accumulation of heavy and biodegraded oil in their 49/19-2 well drilled in 1992. Like most of the other indications of heavy oil this occurred in the Lower Cretaceous succession. The biodegradation of the oil in the basin may be the result of freshwater flushing following Early Cenozoic uplift in the basin. Heavy oil is typically recovered by injecting gas or hot water into the reservoir to lower the viscosity and allow it to flow. While such oil is obviously significantly more expensive to produce than conventional oil, the association in this region of heavy oil with gas, a potential injection material, in the region is a favourable factor in considering such production in the future.

Shale gas: This is an increasingly important source of gas in recent years in many parts of the world, especially in the USA. It is gas that is trapped in natural fractures or pore spaces or is adsorbed onto the organic matter in mudrock-rich, source rock successions. Until the recent advent of horizontal drilling and advanced hydrofracturing techniques, such gas remained largely untapped, with the main focus being on seeking sandstone and limestone reservoirs charged from typically thick source rock mudstones. It is now possible to extract significant quantities of gas from the generally impermeable mudstones by fracturing the rock along horizontal wells, thereby increasing the porosity and permeability and releasing the gas along the resulting connecting fractures as the formation pressure declines. In a number of the Irish offshore basins, notably the North Celtic Sea, Fastnet, Slyne and Erris basins, thick Lower Jurassic mudstones are oil prone (*see* Chapters 10 and 11), while thick Middle and Upper Jurassic oil-prone mudstone units occur in some of these basins and also in the Porcupine Basin. While the large Atlantic margin basins lie in deep water and such exploration is likely to be too costly at present, rising oil prices and increased supply insecurity may result in an increasing level of exploration interest in such shale gas prospects. Onshore in Ireland the source potential in the Upper Palaeozoic shales in the southern half of the country is likely to be exhausted following Variscan orogenic heating. However, as mentioned in Chapters 3 and 4, some gas has been encountered in tight sandstone reservoirs in the Northwest Carboniferous Basin, although successive attempts to prove commercial quantities of gas have been unsuccessful to date. A future phase of exploration for either tight gas in low porosity/permeability sandstones, or in shales, using horizontal wells and hydrofracturing techniques, may be considered in the future.

Coalbed methane: This is another increasingly important non-conventional source of gas. In such plays the methane is adsorbed on the coal and can be released by a drop in pressure. This is achieved by pumping water from the coal-bearing succession – often from unmined or unmineable seams. The resulting pressure drop in the vicinity of the borehole causes methane to bubble into the well bore. As such settings and coal-bearing successions are generally rare in Ireland, with a few exceptions such as in the north of Ireland and a very few areas further south in the country, this is unlikely to be a major new source of gas production onshore in Ireland. However, the Upper Carboniferous coals in the Kish Bank Basin may have a more extensive occurrence throughout the Irish Sea. Perhaps in conjunction with *in situ* coal gasification, there may be some potential for investigating the application of coalbed methane production technology in this region.

Oil shales: These were a major source of oil production in the Midland Valley (West Lothian) region of Scotland from the middle of the nineteenth century until production finally ceased in 1962. At its peak nearly two million tons of shale were being extracted annually, employing 4000 men. The oil was produced by the distillation of very organic-rich Upper Carboniferous shales. Oil shales were also produced in a number of other areas onshore in England from younger rocks, including some units in the Upper Jurassic Kimmeridge Clay Formation at Kimmeridge in Dorset. However, the nature and cost of oil shale production is such that it can only be realistically considered for onshore areas. A very few minor localities have been reported historically in Ireland, with Parnell (1991) reporting that some samples from the Ballyvoy Coals in the Cross borehole are recognized as oil shales giving very high hydrogen indices. These are from a level equivalent to the Upper Oil Shale Group in the Lothian region of Scotland (Whitaker and Butterworth, 1978). Parnell (1991) also mentioned that old borehole records indicate oil shales in the Dinantian (Tournaisian–Viséan) and Namurian (Serpukhovian–early Bashkirian) of North Antrim (*see* Wilson and Robbie, 1966). However, in the Republic of Ireland there are no records of any significant occurrences of oil shales, and it is highly unlikely that there is any realistic prospect of oil shale production here.

Gas hydrates: Methane hydrates (clathrates) are solid compounds in which a large amount of methane is trapped within the crystal structure of water ice. They represent an important unconventional potential source of fossil fuels in certain geological settings. They are restricted to depths of less than 2 km and have been found in polar (permafrost) settings (e.g. Alaska and Siberia) where surface temperatures are less than 0°C and in oceanic settings where water depths are greater than 300 m and where the bottom water is approximately 2°C. They have been recorded along the Atlantic and Pacific margins, including DSDP and ODP drill sites, and have been interpreted and suggested to be present in many regions. The most widely used indicator of gas hydrates is the presence of a bottom simulating reflector (BSR) on seismic reflection data (e.g. Haq, 1998; Kvenvolden 1998). BSRs, generated by the presence of free gas at the base hydrate interval, parallel the seabed but cross-cut stratigraphic reflectors, and are found to be broadly coincident with the base of the gas hydrate interval. However, the absence of a BSR on seismic data does not prove that hydrates are absent, merely that free gas is not present. The presence of hydrates in boreholes may also be indicated by sonic and resistivity wireline log responses (Kvenvolden, 1998). While global estimates of hydrate resources vary considerably, the energy content of methane occurring in hydrates probably exceeds the combined energy content of all other known fossil fuels. However, methane production from hydrates is still largely unproven at a large scale, with a number of small exploratory projects currently under way, notably in Alaska. While methane hydrates have not been observed in the Irish offshore, a body of evidence points to their occurrence. Miles (1995) calculated the extent and thickness of the methane hydrate stability zone along the continental margins of Europe, indicating the stability zone to be present below water depths of about 500 m and with a thickness of up to 300 m. This includes much of the Irish Atlantic margin basins. The long-term sedimentation rates in these regions were probably sufficiently high to support the generation of biogenic methane to form hydrates. The possible presence of free gas in the shallow subsurface of at least part of the Irish Atlantic margin is indicated by seabed pockmarks in the northern Porcupine Seabight (e.g. Games, 2001). The cause of a number of slope failure features in the Porcupine Basin have been interpreted by Henriet et al. (2001) as possible hydrate buildup in glacial times. The presence of hydrates has also been postulated on the basis of geochemical analysis of pore waters at ODP site 982 on the western margin of the Rockall Bank, 50 km north of the UK/Irish boundary (Jansen et al., 1995). Much further work is needed to confirm and assess the likely extent of methane hydrates in the Irish Atlantic margin basins. If located, the biggest challenge is likely to be to develop a safe and efficient method to capture the resource. At present, a solution still seems quite some time away.

Gas storage

The subsurface geological storage of gas – either the short-term storage of natural gas or the long-term storage of greenhouse gases, mainly carbon dioxide – will play an increasing role in the global energy sector during the coming decade. The longer term challenges of identifying possible suitable safe sites for CO_2 storage are dealt with separately below. In this section we address the issue of identifying sites where gas can be stored for a short period before being used to generate electricity or to put into the national gas grid. Such sites could be either dissolution caverns in salt for small, typically onshore or shallow off-shore sites, or larger structures with reservoirs, caps and

trapping structures for offshore sites. Because of the costs involved in transporting, injecting and then exporting gas back to either an electricity generating station or to the national gas grid, it is unlikely that the large Atlantic margin basins will be considered for such prospects. Therefore this discussion is confined to consideration of only the onshore and shallow-water offshore structures.

With a few exceptions, the onshore geology on the island of Ireland is not promising as regards potential subsurface geological storage sites. As detailed in Chapter 4, the Lower Palaeozoic and older rocks, together with the Upper Palaeozoic rocks in the southwest of the country that lie south of the Variscan deformation front, are devoid of significant porosity or permeability and can be excluded from consideration. Studies of regional geology, and the evidence from exploration wells, suggest that the widespread limestone terrain of Central Ireland has low potential for significant subsurface zones of porosity, permeability and seal. Dolomites and fracture zones, which locally – as in the Meelin-1 well in County Cork – may have good porosity, are likely to have variable permeability or poor lateral continuity. Namurian (Pennsylvanian) sandstones with enhanced porosity due to secondary leaching are likely to provide the best reservoir potential, although there are virtually no data regarding subsurface closure or seal. In a very general sense, the reservoir properties of Carboniferous sandstones improve northwards. Gas was discovered in the Northwest Carboniferous Basin (Lough Allen Basin: Chapters 3 and 4) within Mississippian sandstones. Several attempts have been made to establish the commercial viability of the gas, including the use of improved drilling and hydrofracturing techniques, without significant success. The poor porosity and limited permeability of the sandstone horizons would also prove a barrier to any long-term storage scheme in these reservoirs.

Kingscourt, in County Cavan, is the only locality in the Republic where relatively thick (approximately 600 m) Permo-Triassic rocks are preserved onshore (Figure 4.13). The Lower Triassic Kingscourt Sandstone Formation is underlain by the Upper Permian Kingscourt Gypsum Formation. Beneath this sequence lowest Westphalian (<100 m) and Namurian (350–600 m) sandstone–shale sequences are preserved, and the inter-bedded sandstones form potential reservoirs. An almost complete Serpukhovian and Bashkirian (Pennsylvanian) sandstone–shale sequence is preserved within the Kingscourt graben (Jackson, 1965; Sevastopulo, 2009). The Carboniferous and Permo-Triassic rocks of the

Kingscourt area constitute a classic half-graben, with a major bounding fault (the Kingscourt Fault) on the west side (McConnell *et al.*, 2001). There is a possibility of fault-constrained areas of closure in the Upper Carboniferous, with seal provided by the overlying Kingscourt Gypsum Formation. However, the storage potential of the Kingscourt area can only be regarded as theoretical and high-risk.

The only substantial area of Mesozoic on the island is the composite basinal area of the Ulster Basin in Northern Ireland. The Larne and Lough Neagh basins, north of Belfast, contain more than 3000 m of Permo-Triassic rocks. Reservoir quality sandstones have been encountered in onshore wells (Chapters 6 and 12). The seal for Triassic Sherwood sandstone reservoirs is provided by the Mercia Mudstone Group, which in the basin centre contains four thick halite horizons. Major halite horizons are also developed in the Permian. The Belfast–Larne coastal area represents the most promising opportunity onshore on the island of Ireland to develop natural gas storage in caverns within halite horizons or for carbon dioxide sequestration in saline aquifers sealed by mudstones and halite. The salt is at attainable and useful depths along much of the coastal zone. More research is needed to determine the thickness of individual salt beds and to constrain the lateral extent and variability within the succession. From an engineering and solution mining viewpoint, the extent to which shale interbeds within the salt are acceptable will need to be determined. Nevertheless, the geology of this area offers many promising features that are important for a successful and commercial gas storage project. Significantly, application has recently (March 2010) been made by Islandmagee Storage Limited, owned jointly by Infrastrata plc and Mutual Energy Limited (which also operates the gas interconnector between Scotland and Northern Ireland), for a 500 million cubic metre (18 bcf) natural gas storage facility beneath Larne Lough (Figures 6.3 and 12.1A). This is equivalent to more than 60 days of peak gas demand in Northern Ireland. The plan required seven caverns (160 m high x 80 m) to be created by solution mining within Permian halite, which the company believes to be about 200 m thick at this location. Project cost is estimated at £250 million.

As detailed in Chapter 12, good reservoir sandstones exist in the Larne area in the Ormskirk Sandstone Formation (Triassic) and in the Enler Group (Permian), although the former probably has the greater potential. Both levels have overlying mudstone and halite seal. The

sandstones extend across the Larne and Lough Neagh basins, although halite has been intersected in drilling only in the coastal Larne area. As discussed previously, if halite is present in the Lough Neagh basin it is likely to be restricted to the depocentre beneath the Lough. To the north, the Rathlin Basin is less well known, due in part to poor seismic data quality, and no oil exploration wells have been drilled. The sandstones of the Permian Enler Group and the Sherwood Sandstone Group provide reservoir potential, although it is doubtful whether a seal exists for Permian reservoirs. In the Triassic, the thick Mercia Mudstone section should provide an adequate seal, despite a gradational transition with the Sherwood Sandstone Group. Overall, however, the lack of detailed structural data for the Rathlin Basin and associated basins must give them a higher risk rating in terms of storage projects.

Several wells in the Larne and Lough Neagh basins appear to have drilled valid structures, but were dry (Chapter 12). As was the case when considering the exploration potential of the basin, the areas of halite development afford the least risk with respect to seal. Fortunately, this area is close to Belfast, where need for natural gas storage is likely to be centred. Multiple reactivation on faults, that continued into the Cenozoic, was recognised as a significant risk in exploration, and provides the same problem for storage. Outside the halite area the cost of establishing seal viability over any identified structure might prove to be prohibitive. Clearly, four-way closure not dependent on a fault seal would be preferable.

The stratigraphy of the onshore Larne Basin extends offshore into the North Channel and Portpatrick basins (Chapters 6 and 12). However, there is a lack of drill information for the offshore basins. Halokinetic structures and significant tilted fault-block and rollover anticlinal traps have been recognised in seismic studies (Maddox *et al.*, 1997). The risks related to doubts concerning fault seal and trap viability recognised onshore also apply to the offshore basins. Further south in the Irish Sea a number of exploration wells have targeted the Ormskirk Sandstone Formation without success. In the Peel and Solway basins Newman (1999) cited the lack of Carboniferous source rocks beneath the basin as the main reason for the lack of success. The traps are largely fault dependent and trap formation may have post-dated hydrocarbon generation. As with the basins in the northern part of the Irish Sea, the faults have suffered multiple phases of movement, including phases in the Cenozoic, with resultant risk related to present-day trap viability. The Mercia Mudstone Group seal on the main potential reservoir lacks thick annealing halites and, particularly in the Peel Basin, is relatively thin.

The Kish Bank Basin has been the target for four exploration wells (Figure 6.6). The Fina 33/17-1 well demonstrated good porous sandstones (Collyhurst Sandstone) in the Permian, but the overlying Upper Permian Cumbrian Coast Group marls are only 97 m thick, with thin sandstone and siltstone interbeds, and constitute a risk in terms of seal. However, the Triassic Ormskirk Sandstone Formation displays good porosity and the overlying Mercia Mudstone Group, which contains thick halite beds, is considered to afford adequate seal. Dissolution of the thick salt deposits also offers the potential for the creation of a gas storage reservoir similar to that proposed for the Ulster Basin, provided the structural and permeability integrity of the Mercia Mudstone Group mudstones can be confirmed. Despite the probable operation of a petroleum system in the basin (Dunford et al., 2001), the drilled structures have not contained significant hydrocarbons. There has been multiphase reactivation of faults, and particularly, substantial Cenozoic displacement on the Codling Fault and related structures. Whilst it is possible that the structures are now sealed and viable, considerable work would be required to dispel these concerns when considering a gas storage project. The Shell 33/21-1 well (Figure 12.8) drilled a late Cenozoic inversion structure that is less fault-dependent than other structures in the basin. Although it may have post-dated the period of hydrocarbon charge, it, or similar structure, could be viable for gas storage.

Further south, five wells have been drilled in the Irish sector of the Central Irish Sea Basin (Chapter 12), at least three of which tested valid traps (Floodpage et al., 2001). However, most of the potential trapping structures are fault-block closures or footwall rollover anticlines, and as elsewhere in the region there have been multiphase fault movements and inversion episodes. The target Sherwood Sandstone reservoir interval is thinner in the Central Irish Sea Basin than further north, and also is increasingly shale-prone southwards. Overall, however, the sequence is probably capable of serving as an adequate gas reservoir. The Mercia Mudstone Group comprises thick evaporitic mudrocks and interbedded halites and could be anticipated to provide adequate top-seal. However, Maddox et al. (1995) questioned whether sandstones within the basal part of the Mercia Mudstone

Group may form a potential 'thief zone' between the reservoir and the lowest effective halite seal. On balance, the Central Irish Sea Basin has probably a higher risk rating for long-term gas storage than the basins further north in the Irish Sea.

There are a number of hydrocarbon-bearing structures in the Celtic Sea basins, with Jurassic and Cretaceous reservoirs (Chapter 10). The Kinsale and Ballycotton gas fields, producing from the Lower Cretaceous Greensand, are nearing the end of their productive lives. In 2001, the Southwest Kinsale field was converted into Ireland's first gas storage facility while in 2006 the Inch Terminal onshore was converted to enable natural gas from the Bord Gáis Éireann (BGE) system to be injected into the offshore reservoir. This is currently Ireland's only natural gas storage facility and provides approximately 4% of average annual demand. However, Kinsale Energy (part of the Petronas group that purchased Marathon Petroleum) recently announced plans to convert the Ballycotton Gasfield, located c.40 km south of Cork Harbour, to a gas storage facility. This would provide Ireland with the capacity to store approximately 18% of average annual demand, in line with the storage capacities of other EU countries.

With the recent increase in wind turbine clusters throughout Ireland, a challenge that is linked to subsurface gas storage reservoirs is the provision of short-term storage of excess energy. One of the means of achieving such storage is through compressed air energy storage (CAES). While the system has not been applied in Ireland, CAES plants in Alabama (operating since 1991), and Huntorf, Germany (operating since 1978) are located in leached salt caverns, and neither has experienced major geological problems. A number of subsurface settings in the Irish onshore and shallow offshore are potentially amenable to compressed air energy storage. These include Permo-Triassic salt in the Larne and Lough Neagh basins, and in the shallow waters of the Kish Bank Basin. In February 2010, Gaelectric Energy Storage received a mineral prospecting licence for salt which will further advance the research and development of a CAES and gas storage facility in Larne, County Antrim.

CO$_2$ storage/sequestration

Within the past few years carbon capture and storage/sequestration (CCS) has begun to be acknowledged as a potentially major component of carbon abatement strategies in most countries, including Ireland. It has the potential to be an attractive transition technology that will allow the use of fossil fuels while minimising additional CO_2 emissions to the atmosphere as outlined in the Intergovernmental Panel on Climate Change Special Report on carbon dioxide capture and storage (IPCC, 2005). While many sequestration options are potentially available (e.g. mineral, marine and subsurface sequestration) we only consider those subsurface methods that are linked to the sedimentary basins of Ireland (i.e. depleted gas fields and saline aquifers).

In assessing the storage potential of geological structures and basins, different levels of risk need to be applied. Using concepts of increasing certainty, a number of categories of storage capacity are envisaged (after Bachu *et al.*, 2007), as follows:

1. *Theoretical Storage Capacity.* This is the physical limit of the geological reservoir. It assumes that the system's entire capacity to store CO2 in pore space, or dissolved at maximum saturation in formation fluids, is accessible and utilised to its full capacity.

2. *Effective Storage Capacity.* This is the subset of the theoretical capacity obtained by considering that part of the theoretical storage capacity that can be physically accessed and which meets a range of geological and engineering criteria.

3. *Practical Storage Capacity.* This is that subset of the effective capacity obtained by considering technical, legal and regulatory, infrastructural and general economic barriers to CO_2 geological storage. It corresponds to the term 'reserves' used in the petroleum exploration industry.

A recent multidisciplinary study to assess and quantify the potential for geological storage of carbon dioxide on the island of Ireland and its offshore using the above risk categories has been completed and was recently published (Lewis *et al.*, 2008, Lewis 2009). Table 13.1 (after Lewis, 2009) provides a summary of that study. It assessed potential storage that may exist in deep sedimentary basins in depleted gas fields or saline aquifers, in which CO_2 could be stored below 750 m (CO_2 is in a supercritical state and consequently occupies significantly less space below *c.*700 m), and where suitable topseals are present. Well data and published information were used to identify the geological characteristic of potential reservoir/seal pairs and available storage where possible. The results are broadly in line with those to be anticipated from the geological outline given in the sections above. There is little quantified capacity seen in the onshore basins, with the exception of the Northwest Carboniferous Basin and the onshore components of

the basins in northeastern Ireland discussed above. While Ireland has the potential for substantial storage capacity, much of this is in the Theoretical and Effective categories. Some of the Effective capacity must be regarded as near the lower margin of that category. The total quantified CO_2 capacity was estimated at 93000 Mt but the practical capacity was estimated as 1505 Mt. In the Effective/Practical category are three gas fields in Republic of Ireland waters. The East Irish Sea Basin, which constitutes the largest element in this category, lies firmly within UK waters. Whatever the agreement for co-operation on or around the island of Ireland, access to storage in the East Irish Sea Basin would presumably require a separate agreement with the United Kingdom government.

A number of geological factors come into play when planning to use a depleted hydrocarbon reservoir for carbon dioxide storage (Bachu, 2007; Lewis, 2009). These may have the effect of further diminishing available storage space. Where water invasion has occurred from the underlying saline reservoir during depletion, or during water injection in oilfields, some of the original pore capacity may not be available for storage. In some instances gas depletion may disrupt trap integrity. In considering the Kinsale Gasfield as the most promising storage site, Lewis (2009) cited uncertainties with respect to seal efficacy, faulting, gas chimneys, CO_2 reservoir rock interaction and injectivity as requiring additional study and detailed modelling. Existing production wells on the field are also a source of potential leakage and may require recompletions. The cost of assessing the Kinsale field for short-term storage to a probability factor of 90% is judged to be €15 million, but €80 million could be required to bring the field to the same confidence level for long-term storage (Lewis, 2009). Undrilled structures in other basins, even where the geological conditions appear promising, are obviously higher risk, and in consequence will be more costly to assess fully. There are many non-geological factors that affect the economic viability of geological storage schemes, which are beyond the scope of this summary. These include distance to the main point source emission sites, cost and method of carbon capture and cost of infrastructural development. A detailed assessment of these elements can be found in Lewis *et al.* (2008, and Appendices).

It is natural that CO_2 sequestration studies in the region will focus on hydrocarbon-bearing structures, particularly depleted fields. These structures have demonstrated the ability to retain gas over geological time

Table 13.1. Quantified geological storage capacity for CO_2 based on all-island study (Lewis *et al.,* 2008). Table is from Lewis (2009). 'Portpatrick Basin/Larne' on this table includes the North Channel Basin as used in this book.

Basin	Structural Type	Capacity Classification	Storage Capacity (Mt)	Quantified Storage Capacity (Mt)
Kinsale	Gas Field	Effective/Practical	330	1505
South West Kinsale	Gas Field		5	
Spanish Point	Gas Field		120	
East Irish Sea	Oil & Gas Field		1050	
Portpatrick Basin	Sherwood Sandstone selected structures	Effective *(subset of theoretical capacity)*	37	667
Central Irish Sea	Sherwood Sandstone structures		630	
Lough Neagh Basin	Enler Group selected structures	Effective *(additional to theoretical capacity)*	1940	2840
Kish Bank Basin	Sherwood Sandstone structures		270	
East Irish Sea Basin	Ormskirk structures		630	
Celtic Sea	1 structure in the Cretaceous A sand	Theoretical	40	88770
Portpatrick Basin / Larne	Whole basin		2700	
Peel Basin	Sherwood Sandstone whole basin		68000	
NWICB Dowra Basin	Whole basin		730	
Central Irish Sea	Whole basin		17300	
Kish Bank Basin	Carbonifersous sandstone and coal	Theoretical / unquantified		
Rathlin Basin	Sherwood Sandstone, Permian and Carboniferous			
Celtic Sea	Cretaceous A sand			
Porcupine Basin				
Slyne/Erris basins				
Clare Basin				
Rockall Basin				
Gas prospects				
Other onshore basins				
TOTAL (PRACTICAL/EFFECTIVE/THEORETICAL)				93,115

periods – although there are other technical considerations involved (*see* above). While other structures exist in the Celtic Sea basins that were not explicitly considered in the Lewis *et al.* (2008) study, they must be considered to be higher risk than existing fields. In addition to the Lower Cretaceous, other potential reservoir levels also exist in the region – notably in the Triassic and Lower Jurassic – but this potential for sequestration is purely theoretical and unproven. It is difficult to see serious consideration being given to any target in the Atlantic margin region for gas storage or later for CO_2 sequestration in

the foreseeable future, due to harsh operating conditions, inadequate geological data and prohibitive costs. In the event that such areas were considered in the future, the most likely site for consideration would be the Corrib Gasfield, at the northern end of the Slyne Basin, where a massive, presumed Zechstein equivalent evaporite sequence lies beneath the Triassic Sherwood Sandstone. A sandstone-dominant Lower Triassic Sherwood Sandstone Group equivalent (400–420 m) is overlain by Mercia Mudstone Group saliferous and anhydritic mudstones, with a thick basal salt section (Dancer *et al.,* 2005). This, and similar structures, may have future potential for storage when depleted. However, as the first gas still awaits delivery to shore it is probably rather premature to discuss using the field as a site for gas storage or sequestration just yet.

Government and industry research programmes

Within the past 15 years or so there have been a number of research initiatives, funded by the EU, the Irish government and the oil industry, that have helped develop a better understanding of the nature and geological development of Irish sedimentary basins. These have also assisted, directly and indirectly, in assessing the hydrocarbon resource potential of Ireland's petroleum potential and in de-risking onshore and especially offshore petroleum exploration. Several of these are outlined below to indicate the potential benefit such industry–academic research collaboration can have. They can provide a template for the development of future projects that can add further impetus to petroleum-related geoscience research in Ireland.

As part of the Third Frontier Licensing Round of 1997, the Irish Petroleum Affairs Division, Dublin (PAD) established the Petroleum Infrastructure Programme (PIP), with funding support from the companies who were awarded licences. The aim is to promote hydrocarbon exploration and production activities, and this is done by funding and encouraging research into the Irish offshore areas, and providing a forum for co-operation between researchers, especially in Irish universities, and explorationists. A large number of projects have been funded that have provided important new geological and geophysical information from the frontier Atlantic margin basins. Some of these are worth mentioning to illustrate the range of projects and their importance in improving our understanding of the basin development and petroleum potential of the region. These have included the drilling of a set of shallow boreholes (each up to 178 m deep) in 1999 to constrain the Mesozoic and Cenozoic stratigraphic development of the eastern margin of the Rockall Basin west of Ireland. These demonstrated the presence of Jurassic, Cretaceous and Cenozoic strata and provided constraints on the depositional and structural evolution of the region (Haughton *et al.,* 2005). Another project funded the acquisition and interpretation of a set of wide-angle seismic profiles (RAPIDS 3 and 4) across the Rockall Basin, and one across the centre of the Porcupine Basin. These confirmed the presence of continental crust beneath

the basins. This was shown to be very thin (2–6 km) in places (Morewood *et al.,* 2005, O'Reilly *et al.,* 2006) and was explained in terms of differential crustal stretching. This is now known to be a common feature in hyperextended passive margin sedimentary basins in many parts of the world, with significant implications for the thermal and structural development of the basins. The wide-angle data from the Rockall Basin have also indicated the presence of a thick (up to 7 km) sedimentary succession (Morewood *et al.,* 2005) – much greater than could be imaged with confidence from normal-incidence industry seismic profiles, which are negatively impacted by the effects of the extensive Early Cenozoic lavas and sills in the basin. A high-resolution TOBI sidescan sonar survey along the eastern margin of the Rockall Basin in 1998 revealed important details of the slope morphology, including canyon systems, slope failure features and recent deep-water coral mounds, that indicated the complex Neogene to recent sedimentological evolution of the region (Shannon *et al.,* 2001; O'Reilly *et al.,* 2007). A smaller but nonetheless significant pair of PIP-funded projects involved the mapping of the extent of the Irish Atlantic margin basins and the definition of a formal nomenclature for the main structures and geological features in the region (Naylor *et al.,* 1999, 2002). This has been adopted by the Irish licensing authorities and by exploration companies working in the region, and is in widespread use in the recent scientific literature. These names for the basins and features are those used throughout this book.

In 1999 the Geological Survey of Ireland launched its ambitious Irish National Seabed Survey (INSS). The objective of the programme was to map the nature and bathymetry of the seabed through the entire Irish offshore region. The deep-water region was mapped first (1999–2006), while the shallower waters are being mapped under a successor programme (INFOMAR: Integrated Mapping for the Sustainable Development of Ireland's Marine Resources) that has been under way since 2007. To date, the INSS project remains among the largest marine mapping programmes undertaken anywhere in the world. Multibeam swath bathymetric data have yielded high-resolution images of seabed features including canyons, slope failure, contourite drift structures and deep-water coral mounds. These have been used to constrain the Neogene and Holocene evolution of the basins (*see* Chapter 9). As part of the INSS programme the Geological Survey of Ireland, in conjunction with the Petroleum Infrastructure Programme, also funded the acquisition of two long wide-angle seismic profiles in the Hatton Basin. These have helped define the nature of the continental–oceanic boundary, provide a sound understanding of the crustal and sedimentary development of the Hatton region and have indicated the presence of up to 5 km of sediments overlying thinned (15 km) continental crust (Chabert *et al.*, 2008).

Onshore in Northern Ireland, the Tellus project (acquired 2004–2006) is an excellent example of how the acquisition of high-resolution geophysical and geochemical data have helped map lithologies and structures, as well as contributing to the search for onshore oil and gas. The programme was funded by the Department of Enterprise, Trade and Investment. It involved both ground (soil and water geochemistry) and airborne (magnetometer, electromagnetic and gamma-ray spectrometer) geophysical surveys. A brief description and an image of some of the magnetic data are presented in Murphy *et al.* (2009)

A number of EU-funded multinational research projects have also provided an improved understanding of the regional setting and the Neogene to Holocene development of the deep-water basins. The first of these was a project facilitated through the Irish Marine Institute which, in 1996, acquired the first regional sidescan sonar coverage of the Rockall Trough. This provided images of large-scale erosional and depositional features, especially along the basin margins (Unnithan *et al.*, 2001), and laid the foundations for later work, including that of the INSS, outlined above. Another was the STRATAGEM (Stratigraphic Development of the Glaciated European Margin), which ran from 2000 to 2003. This was a study by seven university and research institutions from six European countries. Using an extensive seismic and borehole database, much of it supplied by the oil industry, it provided an integrated, unified stratigraphic framework documenting the detailed Neogene stratigraphy of the northeast Atlantic margin from Lofoten to Porcupine (Evans *et al.*, 2005). Many of the papers published from this research programme have been cited throughout the book, indicating the regional and detailed stratigraphic and tectonic understanding of the Irish offshore basins.

Academic–industry research programmes of the types discussed above are likely to continue to play an important role in helping to provide a better understanding of the Irish sedimentary basins. During the past 15 years or so they have helped to develop and refine models of sedimentary basin development, especially in hyper-extended passive margin settings. They have also provided important new geological and geophysical data that have helped in the task of de-risking exploration. The focus of such collaboration in the future is likely to be in areas of innovative basin modelling, plate reconstruction, and in understanding fluid flow in sedimentologically and structurally complex reservoirs, as well as in working towards maximising petroleum recovery and identifying potential sites for gas storage and long-term sequestration. However, ultimately the costly business of petroleum exploration and development is likely to remain with the large and small exploration companies. The basins of Ireland, both onshore and offshore, have been slow to yield their petroleum secrets. During the past 50 years of exploration much energy and money have been expended, much has been learned about the distribution and geological detail and basin evolution, but much more is still not understood. Exploration in the next 50 years will undoubtedly reveal more geological information and provide a better understanding of Ireland and its offshore. Whether significant discoveries are made remains a matter of conjecture and hope. We remain optimistic about the future.

References

Bachu, S., Bonijoly, D., Bradshaw, J., Burruss, R., Holloway, S., Christensen, N.P., and Mathiassen, O.M. (2008). CO_2 storage capacity estimation: Methodology and gaps. *International Journal of Greenhouse Gas Control*, **1**, 430-443.

Chabert, A., Ravaut, C., O'Reilly, B.M., Readman, P.W., and Shannon, P.M. (2008) Seismic imaging of the crustal and sedimentary structure of the Hatton Basin on the Irish Atlantic Margin. 33rd *International Geological Congress, Oslo. 6–14 August 2008*. STT02729P.

Coney, D., Fyfe, T.B., Retail, P., and Smith, P.J. (1993) Clair appraisal: the benefits of a co-operative approach. *In*: Parker, J.R. (ed.), *Petroleum Geology of Northwest Europe: Proceedings of the 4th Conference*. Geological Society, London, 1409–1420.

Dancer, P.N., Kenyon-Roberts, S.M., Downey, J.W., Baillie, J.M., Meadows, N.S., and Maguire, K. (2005) The Corrib gas field, offshore west of Ireland. *In*: Doré, A.G. and Vining, B.A. (eds), *Petroleum Geology: North- West Europe and Global Perspectives. Proceedings of the 6th Petroleum Geology Conference*. Geological Society, London, 1035–1046.

Dunford, G.M., Dancer, P.N., and Long, K.D. (2001) Hydrocarbon potential of the Kish Bank Basin: integration within a regional model for the Greater Irish Sea Basin. *In*: Shannon, P.M., Haughton, P.D.W., and Corcoran, D.V. (eds), *The Petroleum Exploration of Ireland's Offshore Basins*. Geological Society, London, Special Publications, **188**, 135–154.

Evans, D., Stoker, M.S., Shannon, P.M., and STRATAGEM Partners (2005) The STRATAGEM project: Stratigraphic development of the glaciated European margin. *Marine and*

Petroleum Geology, **22**, 969–976.

Floodpage, J., Newman, P., and White, J. (2001) Hydrocarbon prospectivity in the Irish Sea area: insights from recent exploration in the Central Irish Sea, Solway and Peel basins. *In*: Shannon, P.M., Haughton, P.D.W., and Corcoran, D.V. (eds), *The Petroleum Exploration of Ireland's Offshore Basins.* Geological Society, London, Special Publications, **188**, 107–134.

Games, K.P. (2001) Evidence of shallow gas above the Connemara oil accumulation, Block 26/28, Porcupine Basin. *In*: Shannon, P.M., Haughton, P.D.W., and Corcoran, D.V. (eds), *The Petroleum Exploration of Ireland's Offshore Basins.* Geological Society, London, Special Publications, **188**, 361–373.

Government White Paper (2007) *Delivering a sustainable energy future for Ireland. The Energy Policy Framework 2007–2020.* Department of Communications, Marine and Natural Resources. 68pp.

Haq, B.U. (1998) Gas hydrates: greenhouse nightmare? Energy panacea or pipe dream. *GSA Today*, **8**, 2–6.

Haughton, P., Praeg, D., Shannon, P.M., Harrington, G., Higgs, K., Amy, L., Tyrrell, S., and Morrissey, T. (2005) First results from shallow stratigraphic boreholes on the eastern flank of the Rockall Basin, offshore western Ireland. *In*: Doré, A.G. and Vining, B.A. (eds), *Petroleum Geology: North-West Europe and Global Perspectives – Proceedings of the 6th Petroleum Geology Conference*, Geological Society, London, 1077–1094.

Henriet, J.P., de Mol, B., Vanneste, M., Huvenne, V., van Rooij, D. and The 'Porcupine-Belgica' 97, 98 and 99 Shipboard Parties (2001). *In*: Shannon, P.M., Haughton, P.D.W., and Corcoran, D.V. (eds), *The Petroleum Exploration of Ireland's Offshore Basins.* Geological Society, London, Special Publications, **188**, 375–383.

IPCC (2005). *IPCC Special Report on Carbon Dioxide Capture and Storage.* Metz, B., Davidson, O., de Coninck, Loos, M. and Meyer, L. (eds), Cambridge University Press, 442pp.

Jackson, J.S. (1965) The Upper Carboniferous (Namurian and Westphalian) of Kingscourt, Ireland. *Scientific Proceedings of the Royal Dublin Society*, **A2**, 131–152

Jansen, E., Raymo, M., and Blum, P. (1995) *Ocean Drilling Program, Leg 162 Preliminary Report, North Atlantic Gateways II.* Ocean Drilling Program. Texas A&M University. Preliminary Report **62**, 76pp.

Jenner, J. K. (1981) The structure and stratigraphy of the Kish Bank Basin. *In*: Illing, L.V. and Hobson, G.D. (eds), Petroleum Geology of the Continental Shelf of North-West Europe, Heyden and Son Ltd., London, 426–431.

Kvenvolden, K.A. (1998) A primer on the geological occurrence of gas hydrates. *In*: Henriet, J.-P. & Mienert, J. (eds), *Gas Hydrates: Relevance to World Margin Stability and Climate Change.* Geological Society, London, Special Publications, **137**, 9–30.

Lewis, D., Vernon, R., O'Neill, N., Bentham, M., Cleary. T., Kirk, K., Chadwick, A., Hilditch, D., Michael, K., Allison, G., Neal, P., and Ho, M. (2008) *Assessment of the Potential for Geological Storage of CO$_2$ in the Island of Ireland.* Sustainable Energy Ireland, Dublin, 137pp.

Lewis, D. (2009) Geological storage of CO$_2$ in the island of

Ireland: a review. *European Geologist*, **No. 27**. European Federation of Geologists, Belgium, 8–11.

Maddox, S.J., Blow, R., and Hardman, M. (1995) Hydrocarbon prospectivity of the Central Irish Sea Basin with reference to Block 42/12, offshore Ireland. *In*: Croker, P.F. and Shannon, P.M. (eds), *The Petroleum Geology of Ireland's Offshore Basins.* Geological Society, London, Special Publications, **93**, 59–77.

Maddox, S.J., Blow, R.A., and O'Brien, S.R. (1997) The geology and hydrocarbon prospectivity of the North Channel Basin. *In*: Meadows, N.S., Trueblood, S.P., Hardman, M., and Cowan, G. (eds), *Petroleum Geology of the Irish Sea and Adjacent Areas.* Geological Society, London, Special Publications, **124**, 95–111.

McConnell, B., Philcox, M.E., and Geraghty, M. (2001) *Geology of Meath: a geological description to accompany the bedrock geology 1:100,000 scale map series, Sheet 13, Meath with contributions by J. Morris, W. Cox (Minerals), G. Wright (Groundwater) and R. Meehan (Quaternary).* Geological Survey of Ireland, 78pp.

Miles, P.R. (1995) Potential distribution of methane hydrate beneath the European continental margins. *Geophysical Research Letters*, **22**, 3179–3182.

Morewood, N.C., Mackenzie, G.D., Shannon, P.M., O'Reilly, B.M., Readman, P.W., and Makris, J. (2005) The crustal structure and regional development of the Irish Atlantic margin region. *In*: Doré, A.G. and Vining, B. (eds), *Petroleum Geology: North-West Europe and Global Perspectives – Proceedings of the 6th Petroleum Geology Conference.* Geological Society, London, 1023–1034.

Murphy, T., Jacob, A.W.B., and Sanders, I.S. (2009) Geophysical evidence onshore. *In*: Holland, C.H. and Sanders, I.S. (eds), *The Geology of Ireland.* Second Edition. Dunedin Academic Press, Edinburgh, 461–470.

Naylor, D., Shannon, P.M., and Murphy, N. (1999) *Irish Rockall Basin region – a standard structural nomenclature system.* Petroleum Affairs Division, Dublin, Special Publication **1/99**, 42pp.

Naylor, D., Shannon, P.M., and Murphy, N. (2002) *Porcupine-Goban region – a standard structural nomenclature system.* Petroleum Affairs Division, Dublin, Special Publication **1/02**, 65pp.

Newman, P. (1999) The geology and hydrocarbon potential of the Peel and Solway Basins, East Irish Sea. *Journal of Petroleum Geology*, **22**, 305–324.

O'Reilly, B.M., Hauser, F., Ravaut, C., Shannon, P.M., and Readman, P.W. (2006) Crustal thinning, mantle exhumation and serpentinisation in the Porcupine Basin, offshore Ireland: evidence from wide-angle seismics. *Journal of the Geological Society, London*, **163**, 775–787.

O'Reilly, B.M., Shannon, P.M., and Readman, P.W. (2007) Shelf to slope sedimentation processes and the impact of Plio-Pleistocene glaciations in the northeast Atlantic, west of Ireland. *Marine Geology*, **238**, 21–44.

Parnell, J. (1991) Hydrocarbon potential of Northern Ireland. Part 1. Burial histories and source-rock potential. *Journal of Petroleum Geology*, **14**, 65–78.

Sevastopulo, G.D. (2009) Carboniferous: Serpukhovian and Pennsylvanian. *In*: Holland, C.H. and Sanders, I.S. (eds), *The*

Geology of Ireland. Second Edition. Dunedin Academic Press, Edinburgh, 269–294.

Shannon, P.M., O'Reilly, B.M., Readman, P.W., Jacob, A.W.B., and Kenyon, N. (2001) Slope failure features on the margin of the Rockall Trough. *In*: Shannon, P.M., Haughton, P.D.W., and Corcoran, D.V. (eds), *The Petroleum Exploration of Ireland's Offshore Basins*. Geological Society, London, Special Publications, **188**, 455–464.

Sustainable Energy Ireland (2009) *Energy in Ireland 1990-2008. 2009 Report*. 88pp.

Unnithan, V., Shannon, P.M., McGrane, K., Readman, P.W., Jacob, A.W.B., Keary, R., and Kenyon, N.H. (2001) Slope instability and sediment redistribution in the Rockall Trough: constraints from GLORIA. *In*: Shannon, P.M., Haughton, P.D.W., and Corcoran, D.V. (eds), *The Petroleum Exploration of Ireland's Offshore Basins*. Geological Society, London, Special Publications, **188**, 439–454.

Wilson, H.E. and Robbie, J.A. (1966) *Geology of the country around Ballycastle*. Memoir of the Geological Survey of Northern Ireland. HMSO, Belfast. 370pp.

Index